Textbook of Plant Biology

Textbook of Plant Biology

Edited by **Davis Twomey**

SYRAWOOD
PUBLISHING HOUSE

New York

Published by Syrawood Publishing House,
750 Third Avenue, 9th Floor,
New York, NY 10017, USA
www.syrawoodpublishinghouse.com

Textbook of Plant Biology
Edited by Davis Twomey

International Standard Book Number: 978-1-68286-189-9 (Hardback)

The publisher's policy is to use permanent paper from mills that operate a sustainable forestry policy. Furthermore, the publisher ensures that the text paper and cover boards used have met acceptable environmental accreditation standards.

Trademark Notice: Registered trademark of products or corporate names are used only for explanation and identification without intent to infringe.

Printed in the United States of America.

Contents

Preface   VII

Chapter 1   **Carotenoid accumulation affects redox status, starch metabolism, and flavonoid/anthocyanin accumulation in citrus**   **1**
Hongbo Cao, Jiangbo Wang, Xintian Dong, Yan Han, Qiaoli Ma, Yuduan Ding, Fei Zhao, Jiancheng Zhang, Haijiang Chen, Qiang Xu, Juan Xu and Xiuxin Deng

Chapter 2   **Dof affecting germination 2 is a positive regulator of light-mediated seed germination and is repressed by dof affecting germination 1**   **17**
Silvia Santopolo, Alessandra Boccaccini, Riccardo Lorrai, Veronica Ruta, Davide Capauto, Emanuele Minutello, Giovanna Serino, Paolo Costantino and Paola Vittorioso

Chapter 3   **A novel system for evaluating drought–cold tolerance of grapevines using chlorophyll fluorescence**   **28**
Lingye Su, Zhanwu Dai, Shaohua Li and Haiping Xin

Chapter 4   **Overexpression of a truncated CTF7 construct leads to pleiotropic defects in reproduction and vegetative growth in Arabidopsis**   **40**
Desheng Liu and Christopher A Makaroff

Chapter 5   **Mineral nitrogen sources differently affect root glutamine synthetase isoforms and amino acid balance among organs in maize**   **56**
Bhakti Prinsi and Luca Espen

Chapter 6   **New insights into the evolutionary history of plant sorbitol dehydrogenase**   **69**
Yong Jia, Darren CJ Wong, Crystal Sweetman, John B Bruning and Christopher M Ford

Chapter 7   **Keeping the rhythm: light/dark cycles during postharvest storage preserve the tissue integrity and nutritional content of leafy plants**   **92**
John D Liu, Danielle Goodspeed, Zhengji Sheng, Baohua Li, Yiran Yang, Daniel J Kliebenstein and Janet Braam

Chapter 8   **Impacts of nucleotide fixation during soybean domestication and improvement**   **101**
Shancen Zhao, Fengya Zheng, Weiming He, Haiyang Wu, Shengkai Pan and Hon-Ming Lam

Chapter 9   **High-resolution confocal imaging of wall ingrowth deposition in plant transfer cells: Semi-quantitative analysis of phloem parenchymatransfer cell development in leaf minor veins of Arabidopsis**   **113**
Suong T T Nguyen and David W McCurdy

Chapter 10   **Establishment of *Anthoceros agrestis* as a model species for studying the biology of hornworts**   127
Péter Szövényi, Eftychios Frangedakis, Mariana Ricca, Dietmar Quandt, Susann Wicke and Jane A Langdale

Chapter 11   **Association mapping in sunflower (Helianthus annuus L.) reveals independent control of apical vs. basal branching**   134
Savithri U Nambeesan, Jennifer R Mandel, John E Bowers, Laura F Marek, Daniel Ebert, Jonathan Corbi, Loren H Rieseberg, Steven J Knapp and John M Burke

Chapter 12   **Baseline study of morphometric traits of wild *Capsicum annuum* growing near two biosphere reserves in the Peninsula of Baja California for future conservation management**   146
Bernardo Murillo-Amador, Edgar Omar Rueda-Puente, Enrique Troyo-Diéguez, Miguel Víctor Córdoba-Matson, Luis Guillermo Hernández-Montiel and Alejandra Nieto-Garibay

Chapter 13   **Direct Contact – Sorptive Tape Extraction coupled with Gas Chromatography – Mass Spectrometry to reveal volatile topographical dynamics of lima bean (*Phaseolus lunatus* L.) upon herbivory by *Spodoptera littoralis* Boisd.**   164
Lorenzo Boggia, Barbara Sgorbini, Cinzia M Bertea, Cecilia Cagliero, Carlo Bicchi, Massimo E Maffei and Patrizia Rubiolo

Chapter 14   **Dissection of the style's response to pollination using transcriptome profiling in self-compatible (*Solanum pimpinellifolium*) and self-incompatible (*Solanum chilense*) tomato species**   177
Panfeng Zhao, Lida Zhang and Lingxia Zhao

Chapter 15   **Characterization of *Brachypodium distachyon* as a nonhost model against switchgrass rust pathogen *Puccinia emaculata***   191
Upinder S Gill, Srinivasa R Uppalapati, Jin Nakashima and Kirankumar S Mysore

Chapter 16   **X-ray micro-computed tomography in willow reveals tissue patterning of reaction wood and delay in programmed cell death**   203
Nicholas James Beresford Brereton, Farah Ahmed, Daniel Sykes, Michael Jason Ray, Ian Shield, Angela Karp and Richard James Murphy

Permissions

List of Contributors

Preface

The discipline of plant biology has progressed over time, and various studies have contributed to the better understanding of techniques of growing plants. The understanding of this subject is essential, especially for the students of botany and associated branches of biology. This book strives to provide the readers with a sound knowledge about plants at molecular, cellular, tissue, and organism levels. Aptly suited for students and research scholars alike, this book will succeed in giving a useful insight into the anatomy and physiology of plants.

Significant researches are present in this book. Intensive efforts have been employed by authors to make this book an outstanding discourse. This book contains the enlightening chapters which have been written on the basis of significant researches done by the experts.

Finally, I would also like to thank all the members involved in this book for being a team and meeting all the deadlines for the submission of their respective works. I would also like to thank my friends and family for being supportive in my efforts.

Editor

Carotenoid accumulation affects redox status, starch metabolism, and flavonoid/anthocyanin accumulation in citrus

Hongbo Cao[1,2†], Jiangbo Wang[1,3†], Xintian Dong[2], Yan Han[2], Qiaoli Ma[1], Yuduan Ding[1], Fei Zhao[1], Jiancheng Zhang[1,4], Haijiang Chen[2], Qiang Xu[1], Juan Xu[1] and Xiuxin Deng[1*]

Abstract

Background: Carotenoids are indispensable plant secondary metabolites that are involved in photosynthesis, antioxidation, and phytohormone biosynthesis. Carotenoids are likely involved in other biological functions that have yet to be discovered. In this study, we integrated genomic, biochemical, and cellular studies to gain deep insight into carotenoid-related biological processes in citrus calli overexpressing CrtB (phytoene synthase from *Pantoea agglomerans*). *Fortunella hindsii* Swingle (a citrus relative) and *Malus hupehensis* (a wild apple) calli were also utilized as supporting systems to investigate the effect of altered carotenoid accumulation on carotenoid-related biological processes.

Results: Transcriptomic analysis provided deep insight into the carotenoid-related biological processes of redox status, starch metabolism, and flavonoid/anthocyanin accumulation. By applying biochemical and cytological analyses, we determined that the altered redox status was associated with variations in O_2^- and H_2O_2 levels. We also ascertained a decline in starch accumulation in carotenoid-rich calli. Furthermore, via an extensive cellular investigation of the newly constructed CrtB overexpressing *Fortunella hindsii* Swingle, we demonstrated that starch level reduction occurred in parallel with significant carotenoid accumulation. Moreover, studying anthocyanin-rich *Malus hupehensis* calli showed a negative effect of carotenoids on anthocyanin accumulation.

Conclusions: In citrus, altered carotenoid accumulation resulted in dramatic effects on metabolic processes involved in redox modification, starch degradation, and flavonoid/anthocyanin biosynthesis. These findings provided new perspectives to understand the biological importance of carotenogenesis and of the developmental processes associated with the nutritional and sensory qualities of agricultural products that accumulate carotenoids.

Keywords: Carotenogenesis, Citrus, Redox status, Starch, Chromoplast, Anthocyanin

Background

Carotenoids, which first appeared in bacteria over three billion years ago, belong to a subfamily of isoprenoids that are commonly found in all organisms [1]. In nature, carotenoids originate from the condensation of geranyl-geranyl pyrophosphate (GGPP), which is derived from the synthesis of isoprenoid precursors via the plastid-localized 2-C-methyl-D-erythritol 4-phosphate (MEP) pathway. In the crucial rate-controlling step, phytoene synthase (PSY) mediates the condensation of GGPP, forming the first carotenoid, phytoene [2,3]. Subsequently, different types of carotenoids are generated through various synthetic pathways including desaturation, isomerization, cyclization, hydroxylation, and other modifications [3]. Carotenoids and their derivatives play essential physiological and ecological roles. They are involved in the photosynthetic apparatus, photoprotective pigments, antioxidants, hormone precursors, and attractants for pollinators in plant growth, development, and reproduction [2]. Carotenoids are the precursors of phytohormones such as abscisic acid, strigolactones and the recently identified carlactone, which negatively regulates plant axillary

* Correspondence: xxdeng@mail.hzau.edu.cn
†Equal contributors
[1]Key Laboratory of Horticultural Plant Biology (Ministry of Education), Huazhong Agricultural University, 430070 Wuhan, Hubei, China
Full list of author information is available at the end of the article

outgrowth [4-6]. Ramel et al. [7] recently revealed that a carotenoid endoperoxide that originates from a reaction between β-carotene and reactive oxygen species (ROS) can serve as a stress signal that mediates gene responses to singlet oxygen in *Arabidopsis*. In addition, an epistatic influence on the expression of endogenous carotenogenic genes has been observed in carotenoid-engineered potato tubers, which suggests a feedback regulation of carotenoid metabolites [8,9]. It is reasonable to hypothesize that there are still more biological functions associated with carotenoids or their derivatives that are waiting to be uncovered.

Due to the health promoting function of carotenoids, research on their biosynthesis and accumulation in plants has been a predominant focus. However, knowledge of the effects of carotenoid metabolism on other plant processes is still relatively limited. In natural systems, such as in the fruits of citrus and tomato, carotenoid biosynthesis and accumulation often occur in parallel with the ripening process [10,11]. It is quite certain that the ripening process involves cellular activities related to metabolic networks and to organelle modification, which eventually determine product quality. In addition to playing a role in fruit quality, some metabolites, such as malate and anthocyanin, have newly discovered physiological functions associated with the regulation of fruit metabolism, development, and shelf life [12,13]. Fraser et al. [14] discovered the effects of enhanced carotenoid accumulation on isoprenoids, plastid development, and intermediary metabolism in tomato fruits. In investigations of carotenoid-accumulating citrus mutants, some biological processes associated with carbohydrate metabolism and oxidative stress were found to differ from the wild types [15,16]. Relatively little is known about how the carotenoid accumulation program might contribute to these unintended metabolic changes.

Chromoplast formation is one of the most important cellular changes during the ripening of carotenoid-rich plant tissues; it involves significant carotenoid sequestration and the use of other metabolic pathways, which are all essential for the nutritional and sensory quality of agricultural products [17]. Chromoplasts are generally derived from preexisting plastids such as amyloplasts or chloroplasts, and the chromoplast transition is often associated with tissue and organ development [18]. During chromoplast development, plastoglobules and carotenoid crystals form, starch breakdown occurs, starch granules and thylakoids disappear, and the metabolism of terpenoids and lipids is greatly enhanced [11,17,19,20]. To date, there is limited understanding of how the chromoplast developmental program is established [18], and the associative inner system including metabolic variation and structural remodeling still exhibits intricate behavior. In recent years, a considerable amount of research has shown that chromoplast biogenesis is mediated by crucial factors such as Orange (OR, a DnaJ cysteine-rich domain-

containing protein) and CHRC (chromoplast-specific carotenoid-associated protein) [18,21]. Furthermore, in tobacco floral nectaries and carrot roots, the mutually exclusive relationship between carotenoid accumulation and starch granule development suggests that enhanced carotenogenesis serves as a developmental signal that directs the transition from amyloplasts to chromoplasts [19,22]. Additionally, modification of chromoplast morphology has been previously observed in carotenoid engineered plants, which suggests the possibility that cellular structures can adapt to facilitate the sequestration of newly formed carotenoids [23].

In addition to carotenoid biosynthesis, anthocyanin accumulation is an important biological event during the ripening of some fruits. Anthocyanins are plant phenolic secondary metabolites that are part of the phenylpropanoid pathway [24]. Like carotenoids, they are involved in a series of pivotal biological processes, such as antioxidative protection and the producion of attractants for reproduction, and they are also essential for the protection of human health [13]. Anthocyanins can co-exist with carotenoids in plant tissues and organs, but in some carotenoid-rich organs, there is relatively little anthocyanin accumulation. This phenomenon was observed in *Oncidium* Gower Ramsey flowers and in tomato fruits [25,26]. It is unknown if there is a negative correlation between carotenoid and anthocyanin, and it is a difficult question to answer. Carotenoids and anthocyanins are often involved in concurrent biosynthetic processes that accompany natural development, which makes it difficut to define a causal relationship. Similar observations have also been made for other hypothesized carotenoid-associated biological processes. Recently, genetic manipulation has been used successfully in many plants to modify carotenogenesis and other quality-associated components [27]. In these engineered plants, which have targeted metabolic pathway modifications, various unintended physiological, biochemical, and cellular changes have occurred [12-14]. Engineered systems appear to provide an effective approach for regulating the accumulation of a given metabolite, and they can facilitate the identification of associative biological relationships [12,13].

In our previous study, engineered cell models (ECMs) were established by activating the rate-controlling reaction by overexpressing the CrtB protein (phytoene synthase from *Erwinia herbicola*, now known as *Pantoea agglomerans*) in citrus embryogenic calli [28]. These ECMs exhibit diverse colors and accumulate significant levels of carotenoids. They are useful for understanding not only carotenoid biosynthesis, but also the potential biological processes associated with carotenoid accumulation. In the present study, we use Affymetrix microarrays, biochemistry, and cellular investigation of citrus calli to gain deeper insight into carotenoid-related biological processes, including redox status alternation, starch metabolism, and decreased flavonoid/anthocyanin accumulation. Engineering these pathways in *Fortunella hindsii* Swingle (a citrus relative) and *Malus hupehensis* (a wild apple) calli further validate these results.

Results
ECM transcriptional patterns
Engineered cell models (ECMs) generated by over-expressing 35S:: *CrtB* in citrus embryogenic calli show a striking accumulation of carotenoids [28]. However, relatively little is known about the other biological processes associated with engineered carotenoid accumulation. Thus, to further comprehend the cellular responses to enhanced carotenoid biosynthesis, we used three representative ECMs (M-33, from Marsh grapefruit; RB-4, from Star Ruby grapefruit; and SBT-6, from Sunburst mandarin) and their wild types in Affymetrix microarray analysis. Genes that were up- or down-regulated more than 2-fold (ECMs/WTs, $P < 0.05$) were identified, but a relaxed threshold of 1.5-fold (ECMs/WTs, $P < 0.05$) was used for genes in the M-33/WT microarray data (Additional file 1). This relaxed threshold was utilized because there was a relative lack of differentially expressed genes between M-33 and its wild type, most likely due to strong acclimation to carotenoid accumulation in the Marsh grapefruit genotype. To validate the microarray data, a quantitative real-time PCR (qRT-PCR) experiment was performed on the three representative ECMs and their wild types. A total of 10 genes were selected, and gene-specific primers were designed. Linear regression analysis showed an overall correlation coefficient of $R^2 = 0.6605$ between the qRT-PCR and microarray data, which confirmed that the microarray data were reliable (Additional file 2).

MapMan Bin indicated that the three representative ECMs had similar transcriptional responses to carotenoid accumulation (Figure 1). We further examined the differentially expressed genes that were annotated as being involved in stress, redox, hormone metabolism, and secondary metabolism; these groups represented the major transcriptional responses in the ECMs. As shown in Additional file 3, many stress- and redox-response genes were up-regulated in the ECMs. Hormone metabolism genes for ABA, auxin, ethylene, gibberellin, jasmonate (JA), and salicylic acid (SA) showed significantly higher transcription levels in the ECMs than in the wild types. Moreover, many genes involved in the synthesis of phenylpropanoids, alkaloids, wax, and simple phenols were up-regulated in the ECMs, whereas the majority of the genes related to flavonoid/anthocyanin and isoprenoid synthesis were down-regulated. Notably, a MapMan Bin for major CHO metabolism contained one gene that encoded the α-amylase. This gene was up-regulated in

the ECMs compared with the wild types (Figure 2A). This result was supported by qRT-PCR analysis and was in accordance with the consistently up-regulated transcriptional pattern of three α-amylase genes (Figure 2B), and their significantly increased enzymatic activity, as shown by enzyme activity analysis (Figure 2C).

Furthermore, gene annotation revealed that in the three ECMs, a significant number of up-regulated genes were annotated as encoding peroxidases (PODs), glutathione S-transferase (GST), and hydroxyproline-rich glycoprotein family proteins. The down-regulated genes primarily encoded protein kinases, zinc finger family proteins, glycine-rich proteins, and senescence-related factors (Additional files 1 and 4).

Redox status was significantly altered in the ECMs
POD, GST, hydroxyproline-rich glycoprotein family proteins, heat shock proteins, and universal stress protein (USP) family proteins have often been regarded as important stress response factors in plants [29-31]. Interestingly, our microarray showed that these stress-related genes, especially *PODs*, were significantly induced in the ECMs (Additional files 1 and 4). Prediction of subcellular localization and class of the differentially expressed PODs suggested that they were extracellular class III peroxidases (Additional file 5). Engineered carotenoid modification can disturb ABA levels [32], which could be associated with the stress and redox responses observed in the ECMs. However, comparative analysis showed that, compared with the corresponding wild type, ABA content was higher only in the RB ECM and not in the M and SBT ECMs (Additional file 6). Furthermore, several differentially expressed genes from the microarray data were verified by RT-PCR analysis in the calli, these genes included WRKY75, a protease inhibitor gene, a hydroxyproline-rich glycoprotein family protein gene, and a USP family protein gene (Additional file 7). Previous reports have shown that these four investigated genes are induced by ROS [29-31,33,34]. This information suggested the possibility that engineered carotenoid accumulation modified the ROS levels in the ECMs. Therefore, we investigated the ROS levels in the calli. NBT staining showed that superoxide radical (O_2^-) levels in the ECMs were markedly reduced compared with the wild types (Figure 3A). In contrast, hydrogen peroxide (H_2O_2) levels showed an unexpected increase (Figure 3B). This result corroborated the microarray data, which revealed that many stress- and redox-responsive genes were up-regulated; H_2O_2 has a signaling role in stress responses [29].

To further identify ROS changes in the ECMs, we used M-33 and its wild-type control to analyze the activities of ROS-related enzymes, including NADPH oxidase (NOX), which is requried for O_2^- production, as well as

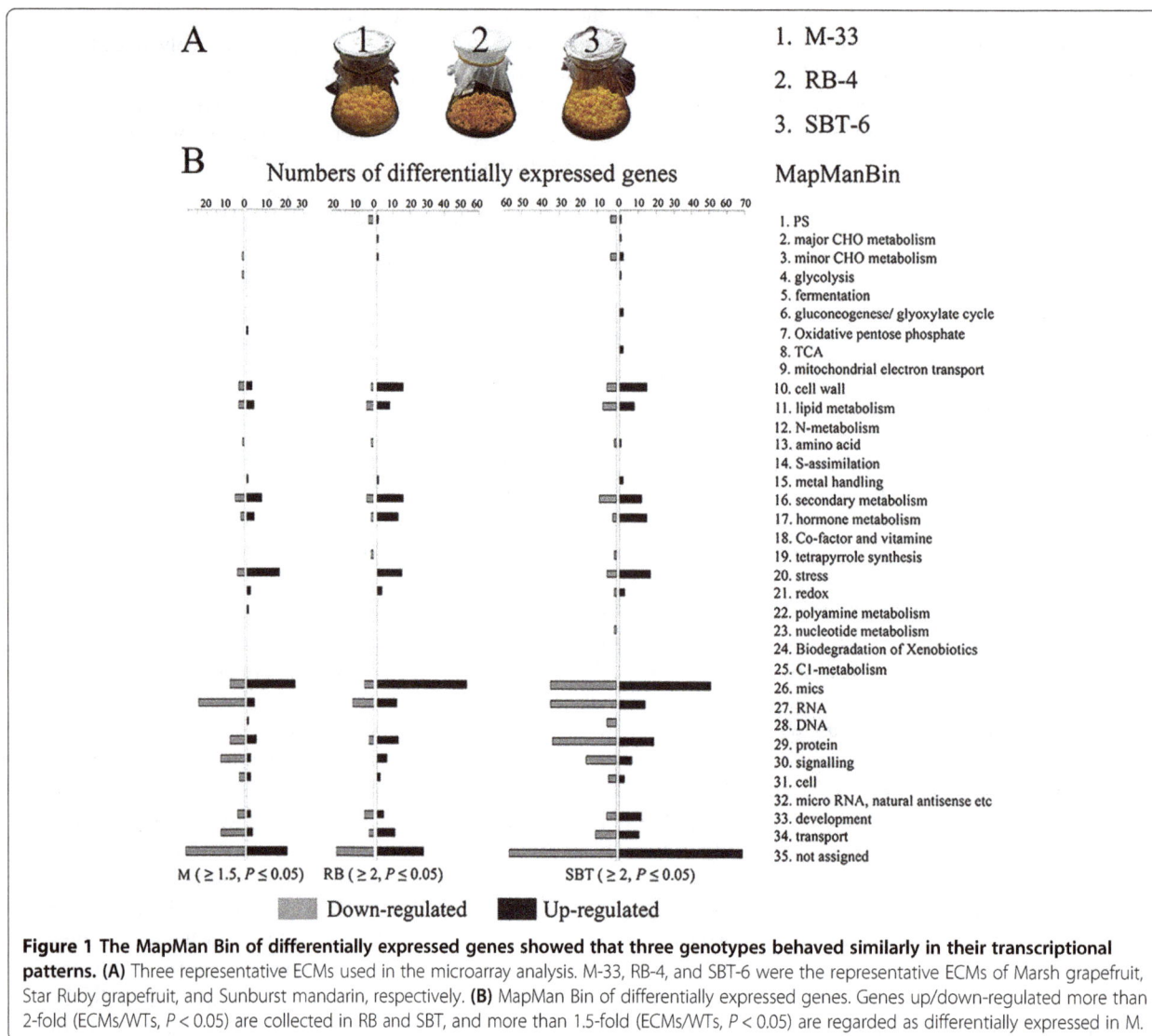

Figure 1 The MapMan Bin of differentially expressed genes showed that three genotypes behaved similarly in their transcriptional patterns. (A) Three representative ECMs used in the microarray analysis. M-33, RB-4, and SBT-6 were the representative ECMs of Marsh grapefruit, Star Ruby grapefruit, and Sunburst mandarin, respectively. **(B)** MapMan Bin of differentially expressed genes. Genes up/down-regulated more than 2-fold (ECMs/WTs, P < 0.05) are collected in RB and SBT, and more than 1.5-fold (ECMs/WTs, P < 0.05) are regarded as differentially expressed in M.

superoxide dismutase (SOD) and catalase (CAT), which are involved in H_2O_2 production and scavenging, respectively. As shown in Figure 3C, NOX activity was slightly elevated in M-33, but the difference was statistically insignificant. Similarly, no significant difference in SOD activity was observed between M-33 and the wild-type control. It is noteworthy that CAT activity was significantly lower in M-33 compared with the wild type.

Carotenoid accumulation altered starch metabolism in ECMs

The clear induction of the α-amylase genes and their elevated enzymatic activity (Figures 1 and 2) suggested a modification of starch metabolism in the ECMs. This provided a key clue in explaining some of the phenomena observed in the ECMs. For example, they generally have reduced dry weight (Additional file 8), and their cytological profiles often showed fewer starch granules than the wild types (Figure 4A). Therefore, we investigated

starch levels in the calli. Compared with the corresponding wild types, the ECMs had reduced starch content (Figure 4B), which was consistent with the results of iodine staining (Figure 4C). Furthermore, we isolated the starch granules and noted that there are substantially more in the wild types than in the ECMs (Figure 4C). Subsequent SEM investigation demonstrated the absence of large starch granules in ECM cells (Figure 4C). We also discovered that the ECMs had higher soluble sugar (fructose, glucose, and sucrose) content in the ECMs (Additional file 9), which suggested a biochemical alteration of starch metabolism in the ECMs.

Reduced starch content occurred in parallel with significant carotenoid accumulation in citrus

We generated transgenic plants by overexpressing 35S:: *CrtB* in an early flowering citrus relative, Hongkong kumquat (*F. hindsii* Swingle). Overexpression of the

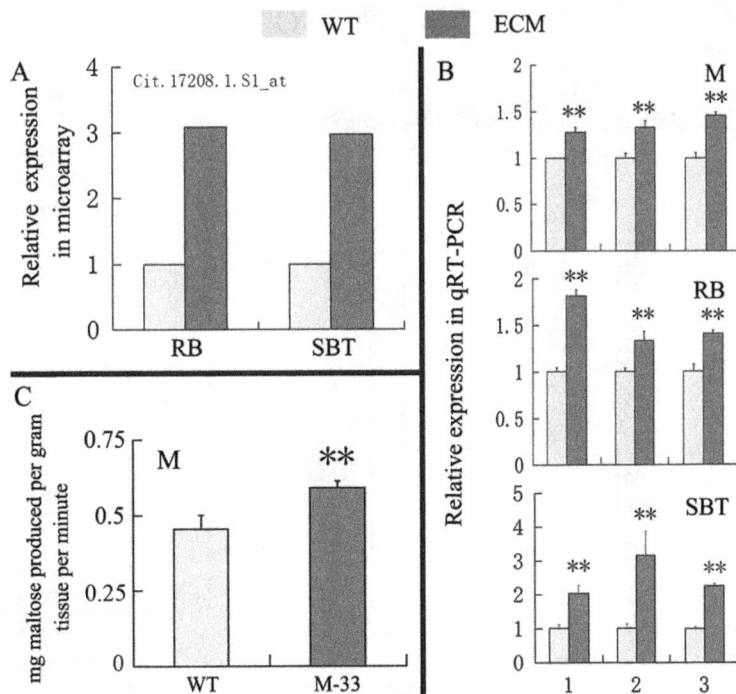

Figure 2 Validation of the up-regulated transcriptional pattern of the α-amylase gene from microarray data via qRT-PCR and enzyme activity analysis. (A) Up-regulated transcriptional pattern of Cit.17208.1.S1_at in microarray data, and this gene was annotated as α-amylase gene. **(B)** qRT-PCR analysis of α-amylase genes in M, RB, and SBT. Three α-amylase genes used for qRT-PCR analysis were from a previous report [25] and the NCBI. *AMY*, citrus sinensis alpha-amylase-like gene (accession number: XM_006473264); *SD1*, α-amylase gene (accession number: JN793456); *SD2*, α-amylase 3 gene (accession number: JN793457). Transcript levels are expressed relative to WT (wild type). **(C)** Enzyme activity analysis of α-amylase in wild-type M and M-33. α-amylase activity was expressed as mg maltose produced per gram tissue per minute. Columns and bars represent the means and ± SD, respectively (n = 3 replicate experiments). **Indicates that the values are significantly different compared with wild type at the significance level of P < 0.01. M, RB, and SBT represent Marsh grapefruit, Star Ruby grapefruit, and Sunburst mandarin, respectively. M-33 represents the ECM line of Marsh grapefruit.

CrtB gene led to orange pigmentation of the flower petals, roots, and other tissues (Figure 5A-D), which demonstrated increased carotenoid accumulation. Cellular inspection using light and electron microscopy revealed significant modifications of the plastids in the orange tissues, which led the preexisting plastids to show a chromoplast-like profile. For instance, in the wild-type tissues, amyloplasts were found in the flowers and dark-grown roots, etioplasts were found in dark-grown embryoids, and chloroplasts in old petioles and light-grown roots, and all of these plastids showed significant starch granule deposition (Figure 5E, J, I, K, M, O; Additional file 10A). However, in the corresponding transgenic tissues, the plastids showed an entirely different morphology: abundant plastoglobules and crystal structures could be observed, but starch granules and thylakoid membranes were scarce (Figure 5F, H, J, L, N, P; Additional file 10A). We also investigated the expression of genes related to starch metabolism in the roots. The results confirmed that α-amylase genes were significantly up-regulated in transgenic roots (Additional file 10B). These observations demonstrated that the

reduction in starch occurred in parallel with significant carotenoid accumulation in citrus, and they also suggested that carotenoid accumulation could induce similar transcriptional perturbation in the roots and in citrus calli. Moreover, six differentially expressed genes involved in stress and senescence that were identified in the microarray data from citrus calli were analyzed in the roots of the transgenic Hongkong kumquat and in the wild-type control. qRT-PCR analysis demonstrated that five of six genes, including those encoding WRKY75, the USP family protein, the hydroxyproline-rich glycoprotein family protein, the senescence-related factor, and the plastocyanin-like domain-containing protein, showed significant transcriptional alterations in the carotenoid-rich roots compared with the wild-type control. Importantly, the changes were consistent with those seen in the ECMs (Additional file 10C).

Flavonoid/anthocyanin biosynthesis was negatively affected by carotenoid accumulation

In the ECMs, transcriptional changes associated with secondary metabolism were obvious (Additional file 3).

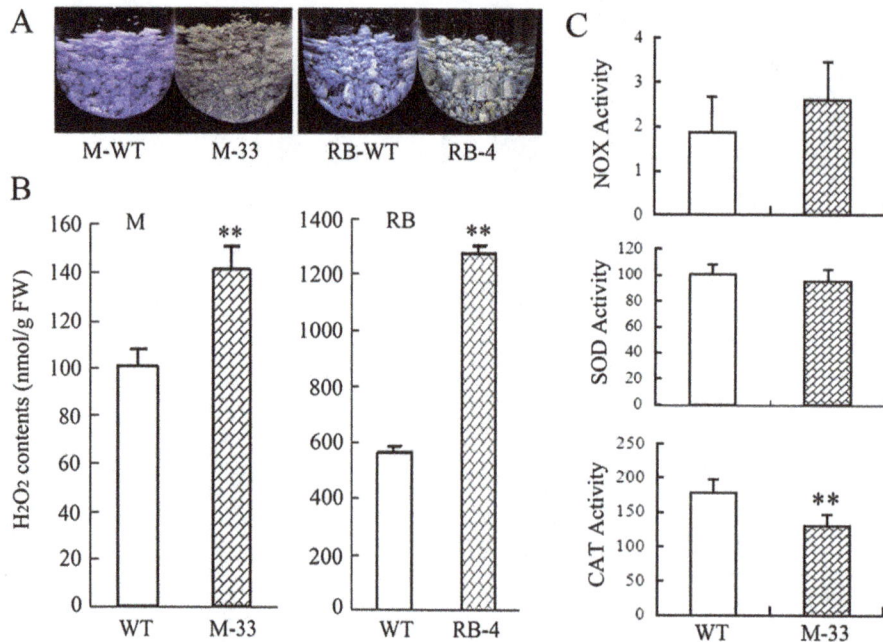

Figure 3 Determination of ROS levels and the activities of related enzymes in calli. M-33, and RB-4 represent the ECM lines of Marsh grapefruit and Star Ruby grapefruit, respectively. WT represents wild-type calli. **(A)** O_2^- is detected by histochemical staining with NBT. M and RB are shown as representatives for O_2^- levels. **(B)** H_2O_2 determination is based on the titanium reagent method. **(C)** NOX activity (nmol NADPH/min/g FW), SOD activity (U/g FW) and CAT activity (U/g FW) were analyzed in the M-33 and its wild-type control. The columns and bars represent the means and ± SD, respectively (n = 3 replicate experiments). **Indicates that the values are significantly different compared with wild type at the significance level of $P < 0.01$.

In particular, the transcription of flavonoid/anthocyanin biosynthetic genes was repressed in the ECMs. Additionally, we tested the expression of chalcone synthase (CHS) and chalcone isomerase (CHI) (the key enzymes of flavonoid biosynthesis) in the roots of the transgenic Hongkong kumquat and its control. qRT-PCR analysis showed that the *CHS* and *CHI* genes were both expressed at lower levels in the orange roots of the transgenic Hongkong kumquats than in the wild-type control (Additional file 10D). These results suggested that carotenoids have a negative effect on flavonoid/anthocyanin accumulation. Citrus embryogenic calli and roots are anthocyanin-free, making it difficult to investigate this phenotype in these tissues. To study this negative correlation, we established an apple ECM by engineering an anthocyanin-rich *M. hupehensis* callus using a 35S::CrtB construct. Overexpression of the *CrtB* gene led to a 3.68-fold increase in total carotenoid levels in the apple ECM (Figure 6A; Additional file 11). Abundant carotenoid accumulation could also be observed through cellular inspection, which showed that apple ECM cells formed many plastoglobules in the plastids, but that the plastids in the wild-type cells were filled with starch granules (Figure 6B). Notably, under visible light, the wild-type callus contained abundant anthocyanins, but the apple ECM showed minimal anthocyanin accumulation (Figure 6A). Additionally, qRT-PCR analysis of anthocyanin biosynthetic

genes showed that they were repressed in the apple ECM (Figure 6C). To further understand the negative effect of carotenoids on anthocyanin accumulation in the apple callus system, norflurazon (phytoene desaturase inhibitor) treatment was performed to block colored carotenoid accumulation. After a twenty-day culture with 10 μM norflurazon treatment under visible light, the apple ECM, which was initially yellow, turned red. Anthocyanin analysis revealed that norflurazon treatment could partially rescue anthocyanin accumulation in apple ECM (Figure 6D).

Discussion

The callus model system has provided a unique opportunity to investigate the effect of carotenoid metabolic engineering in plants [35]. Using our previously constructed, carotenoid-engineered cell models in citrus [28], we examined carotenoid-related biological processes. Despite the genotypic diversity of the representative ECMs, they had similar transcriptional patterns (Figure 1; Additional file 2). Our results revealed new aspects of carotenoid-induced redox modification, starch metabolism, and anthocyanin loss. The present discoveries of carotenoid-related biological processes also support previous studies of carotenoid-associatied plastid development and key metabolic modifications [14,22,23].

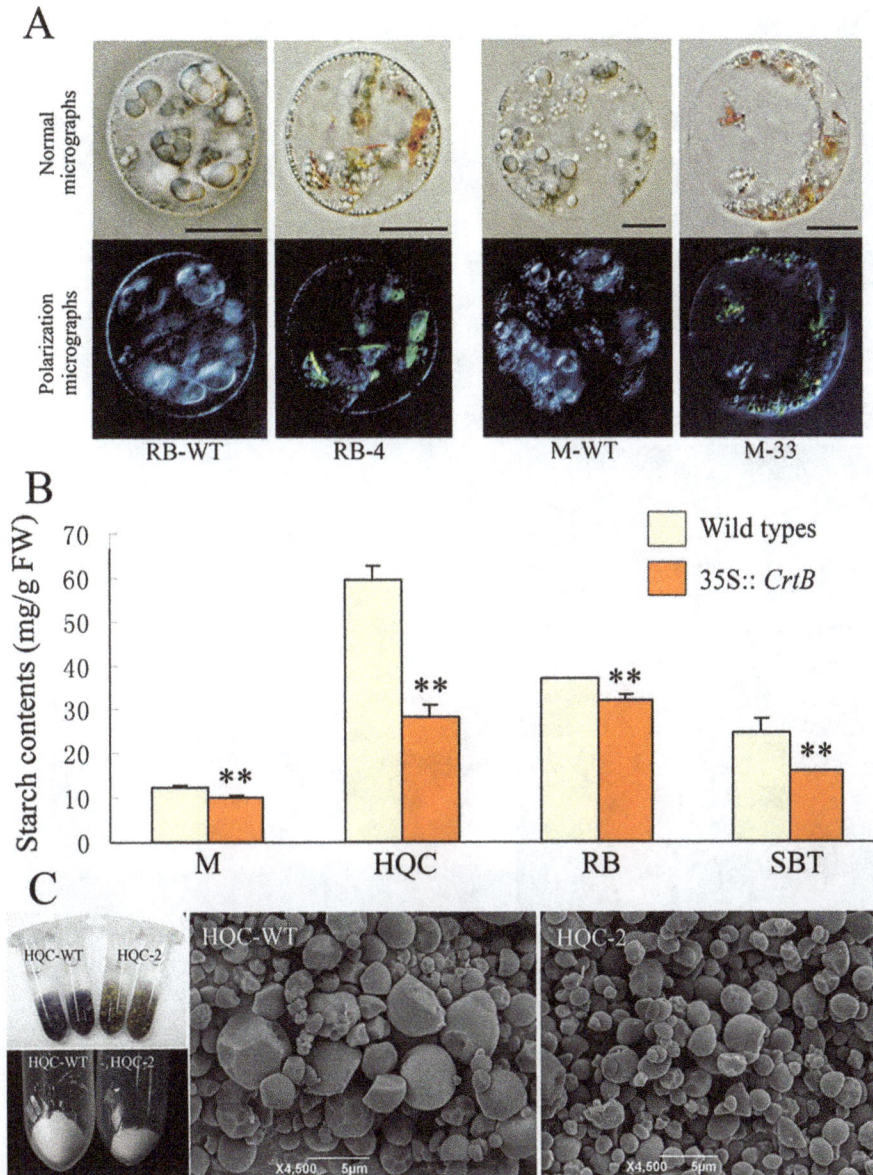

Figure 4 Comparison of starch accumulation between the ECMs and wild-type calli. (A) Light microscopic inspection of starch granule deposition in the callus cells. Protoplasts from wild-type calli and representative ECMs, respectively, are observed under a normal light field and a polarization microscope. The polarization microscope shows the characteristic birefringences of starch granules and carotenoid crystals, respectively. **(B)** Starch content analysis. Columns and bars represent the means and ± SD, respectively (n = 3 technical triplicate experiments). **Indicates that the values are significantly different compared with ECMs at the significance level of $P < 0.01$. **(C)** I_2/KI staining analysis and isolation of starch granules for SEM inspection. HQC is shown as a representative. The image in the lower left corner shows the yields of starch granules from equal weight of calli. HQC-WT and HQC-2 represent the ECM line and wild type, respectively. The right images show the starch granule morphology magnified 4500 times via SEM.

Altered carotenoid accumulation changed the redox status of ECMs

Transcriptomic data showed that *POD* and *GST* genes were significantly induced (Additional file 4). Additionally, genes involved in phenylpropanoid metabolism and hormone metabolism (involved in ABA, JA, and SA) were expressed at higher levels in the ECMs than in the wild types (Additional file 3). These transcriptional characteristics conferred a clear stress response pattern in the ECMs,

but this pattern could not be explained by ABA levels (Additional file 6). A previous study revealed that carotenoids can react with ROS to suppress oxidative stress in *Arabidopsis* leaves under high light [7]. In our study, O_2^- levels were markedly reduced in the ECMs compared with the wild types, but NOX and SOD, which are required for O_2^- production and scavenging, respectively, showed insignificant changes in activity in the M-33 ECM (Figure 3A, C). Therefore, it was hypothesized that

Figure 5 Phenotypes of transgenic and wild-type Hongkong kumquats and cellular investigation. (A) Two-year-old flowering Hongkong kumquats (left, wild type; right, 35S:: *CrtB*, which represents transgenic Hongkong kumquat). **(B)** Phenotypes of flowers (left, wild type; right, 35S:: *CrtB*). **(C)** Phenotype of senescent leaf (top, wild type; lower, 35S:: *CrtB*). **(D)** Nucellar seedlings under light-grown for 60 d (left, wild type; right, 35S:: *CrtB*). **(E, G,** and **I)** Frozen sectioning investigation of the petal, leafstalk, and root of the wild-type Hongkong kumquat, respectively. **(F, H,** and **J)** Frozen sectioning investigation of the petal, leafstalk, and root of the transgenic Hongkong kumquat, respectively. **(K, M,** and **O)** Ultrastructural inspection of the petal, leafstalk, and root of the wild-type Hongkong kumquat, respectively. **(L, N,** and **P)** Ultrastructural inspection of petal, leafstalk, and root of the transgenic Hongkong kumquat, respectively. s, starch granules; p, plastoglobules; th, thylakoids; c, carotenoid crystal and characteristic internal membrane. The bars represent 10 μm in light microscopy and represent 1 μm in transmission electron microscopy.

carotenoids participated in the elimination of O_2^- in the ECMs. However, carotenoid accumulation did not lead to a similar decrease in H_2O_2 levels; instead, the ECMs had more H_2O_2 than the wild types (Figure 3). This result provided an explanation for the up-regulated stress response in the ECMs, as H_2O_2 plays an important signaling role in plant protection systems and could induce a transcriptional pattern similar to that found in the ECMs [29,30]. This hypothesis was supported by qRT-PCR analysis of several ROS-induced genes in the calli (Additional file 7). Increased H_2O_2 levels was an unexpected example of ROS modification in the ECMs. One possibility is that significant degradation of O_2^- enhanced H_2O_2 production in the

ECMs [30,36]. In addition, the reduction of CAT activity in the M-33 ECM provided evidence of a weakened H_2O_2 scavenging system, which could lead to increased H_2O_2 levels [37].

Similar stress responses were also observed in other plant organs with altered carotenoid accumulation, such as in *OR* transgenic potato and in red-fleshed mutant citrus fruits [16,18]. It is unclear if there is a common mechanism mediating the carotenoid-associated stress response in plants. Recently, carotenoid oxidation products, such as β-cyclocitral, which is generated from the oxidation of β-carotene by singlet oxygen, have been shown to be signals mediating stress responses in *Arabidopsis* [7]. Thus,

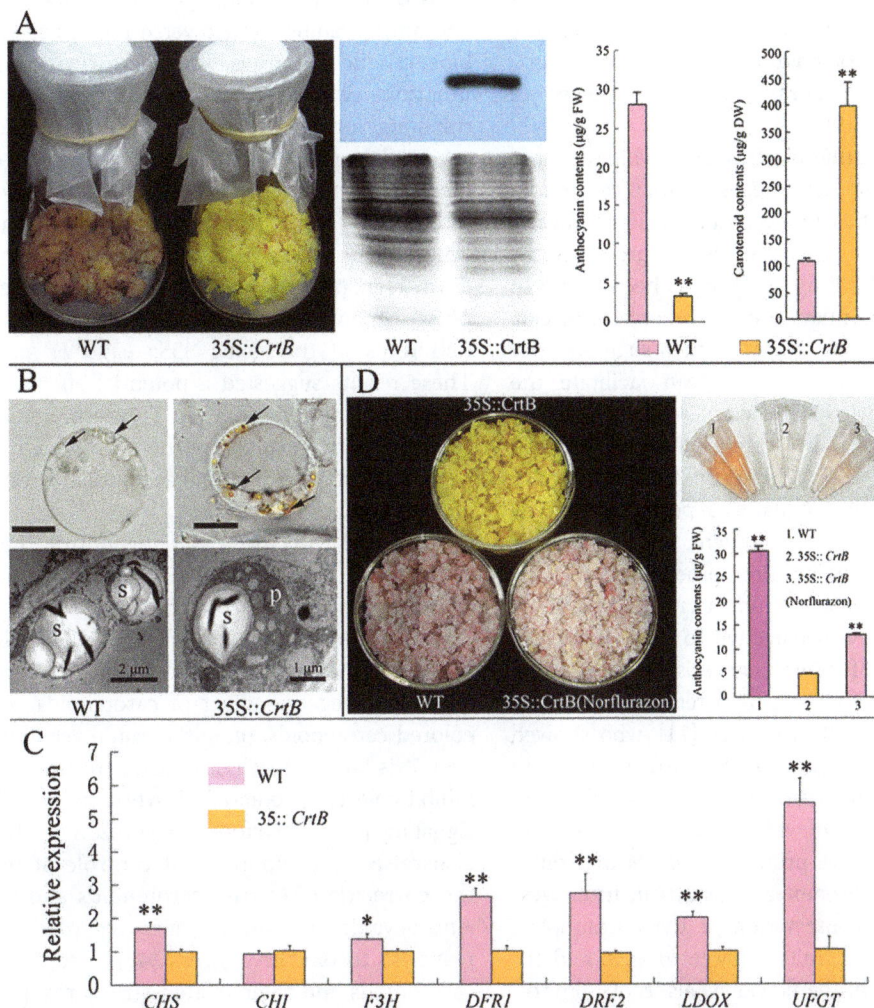

Figure 6 Pigment accumulation in the transgenic *M. hupehensis* callus with overexpressive *CrtB* gene. (A) Left image shows the phenotypes of the *M. hupehensis* calli cultured under visible light condition; middle image is the western blotting confirmation of CrtB protein, coomassie blue staining for loading control; right image displays the results of pigment (anthocyanins and carotenoids) levels analysis, columns and bars represent the means and ± SD, respectively (n = 3 replicate experiments); 35S:: *CrtB* represents the apple ECM. **(B)** Cytological investigation of *M. hupehensis* callus shows the amyloplasts in the wild-type cell, and the chromoplast-like structures in the ECM cell (arrows shown); s, starch granule; p, plastoglobule; m, mitochondrion. Bars represent 10 μm. **(C)** qRT-PCR analysis of anthocyanin related genes in the *M. hupehensis* calli. UFGT, uridine diphosphate (UDP)-glucose: flavonoid 3-O-glycosyltransferase; LDOX, leucoanthocyanidin dioxygenase; DFR1/2, dihydroflavonol 4-reductase 1/2; F3H, flavanone 3 β-hydroxylase; CHI, chalcone isomerase; CHS, chalcone synthase. Columns and bars represent the means and ± SD, respectively (n = 3 replicate experiments). 35S:: *CrtB* represents the light-cultured apple ECM; WT represents the light-cultured wild-type apple callus. **(D)** Phenotypes and pigment analyses under norflurazon treatment for 20 days. The medium contained 10 μM norflurazon. Columns and bars represent the means and ± SD, respectively (n = 3 replicate experiments). Transcript levels are expressed relative to the apple ECM (35S:: *CrtB*). * and ** indicate that the values are significantly different at the significance levels of $P < 0.05$ and $P < 0.01$, respectively.

β-cyclocitral-associated stress response may exist in the ECMs.

Carotenoid accumulation mediated starch metabolism

Our cytological and biochemical observations (Figure 4 and Additional file 8) reveal a very interesting correlation and suggest that carotenoids might regulate carbohydrate metabolism in plants. The discovery of this process clarifies the developmental processes associated with the nutritional and sensory qualities of agricultural products that accumulate carotenoids. Carotenoid biosynthesis requires a carbohydrate supply for assembling the carotenoid molecular backbone. The plastids are the only organelles involved in carotenoid biosynthesis, and they are also the sites for sugar and starch carbohydrate metabolism [11,18,23]. Thus, a feedback mechanism for maintaining a carbon supply for carotenogenesis could be involved in the correlation between carotenoid accumulation and starch degradation. This hypothesis is supported by a previous investigation that demonstrates a mutually exclusive relationship between carotenoid accumulation and starch deposition during the natural ripening processes of tobacco floral nectaries [19].

In addition, by comparing plastids from white and red carrot roots, Kim *et al.* [22] suggested that carotenoid accumulation might act as a developmental signal directing plastid modification. Carotenogenesis promoted by the overexpression of CrtB or by light has been found to coincide with the differentiation of chromoplasts in carrot roots [38,39]. A recent study further suggests that the adaptation of plastid structures can facilitate the sequestration of the newly formed carotenoids [23]. In the present study, chromoplast-like profiles were observed in various tissues of the 35S:: *CrtB* transgenic *F. hindsii* Swingle, as well as in the ECMs, as reported previously [28]. Therefore, reduced starch content was presumably related to the plastid modification process induced by significant carotenoid accumulation in citrus. This interpretation provides a new perspective to understand the feedback mechanism of carotenogenesis. However, the up-regulation of α-amylase is in apparent contrast with the proteomic analysis of Barsan et al. [11], who showed that proteins involved in starch metabolism decrease in abundance during chromoplastogenesis. Presumably, plastid modifications associated with engineered carotenoid accumulation might involve protein dynamics that differ from those of natural chromoplastogenesis in fruit. Despite the high level of conservation of the chromoplast proteome in the ripening fruits of sweet orange and tomato [40], the plastids in the flower petals, roots, embryoids, petioles, and callus systems of citrus are never involved in chromoplastogenesis during natural developmental processes, and they are distinct from those in citrus and tomato fruits. Additionally, engineered carotenoid accumulation could alter the redox status of ECMs, and this observation suggested that redox-regulated starch degradation occurred in the engineered carotenoid-rich tissues [41]. Further studies are required to identify the direct causal link between carotenoid accumulation and starch reduction.

Carotenoid accumulation negatively regulates anthocyanin biosynthesis

Carotenoids and anthocyanins are both biological pigments and can co-exist in plant tissues. However, there is little to no anthocyanin accumulation in some carotenoid-rich tissues [25,26]. This phenomenon is not absolute, but it seems to be prevalent in nature. For example, the ripening flesh of tomatoes and apricots accumulates abundant carotenoids but has little to no anthocyanins [42,43]. In addition, the negative correlation between the accumulation of carotenoids and anthocyanins was observed in the

peels of five apple genotypes [44]. Although carotenoids and anthocyanins show diverse molecular structures and biosynthetic pathways, they perform similar biological functions, including acting as antioxidants, and as attractants for pollinators [24]. Perhaps, the alternative accumulation of pigments represents an evolutionary mechanism to escape functional redundancy. However, to date, there is no evidence supporting this hypothetic evolutionary mechanism. Our present study found many down-regulated flavonoid/anthocyanin genes in ECMs. This phenomenon was also observed in the carotenoid-rich roots of transgenic 35S:: *CrtB F. hindsii* Swingle. These results suggested a potential effect of carotenoids on anthocyanin biosynthesis. Furthermore, we utilized an apple ECM overexpressing the *CrtB* gene to confirm the negative effect of carotenoid accumulation on anthocyanin biosynthesis.

Compared with the wild type control, the carotenoid-rich apple ECM had minimal anthocyanin accumulation. Additionally, norflurazon could partially rescue anthocyanin accumulation in the apple ECM. These results supported the transcriptional data from citrus, indicating a possible negative effect of carotenoids, and especially colored carotenoids, on anthocyanin accumulation. However, it is known that norflurazon treatment can not only inhibit colored carotenoid biosynthesis, but also alter ROS signaling [45]. Therefore, norflurazon treatment analysis also raised a question about the role of redox state in the correlation between carotenoids and anthocyanins. Anthocyanin accumulation is regarded as a positive response to oxidative stress [46]. The accumulation of carotenoids and their related structures, plastoglobules, may provide a more effective approach than anthocyanin accumulation to suppress oxidation. This interpretation is supported by previous studies showing that increasing levels of xanthophylls or plastoglobules could enhance the photooxidative tolerance and reduce anthocyanin accumulation in *Arabidopsis* and apple leaves [47,48]. Moreover, a recent study has probed into anthocyanin biosynthesis with an early redox signaling control upstream of the known transcription factors [49]. qRT-PCR analysis revealed that most the anthocyanin biosynthetic genes were consistently suppressed in the apple ECM, which suggested that there is significant transcriptional regulation involved in the negative effect of carotenoid accumulation on anthocyanin biosynthesis. However, the redox signaling-based regulatory mechanism that could mediate the link between carotenoid accumulation and anthocyanin biosynthesis is still unclear and requires further study. In addition, in strawberry fruit, competitive regulation via the peroxidase FaPRX27 has recently been proposed; FaPRX27 diverts phenolic flux from anthocyanins to lignin [50]. The existence of such competitive regulation in the apple ECM warrants future study.

Conclusions

Our studies on the transcriptional patterns of citrus calli linked carotenoid accumulation with redox state, starch metabolism, and flavonoid/anthocyanin accumulation. The existence of these physiological processes was further elucidated using biochemical and cytological analyses, as well as genetic manipulation of carotenoid biosynthesis in citrus calli, *F. hindsii* Swingle, and *M. hupehensis* calli. Our findings provide a new perspective on the complexity of carotenoid accumulation and its associated biological processes. For example, we confirmed that significant carotenoid accumulation could induce starch degradation in callus systems and in tissues such as flower petals and roots. The data generated from these model systems provide important information that could promote the understanding of starch metabolism and carotenoid accumulation during the ripening process in other plant systems. Equally importantly, our discoveries have significant implications for carotenoid metabolic engineering by providing the knowledge needed to give close consideration to a wider range of characteristics, such as plant resistance and systematic metabolic modification. In particular, the decreased anthocyanin levels associated with carotenoid accumulation should be avoided in carotenoid metabolic engineered plants. Anthocyanins are an important source of hydrophilic dietary antioxidants, and fruits and vegetables rich in both soluble and lipophilic antioxidants are considered to offer the best health protection [25].

Methods

Plant materials

Engineered cell models (ECMs) were established by over-expressing 35S:: *CrtB* (*tp–rbcS–CrtB*) (CrtB protein, phytoene synthase from *Erwinia herbicola*, now known as *Pantoea agglomerans*, containing a Pea rbcS transit peptide) in citrus embryogenic calli [28]. The ECMs and wild-type embryogenic calli were obtained from four citrus genotypes, Star Ruby grapefruit (*C. paradise* Macf.), Marsh grapefruit (*C. paradise* Macf.), Cara Cara navel orange [*C. sinensis* (L.) Osb.], and Sunburst mandarin [*C. reticulata* Blanco × (*C. paradisi* Macf. × *C. reticulata*)] designated as RB, M, HQC and SBT, respectively. The calli were propagated on MT medium in dark and kept at 25 ± 1°C. MT medium, which is typically used for citrus culture in vitro, was prepared according to Murashige and Tucker [51]. Twenty-day-old calli were harvested and used for immediate cellular and biochemical analyses or stored at -80°C for later molecular analysis.

Transgenic Hongkong kumquats (*Fortunella hindsii* Swingle), an early-flower citrus relative, were recovered through *Agrobacterium*-mediated transformation using 35S:: *CrtB* (*tp–rbcS–CrtB*) construct according to the method of Zhang *et al.* [52]. Regenerated resistant shoots were rooted directly on rooting medium (1/2MT medium

supplemented with 0.5 mg/L 1-naphthylacetic acid, 0.1 mg/L indolebutyric acid, 25 g/L sucrose, 0.5 g/L activated charcoal, and 8 g/L agar; pH 5.8). Rooted plantlets were transplanted into pots containing commercial substrates with organic matter and were placed in greenhouse facilities. Nucellar seedlings were recovered through cultivating mature seeds of transgenic and wild-type Hongkong kumquats in solidified MT basal medium containing 20 g/L sucrose.

The apple calli were initiated from the young embryo of the *Malus hupehensis* (a wild apple). They can be grown well on MT medium at 25°C under visible light (40 μmol m^{-2} s^{-1}) and display a typical character of anthocyanin accumulation, however, it must be supplemented with 0.1 mg·L^{-1} naphthalene acetic acid (NAA) and 0.5 mg·L^{-1} 6-benzylaminopurine (6-BA). The apple calli were used for genetic transformation using a 35S:: *CrtB* (*tp–rbcS–CrtB*) construct as detailed in our previous paper [28]. Explants preparation and transformation were performed according to the citrus callus transformation protocol described by Cao et al. [28] with a minor modification in which the transgenic calli were selected with 20 mg/L kanamycin. Fifteen-day-old calli were harvested and used for immediate cellular and biochemical analyses or stored at -80°C for molecular analysis. The calli used for extraction of carotenoids were lyophilized and stored at -80°C until use.

Norflurazon (an inhibitor of phytoene desaturase, Sigma, St. Louis, MO, USA) treatment with 10 μM norflurazon (dissolved in acetone) was performed on solid medium for apple calli. Control plates received equivalent acetone. After a twenty-day culture at 25°C under visible light (40 μmol m^{-2} s^{-1}), the yellow transgenic apple calli turned a red color, then all samples were collected for anthocyanin analysis.

Quantitative analysis of gene expression

Total RNA of citrus samples was extracted using a modified Trizol extraction protocol, as described previously [53]. Due to high contents of polyphenol compounds, a CTAB protocol was used to extract the total RNA from apple calli according to Hu *et al.* [54]. First-strand cDNA was synthesized from 1 μg of total RNA isolated from calli and roots using the RevertAid M-MuLV KIT (MBI, Lithuania) according to the manufacturer's instructions. The primer pairs used in the present study were as listed in previous reports or designed using the Primer Express software (Applied Biosystems, Foster City, CA, USA) (Additional file 12). *UBF5* [a suitable reference gene for qRT-PCR analysis using embryogenic callus culture] and *Actin* were used as the endogenous control to normalize expression in citrus calli and the roots of *F. hindsii* Swingle, respectively [15,55]. In apple calli, *MdActin* was used as the endogenous control [56]. qRT-PCR was performed using ABI 7500 Real Time System (PE Applied Biosystems; Foster City, CA, USA).

Microarray analysis

Affymetrix GeneChip Citrus Genome Arrays (Affymetrix, Santa Clara, CA, USA) were used for detecting transcriptional diversities between wild-type calli and ECMs. For each sample, RNA was extracted from two biological replicates. A total of 10 µg of fragmented cRNA from each sample was used for hybridization. The procedure for GeneChip Citrus Arrays (hybridization, washing, staining, and scanning with a GeneChip Scanner 3000) was followed carefully according to the Affymetrix GeneChip Expression Analysis Technical Manual.

Scanned images from GeneChip Citrus Arrays were analyzed using GeneChip Operating Software (GCOS 1.4; Affymetrix) with its default settings to generate raw data, which were saved as CEL files. The raw data were normalized using a robust multichip analysis approach implemented in the Affy package [57,58]. Analysis of variance (ANOVA) was used to look for significant differences between samples, using transformation and wild type as factors. The probe sets were filtered for a 2-fold or greater change in expression in RB and SBT, then filtered for a 1.5-fold expression level difference in M. Differentially expressed genes were ranked by P values, and genes with a P value of ≤0.05 were considered differentially expressed at a statistically significant level. Gene annotation was carried out based on similarity scores in BLASTX comparisons against sequences contained in the Harvest: Citrus database (http://harvest.ucr.edu/). Differentially expressed genes were further analyzed using Map-Man Bin (http://ppdb.tc.cornell.edu/default.aspx). The subcellular localization of differentially expressed peroxidase genes was predicted using TargetP (http://www.cbs.dtu.dk/services/TargetP/) and SUBA3 (http://suba.plantenergy.uwa.edu.au/). Peroxidase classification was based on PeroxiBase analysis (https://peroxibase.toulouse.inra.fr/tools/peroxiscan.php).

Starch analysis

Starch contents in various calli (20-day-old) were detected by the anthrone reagent method according to Chen et al. [59]. The procedure of starch granules isolation was based on the method described by Ritte et al. [60] with minor modifications. Five grams from each callus were mixed with 10 ml extraction buffer [100 mM N-2-hydroxyethylpiperazine-N-2-ethanesulfonic acid (HEPES)-KOH (pH 8.0), 1 mM ethylenediaminetetraacetic acid (EDTA), and 0.05% (v/v) Triton-X-100] and homogenized for 20 s using a Waring blender. The homogenate was filtered through 3 layers of Micracloth (Calbiochem), and the pooled filtrates were subsequently centrifuged for 5 min at 1000 g. The supernatant is referred to as the soluble fraction. The

remaining pellet was then suspended in 5 ml of extraction buffer. The homogenate was centrifuged for 5 min at 1000 g. The supernatant was discarded, and the pellet was suspended in 2 ml of extraction buffer. Subsequently, the starch suspension was layered on the top of a 5 ml cushion consisting of 90% (v/v) Percoll (GE Healthcare Bio-Sciences AB, Uppsala, Sweden) and 10% (v/v) extraction buffer, the mixture was centrifuged for 15 min at 400 g. The pelleted granules were washed twice in extraction buffer, dried under vacuum condition, and stored at $-80°C$ until use.

α-Amylase activity was assayed by testing for the release of reducing sugars from soluble starch according to a previously described method [61] with appropriate modifications. The assay buffer consisted of 50 mM Na-acetate and 10 mM $CaCl_2$, pH 5.2. Heat-treated extracts ($70°C$ for 15 min) were used to inactivate β-amylase. The substrate was 1% boiled soluble starch and incubation ($40°C$) lasted for up to 5 min. An aliquot (200 µl) was taken from the assay mixture, treated with 2 ml of 3,5-dinitrosalicylic acid (DNS) solution (40 mM DNS, 400 mM NaOH, and 1 M K-Na tartrate), then heated for 10 min at $100°C$. After dilution with distilled water (up to 5 ml), the A520 was taken, and the reducing power was evaluated using a standard curve obtained using maltose. α-Amylase activity was expressed as mg of maltose produced per gram of tissue per minute.

Soluble sugar content measurement

Twenty-day-old calli were were washed 5 times using distilled water to remove the soluble sugar from the medium. Soluble sugar contents were quantified using gas chromatography. Two grams of fresh calli were homogenized and reconstituted in 80% (v/v) methanol for 30 min at $70°C$. After centrifugation at 4000 g for 10 min, the supernatant was withdrawn and diluted to a volume of 10 ml; 0.2 ml of methyl-α-D-glucopyranoside and phenyl β-D-glucopyranoside was added as an internal standard. The procedure for derivatization was performed as described by Bartolozzi et al. [62]. The derivatized samples were injected into an Agilent 6890 N gas chromatograph (Agilent, Palo Alto, CA, USA) using an Agilent 7683 autosampler.

Measurement of ABA levels

Various calli for abscisic acid (ABA) quantification were prepared according to the method described by Pan et al. [63] with some modifications. Calli (0.8 g per sample) were ground into powder in liquid nitrogen, and each sample was transferred to 10 ml screw-cap tubes. Two microliters of extraction solvent [2-propanol: H_2O: concentrated HCl (2: 1: 0.002, v/v/v)] was added to each tube and shaken at 200 rpm for 30 min at $4°C$. Subsequently, dichloromethane (4 ml) was added to each sample and the mixture was continually shaken for 30 min at 4°C. The mixtures were centrifuged at 13000 g for 5 min, then the lower phase was transferred into a screw-cap tube and concentrated using the nitrogen. The samples were redissolved in 0.2 ml of methanol and filtered with 0.22 μm organic membrane filters for analysis via HPLC electrospray ionization tandem mass spectrometry (HPLC-ESI-MS/ MS). An Agilent 1100 HPLC (Agilent Technologies, Palo Alto, CA, USA), a Waters C18 column (150 × 2.1 mm, 5 μm), and the API3000 MS-MRM (Applied Biosystems, Foster City, CA, USA) were used for ABA measure. The reaction monitoring acquisition of the transition 263/153 was used for quantitation of ABA extracts.

ROS levels and the activities of related enzymes

Twenty-day-old calli were harvested for ROS analysis. *In situ* accumulation of O_2^- was examined based on histochemical staining by nitroblue tetrazolium (NBT) [64]. H_2O_2 determination was based on the fact that hydroperoxides form a specific complex with titanium ($Ti4^+$) that can be measured by colorimetry, as described by Brennan and Frekel [65].

NADPH oxidase (NOX, EC 1.6.3.1) activity in the callus samples was determined using a commercial plant NADPH oxidase detection kit (GMS50096.3, Genmed Scientifics Inc. USA) according to the manufacturer's instructions. Extraction of superoxide dismutase (SOD, EC 1.15.1.1) and catalase (CAT, EC 1.11.1.6) was conducted as previously described [66]. SOD activity was spectrophotometrically measured using a photochemical assay system based on the inhibition of NBT reduction, and one unit of SOD was defined as the enzyme quantity that inhibited NBT photoreduction by 50% [67]. The CAT activity was assessed by monitoring the decrease in absorbance at 240 nm resulting from H_2O_2 consumption, and one unit of CAT activity was defined as a 0.01 reduction in absorbance units per min [66].

Western blot analysis

Total proteins of *M. hupehensis* calli were prepared via the phenol extraction protocol described by Pan *et al.* [16]. The total proteins were quantified using a Bio-Rad protein assay kit (Bio-Rad, Hercules, CA, USA) based on the Lowry method using bovine serum albumin (BSA) as standard. Anti-CrtB antibodies were generated through immunizing rabbits using a peptide that contains 117 amino acids in the C-terminus of CrtB [28]. Subsequent protein separation and Western blot analysis were performed accordingly to Cao *et al.* [28].

Pigment analyses

Carotenoid extraction and analysis using reversed-phase high-performance liquid chromatography (RP-HPLC) was conducted as previously described [28]. Because of

carotenoid esters in *M. hupehensis* calli, the extracts were saponified with 15% (w/v) KOH: methanol. The carotenoids were identified by their characteristic absorption spectra and typical retention time which were based on the literature and standards of the CaroNature Co. (Bern, Switzerland). The quantification of the carotenoids was achieved using calibration curves for violaxanthin, lutein, antheraxanthin, phytoene, α-carotene, β-carotene, and lycopene; phytofluene was quantified as phytoene, luteoxanthin was quantified as lutein, and zeaxanthin was quantified as antheraxanthin.

Total anthocyanins were measured using a spectrophotometric differential pH method following the procedure of Yuan *et al.* [68] with a minor modification. Frozen samples (400 mg) were crushed into powder and extracted separately with 2 ml of pH 1.0 buffer containing 50 mM KCl and 150 mM HCl as well as with 2 ml of pH 4.5 buffer containing 400 mM sodium acetate and 240 mM HCl. The mixtures were centrifuged at 12000 g for 15 min at 4°C. Supernatants were collected and diluted for direct measurement of absorbance at 510 nm. Total anthocyanin content was calculated using the following equation: amount (μg/g FW) = $(A_{pH1} - A_{pH4.5}) \times 1000 \times 484.8/24825 \times 6$. The number 484.8 is the molecular mass of cyanidin-3-glucoside chloride and 24825 is its molar absorptivity at 510 nm. Six is the dilution factor in this experiment.

Microscopy analyses

Protoplasts from the calli were generated as previously described [69], then protoplast suspensions were dropped onto microscope slides to observe the plastid modes. Light microscopy of various orange tissues of Hongkong kumquats was performed using a frozen sectioning technique with a Leica CM1900 (Leica, Germany). An optical microscope (BX61, Olympus) equipped with a DP70 camera was used in tandem with a differential interference contrast (DIC) technique.

Transmission electron microscopy (TEM) analysis was performed according to Cao *et al.* [28]. Samples were prepared using a normal fixation process with 2.5% glutaraldehyde adjusted to pH 7.4, and a 0.1 M phosphate buffer with 2% OsO_4. The preparations were dehydrated and embedded in epoxy resin and SPI-812, respectively. Ultrathin sections obtained with a Leica UC6 ultramicrotome were stained with uranyl acetate and subsequently with lead citrate. Image recording was performed with a HITACHI H-7650 transmission electron microscope at 80 KV and a Gatan 832 CCD camera.

Starch granule morphology was examined with a scanning electron microscope (SEM). The samples were mounted on studs, sputter coated with gold (Balzers, JFC-1600), and examined under a JSM-6390LV SEM (JEOL, Japan).

Statistical analysis

The SAS statistical software was used to compare the statistical difference based on the Student-Newman-Keuls' multiple range test at significance levels of $P < 0.05$ (*) and $P < 0.01$ (**), respectively. A linear regression calculation was implemented in a Microsoft Excel® spreadsheet.

Availability of supporting data

The raw data sets supporting the results of this article are available in the Gene Expression Omnibus (GEO) repository under accession No. GSE61633 at website: http://www.ncbi.nlm.nih.gov/geo/query/acc.cgi?acc = GSE61633, and LabArchives (doi:10.6070/H4XW4GRZ).

Additional files

Additional file 1: List of differentially expressed genes (including annotation and MapMan Bin) between the wild types and ECMs.

Additional file 2: Validation of the microarray expression data using qRT-PCR. (A) Relative transcript levels of 10 genes; M, RB, and SBT represent Marsh grapefruit, Star Ruby grapefruit, and Sunburst mandarin, respectively. Transcript levels are expressed relative to WT (wild type). (B) Microarray probes used for qRT-PCR validation and the annotation. (C) Comparison between the gene expression ratios obtained from microarray data and qRT-PCR.

Additional file 3: Number of differentially expressed genes involved in stress and redox, hormone metabolism, and secondary metabolism. M, RB, and SBT represent Marsh grapefruit, Star Ruby grapefruit, and Sunburst mandarin, respectively. Positive axes represent the number of genes up-regulated, and negative axes represent the number of genes down-regulated.

Additional file 4: Function annotation of predominantly detected genes. The graphs show number of genes annotated as the same function. The predominant functions are listed on the right of the graphs. M, RB, and SBT represent Marsh grapefruit, Star Ruby grapefruit, and Sunburst mandarin, respectively.

Additional file 5: Subcellular localization and class of differentially expressed peroxidase genes in this study.

Additional file 6: ABA contents in the ECMs and their wild-type controls. Columns and bars represent the means and ± SD, respectively (n = 3 replicate experiments). **indicates that the values are significantly different compared with wild type at the significance level of P < 0.01. M-33, RB-4, and SBT-6 represent the ECM lines of Marsh grapefruit, Star Ruby grapefruit, and Sunburst mandarin, respectively.

Additional file 7: Differentially expressed ROS-induced genes from the microarray data were verified in the calli via RT-PCR analysis. M, RB, and SBT represent Marsh grapefruit, Star Ruby grapefruit, and Sunburst mandarin, respectively. Trangenic calli (35S:: CrtB) were the representative ECMs, M-33, RB-4, and SBT-6, which were also used for Affymetrix microarray analysis.

Additional file 8: Dry weight analysis of the wild types and ECMs. Columns and bars represent the means and ± SD, respectively (n = 3 biological replicate experiments). **indicates that the values are significantly different at the significance level of P < 0.01.

Additional file 9: Soluble sugar (fructose, glucose, and sucrose) contents in the ECMs and wild types. 35S:: CrtB represents the ECM lines. RB, M, HQC, and SBT represent Star Ruby grapefruit, Marsh grapefruit, Cara Cara navel orange, and Sunburst mandarin, respectively. Columns and bars represent the means and ± SD, respectively (n = 3 biological replicate experiments). * and ** indicate that the values are significantly different compared with wild type at the significance levels of P < 0.05 and P < 0.01, respectively.

Additional file 10: Cellular investigation and qRT-PCR analysis of the roots of Hongkong kumquats (*F. hindsii* Swingle). (A) Cellular investigation of Hongkong kumquats. Ultrastructural inspection of dark-grown roots and embryoids. 35S:: CrtB represents the transgenic line. s, starch granules; p, plastoglobules; th, thylakoids; c, carotenoid crystal and characteristic internal membrane; ch, chromoplast; am, amyloplast. (B) qRT-PCR analysis of starch related genes in the roots of Hongkong kumquats. AMY, citrus sinensis alpha-amylase-like; SD1, α-amylase; SD2, α-amylase 3. (C) Expression levels of 6 stress-related and senescence-related genes that had been identified as differentially expressed between the ECMs and their wild types in microarray and qRT-PCR analyses. 1, WRKY75 (Cit.341.1.S1_s_at); 2, Protease inhibitor (Cit.16616.1.S1_at); 3, Universal stress protein (USP) family protein (Cit.14892.1.S1_at); 4, Hydroxyproline-rich glycoprotein family protein (Cit.37479.1.S1_at); 5, Senescence-related gene (Cit.14916.1.S1_at); 6, Plastocyanin-like domain-containing protein (Cit.5498.1.S1_at). (D) qRT-PCR analysis of key flavonoid biosynthetic genes in the roots of transgenic Hongkong kumquat and its control. CHS, chalcone synthase; CHI, chalcone isomerase. All transcript levels are expressed relative to WT (wild type). * and ** indicate that values are significantly different compared with wild type at the significance levels of P < 0.05 and P < 0.01, respectively.

Additional file 11: Contents of various carotenoids in the *M. hupehensis* calli. Vio., Violaxanthin; Luteo., Luteoxanthin; Lut., lutein; Phy., Phytoene; Phytof., Phytofluene; Anth., Antheraxanthin; Zea., Zeaxanthin; β-Car., β-Carotene; Lcy., Lycopene. WT represents the light-cultured wild-type apple callus and 35S:: CrtB represents the light-cultured transgenic apple callus with overexpression of CrtB. Columns and bars represent the means and ± SD, respectively (n = 2 replicate experiments). * and ** indicate that the values are significantly different compared with wild type at the significance levels of P < 0.05 and P < 0.01, respectively.

Additional file 12: Specific primer pairs used for qRT-PCR analysis in the present study.

Competing interests

The authors declare that they have no competing interests.

Authors' contributions

HC was responsible for generating the Microarray data and for the interpretation of the data. JW carried out qRT-PCR experiments. HC, XD, YH, QM, and FZ performed cytological analysis and measured starch, sugar, ROS contents. YD participated in the statistical analyses. HC established the transgenic plants and calli. HC, JW, HC, JZ, QX, JX, and XD interpreted the experimental dada and participated in writing the manuscript. XD supervised the research. All authors read and approved the final manuscript.

Acknowledgements

This research was supported by National Basic Research Program of China (No. 2011CB100601) and National Natural Science Foundation of China (31401841). We thank Professor Shih-Tung Liu (Taiwan, China) for providing the *Erwinia herbicola* CrtB gene. We thank Professor Li Li (Cornell University, USA) for her critical reading of this paper. We also thank Junli Ye (Huazhong Agricultural University, China), Jianbo Cao (Huazhong Agricultural University, China), Dongqin Li (Huazhong Agricultural University, China), and Baoping Chen (Medical College, Wuhan University) for technical assistance.

Author details

[1]Key Laboratory of Horticultural Plant Biology (Ministry of Education), Huazhong Agricultural University, 430070 Wuhan, Hubei, China. [2]College of Horticulture, Agricultural University of Hebei, 071001 Baoding, Hebei, China. [3]Present address: College of Plant Science, Tarim University, 843300 Alar, China. [4]Present address: Shanxi Agricultural University, 030801 Taigu, Shanxi, China.

References

1.   Johnson JD. Do carotenoids serve as transmembrane radical channels? Free Radical Bio Med. 2009;47:321–3.

2. Cazzonelli CI, Pogson BJ. Source to sink: regulation of carotenoid biosynthesis in plants. Trends Plant Sci. 2010;15:266–74.

3. Moise AR, Al-Babili S, Wurtzel ET. Mechanistic aspects of carotenoid biosynthesis. Chem Rev. 2013;114:164–93.

4. Cazzonelli CI, Cuttriss AJ, Cossetto SB, Pye W, Crisp P, Whelan J, et al. Regulation of carotenoid composition and shoot branching in Arabidopsis by a chromatin modifying histone methyltransferase, SDG8. Plant Cell. 2009;21:39–53.

5. de Saint GA, Bonhomme S, Boyer FD, Rameau C. Novel insights into strigolactone distribution and signalling. Curr Opin Plant Biol. 2013;16:583–9.

6. Alder A, Jamil M, Marzorati M, Bruno M, Vermathen M, Bigler P, et al. The Path from β-carotene to carlactone, a strigolactone-like plant hormone. Science. 2012;335:1348–51.

7. Ramel F, Mialoundama AS, Havaux M. Nonenzymic carotenoid oxidation and photooxidative stress signalling in plants. J Exp Bot. 2013;64:799–805.

8. Diretto G, Welsch R, Tavazza R, Mourgues F, Pizzichini D, Beyer P, et al. Silencing of beta-carotene hydroxylase increases total carotenoid and beta-carotene levels in potato tubers. BMC Pant Biol. 2007;7:11.

9. Diretto G, Al-Babili S, Tavazza R, Scossa F, Papacchioli V, Migliore M, et al. Transcriptional-metabolic networks in β-carotene-enriched potato tubers: the long and winding road to the Golden phenotype. Plant Physiol. 2010;154:899–912.

10. Kato M, Ikoma Y, Matsumoto H, Sugiura M, Hyodo H, Yano M. Accumulation of carotenoids and expression of carotenoid biosynthetic genes during maturation in citrus fruit. Plant Physiol. 2004;134:824–37.

11. Barsan C, Zouine M, Maza E, Bian WP, Egea I, Rossignol M, et al. Proteomic analysis of chloroplast-to-chromoplast transition in tomato reveals metabolic shifts coupled with disrupted thylakoid biogenesis machinery and elevated energy-production components. Plant Physiol. 2012;160:708–25.

12. Centeno DC, Osorio S, Nunes-Nesi A, Bertolo ALF, Carneiro RT, Araujo WL, et al. Malate plays a crucial role in starch metabolism, ripening, and soluble solid content of tomato fruit and affects postharvest softening. Plant Cell. 2011;23:162–84.

13. Zhang Y, Butelli E, De Stefano R, Schoonbeek HJ, Magusin A, Pagliarani C, et al. Anthocyanins double the shelf life of tomatoes by delaying overripening and reducing susceptibility to gray mold. Curr Biol. 2013;23:1094–100.

14. Fraser PD, Enfissi EM, Halket JM, Truesdale MR, Yu D, Gerrish C, et al. Manipulation of phytoene levels in tomato fruit: effects on isoprenoids, plastids, and intermediary metabolism. Plant Cell. 2007;19:3194–211.

15. Liu Q, Xu J, Liu YZ, Zhao XL, Deng XX, Guo LL, et al. A novel bud mutation that confers abnormal patterns of lycopene accumulation in sweet orange fruit (Citrus sinensis L. Osbeck). J Exp Bot. 2007;58:4161–71.

16. Pan ZY, Liu Q, Yun Z, Guan R, Zeng WF, Xu Q, et al. Comparative proteomics of a lycopene-accumulating mutant reveals the important role of oxidative stress on carotenogenesis in sweet orange (Citrus sinensis [L.] Osbeck). Proteomics. 2009;9:5455–70.

17. Wang YQ, Yang Y, Fei Z, Yuan H, Fish T, Thannhauser TW, et al. Proteomic analysis of chromoplasts from six crop species reveals insights into chromoplast function and development. J Exp Bot. 2013;64:949–61.

18. Li L, Yuan H. Chromoplast biogenesis and carotenoid accumulation. Arch Biochem Biophy. 2013;539:102–9.

19. Horner HT, Healy RA, Ren G, Fritz D, Klyne A, Seames C, et al. Amyloplast to chromoplast conversion in developing ornamental tobacco floral nectaries provides sugar for nectar and antioxidants for protection. Am J Bot. 2007;94:12–24.

20. Tranbarger TJ, Dussert S, Joet T, Argout X, Summo M, Champion A, et al. Regulatory mechanisms underlying oil palm fruit mesocarp maturation, ripening, and functional specialization in lipid and carotenoid metabolism. Plant Physiol. 2011;156:564–84.

21. Leitner-Dagan Y, Ovadis M, Shklarman E, Elad Y, David DR, Vainstein A. Expression and functional analyses of the plastid lipid-associated protein CHRC suggest its role in chromoplastogenesis and stress. Plant Physiol. 2006;142:233–44.

22. Kim J, Rensing K, Douglas C, Cheng K. Chromoplasts ultrastructure and estimated carotene content in root secondary phloem of different carrot varieties. Planta. 2010;231:549–58.

23. Nogueira M, Mora L, Enfissi EM, Bramley PM, Fraser PD. Subchromoplast sequestration of carotenoids affects regulatory mechanisms in tomato lines expressing different carotenoid gene combinations. Plant Cell. 2013;25:4560–79.

24. Tanaka Y, Sasaki N, Ohmiya A. Biosynthesis of plant pigments: anthocyanins, betalains and carotenoids. Plant J. 2008;54:733–49.

25. Butelli E, Titta L, Giorgio M, et al. Enrichment of tomato fruit with health-promoting anthocyanins by expression of select transcription factors. Nat Biotechnol. 2008;26:1301–8.

26. Chiou CY, Yeh KW. Differential expression of MYB gene (OgMYB1) determines color patterning in floral tissue of Oncidium Gower Ramsey. Plant Mol Biol. 2008;66:379–88.

27. Martin C. The interface between plant metabolic engineering and human health. Curr Opinn Biotech. 2012;24:344–53.

28. Cao HB, Zhang JC, Xu JD, Ye JL, Yun Z, Xu Q, et al. Comprehending crystalline β-carotene accumulation by comparing engineered cell models and the natural carotenoid-rich system of citrus. J Exp Bot. 2012;63:4403–17.

29. Desikan R, Soheila AH, Hancock JT, Neill SJ. Regulation of the Arabidopsis transcriptome by oxidative stress. Plant Physiol. 2001;127:159–72.

30. Vandenabeele S, Van Der Kelen K, Dat J, Gadjev I, Boonefaes T, Morsa S, et al. A comprehensive analysis of hydrogen peroxide-induced gene expression in tobacco. Proc Natl Acad Sci U S A. 2003;100:16113–8.

31. Pucciariello C, Parlanti S, Banti V, Novi G, Perata P. Reactive oxygen species-driven transcription in Arabidopsis under oxygen deprivation. Plant Physiol. 2012;159:184–96.

32. Lindgren LO, Stalberg KG, Hoglund AS. Seed-specific overexpression of an endogenous Arabidopsis phytoene synthase gene results in delayed germination and increased levels of carotenoids, chlorophyll, and abscisic acid. Plant Physiol. 2003;132:779–85.

33. Inzé A, Vanderauwera S, Hoeberichts FA, Vandorpe M, Van Gaever TIM, Van Breusegem F. A subcellular localization compendium of hydrogen peroxide-induced proteins. Plant Cell Environ. 2012;35:308–20.

34. Orozco-Cárdenas ML, Narváez-Vásquez J, Ryan CA. Hydrogen peroxide acts as a second messenger for the induction of defense genes in tomato plants in response to wounding, systemin, and methyl jasmonate. Plant Cell. 2001;13:179–91.

35. Bai C, Rivera SM, Medina V, et al. An in vitro system for the rapid functional characterization of genes involved in carotenoid biosynthesis and accumulation. Plant J. 2014;77:464–75.

36. Asada K. Production and scavenging of reactive oxygen species in chloroplasts and their functions. Plant Physiol. 2006;141:391–6.

37. Willekens H, Chamnongpol S, Davey M, Schraudner M, Langebartels C, et al. Catalase is a sink for H_2O_2 and is indispensable for stress defense in C3 plants. EMBO J. 1997;16:4806–16.

38. Maass D, Arango J, Wust F, Beyer P, Welsch R. Carotenoid crystal formation in Arabidopsis and carrot roots caused by increased phytoene synthase protein levels. PLoS One. 2009;4:e6373.

39. Fuentes P, Pizarro L, Moreno JC, Handford M, Rodriguez-Concepcion M, Stange C. Light-dependent changes in plastid differentiation influence carotenoid gene expression and accumulation in carrot roots. Plant Mol Biol. 2012;79:47–59.

40. Zeng YL, Pan ZY, Ding YD, Zhu AD, Cao HB, Xu Q, et al. A proteomic analysis of the chromoplasts isolated from sweet orange fruits [Citrus sinensis (L.) Osbeck]. J Exp Bot. 2011;62:5297–309.

41. Sparla F, Costa A, Schiavo FL, Pupillo P, Trost P. Redox regulation of a novel plastid-targeted β-amylase of Arabidopsis. Plant Physiol. 2006;141:840–50.

42. Sass-Kiss A, Kiss J, Milotay P, Kerek MM, Toth-Markus M. Differences in anthocyanin and carotenoid content of fruits and vegetables. Food Res Int. 2005;38:1023–9.

43. Bureau S, Renard CM, Reich M, Ginies C, Audergon JM. Change in anthocyanin concentrations in red apricot fruits during ripening. LWT-Food Sci Technol. 2009;42:372–7.

44. Lancaster JE, Grant JE, Lister CE, Taylor MC. Skin color in apples—influence of copigmentation and plastid pigments on shade and darkness of red color in five genotypes. J Am Soc Hortic Sci. 1994;119:63–9.

45. Kim C, Apel K. $^1O_2$-mediated and EXECUTER dependent retrograde plastid-to-nucleus signaling in norflurazon-treated seedlings of Arabidopsis thaliana. Mol Plant. 2013;6:1580–91.

46. Li YY, Mao K, Zhao C, Zhao XY, Zhang HL, Shu HR, et al. MdCOP1 ubiquitin E3 ligases interact with MdMYB1 to regulate light-induced anthocyanin biosynthesis and red fruit coloration in apple. Plant Physiol. 2012;160:1011–22.

47. Davison P, Hunter C, Horton P. Overexpression of beta-carotene hydroxylase enhances stress tolerance in Arabidopsis. Nature. 2002;418:203–6.

48. Singh D, Maximova S, Jensen P, Lehman B, Ngugi H, McNellis T. FIBRILLIN 4 is required for plastoglobule development and stress resistance in apple and Arabidopsis. Plant Physiol. 2010;154:1281–93.

49. Page M, Sultana N, Paszkiewicz K, Florance H, Smirnoff N. The influence of ascorbate on anthocyanin accumulation during high light acclimation in Arabidopsis thaliana: further evidence for redox control of anthocyanin synthesis. Plant Cell Environ. 2011;35:388–404.

50. Ring L, Yeh SY, Hücherig S, et al. Metabolic interaction between anthocyanin and lignin biosynthesis is associated with peroxidase FaPRX27 in strawberry fruit. Plant Physiol. 2013;163:43–60.

51. Murashige T, Tucker DH. Growth factor requirements of *Citrus* tissue culture. Proc First Int Citrus Symp. 1969;3:1155–61.

52. Zhang JC, Tao NG, Xu Q, Zhou WJ, Cao HB, Xu JA, et al. Functional characterization of Citrus PSY gene in Hongkong kumquat (*Fortunella hindsii* Swingle). Plant Cell Rep. 2009;28:1737–46.

53. Liu YZ, Liu Q, Tao NG, Deng XX. Efficient isolation of RNA from fruit peel and pulp of ripening navel orange (*Citrus sinensis* Osbeck). J Huazhong Agric Univ. 2006;25:300–4.

54. Hu CG, Honda C, Kita M, Zhang Z, Tsuda T, Moriguchi T. A simple protocol for RNA isolation from fruit trees containing high levels of polysaccharides and polyphenol compounds. Plant Mol Biol Rep. 2002;20:69a–g.

55. Liu Z, Ge XX, Wu XM, Kou SJ, Chai LJ, Guo WW. Selection and validation of suitable reference genes for mRNA qRT-PCR analysis using somatic embryogenic cultures, floral and vegetative tissues in citrus. Plant Cell Tiss Org Cult. 2013;113:469–81.

56. Espley RV, Hellens RP, Putterill J, Stevenson DE, Kutty-Amma S, Allan AC. Red colouration in apple fruit is due to the activity of the MYB transcription factor, MdMYB10. Plant J. 2007;49:414–27.

57. Bolstad BM, Irizarry RA, Åstrand M, Speed TP. A comparison of normalization methods for high density oligonucleotide array data based on variance and bias. Bioinformatics. 2003;19:185–93.

58. Gautier L, Cope L, Bolstad BM, Irizarry RA. affy—analysis of Affymetrix GeneChip data at the probe level. Bioinformatics. 2004;20:307–15.

59. Chen MH, Liu LF, Chen YR, Wu HK, Yu SM. Expression of α-amylases, carbohydrate metabolism, and autophagy in cultured rice cells is coordinately regulated by sugar nutrient. Plant J. 1994;6:625–36.

60. Ritte G, Lorberth R, Steup M. Reversible binding of the starch-related R1 protein to the surface of transitory starch granules. Plant J. 2000;21:387–91.

61. Guglielminetti L, Yamaguchi J, Perata P, Alpi A. Amylolytic activities in cereal seeds under aerobic and anaerobic conditions. Plant Physiol. 1995;109:1069–76.

62. Bartolozzi F, Bertazza G, Bassi D. Simultaneous determination of soluble sugars and organic acids as their trimethylsilyl derivatives in apricot fruits by gas-liqiud chromatography. J Chromatogr A. 1997;758:99–107.

63. Pan X, Welti R, Wang X. Quantitative analysis of major plant hormones in crude plant extracts by high-performance liquid chromatography-mass spectrometry. Nat Protoc. 2010;5:986–92.

64. Huang XS, Luo T, Fu XZ, Fan QJ, Liu JH. Cloning and molecular characterization of a mitogen-activated protein kinase gene from Poncirus trifoliata whose ectopic expression confers dehydration/drought tolerance in transgenic tobacco. J Exp Bot. 2011;62:5191–206.

65. Brennan T, Frenkel C. Involvement of hydrogen peroxide in the regulation of senescence in pear. Plant Physiol. 1977;59:411.

66. Huang XS, Liu JH, Chen XJ. Overexpression of PtrABF gene, a bZIP transcription factor isolated from Poncirus trifoliata, enhances dehydration and drought tolerance in tobacco via scavenging ROS and modulating expression of stress-responsive genes. BMC Plant Biol. 2010;10:230.

67. Beauchamp C, Fridovich I. Superoxide dismutase: improved assays and an assay applicable to acrylamide gels. Ana Biochem. 1971;44:276–87.

68. Yuan Y, Chiu LW, Li L. Transcriptional regulation of anthocyanin biosynthesis in red cabbage. Planta. 2009;230:1141–53.

69. Grosser J, Gmitter F. Protoplast fusion and citrus improvement. Plant Breed Rev. 1990;8:339–74.

DOF AFFECTING GERMINATION 2 is a positive regulator of light-mediated seed germination and is repressed by DOF AFFECTING GERMINATION 1

Silvia Santopolo[1], Alessandra Boccaccini[1,2], Riccardo Lorrai[1,2], Veronica Ruta[1], Davide Capauto[1], Emanuele Minutello[1], Giovanna Serino[1], Paolo Costantino[1] and Paola Vittorioso[1,2*]

Abstract

Background: The transcription factor DOF AFFECTING GERMINATION1 (DAG1) is a repressor of the light-mediated seed germination process. DAG1 acts downstream PHYTOCHROME INTERACTING FACTOR3-LIKE 5 (PIL5), the master repressor, and negatively regulates gibberellin biosynthesis by directly repressing the biosynthetic gene AtGA3ox1. The Dof protein DOF AFFECTING GERMINATION (DAG2) shares a high degree of aminoacidic identity with DAG1. While DAG1 inactivation considerably increases the germination capability of seeds, the dag2 mutant has seeds with a germination potential substantially lower than the wild-type, indicating that these factors may play opposite roles in seed germination.

Results: We show here that DAG2 expression is positively regulated by environmental factors triggering germination, whereas its expression is repressed by PIL5 and DAG1; by Chromatin Immuno Precipitation (ChIP) analysis we prove that DAG1 directly regulates DAG2. In addition, we show that Red light significantly reduces germination of dag2 mutant seeds.

Conclusions: In agreement with the seed germination phenotype of the dag2 mutant previously published, the present data prove that DAG2 is a positive regulator of the light-mediated seed germination process, and particularly reveal that this protein plays its main role downstream of PIL5 and DAG1 in the phytochrome B (phyB)-mediated pathway.

Keywords: DAG2, Seed germination, DAG1, Arabidopsis thaliana

Background

The DNA BINDING WITH ONE FINGER (Dof) proteins are a family of plant-specific transcription factors characterised by a single zinc-finger DNA-binding domain. So far Dof proteins have been identified in *Chlamydomonas reinharditii*, where only one *Dof* gene is present, in ferns, mosses and in higher plants [1-3].

The number of *Dof* genes varies depending on the species; bioinformatic analysis of the *Arabidopsis* and rice genome predicts 36 and 30 *Dof* genes, respectively [1], while 26 are present in barley [2], 31 in wheat [4], and 28 in sorghum [5]. Members of this family have been found to be involved in the regulation of diverse plant-specific processes. Although the biological role of many Dof proteins has not been clarified yet, a number of them has been shown to be involved in responses to light and phytohormones, as well as in seed development and germination [6-15].

Seed germination is regulated by environmental factors such as light, temperature and nutrients, and by phytohormones, particularly gibberellins (GA) and abscissic acid (ABA) [16]. The effect of light is mediated mainly by the photoreceptor phytochrome B (phyB) [17], and light modulates in opposite ways the levels of GA and ABA, as it induces GA biosynthesis and causes a reduction in ABA levels [18,19]. Among the factors involved in phyB-mediated GA-induced seed germination, the bHLH

* Correspondence: paola.vittorioso@uniroma1.it
[1]Dipartimento di Biologia e Biotecnologie "C. Darwin", Sapienza Università di Roma, Piazzale Aldo Moro 5, 00185 Rome, Italy
[2]Istituto Pasteur Fondazione Cenci Bolognetti, Dipartimento di Biologia e Biotecnologie "C. Darwin", Sapienza Università di Roma, Piazzale Aldo Moro 5, 00185 Rome, Italy

transcription factor PHYTOCHROME INTERACTING FACTOR 3-LIKE 5 (PIL5) represents the master repressor of this process in *Arabidopsis* [20].

We have previously shown that inactivation of the Dof proteins DAG1 and DAG2 affects in opposite ways seed germination: *dag2* mutant seeds required more light and GA than wild-type seeds to germinate, whereas germination of *dag1* seeds was less dependent on these factors [7,8,21].

Recently, we have also pointed out that DAG1 acts as a negative regulator in the phyB-mediated pathway: *DAG1* gene expression is reduced in seeds irradiated for 24 hours with Red light, and this reduction is dependent on PIL5; in *pil5* mutant seeds *DAG1* expression is reduced irrespective of light conditions, indicating that DAG1 acts downstream of PIL5. Moreover, DAG1 negatively regulates GA biosynthesis by directly repressing the GA biosynthetic gene *AtGA3ox1* [22]. Very recently we showed that in repressing *AtGA3ox1* DAG1 directly interacts with the GA INSENSITIVE (GAI) DELLA protein [23]. Furthermore, we pointed out that DAG1 plays a role also in embryo development, as inactivation of *DAG1* results in a significant number of embryo abnormalities [7,24], and simultaneous inactivation of both *DAG1* and *GAI* results in an embryo-lethal phenotype. Here, we provide evidence suggesting that DAG2, opposite to DAG1, functions as a positive regulator in the molecular pathway controlling seed germination, and that it is negatively regulated by DAG1.

Differently from DAG1, DAG2, although it is expressed during embryo development, is not likely to play a role in this process, as *dag2* mutant embryos develop similarly to wild-type embryos.

Results

DAG2 inactivation affects phyB-dependent seed germination

We have previously demonstrated that *dag2* mutant seeds have a reduced germination potential, as they are substantially more dependent than the wild-type on the stimuli that promote germination [8]. This germination phenotype is opposite to that of *dag1* mutant seeds. As we have recently shown that DAG1 is a component of the phyB-mediated pathway controlling seed germination in *Arabidopsis* [22,23], we set up to verify whether DAG2 is also a component of this regulatory network.

Since seed germination, although promoted mainly by phyB, may be induced also by phyA under very low light fluences [17], we checked whether Red (R) or Far Red (FR) light may control expression of the *DAG2* gene. Analysis of wild-type seeds exposed to phyB- or phyA-dependent conditions, according to Oh *et al*. 2006 [25], revealed that the *DAG2* gene is induced by exposure to R light (Figure 1A), whereas *DAG2* expression in seeds

exposed to FR light was not significantly different than in seeds kept in the dark (Figure 1B). To assess whether DAG2 plays its role under R light, we analysed seed germination under phyB-dependent conditions [22] using the *dag2* mutant previously characterised [8], compared to the corresponding wild-type (Ws-4). Germination of *dag2* mutant seeds was significantly lower than that of wild-type seeds (30% and 90%, respectively - Figure 1C), thus confirming that DAG2 plays a positive role in seed germination and showing that it acts in the phyB-mediated pathway.

Since water uptake is a fundamental requirement for seed germination, we verified whether expression of *DAG2* was regulated during imbibition. We performed RT-qPCR assays on wild-type (Ws-4) dry seeds, and on seeds imbibed under White (W) and R light or in the dark for 12 and 24 hrs. Figure 2A shows that, compared to the low amount present in dry seeds, *DAG2* expression in seeds was much increased following water uptake in the dark (2 and 4 fold, respectively, at 12 and 24 hrs). Interestingly, the increase in *DAG2* mRNA level in seeds exposed to W or R light was even higher, probably due to the effect of both light and imbibition (3.7 and 7.8 fold in W light and 4 and 7-fold in R light, at 12 and 24 hrs, respectively - Figure 2A). GUS histological assays, performed on seeds of the *DAG2:GUS* transgenic line [8], dry or imbibed 12 hours under W light or in the dark respectively, showed that the *DAG2* promoter was active only in the vascular tissue (Figure 2B).

DAG2 is directly regulated by DAG1

We have previously investigated the genetic interactions between the *DAG2* and *DAG1* genes by isolating the *dag2dag1* double mutant, and showed that *DAG1* is epistatic over *DAG2* [8]. Since the function of DAG2 appears to be opposite to that of DAG1, we verified whether DAG1 and DAG2 would mutually affect their expression, by performing an RT-qPCR analysis in *dag1* and *dag2* mutant seeds imbibed for 12 hours in the dark or under R light. As shown in Figure 3A, expression of *DAG2* is significantly (approximately 3-fold) increased by lack of DAG1, irrespective of light conditions. Conversely, *DAG1* expression level in wild-type and *dag2* mutant seeds was comparable, both in the dark and under R light (Figure 3B).

To assess whether DAG1 regulates *DAG2* by directly binding to the *DAG2* promoter *in vivo*, we performed chromatin immunoprecipitation (ChIP) assays, utilizing the *dag1DAG1-HA* line previously reported [22,23]. A scheme of the *DAG2* promoter is reported in Figure 3C, showing the positions of the PCR fragments amplified for the ChIP assays, each containing different numbers of Dof binding sites: 0 (a, b), 4 (c) and 7 sites (d). Consistently, anti-HA antibodies revealed that the amplification of

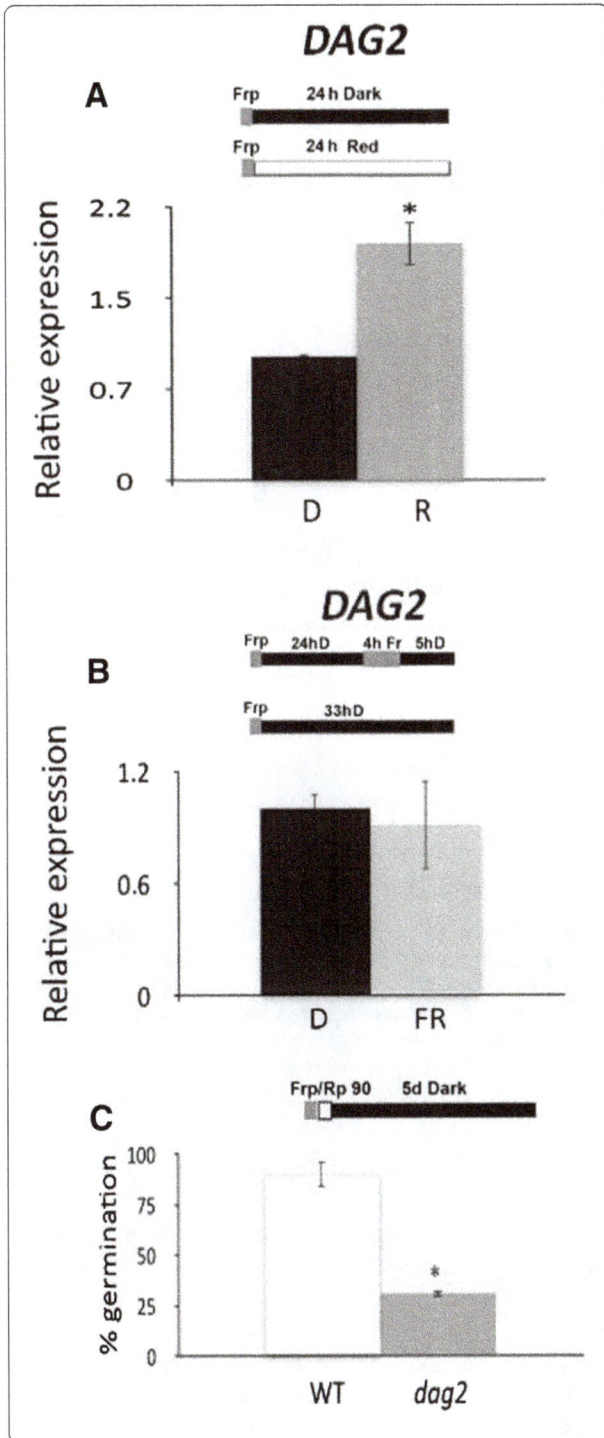

DAG2

A

DAG2

B

C

Figure 1 Mutation of *DAG2* affects seed germination under R light. Relative expression level of *DAG2* in wild-type seeds imbibed 24 hours in the dark (D), or under phyB-dependent conditions, **(A)**, and in the dark or under phyA-dependent conditions **(B)**. Relative expression levels were normalized with that of the *UBQ10* (*At4g05320*) gene, and are presented by the ratio of the corresponding mRNA level in Dark, which was set to 1. Similar results were obtained from three independent experiments, and a typical result is presented with SD values. Germination rates of wild-type and *dag2* mutant seeds, grown 5 days under phyB-dependent germination conditions **(C)**. Error bars = SEM. The diagram at top depicts the light treatment scheme for the experiment. FRp, Far Red pulse (40 µmol m^{-2} s^{-1}); Rp, Red pulse (90 µmol m^{-2} s^{-1}). Significative differences were analyzed by *t*-test (*P ≤ 0,05).

fragments c and d were the most efficient, compared to the positive control, the fragment B3 of the *AtGA3ox1* promoter bound by DAG1-HA, as previously reported [23]. On the contrary, the signal for fragments a and b was quite faint. No PCR product was present for any of the fragments in the sample precipitated without antibodies as

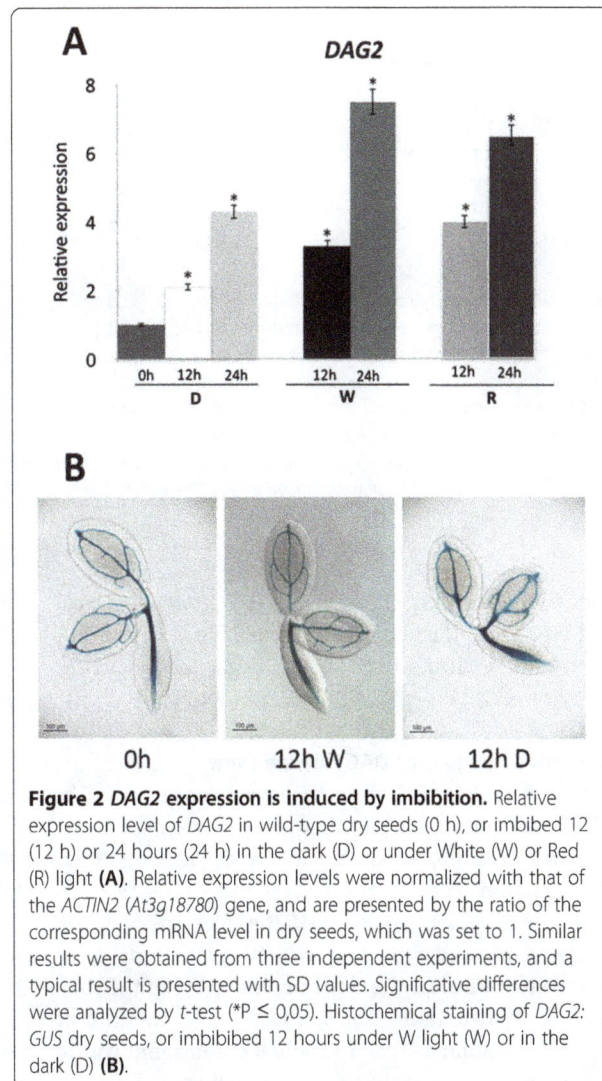

Figure 2 *DAG2* expression is induced by imbibition. Relative expression level of *DAG2* in wild-type dry seeds (0 h), or imbibed 12 (12 h) or 24 hours (24 h) in the dark (D) or under White (W) or Red (R) light **(A)**. Relative expression levels were normalized with that of the *ACTIN2* (*At3g18780*) gene, and are presented by the ratio of the corresponding mRNA level in dry seeds, which was set to 1. Similar results were obtained from three independent experiments, and a typical result is presented with SD values. Significative differences were analyzed by *t*-test (*P ≤ 0,05). Histochemical staining of *DAG2:GUS* dry seeds, or imbibed 12 hours under W light (W) or in the dark (D) **(B)**.

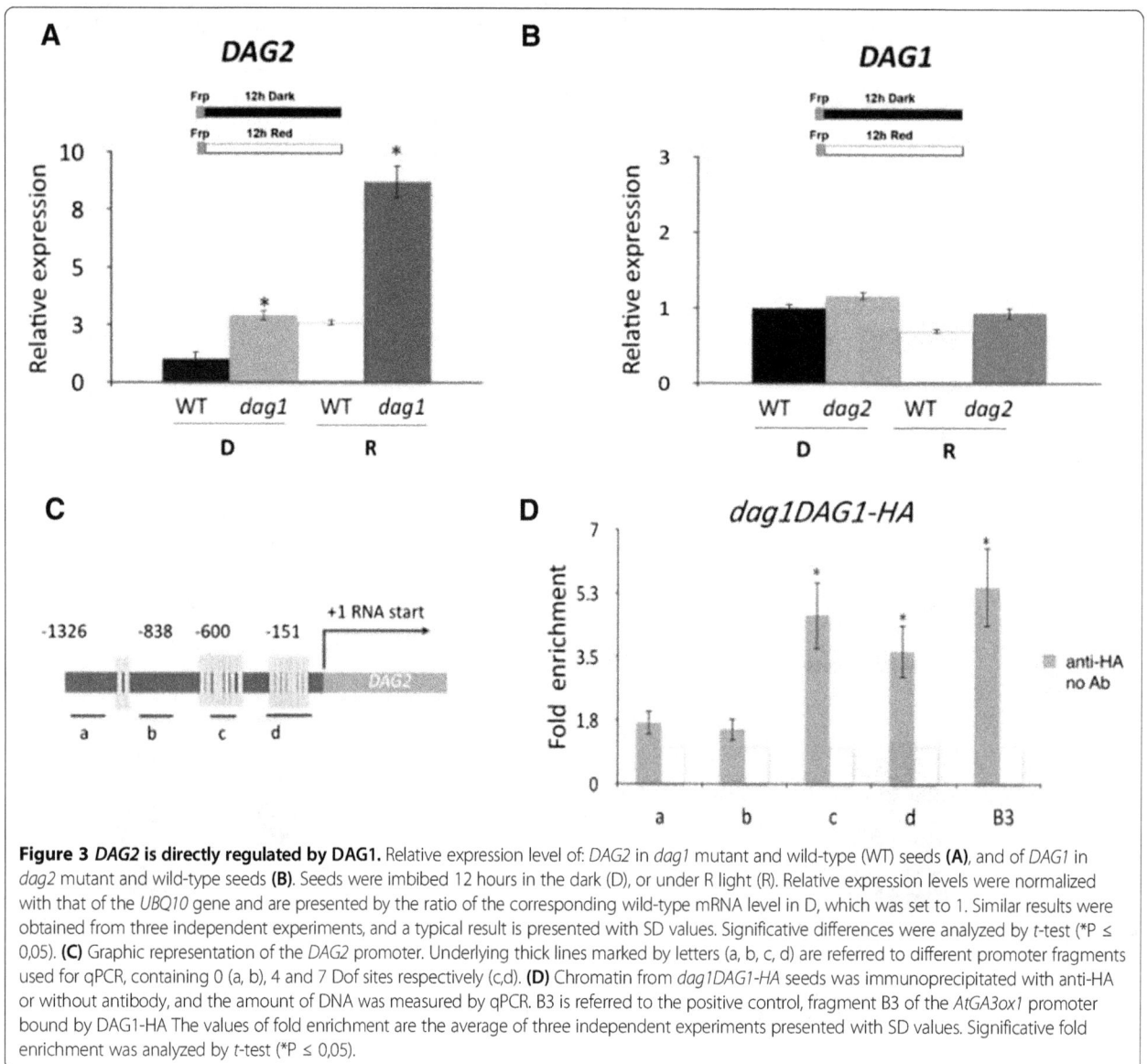

Figure 3 *DAG2* **is directly regulated by DAG1.** Relative expression level of: *DAG2* in *dag1* mutant and wild-type (WT) seeds **(A)**, and of *DAG1* in *dag2* mutant and wild-type seeds **(B)**. Seeds were imbibed 12 hours in the dark (D), or under R light (R). Relative expression levels were normalized with that of the *UBQ10* gene and are presented by the ratio of the corresponding wild-type mRNA level in D, which was set to 1. Similar results were obtained from three independent experiments, and a typical result is presented with SD values. Significative differences were analyzed by *t*-test (*P ≤ 0,05). **(C)** Graphic representation of the *DAG2* promoter. Underlying thick lines marked by letters (a, b, c, d) are referred to different promoter fragments used for qPCR, containing 0 (a, b), 4 and 7 Dof sites respectively (c,d). **(D)** Chromatin from *dag1DAG1-HA* seeds was immunoprecipitated with anti-HA or without antibody, and the amount of DNA was measured by qPCR. B3 is referred to the positive control, fragment B3 of the *AtGA3ox1* promoter bound by DAG1-HA The values of fold enrichment are the average of three independent experiments presented with SD values. Significative fold enrichment was analyzed by *t*-test (*P ≤ 0,05).

a negative control, and not even for the negative control on wild-type seeds (Figure 3D; Additional file 1: Figure S1). These results indicate that DAG1 negatively regulates *DAG2* by directly binding the *DAG2* promoter.

PIL5 negatively regulates DAG2 in the Dark

Since DAG1 and DAG2 seem to have opposite roles in the phyB-mediated seed germination pathway, we wondered whether PIL5, which positively regulates *DAG1*, might negatively control the expression of *DAG2*. To verify this hypothesis, we analysed the expression of the *DAG2* gene in wild-type and *pil5* mutant seeds after 12 hours imbibition in the dark or under R light. Interestingly, as shown in Figure 4, the relative amount of *DAG2* in *pil5* mutant seeds in the dark was significantly higher than in the wild-type, suggesting that PIL5

negatively regulates the expression of *DAG2* in the dark. On the other hand, *DAG2* expression level in R light does not depend on PIL5, as it is degraded following interaction with phyB.

The DELLA proteins GAI and RGA are negative regulators of seed germination, acting downstream of PIL5 [26]. In particular, we have recently shown that GAI and DAG1 mutually regulate their expression level and directly interact with each other [23]. Thus, we set to assess whether GAI and/or RGA might control the expression level of *DAG2*. Analysis of *DAG2* expression on *gai-t6* and *rga28* mutant seeds and on the corresponding Col-0 wild-type seeds, imbibed 12 hours in the dark or under R light, revealed that neither GAI nor RGA control *DAG2* expression, as the relative amount of *DAG2* mRNA was similar in the *gai-t6* and *rga28* single mutants compared

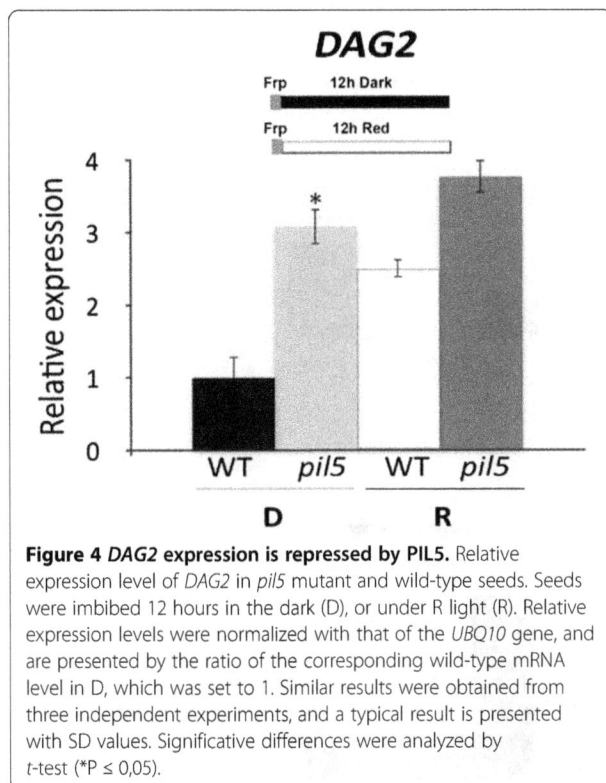

Figure 4 DAG2 expression is repressed by PIL5. Relative expression level of *DAG2* in *pil5* mutant and wild-type seeds. Seeds were imbibed 12 hours in the dark (D), or under R light (R). Relative expression levels were normalized with that of the *UBQ10* gene, and are presented by the ratio of the corresponding wild-type mRNA level in D, which was set to 1. Similar results were obtained from three independent experiments, and a typical result is presented with SD values. Significative differences were analyzed by *t*-test (*P ≤ 0,05).

to the wild-type, under both light conditions (Figure 5A, B). To verify whether DAG2 might regulate expression of these DELLA proteins, we analysed the expression of *GAI* and *RGA* in *dag2* mutant seeds compared to the wild-type. As shown in Figure 5C, the expression of the *RGA* gene was significantly increased in *dag2* mutant seeds, whereas *GAI* expression in wild-type and *dag2* mutant seeds was not significantly different, both in the dark and under R light (Figure 5D).

These results point to DAG2 as a positive component of light-mediated signalling pathway, downstream of PIL5 and in turn controlling the DELLA protein RGA in the phyB signalling pathway.

We then verified whether expression of some factors known to be involved in the phyA-signalling pathway [27] may be affected in *dag2* mutant seeds. In particular, we analysed expression of the FR light-regulated *ARABIDOPSIS THALIANA HOMEOBOX PROTEIN 2 (ATHB2)* and *PHYTOCHROME INTERACTING FACTOR 3-LIKE 2 (PIL2)* genes, of the GA-regulated *GA-STIMULATED ARABIDOPSIS 4* and *6 (GASA4* and *GASA6)*, and of the ABA signalling gene *ABA INSENSITIVE 4 (ABI4)*. Our data revealed that under phyA-dependent conditions neither expression of *ATHB2* and *PIL2*, nor that of *ABI4* were affected in the *dag2* mutant (Additional file 2: Figure S2) whereas expression of both *GASA4* and *GASA6* was downregulated, thus opening the possibility that DAG2 may also play a role in phyA signalling.

The *dag2* mutation alters GA metabolism

It has been shown that phyB controls the ratio of GA and ABA levels during seed germination by altering the expression of different GA and ABA metabolic genes through PIL5 [18,26]. In particular, DAG1 directly represses the GA biosynthetic gene *AtGA3ox1* in cooperation with GAI [22,23], and its inactivation affects expression of the ABA metabolic genes *ABA1*, *ABA2* and *CYP707A2* [22].

As DAG2 seems to have a role opposite to DAG1 in seed germination, we investigated whether DAG2 would regulate the expression of GA and ABA metabolic genes in germinating seeds. We performed a RT-qPCR analysis of the expression of the GA biosynthetic genes *AtGA3ox1*, *AtGA3ox2* and of the catabolic gene *AtGA2ox2* in *dag2* and wild-type seeds imbibed 12 hours in the dark or under R light. As shown in Figure 6A, the expression of both GA biosynthetic genes was significantly reduced in *dag2* mutant seeds irrespective of light conditions, whereas the catabolic gene *AtGA2ox2* was expressed similarly in *dag2* and wild-type seeds (Figure 6A).

As for ABA metabolism, we analysed the expression level of the biosynthetic genes *ABA1*, *ABA2*, *NCED6* and *NCED9*, and of the catabolic gene *CYP707A2*, on *dag2* and wild-type seeds imbibed 12 hours in the dark or under R light. The expression profile of the biosynthetic genes, as well as of the catabolic gene *CYP707A2* did not show significant differences in *dag2* and wild-type seeds (Figure 6B).

We have previously shown that the sensitivity of seeds to GA is affected by mutation of the *DAG2* gene: a concentration of GA 10-fold higher than for wild-type seeds was needed for *dag2* mutant seeds to attain 50% germination [8]. To verify whether GA affect *DAG2* expression, we carried out an RT-qPCR analysis on wild-type seeds imbibed 24 hours in the presence of GA or of paclobutrazol, an inhibitor of GA biosynthesis. Since GA metabolism is controlled by the ABA level [18], we also checked *DAG2* expression on wild-type seeds imbibed 24 hours in the presence of ABA. The results of this analysis did not show any significant difference in *DAG2* transcript levels in all conditions tested, clearly showing that the *DAG2* gene is not regulated by GA nor by ABA irrespective of light conditions (Figure 7).

Inactivation of the *DAG2* gene does not affect embryo development

We have recently shown that *DAG1* is expressed during embryo development, and that lack of DAG1 affects this process [24]. Thus, we set to assess whether also DAG2 is required for embryo development. We first analyzed the expression of *DAG2* during embryo development by histochemical GUS analysis of seeds of the *DAG2:GUS* transgenic line. GUS activity was observed in embryos at

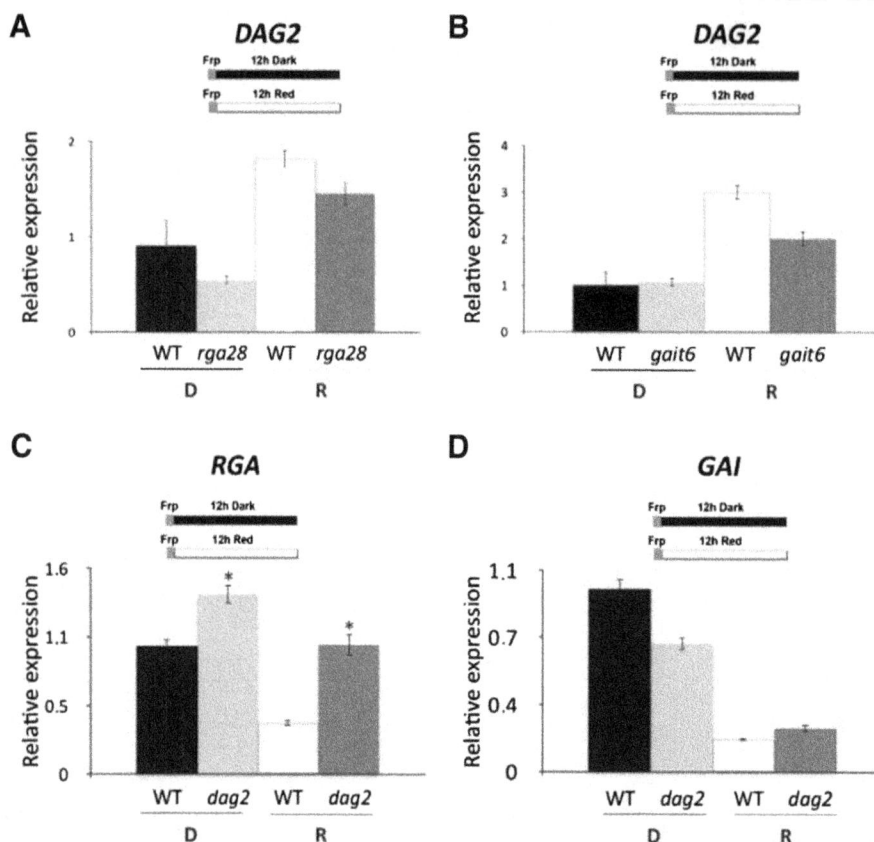

Figure 5 *DAG2* expression is regulated by RGA and GAI. Relative expression level of: *DAG2* in *rga28* **(A)**, and *gai-t6* mutant seeds **(B)**, and of *RGA* **(C)** or *GAI* **(D)** in *dag2* mutant seeds, compared to wild-type seeds. Seeds were imbibed 12 hours in the dark (D), or under R light (R). Relative expression levels were normalized with that of *PP2A* (*At1g13320*) **(A, B)**, or of *UBQ10* **(C, D)**, and are presented by the ratio of the corresponding wild-type mRNA level in D, which was set to 1. Similar results were obtained from three independent experiments, and a typical result is presented with SD values. Significative differences were analyzed by *t*-test (*P ≤ 0,05).

the heart, torpedo, and bent-cotyledon stages. Interestingly, GUS staining was extended to all cells at the heart stage, whereas from the torpedo stage on it was restricted to the procambium (Figure 8A). These results were confirmed and extended to later seed development stages by a RT-qPCR analysis on wild-type embryos at 13, 16 and 19 Days After Pollination (DAP), compared to mature seeds, to verify whether the *DAG2* gene was expressed also during seed maturation.

Expression of *DAG2*, at 13 and 16 DAP was extremely high (63- and 57-fold the basal level, respectively), and gradually decreased at 19 DAP (24-fold) compared to mature seeds (Figure 8B).

Despite the high expression level of the *DAG2* gene during embryo and seed development, microscopic analysis of *dag2* mutant embryos did not reveal any noticeable phenotypical alteration (Figure 8C).

Discussion

We had previously shown that the *dag2* mutant has seeds which require higher light fluences and higher GA

levels than wild-type ones to germinate [8], suggesting a positive role of the Dof transcription factor DAG2 in the regulation of seed germination.

Here, we have expanded our analysis of the function of DAG2 and we confirm the positive role of DAG2 in seed germination and provide molecular and genetic evidences that assign this protein to the phyB/PIL5 pathway.

To date the molecular pathway controlling seed germination has been partially elucidated. In this model PIL5 acts as the master repressor, which inhibits seed germination in the dark partly by activating the expression of the genes encoding the DELLA proteins RGA and GAI - which repress germination acting as negative GA signaling components - and of the transcription factors *ABA INSENSITIVE 3* and *5* (*ABI3* and *ABI5*) - which function as positive ABA signaling molecules [26,28]. Other transcription factors acting as repressors have been added in this pathway: the bHLH transcription factor SPATULA (SPT) [29], the C3H-type zinc finger protein SOMNUS (SOM) [30], and the Dof transcription factor DAG1, which we have shown to directly regulate the GA

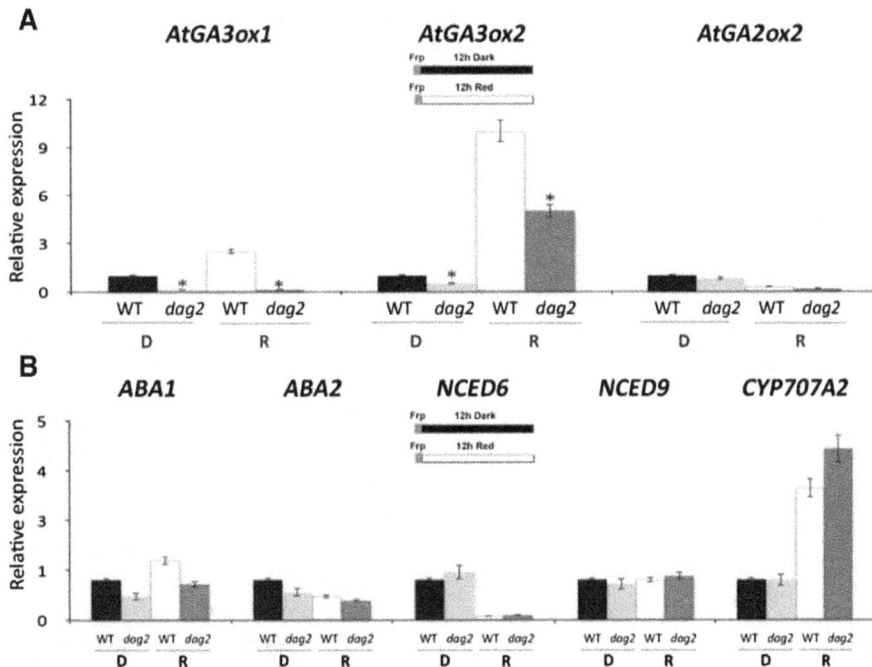

Figure 6 Mutation of the DAG2 gene affects GA biosynthesis. Relative expression level of *AtGA3ox1*, *AtGA3ox2* and *AtGA2ox2* **(A)**, and of *ABA1*, *ABA2*, *NCED6*, *NCED9* and *CYP707A2* **(B)** in *dag2* mutant seeds compared to wild-type seeds. Seeds were imbibed 12 hours in the dark (D), or under R light (R). Relative expression levels were normalized with that of the *UBQ10* gene, and are presented by the ratio of the corresponding wild-type mRNA level in D, which was set to 1. Similar results were obtained from three independent experiments, and a typical result is presented with SD values. Significative differences were analyzed by *t*-test (*P ≤ 0,05).

biosynthetic gene *AtGA3ox1*, with the cooperation of GAI [22,23].

DAG1 and DAG2 share 77% overall aminoacidic identity, with 100% identity in the Dof domain and, based on the opposite germination properties of *dag1* and *dag2* mutant seeds, we had assumed that the function of these two Dof proteins was opposite. This was also supported by *DAG1* overexpression, which caused phenotypes similar to mutation of *DAG2* [8]. Consistently, germination of *dag2* mutant seeds in phyB-dependent conditions (i.e.

under R light) was significantly reduced compared to wild-type seeds, whereas *dag1* seeds showed a higher germination frequency [21,22]. In addition, *DAG2* expression is induced by exposure to R light, as opposed to *DAG1*, whose transcript level is lower in R light than in the dark [22].

Analysis of the germination properties of *dag2dag1* double mutant seeds revealed that the *dag1* mutation is epistatic over the *dag2* one [8]. Consistent with these previous reports, here we showed that *DAG2* expression

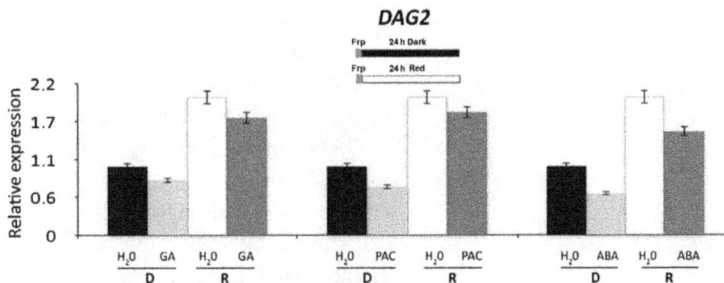

Figure 7 DAG2 expression is not altered by ABA or GA. Relative expression level of *DAG2* in wild-type seeds imbibed 24 hours in the presence of GA, of Paclobutrazol, an inhibitor of GA biosynthesis, or of ABA in the dark (D), or under R light (R), compared to seeds imbibed in water as a control (H₂O). Relative expression levels were normalized with that of the *UBQ10* gene, and are presented by the ratio of the corresponding mRNA level in seeds imbibed in water, which was set to 1. Similar results were obtained from three independent experiments, and a typical result is presented with SD values.

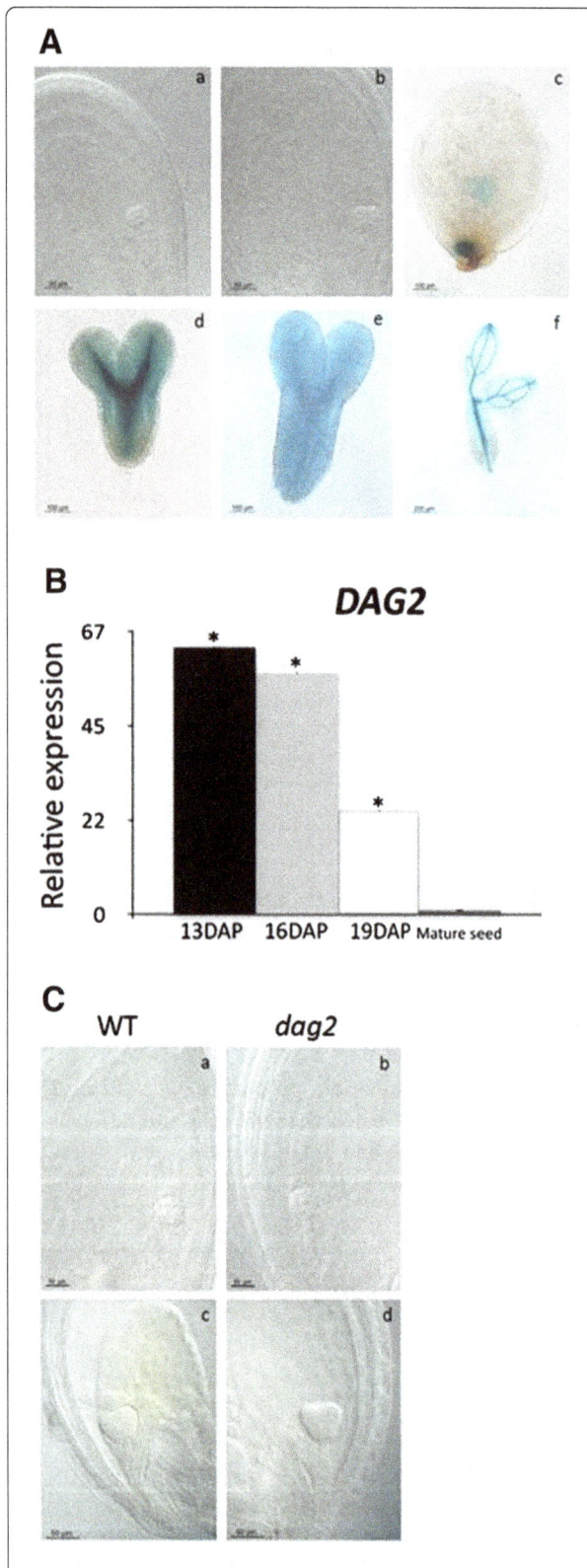

Figure 8 *DAG2* inactivation does not affect embryo development. Histochemical staining of *DAG2:GUS* during embryogenesis, in early globular, globular, heart, late heart, torpedo and mature embryo **(A)**.Relative expression level of *DAG2* in wild-type seeds at 13, 16 and 19 Days After Pollination (DAP), and in mature seeds (28 DAR). Relative expression levels were normalized with that of the *UBQ10* gene, and are presented by the ratio of the corresponding mRNA level in mature seeds, which was set to 1. Similar results were obtained from three independent experiments, and a typical result is presented with SD values. Significative differences were analyzed by *t*-test (*P ≤ 0,05) **(B)**. Phenotypes of wild-type (a, c) and *dag2* mutant (b, d) embryos, at globular (a, b) and heart stage (c, d) **(C)**.

is negatively controlled by DAG1, and that DAG1 directly binds the *DAG2* promoter as demonstrated by ChIP assay.

This provides molecular support to the genetic evidence of the epistatic relationship between these two Dof proteins shown in previous work [8]. We show here that *DAG2* is also repressed by PIL5, since the *DAG2* mRNA level is significantly increased in *pil5* mutant seeds in the dark but not under R light, where PIL5 is degraded following interaction with phyB in its activated form (Pfr).

Since DAG1 directly interacts with GAI, and cooperates with this DELLA protein in repressing the GA biosynthetic gene *AtGA3ox1* [23], and in the light of the opposite role of DAG2 in this molecular pathway, one could hypotesize a relationship of DAG2 with RGA or GAI. Interestingly, our results revealed that expression of *RGA*, but not of *GAI*, is significantly affected in *dag2* mutant seeds exposed to R light, suggesting that DAG2 may negatively regulate this *DELLA* gene, whereas expression of *DAG2* is not likely to be controlled by both RGA and GAI, as *DAG2* transcript levels are similar in *rga28* and in *gai-t6* mutant seeds compared to wild-type seeds, in both light conditions.

Our expression analysis under phyA-dependent conditions further supports the notion that DAG2 acts in the phytochrome-mediated seed germination. In fact, of the marker genes of the phyA-dependent germination pathway we analyzed, *PIL2*, *ATHB2*, and *ABI4* remained unaffected in the *dag2* mutant, while *GASA4* and *GASA6* were severely downregulated - consistent with the role of DAG2 in the positive control of GA biosynthesis - opening the interesting possibility that DAG2 participates also in phyA signalling.

It should be noted that GASA4 has been previously characterised as a regulatory protein, induced by GA and involved in seed development and germination, independently of light conditions [31,32].

Phytochromes promote seed germination partly through GA. Red light induces the expression of the two GA anabolic genes *GA3-oxidase* genes *GA3ox1* and *GA3ox2*, whereas it represses the GA catabolic gene *GA2ox2* [33,34]. Consistent with a positive role of DAG2 in seed germination, mutation of the *DAG2* gene severely affects expression of both *AtGA3ox1* and *AtGA3ox2*, although it does not alter the expression level of *AtGA2ox2*. Unlike DAG1, DAG2 does not seem to play its function through regulation of ABA metabolism, as the expression profile of the ABA metabolic genes tested is quite similar in *dag2* and wild-type seeds [22].

In recent years, the molecular mechanisms underlying light-mediated seed germination has been partly elucidated; however, it still remains an open question which are the positive regulators of this process. In fact, so far only LONG HYPOCOTYL IN FAR RED1 (HFR1) has been identified as a positive regulator of seed germination: HFR1 acts upstream of PIL5 and interacts directly with PIL5 thus sequestering it to prevent it from binding to its target genes [35]. Interestingly, germination of *hfr1* mutant seeds under phyB-dependent germination conditions is very similar to that of *dag2* mutant seeds, strengthening the notion that DAG2 is also a positive regulator in the phyB-dependent seed germination pathway.

As previously reported, *DAG2* and *DAG1* show a very similar expression profile, restricted to the vascular tissue [8], and we showed that during embryo development, *DAG1* is expressed from late globular stage [22,24]. We also showed that *dag1* mutant embryos displayed abnormal cell divisions at globular stage, altering the radial symmetry of the embryo axis [24].

Here we showed that, in contrast with DAG1, although also *DAG2* is expressed during embryo development, its absence does not produce obvious embryo phenotypes.

Conclusions

Our genetic and molecular data indicate that DAG2 is a new positive factor of the phyB/PIL5-mediated seed germination pathway. DAG2 is located downstream PIL5 and DAG1, which directly represses *DAG2* expression. Consistent with previous genetic data, DAG2 plays an opposite role to DAG1, although our results indicate that DAG2 acts on GA, but not on ABA, metabolism.

Methods

Plant material and growth conditions

dag2 is the allele described in Gualberti *et al.* [8] in Ws-4 ecotype.

All *Arabidopsis thaliana* lines used in this work were grown in a growth chamber at 24/21°C with 16/8-h day/night cycles and light intensity of 300 μmol/m^{-2} s^{-1} as previously described [7,22].

Seed germination assays

All seeds used for germination tests were harvested from mature plants grown at the same time, in the same conditions, and stored for the same time (28 Days After Ripening, DAR) under the same conditions. Germination assays were performed according to Gabriele *et al.* [22]. For phyB-dependent germination experiments, seeds were exposed to a pulse of FR light (40 μmol m^{-2} s^{-1}), then a pulse of R light (90 μmol m^{-2} s^{-1}) and subsequently kept in the dark for 5 days. Germination assays were repeated with three seed batches, and one representative experiment is shown. Bars represent the mean ± SEM of three biological repeats (25 seeds per biological repeat). P values were obtained from a Student's unpaired two-tail *t* test comparing the mutant with its control (* = p ≤ 0,05).

Expression analysis

For expression analysis, seeds were imbibed for 12 or 24 hours, on five layers of filter paper, soaked with 5 ml water, exposed to a pulse of FR (40 μmol m^{-2} s^{-1}), then incubated in the dark or under R light (90 μmol m^{-2} s^{-1}), in the presence of PAC (100 μM) to prevent de-novo GA biosynthesis in response to light [26]. For phyA-dependent conditions, seeds were treated according to Oh *et al.*, 2006 [25]. RNA extraction and RT-qPCR were performed according to Gabriele *et al.* [22]. Quantification of gene expression was expressed in comparison to the reference gene (See legends of figures), and relative expression ratio was calculated based on the qRT-PCR efficiency (E) for each gene and the crossing point (CP) deviation of our target genes versus a control [36]. The expression analyses were repeated in comparison with a second reference gene (Additional file 3: Figure S3).

Three independent biological replicates were performed, and one representative experiment is reported. Significative differences were analyzed by *t*-test (*P ≤ 0,05). The primers used for the assays are listed in Additional file 4: Table S1.

ChIP analysis

The *dag1DAG1-HA* line is the one previously described in Gabriele *et al.* [22]. ChIP was performed as previously described [22], with 12 hours imbibed seeds. Antibodies against HA tag (Santa Cruz, CA, USA) were used for immunoprecipitation. Equal amounts of starting material and ChIP products were used for qPCR reaction. The primers used are listed in Table S1. Three independent biological replicates were performed. Significative differences were analyzed by *t*-test (*P ≤ 0,05).

Microscopy and GUS analysis

Analysis of *dag2* and wild-type embryos was performed under an Axioskop 2 plus microscope (Zeiss).

The *DAG2:GUS* line is the one described in Gualberti *et al.* [8]. Histochemical staining and microscopic analysis were carried out according to Blazquez *et al.* [37]. Stained embryos (after washing in 70% ethanol) were analysed and photographed under an Axioskop 2 plus microscope (Zeiss).

Availability of supporting data

All the supporting data of this article are included as additional files (Additional files 1, 2 and 3: Figures S1-S3; Additional file 4: Table S1).

Additional files

Additional file 1: ChIP analysis of wild-type (WS) seeds immunoprecipitated with anti-HA antibody or without antibody.

Additional file 2: Relative expression levels of *ATHB2, PIL2, GASA4, GASA6* and *ABI4* in wild-type (WT) and *dag2* mutant seeds. Relative expression levels were normalized with that of the *UBQ10* gene.

Additional file 3: Expression analysis with a second reference gene. Relative expression levels of *DAG2* in wild-type seeds in the dark (D), Red (R) (A), or Far Red (FR) (B) light, in dry seeds (0h), or imbibed 12 (12h), 24 hours (24h) in the dark, under White (W) or Red light (C), normalized with the *PP2A* gene. Relative expression levels of *DAG2* in *dag1* (D), *pil5* (F), *rga28* (G), *gai-t6* (H) seeds compared to WT. Relative expression levels of *DAG1* (E), *RGA* (I), GAI (L), in *dag2* seeds compared to WT. The expression levels were normalized with *PP2A* (D, F, E, I, L), or with *eIF1a* (At5g60390) (G, H). Relative expression levels of GA (M) or ABA (N) metabolic genes in *dag2* seeds compared to WT, normalized with *PP2A* gene. Relative expression levels of *DAG2* in WT seeds imbibed in the presence of ABA, or GA, or PAC (O), in WT embryos at 13, 16, 19 DAP (P), normalized with *PP2A* gene. Relative expression levels of *ATHB2, PIL2, GASA4, GASA6* and *ABI4* in *dag2* seeds compared to WT (Q), normalized with *PP2A*.

Additional file 4: Table S1: List of the primers used for expression analyses and for the ChIP assays.

Abbreviations

DOF: DNA Binding With One Finger; DAG1: Dof AFFECTING GERMINATION 1; DAG2: Dof AFFECTING GERMINATION 2; phyB: Phytochrome B; PIL5: PHYTOCHROME INTERACTING FACTOR3-LIKE 5; GAI: GA INSENSITIVE; RGA: REPRESSOR OF *ga1-3*; ABI3: ABA INSENSITIVE 3; ABI5: ABA INSENSITIVE 5; SPT: SPATULA; SOM: SOMNUS; HFR1: LONG HYPOCOTYL IN FAR RED1; ATHB2: ARABIDOPSIS THALIANA HOMEOBOX PROTEIN 2; GASA4: 6, GA-STIMULATED ARABIDOPSIS 4, 6; PIL2: PHYTOCHROME INTERACTING FACTOR3-LIKE 2; ABI4: ABA Insensitive 4,ABA, Abscissic Acid; GA: Gibberellins; PAC: Paclobutrazol; ChIP: Chromatin Immuno Precipitation; RT-qPCR: quantitative reverse transcriptase-polymerase chain reaction; W light: White light; R light: Red light; D: Dark; FR Light: Far Red Light; DAP: Days After Pollination; DAR: Days After Ripening; GUS: β-glucuronidase; HA: Heme Agglutinin.

Competing interests

The authors declare they have no competing interests.

Authors' contributions

PV designed the research. SS, AB and GS contributed to the experimental design and to analysis of the results. SS, AB, RL, VR, DC, EM performed the experiments. All authors analyzed and discussed the data. SS prepared the figures and PV wrote the article. PC supervised the research and the writing of the manuscript. All authors read and approved the final manuscript.

Acknowledgments

This work was partially supported by research grants from Ministero dell'Istruzione, Università e Ricerca, Progetti di Ricerca di Interesse Nazionale, and from Sapienza Università di Roma to PC, and from Istituto Pasteur Fondazione Cenci Bolognetti to PV.

References

1. Lijavetzky D, Carbonero P, Vicente-Carbajosa J. Genome-wide comparative phylogenetic analysis of the rice and Arabidopsis Dof gene families. BMC Evol Biol. 2003;3:17.
2. Moreno-Risueno MA, Martinez M, Vicente-Carbajosa J, Carbonero P. The family of DOF transcription factors: from green unicellular algae to vascular plants. Mol Genet Genomics. 2007;277(4):379–90.
3. Hernando-Amado S, Gonzalez-Calle V, Carbonero P, Barrero-Sicilia C. The family of DOF transcription factors in Brachypodium distachyon: phylogenetic comparison with rice and barley DOFs and expression profiling. BMC Plant Biol. 2012;12:202.
4. Shaw LM, McIntyre CL, Gresshoff PM, Xue GP. Members of the Dof transcription factor family in Triticum aestivum are associated with light-mediated gene regulation. Funct Integr Genomics. 2009;9(4):485–98.
5. Kushwaha H, Gupta S, Singh VK, Rastogi S, Yadav D. Genome wide identification of Dof transcription factor gene family in sorghum and its comparative phylogenetic analysis with rice and Arabidopsis. Mol Biol Rep. 2011;38(8):5037–53.
6. Yanagisawa S, Sheen J. Involvement of maize Dof zinc finger proteins in tissue-specific and light-regulated gene expression. Plant Cell. 1998;10(1):75–89.
7. Papi M, Sabatini S, Bouchez D, Camilleri C, Costantino P, Vittorioso P. Identification and disruption of an Arabidopsis zinc finger gene controlling seed germination. Genes Dev. 2000;14(1):28–33.
8. Gualberti G, Papi M, Bellucci L, Ricci I, Bouchez D, Camilleri C, et al. Mutations in the Dof zinc finger genes DAG2 and DAG1 influence with opposite effects the germination of Arabidopsis seeds. Plant Cell. 2002;14(6):1253–63.
9. Isabel-LaMoneda I, Diaz I, Martinez M, Mena M, Carbonero P. SAD: a new DOF protein from barley that activates transcription of a cathepsin B-like thiol protease gene in the aleurone of germinating seeds. Plant J. 2003;33(2):329–40.
10. Park DH, Lim PO, Kim JS, Cho DS, Hong SH, Nam HG. The Arabidopsis COG1 gene encodes a Dof domain transcription factor and negatively regulates phytochrome signaling. Plant J. 2003;34(2):161–71.
11. Imaizumi T, Schultz TF, Harmon FG, Ho LA, Kay SA. FKF1 F-box protein mediates cyclic degradation of a repressor of CONSTANS in Arabidopsis. Science. 2005;9(5732):3–7.
12. Ward JM, Cufr CA, Denzel MA, Neff MM. The Dof transcription factor OBP3 modulates phytochrome and cryptochrome signaling in Arabidopsis. Plant Cell. 2005;17(2):475–85.
13. Iwamoto M, Higo K, Takano M. Circadian clock- and phytochrome-regulated Dof-like gene, Rdd1, is associated with grain size in rice. Plant Cell Environ. 2009;32(5):592–603.
14. Rueda-Romero P, Barrero-Sicilia C, Gomez-Cadenas A, Carbonero P, Onate-Sanchez L. Arabidopsis thaliana DOF6 negatively affects germination in non-after-ripened seeds and interacts with TCP14. J Exp Bot. 2012;63 (5):1937–49.
15. Noguero M, Atif RM, Ochatt S, Thompson RD. The role of the DNA-binding One Zinc Finger (DOF) transcription factor family in plants. Plant science : an international journal of experimental plant biology. 2013;209:32–45.
16. Bewley JD. Seed Germination and Dormancy. Plant Cell. 1997;9(7):1055–66.
17. Shinomura T, Nagatani A, Chory J, Furuya M. The Induction of Seed Germination in Arabidopsis thaliana Is Regulated Principally by Phytochrome B and Secondarily by Phytochrome A. Plant Physiol. 1994;104(2):363–71.
18. Seo M, Hanada A, Kuwahara A, Endo A, Okamoto M, Yamauchi Y, et al. Regulation of hormone metabolism in Arabidopsis seeds: phytochrome regulation of abscisic acid metabolism and abscisic acid regulation of gibberellin metabolism. Plant J. 2006;48(3):354–66.
19. Yamaguchi S, Sun T, Kawaide H, Kamiya Y. The GA2 locus of Arabidopsis thaliana encodes ent-kaurene synthase of gibberellin biosynthesis. Plant Physiol. 1998;116(4):1271–8.
20. Oh E, Kim J, Park E, Kim JI, Kang C, Choi G. PIL5, a phytochrome-interacting basic helix-loop-helix protein, is a key negative regulator of seed germination in Arabidopsis thaliana. Plant Cell. 2004;16(11):3045–58.

DOF AFFECTING GERMINATION 2 is a positive regulator of light-mediated seed germination...

27

21. Papi M, Sabatini S, Altamura MM, Hennig L, Schafer E, Costantino P, et al. Inactivation of the phloem-specific Dof zinc finger gene DAG1 affects response to light and integrity of the testa of Arabidopsis seeds. Plant Physiol. 2002;128(2):411–7.

22. Gabriele S, Rizza A, Martone J, Circelli P, Costantino P, Vittorioso P. The Dof protein DAG1 mediates PIL5 activity on seed germination by negatively regulating GA biosynthetic gene AtGA3ox1. Plant J. 2010;61(2):312–23.

23. Boccaccini A, Santopolo S, Capauto D, Lorrai R, Minutello E, Serino G, et al. The DOF Protein DAG1 and the DELLA Protein GAI Cooperate in Negatively Regulating the AtGA3ox1 Gene. Mol Plant. 2014;7(9):1486–9.

24. Boccaccini A, Santopolo S, Capauto D, Lorrai R, Minutello E, Belcram K, et al. Independent and interactive effects of DOF affecting germination 1 (DAG1) and the Della proteins GA insensitive (GAI) and Repressor of. BMC Plant Biol. 2014;14(1):200.

25. Oh E, Yamaguchi S, Kamiya Y, Bae G, Chung WI, Choi G. Light activates the degradation of PIL5 protein to promote seed germination through gibberellin in Arabidopsis. Plant J. 2006;47(1):124–39.

26. Oh E, Yamaguchi S, Hu J, Yusuke J, Jung B, Paik I, et al. PIL5, a phytochrome-interacting bHLH protein, regulates gibberellin responsiveness by binding directly to the GAI and RGA promoters in Arabidopsis seeds. Plant Cell. 2007;19(4):1192–208.

27. Ibarra SE, Auge G, Sanchez RA, Botto JF. Transcriptional programs related to phytochrome A function in Arabidopsis seed germination. Mol Plant. 2013;6(4):1261–73.

28. Park J, Lee N, Kim W, Lim S, Choi G. ABI3 and PIL5 collaboratively activate the expression of SOMNUS by directly binding to its promoter in imbibed Arabidopsis seeds. Plant Cell. 2011;23(4):1404–15.

29. Penfield S, Josse EM, Kannangara R, Gilday AD, Halliday KJ, Graham IA. Cold and light control seed germination through the bHLH transcription factor SPATULA. Curr Biol. 2005;15(22):1998–2006.

30. Kim DH, Yamaguchi S, Lim S, Oh E, Park J, Hanada A, et al. SOMNUS, a CCCH-type zinc finger protein in Arabidopsis, negatively regulates light-dependent seed germination downstream of PIL5. Plant Cell. 2008;20(5):1260–77.

31. Aubert D, Chevillard M, Dorne AM, Arlaud G, Herzog M. Expression patterns of GASA genes in Arabidopsis thaliana: the GASA4 gene is up-regulated by gibberellins in meristematic regions. Plant Mol Biol. 1998;36(6):871–83.

32. Roxrud I, Lid SE, Fletcher JC, Schmidt ED, Opsahl-Sorteberg HG. GASA4, one of the 14-member Arabidopsis GASA family of small polypeptides, regulates flowering and seed development. Plant Cell Physiol. 2007;48(3):471–83.

33. Yamaguchi S, Smith MW, Brown RG, Kamiya Y, Sun T. Phytochrome regulation and differential expression of gibberellin 3beta-hydroxylase genes in germinating Arabidopsis seeds. Plant Cell. 1998;10(12):2115–26.

34. Yamauchi Y, Takeda-Kamiya N, Hanada A, Ogawa M, Kuwahara A, Seo M, et al. Contribution of gibberellin deactivation by AtGA2ox2 to the suppression of germination of dark-imbibed Arabidopsis thaliana seeds. Plant Cell Physiol. 2007;48(3):555–61.

35. Shi H, Zhong S, Mo X, Liu N, Nezames CD, Deng XW. HFR1 sequesters PIF1 to govern the transcriptional network underlying light-initiated seed germination in Arabidopsis. Plant Cell. 2013;25(10):3770–84.

36. Pfaffl MW. A new mathematical model for relative quantification in real-time RT-PCR. Nucleic Acids Res. 2001;29(9):e45.

37. Blazquez MA, Soowal LN, Lee I, Weigel D. LEAFY expression and flower initiation in Arabidopsis. Development. 1997;124(19):3835–44.

A novel system for evaluating drought–cold tolerance of grapevines using chlorophyll fluorescence

Lingye Su[1,2], Zhanwu Dai[3], Shaohua Li[1,4*] and Haiping Xin[4*]

Abstract

Background: Grape production in continental climatic regions suffers from the combination of drought and cold stresses during winter. Developing a reliable system to simulate combined drought–cold stress and to determine physiological responses and regulatory mechanisms is important. Evaluating tolerance to combined stress at germplasm level is crucial to select parents for breeding grapevines.

Results: In the present study, two species, namely, *Vitis amurensis* and *V. vinifera* cv. 'Muscat Hamburg', were used to develop a reliable system for evaluating their tolerance to drought–cold stress. This system used tissue –cultured grapevine plants, 6% PEG solution, and gradient cooling mode to simulate drought–cold stress. *V. amurensis* had a significantly lower LT50 value (the temperature of 50% electrolyte leakage) than 'Muscat Hamburg' during simulated drought–cold stress. Thus, the former had higher tolerance than the latter to drought–cold stress based on electrolyte leakage (EL) measurements. Moreover, the chlorophyll fluorescence responses of *V. amurensis* and 'Muscat Hamburg' were also analyzed under drought–cold stress. The maximum photochemical quantum yield of PS II (*Fv/Fm*) exhibited a significant linear correlationship with EL. The relationship of EL with *Fv/Fm* in the other four genotypes of grapevines under drought–cold stress was also detected.

Conclusions: A novel LT50 estimation model was established, and the LT50 values can be well calculated based on *Fv/Fm* in replacement of EL measurement. The *Fv/Fm*–based model exhibits good reliability for evaluating the tolerance of different grapevine genotypes to drought–cold stress.

Keywords: Drought–cold stress, Electrolyte leakage, *Fv/Fm*, Grapevine, LT50

Background

Abiotic stresses are major factors that affect the growth, development, and productivity of crops. Most studies have mainly focused on individual stresses, such as cold, drought, and high salinity [1-3]. However, different stresses might occur simultaneously in the field; thus, crops can suffer from the superimposition of these stresses [4,5]. Hence, cross–breeding or marker–assisted breeding, which targets single abiotic stress, might be insufficient for enhancing the performance of crops in the field.

Therefore, the combination of different stresses should be considered in evaluating tolerance and stress–related molecular mechanism [4,6].

Summer drought with heat waves has been noticed in grape–producing regions [7-9]. The mechanisms of drought–heat effects have also been reported in different plants [6,10]. In addition to summer drought, grapevine routinely suffers from dry winter; during this season, regions such as North China with extremely continental climate experience a low temperature and air humidity with little snow [11,12]. Frozen water in the soil support-ing the main roots results in limited water use in the soil by grapevine plants during winter, on the contrary, tran-spiration by woody tissues (cuticular transpiration and lenticular transpiration) from grapevine canes is relatively high due to low humidity. All *Vitis vinifera* cultivars can't be survival under natural condition in the main Chinese

---

* Correspondence: shhli@ibcas.ac.cn; xinhaiping215@hotmail.com
[1]Beijing Key Laboratory of Grape Sciences and Enology and CAS Key Laboratory of Plant Resources, Institute of Botany, Chinese Academy of Sciences, Beijing 100093, China
[4]Key Laboratory of Plant Germplasm Enhancement and Specialty Agriculture, Wuhan Botanical Garden, Chinese Academy of Sciences, Wuhan 430074, China
Full list of author information is available at the end of the article

grape–producing areas in North China. To have economy income, all grapevine canes should be buried during winter, even if the temperature is higher than −10°C. This process requires more labor, and thus, increases product cost. Generally, extremely low temperature could damage the bud and cane of grapevines [13]. Moreover, the combination of drought–cold stress in winter in North China might result in death of shoots, even death of young trees such as in apple trees which can be survival under individual cold stress [14]. Even a special term 'choutiao' in Chinese is given for the phenomenon concerning death of shoots or whole trees due to drought stress under cold winter and some special culture management were developed to overcome drought–cold stress in apple trees [14].

Various evaluation methods are available for quantifying the tolerances to individual drought or cold stress in the laboratory [15]. Measuring electrolyte leakage (EL) is one of the most frequently used methods to assess plant tolerance in response to drought and low temperature [16,17]. Abiotic stresses induce cell membrane injury, leading to intracellular ion efflux. EL measurement can reflect the change of ion exosmosis, and determine the cell damage level. Half–lethal temperature (LT50) is widely considered to represent the low–temperature tolerance in plants. The LT50 value can be generally calculated by EL measurement defined as the temperature at which EL decreases to 50% of that under optimal growth conditions [18]. However, this method is time consuming [19]. Moreover, severe stress (e.g., freezing environment) could seriously damage the membrane structure and cause secondary stress to the samples, thus affecting the accuracy of the method [20]. Few studies have focused on the combination of the two stresses. However, the damages induced by drought and cold have several common characteristics. Both stresses may cause cell dehydration and accumulation of reactive oxygen species, resulting in damaged membrane and photosynthesis system at cellular level [21,22]. Consequently, tolerance to combined stress could be quantified through methodologies similar to those for each individual stress.

The negative impacts on photosynthesis have been widely studied under abiotic stresses, and chlorophyll fluorescence measurement has been proven as an efficient and reproducible tool for evaluating plant susceptibility index to drought [23,24] or low temperature [20,25] stresses. This method reflects the susceptibility to the damages of the photo system II (PSII) in the photosynthesis electron transport chains [26]. As a nondestructive diagnostic tool, chlorophyll fluorescence method shows more benefits compared with EL measurement, especially the more rapid process induces less secondary stresses to the samples. Moreover, different parameters (e.g., Fo, Fv/Fm, and qP) can be measured [25,27].

In the present study, we mimicked a drought–cold stress condition by coupling polyethylene glycol (PEG)–induced water–deficit hydroponic culture system with cooling environment. Fluorescence parameters were determined to evaluate the tolerance of grapevine to combined drought–cold stress. We established a novel model to estimate LT50 values using Fv/Fm measurement based on the correlation between the EL and chlorophyll fluorescence parameters of the grape leaves exposed under combined drought–cold stress condition. This model simplifies the evaluation of the damages caused by drought–cold stress. The proposed model can be readily applied to determine the tolerance of the grape germplasm and cross–progeny individuals to breed drought–cold–tolerant grapevines.

Results

Individual drought and cold tolerance of V. amurensis and 'Muscat Hamburg'

After exposure to PEG–simulated drought stress for 1 d, *V. amurensis* showed significantly lower EL than 'Muscat Hamburg' (*V. vinifera*) at all PEG levels (Figure 1a). *V. amurensis* showed a lower increase in EL than that of 'Muscat Hamburg' (12.2 vs 18.3 times) at 10% PEG compared with the controls. The EL difference between *V. amurensis* and 'Muscat Hamburg' increased as PEG concentration increased. Moreover, leaf relative water content (RWC) was lower in 'Muscat Hamburg' than that in *V. amurensis* under PEG stress, particularly at high PEG concentration (Additional file 1: Figure S1). RWC (75.8%, 68.0%, and 31.8%) was significantly lower in 'Muscat Hamburg' than that in *V. amurensis* under 6%, 8%, and 10% PEG treatments, respectively. The effect of the transpiration volume of the plantlets on the water potential of nutrient solution was also investigated. We filled the solution with distilled water to the initial volume every 12h after treatment. The two grape species exhibited significant phenotypic differences (Additional file 2: Figure S2).

To determine cold tolerance, we examined LT50 values calculated based on the measured ELs in both species. The LT50 values of *V. amurensis* and 'Muscat Hamburg' were −10.77 and −5.35°C, respectively; and they were significantly different in LT50 values between the previous two genotypes (Figure 1b). The different tolerances of *V. amurensis* and 'Muscat Hamburg' to the two individual stresses could be used as foundation for subsequent combined studies.

Tolerances to drought–cold stress evaluated using EL–based LT50 value

To establish optimal conditions for combined stress, we performed a series of preliminary examinations for drought and low–temperature treating modes. A suitable PEG concentration should immediately trigger plant physiological

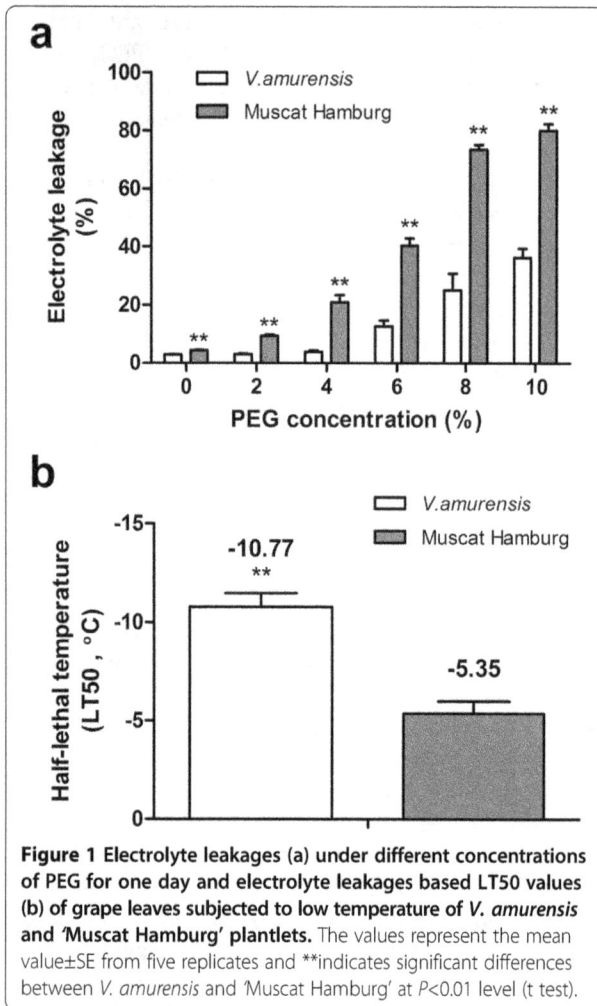

Figure 1 Electrolyte leakages (a) under different concentrations of PEG for one day and electrolyte leakages based LT50 values (b) of grape leaves subjected to low temperature of *V. amurensis* and 'Muscat Hamburg' plantlets. The values represent the mean value±SE from five replicates and **indicates significant differences between *V. amurensis* and 'Muscat Hamburg' at $P<0.01$ level (t test).

'NAF', Figure 2b) modes. Low temperature significantly increased the EL values in both species; the increase in EL was higher in 'Muscat Hamburg' than that in *V. amurensis* (Figure 2c and d). EL was significantly different between the two genotypes from −4°C to −7°C under NAF mode, while the significant differences under GC mode were only observed at −4°C and −5°C. Under GC mode, the EL values in both genotypes slightly increased at initial degrees, whereas inflection point increased at high temperature in 'Muscat Hamburg' (−4°C, 5.38–fold increase) compared with that in *V. amurensis* (−6°C, 7.23–fold increase). Moreover, EL slowly increased under NAF mode compared with that under GC mode. LT50 values were calculated based on the EL data. As shown in Figure 2e, the LT50 values of *V. amurensis* and 'Muscat Hamburg' were −5.61±0.19°C and −3.72±0.42°C under GC mode and −6.88±0.34°C and −4.84±0.13°C under NAC mode, respectively.

Chlorophyll fluorescence response

As shown in Figure 3, we examined three chlorophyll fluorescence parameters (*Fo*, *Fv/Fm*, and *Fv/Fo*) under drought–cold stress at the two cooling modes. *Fo* rapidly increased at temperatures lower than −4°C under both GC (Figure 3a) and NAF (Figure 3d) modes. Moreover, *Fo* was significantly higher in 'Muscat Hamburg' than that in *V. amurensis* at −6°C or/and −7°C. The *Fv/Fm* (Figure 3b and e) and *Fv/Fo* (Figure 3c and f) values decreased as temperature decreased under both cooling modes. In addition, a more rapid decrease of their values in 'Muscat of Hamburg' was observed than those in *V. amurensis* and significant difference was observed at −5°C at both cooling modes.

To establish an LT50 estimation model based on chlorophyll fluorescence responses, we should ensure a good correlation between EL and the candidate parameters. All the three chlorophyll fluorescence parameters were significantly correlated with EL under both cooling modes (Figure 4). Interestingly, the cooling modes affected the coefficient of correlation for different chlorophyll fluorescence-to-EL pairs. *Fv/Fm* and *Fv/Fo* showed higher correlations with EL under GC than those under NAF. The low correlation under NAF was mainly caused by the non–synchronous variation in the responses of chlorophyll fluorescence and EL to the decreasing temperatures. *Fv/Fm* and *Fv/Fo* reached their higher limits when EL was approximately 20%; thereafter, any further increase in EL (from 20% to 60%) was not accompanied by a proportional decrease in the two chlorophyll florescence parameters (Figure 3e and f). The *Fv/Fm* under GC showed the highest correlation with EL (r^2=0.9772) among the three candidate parameters, and the two genotypes exhibited a unique regression line (Additional file 4: Table S1); thus, *Fv/Fm* was selected as the model for further analysis.

responses and effectively discriminate drought tolerance among genotypes. However, the nutrient solution should remain unfrozen under the given freezing condition; freezing causes lower water potential [15] and therefore decreases the accuracy of the PEG concentration. According to these criteria and the results in Figure 1a, we selected 4%, 6%, and 8% as the candidate PEG concentrations. We then assessed the freezing pattern of the three PEG solutions at −6°C based on the pre–experiment, which showed that even *V. amurensis* exhibited severe water–soaking damage and EL almost reached the upper limit in all PEG concentrations at temperatures lower than −6°C. Moreover, 6% and 8% PEG remained unfrozen in the solution for 2 h (Additional file 3: Figure S3), whereas the solutions without PEG or with 4% PEG became frozen. Finally, 6% PEG, which induced moderate stress compared with 8% PEG, was selected for subsequent experiments.

EL was measured in both genotypes under 6% PEG coupled with simultaneous cooling treatment in both gradient cooling (hereafter referred to as 'GC', Figure 2a) and non–acclimated freezing (hereafter referred to as

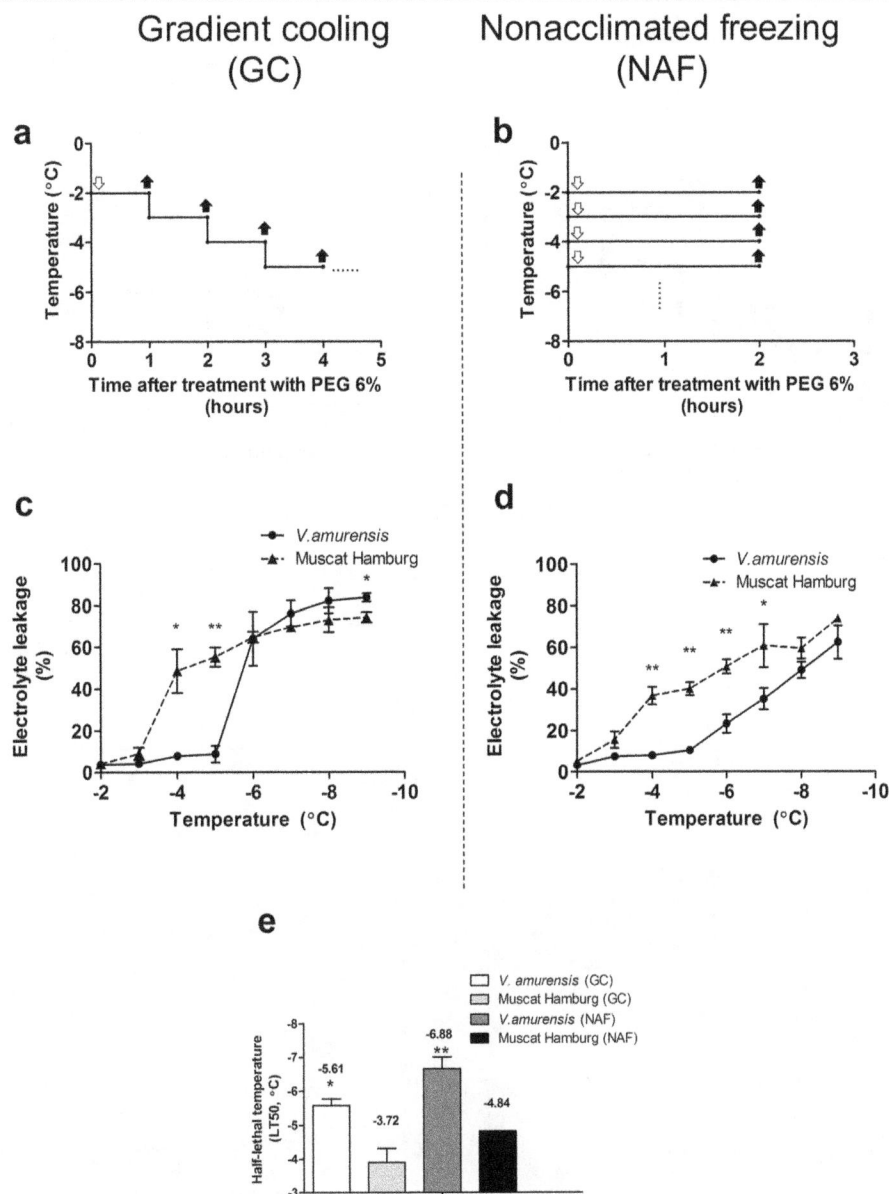

Figure 2 Electrolyte leakages and LT50 values of *V. amurensis* and 'Muscat Hamburg' plantlets under combined drought–cold systems. **(a)** and **(b)** represent the pattern diagrams of different cooling modes. **(a)** Gradient cooling (GC) combined PEG 6% and continuous temperature decreased at a rate of 1°C/h from −2°C; **(b)** non–acclimated freezing (NAF) combined PEG 6% and directly frozen to each given temperature for 2 h. The feint arrows indicate the points when the plantlets began to subject cold stress, while the solid arrows represent the sample time at the end of each defined temperatures. **(c)** and **(d)** show electrolyte leakages under GC and NAF modes, respectively. **(e)** LT50 values of GC and NAF modes in *V. amurensis* and 'Muscat Hamburg'. The values represent the mean value±SE from three to five replicates, * and ** indicate significant differences between *V. amurensis* and 'Muscat Hamburg' at *P*<0.05 and *P*<0.01 level (t test), respectively.

LT50 estimation model under drought–cold stress based on chlorophyll fluorescence

To confirm the reliability of our "PEG 6%+GC" system and the use of *Fv/Fm* as an alternative indicator of cold tolerance, we applied these parameters in the four other grape genotypes. Figure 5 shows the comparison between the LT50 values obtained from EL in the four newly investigated genotypes with those of the two genotypes used

during system establishment under drought–cold stress. The lowest LT50 values were observed in *V. amurensis* at −5.61°C, whereas the highest in 'Cardinal' at −3.71°C.

As shown in Figure 6, high correlations (r^2>0.97, Additional file 5: Table S2) were observed between EL and *Fv/Fm* under GC mode for all the tested cultivars. In addition, all cultivars presented similar linear regression slope between *Fv/Fm* and EL; however, some differences

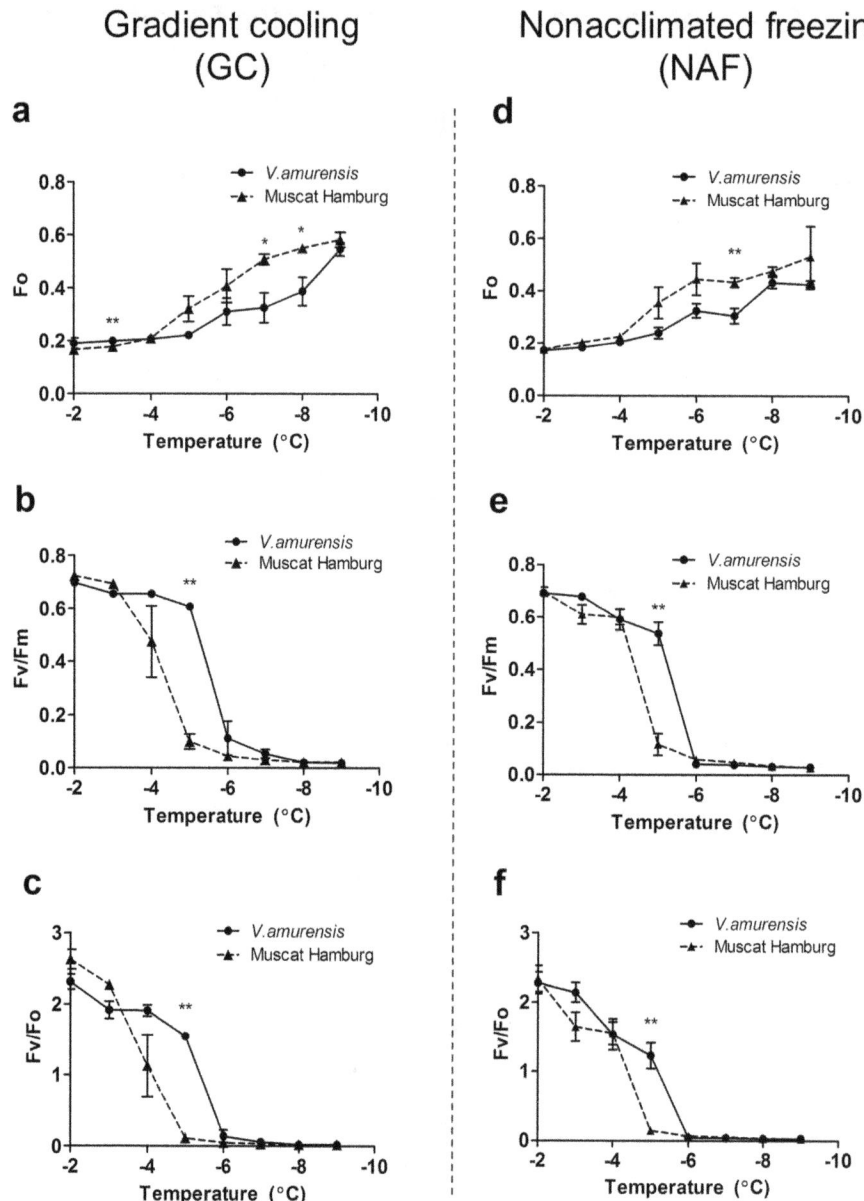

Figure 3 Chlorophyll fluorescence response of *V. amurensis* and 'Muscat Hamburg' under two combined drought–cold stress modes.
(a)–(c) indicate the response of *Fo* **(a)**, *Fv/Fm* **(b)** and *Fv/Fo* **(c)** to gradient cooling (GC) mode, while **(d)–(f)** represent the response of *Fo* **(d)**,
Fv/Fm **(e)** and *Fv/Fo* **(f)** to non–acclimated freezing (NAF) mode. The values were the mean value±SE of three replicates, and * and ** indicate
significant differences between *V. amurensis* and 'Muscat Hamburg' at *P*<0.05 and *P*<0.01 level (t test), respectively.

were observed in their intercepts (Additional file 4: Table S1). This synchronization between the responses of *Fv/Fm* to EL under GC mode confirms the reliability of *Fv/Fm* as an effective indicator of cold tolerance. Therefore, we compared the LT50 estimated from *Fv/Fm* with the values estimated from classic EL values.

Figure 7 and S4 demonstrate the comparison of the LT50 obtained from EL with those obtained from *Fv/Fm* under GC mode. A close correlation was observed between LT50–EL and LT50–*Fv/Fm* for all genotypes. The

values of LT50–*Fv/Fm* were consistent with those of LT50–EL. A minor absolute difference of 0.3°C (RMSE), a low relative difference of 7.1% (RRMSE), and a very high agreement index of 93.4% were obtained. All these indexes indicate that LT50–*Fv/Fm* provides a reliable and precise representation of LT50–EL. Paired t tests have revealed that LT50–EL and LT50–*Fv/Fm* values were significantly different in the two genotypes (*V. amurensis* and 'Muscat Hamburg') under NAF, whereas no difference was observed under GC mode (Additional file 6: Table S3). This

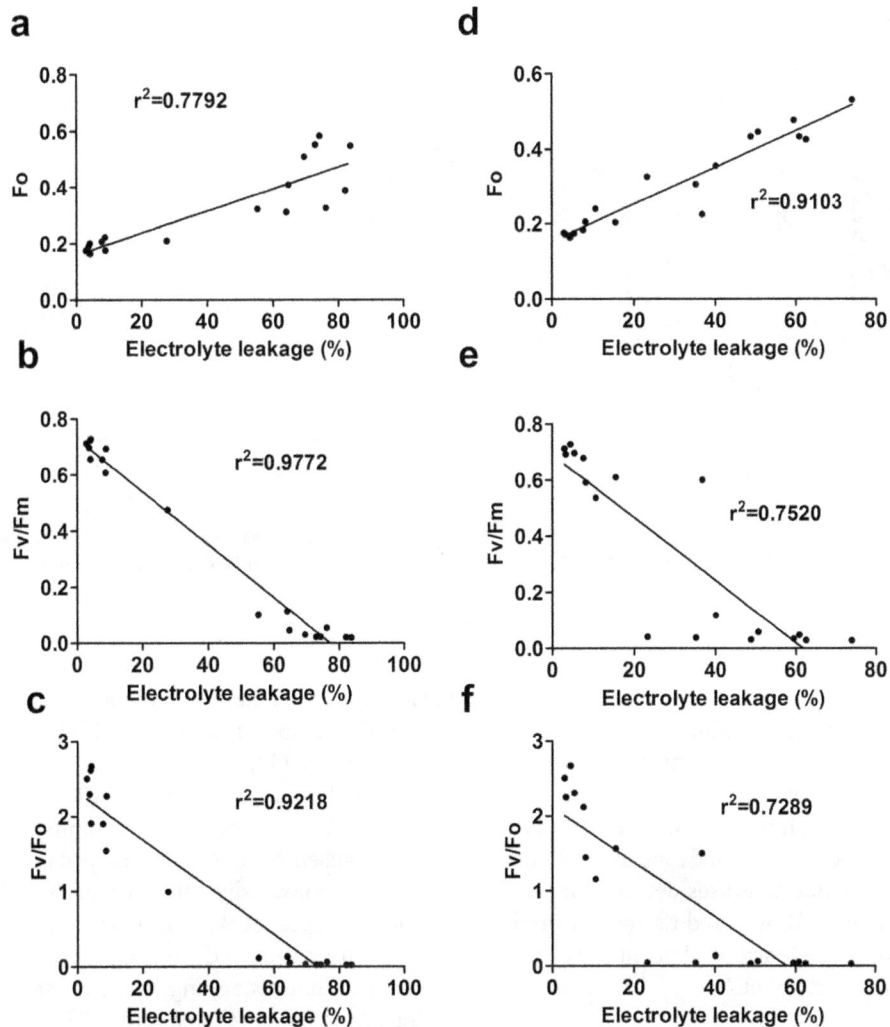

Figure 4 Correlations between electrolyte leakage (EL) and three chlorophyll fluorescence parameters under two different drought–cold systems in V. amurensis and 'Muscat Hamburg'. (a)–(c): Correlation between EL and Fo **(a)**, Fv/Fm **(b)** and Fv/Fo **(c)** under gradient cooling (GC) mode; **(d)–(f):** Correlation between EL and Fo **(d)**, Fv/Fm **(e)** and Fv/Fo **(f)** under non–acclimated freezing (NAF) mode. Data were from those shown in Figures 2 and 3 as well as controls.

finding indicates that GC mode provided more consistent results between LT50–EL and LT50–Fv/Fm, and thus, more suitable for this system.

Discussion

Experimental system of combined stress

Mittler [4,28] emphasized that combined stress is not merely an addition of two individual stresses; the physiological and molecular mechanisms of combined stress should be studied and regarded as a novel stress. Some studies have elucidated the plant tolerance mechanisms to drought–heat [28], salinity–heat [29], drought–ozone [30], and drought–heat–virus [31]. However, the combination of drought and low–temperature stresses has been rarely reported except for the study on wheat [32]. This unique stress combination should be considered for

actual fruit production. An accurate and simple method for evaluation is crucial for subsequent physiological and molecular research. The parents for breeding new cultivars with high resistance to the combined stress should be selected through stress evaluation at the germplasm level.

Establishing a suitable experimental platform for stress mimic is the prerequisite for evaluating drought–cold stress. In this study, the grape plant tissues cultured with 6% PEG solution under GC mode were subjected to a simulated drought–cold stress. PEG–induced hydroponic culture results in decreased water utilization by plants and is used for stable drought simulation because it is quantifiable and can be easily maintained. This culture condition is comparable with dry soil in winter; in which the frozen state causes unavailability of water in the upper soil layer,

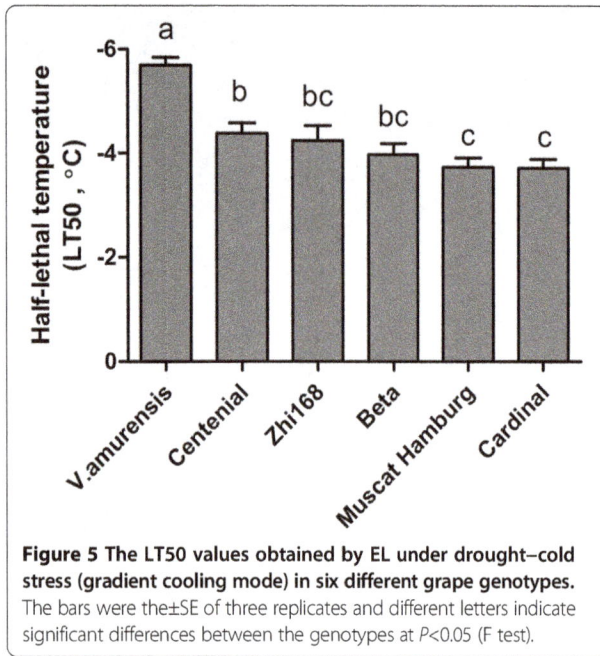

Figure 5 The LT50 values obtained by EL under drought–cold stress (gradient cooling mode) in six different grape genotypes. The bars were the ±SE of three replicates and different letters indicate significant differences between the genotypes at $P<0.05$ (F test).

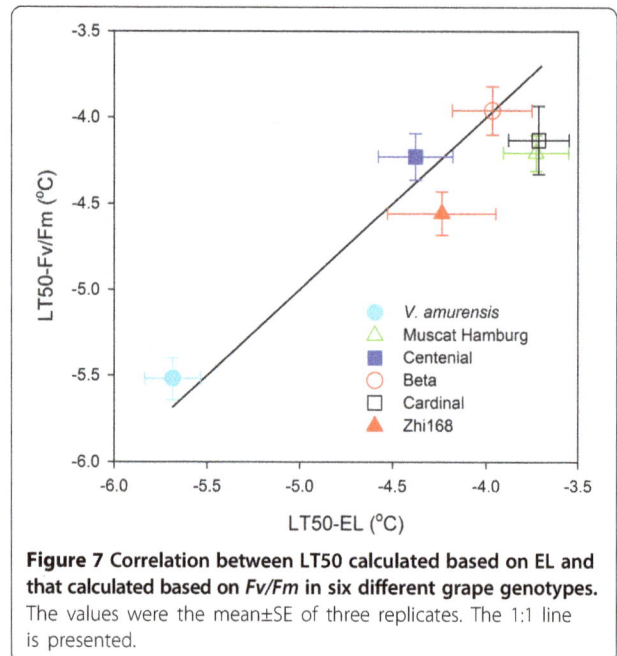

Figure 7 Correlation between LT50 calculated based on EL and that calculated based on *Fv/Fm* in six different grape genotypes. The values were the mean±SE of three replicates. The 1:1 line is presented.

where most grapevine roots are distributed. In addition, *in vitro* grapevine hydroponic system exhibits rapid and easily reproducible abilities; this finding has also been observed on some other horticultural crops, such as apple [33], banana [34], sugar beet [35], and poplar [36]. By contrast to the classic method that uses detached leaves to evaluate cold tolerance [15], we used tissue–cultured grape plants to ensure consistency of plant material for investigating the whole–plant level.

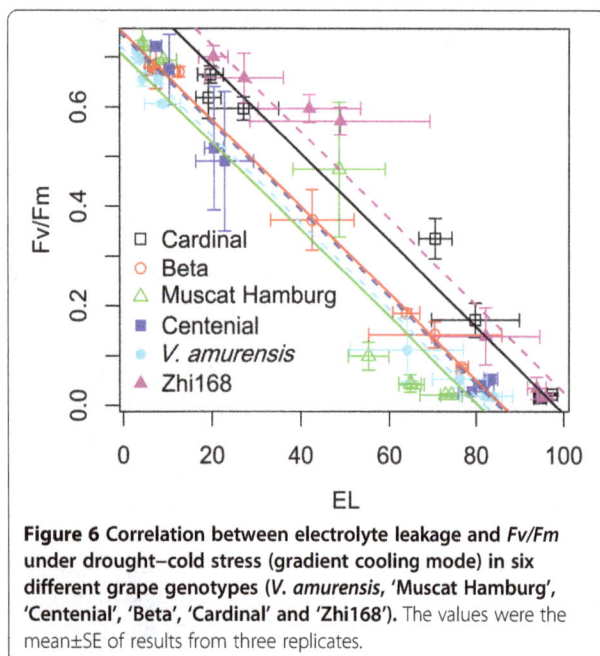

Figure 6 Correlation between electrolyte leakage and *Fv/Fm* under drought–cold stress (gradient cooling mode) in six different grape genotypes (*V. amurensis*, 'Muscat Hamburg', 'Centenial', 'Beta', 'Cardinal' and 'Zhi168'). The values were the mean±SE of results from three replicates.

Tolerances to individual and combined stresses

V. amurensis is one of the most cold–tolerant species in the *Vitis* genus [37,38]; our present study confirmed this finding based on the lower LT50 value obtained (Figure 1b). *V. amurensis* also exhibits better drought tolerance caused by less membrane damage and water loss. To our knowledge, the comparison between the drought tolerance of *V. amurensis* with other grapes has been rarely reported [39], the result of which may broaden our understanding on the use of this genotype for evaluating combined stress. The combined stress was investigated under two different cooling modes with 6% PEG solution. EL assays showed that *V. amurensis* had significantly lower LT50 values than 'Muscat Hamburg' under both modes (Figure 2c and d). This finding indicates that *V. amurensis* had high tolerance to drought–cold stress. Remarkably, the LT50 values between *V. amurensis* and 'Muscat Hamburg' were similar under both cooling modes (about 2°C, Figure 2e). Thus, this evaluation method could significantly distinguish the tolerance of the two genotypes to the combined stress condition. The LT50 values of both genotypes under the combined stress were higher than the values under individual cold stress (Figure 1b). This result could be attributed to increased stress effect by drought stress. Adding PEG to induce drought stress might damage membrane stability, as indicated by the increasing LT50 values.

To determine the effects of the two different cooling modes combined with a fixed PEG concentration of 6%, we emphasized the relationship between EL and chlorophyll fluorescence measurements (*Fo*, *Fv/Fo*, and *Fv/Fm*) under both cooling modes. *Fv/Fo* and *Fv/Fm* exhibited significant higher correlations with EL under GC mode than those under NAF mode, indicating that GC mode was more suitable than NAF mode for drought–cold treatment. The changed trends of chlorophyll fluorescence data under two modes were similar (Figure 3), so the correlation differences possibly originated from the EL measurement results. This finding could be attributed to the insufficient time for increasing membrane damage within short–term freezing at defined temperature without pre–chilling accumulation; hence, EL did not exhibit a "steep–rise" at one inflection point under NAF mode, as opposed to that under GC mode. Under field conditions, the chilling temperatures in winter routinely and gradually decrease; therefore, GC mode could better mimic the natural environment than NAF mode. Indeed, an exponential decay regression might be better for correlating Fv/Fm and Fv/Fo to EL, particularly under the mode NAF (Figure 4e and f). However, the correlation Fv/Fm and EL under GC is clearly linear and an exponential decay regression may cause overfitting. Therefore, the linear regression was applied for all the correlations, which highlight the differences between the two cooling modes and provide support for better choice of GC mode.

In this study, we selected the chlorophyll fluorescence parameters from three designated indexes. Other parameters (e.g., coefficient of photochemical fluorescence quenching (qP), effective photochemical quantum yield of PS II, (ϕ_{PSII}), and Stern-Volmer type non-photochemical fluorescence quenching (NPQ)) measured using chlorophyll fluorescence could also be used for model establishment [24]. However, compared with other parameters, the three candidate indexes could be easily obtained after dark adaptation without requiring actinic light adaption or far–red light illumination. Using these three parameters could provide a more convenient process that is not destructive to the leaf samples, and thus, is more advantageous for large–scale grapevine production and resistant breeding.

There are few points on all correlations of Figure 4 between 20% and 60% EL. This lack of evenly scattered points between 20% and 60% EL was a result of the sharp burst of cell damage occurred between –3 to –5°C (Additional file 7: Figure S4). Some suggestions to improve the system accuracy include to take an even smaller temperature decreasing gradient, e.g. changing from 1°C/h to 0.5°C/h between –3 to –6°C in GC mode. However, this will not only double the number of measurements but also challenge the accuracy of the cooling instruments. Consequently, the balance between the gain of accuracy and increase of manpower needs to be checked when applying the updated system to large scale drought–cold tolerance screening at population or germplasm level.

Fv/Fm is one of the most commonly used indexes for tolerance evaluation. However, using Fv/Fm mainly focuses on the tolerance to individual stress, such as pathogen [40], drought [41], freezing [20], and heat [42]. Some previous studies have reported that water deficit minimally affects Fv/Fm [23,43]; however, our preliminary experiments showed that Fv/Fm, as well as Fv/Fo, qP, and ϕ_{PSII}, significantly decreased under individual drought stress (Additional file 8: Figure S5). The discrepancies in these studies could be attributed to the different growth conditions of the plant; the plants in the hydroponic system are more sensitive to drought than those grown in soil. The possible reason is the water potential may gradually decrease in soil dry, while the plants suffer from continuous given low water potential stress in the PEG–added hydroponic system during the whole treatment process, which leads to more rapid and severe damages in electron transport chains. Moreover, the findings suggest that the experimental system used in the present study could ensure that drought and cold stress, which were used as combined stress, individually affected the chlorophyll fluorescence results.

LT50 estimation based on chlorophyll fluorescence parameter

The LT50 value is an easily comparable parameter for quantifying tolerance to drought and cold stresses [15]. However, classic LT50 calculation by measuring EL is time consuming and less accurate, and thus, unsuitable for large–scale screenings of drought-or cold–tolerant grapes. Hence, we established a suitable model to estimate the LT50 values without EL measurement. Determining chlorophyll fluorescence is a good alternative for EL measurement because of its non–invasiveness and rapidness, as well as its potential for estimating LT50 according to the high correlations between EL and given chlorophyll fluorescence parameters.

This study also reported a significant correlation between LT50–EL and LT50–Fv/Fm across different grape genotypes under simultaneous drought–cold stress; this correlation is beneficial for estimating LT50 without EL measurement. Moreover, investigating two markedly resistance–different genotypes, namely, $V.$ $amurensis$ and 'Muscat Hamburg', and four genotypes increases the coverage in the spectrum of the natural drought–cold resistance of grapevine. The synchronization between the responses of Fv/Fm to EL under GC mode confirms the reliability of using Fv/Fm as an effective indicator of drought–cold resistance.

The significant correlation between LT50–EL and LT50–Fv/Fm has been observed under freezing condition in $Arabidopsis$ ($Arabidopsis$ $thaliana$) [20] and grape [19,44]. Interestingly, Ehlert et $al.$ [20] emphasized that LT50–Fv/Fm is slightly lower than the LT50 value in $Arabidopsis$ leaves. Jiang et $al.$ [19] concluded that the Fv/Fm inflection point is higher than the LT50 of grape woody tissues. In the present study, LT50–EL and LT50–*Fv/Fm* are approximately equal, which may be due to the different cold sensitivities of plant tissues and the additive effect of drought and cold stresses.

The proposed evaluation system provides a more convenient and reliable tool for determining drought–cold resistance in the laboratory and for large–scale screening in the field. The system should be further improved before use for actual grapevine breeding.

Conclusions

In the present study, we established and validated a novel experimental system for evaluating the resistance of grapevines against drought–cold stress. This system used tissue–cultured grape plants and 6% PEG solution under GC mode to simulate drought–cold stress. The resistance against drought–cold stress was evaluated in six different representative germplasms based on EL and chlorophyll fluorescence parameters, particularly *Fv/Fm*. A high correlation was observed between EL and *Fv/Fm*. Therefore, LT50 values can be well calculated based on *Fv/Fm* using the present system to evaluate the resistance of grapevine germplasms against drought–cold stress.

Methods

Plant material and culture conditions

Six genotypes of grape were pre–cultured on 1/2 B5 medium [45]. These genotypes included Chinese wild species *V. amurensis* (strongly tolerant to combined stress); three cultivars from *V. vinifera*, namely, 'Muscat Hamburg' (moderately tolerant to combined stress) [38], 'Centenial', and 'Cardinal'; and two interspecific hybrids, namely, 'Zhi168' (*V. monticola* × *V. riparia*) and 'Beta' (*V. labrusca* × *V. riparia*). The plantlets with heights of 5–8 cm were transferred to 1/2 Hoagland nutrient solution in hydroponic boxes (37 cm × 8 cm × 5 cm) with continuous aeration. Culture conditions were 23±1°C and 60% relative humidity with 16–h light (120 μmolm^{-2}s^{-1})/ 8–h dark photoperiod. After two weeks, the first three fully expanded leaves near the shoot apex were used for subsequent analysis.

Evaluation of the individual resistance of V. amurensis and 'Muscat Hamburg' against drought and cold stresses

Individual resistance against drought and cold stresses was evaluated using the micropropagated plantlets of *V. amurensis* and 'Muscat Hamburg', which were acclimated in 1/2 Hoagland nutrient solution for two weeks. PEG–6000 was added into the solution to decrease water potential for mimicking drought stress. *V. amurensis* and 'Muscat Hamburg' were subjected to five different concentrations of PEG (2%, 4%, 6%, 8%, and 10%) for 1 d, whereas control plants were grown in a solution without

PEG (CK). The third fully expanded leaf of each plantlet was sampled. The leaf samples were divided into two groups, which were subjected to EL (approximately 0.1 g) and RWC measurement. Moreover, chlorophyll fluorescence responses of the plantlets of both genotypes subjected to 6% PEG were evaluated using the third fully expanded leaf. All set of data had five replicates.

A classic method was used to assess the tolerance to a single cold stress [15]. Three leaf discs (6 mm in diameter) from the third fully expanded leaf were added into one tube containing 100 μL of distilled water. The tubes were transferred to a low–temperature incubator. After equilibrium at 0°C for 1 h, the temperature was decreased at a rate of 2°C/h from –2°C to –16°C. The samples were collected at defined temperatures to measure EL and LT50 values.

Combined drought–cold treatments

To establish the drought–cold treatment system, we selected a suitable cooling mode combined with a fixed PEG concentration. To determine the PEG concentration, we added 100 mL of 1/2 Hoagland nutrient with different PEG concentrations (0%, 4%, 6%, and 8%). The solutions were distributed into flasks and placed in a specific freezing environment. The optimal PEG concentration was selected to effectively distinguish the drought tolerance among genotypes; PEG should be non–frozen at a given temperature. The plantlets in the selected PEG concentration (6%) were subjected to two different cooling modes for mimicking combined drought–cold treatments. The two cooling modes were as follows: (1) gradient cooling ('GC', Figure 2a): a given low temperature was maintained for 1 h and rapidly reduced by 1°C; the procedure was repeated from –2°C to –9°C (each temperature point had three replicates); and (2) non–acclimated freezing ('NAF', Figure 2b): direct freezing from normal growth temperature (23°C) to a given low temperature as GC mode and then maintained at the low temperature for 2 h (each temperature point had five replicates). The third fully expanded leaf attached to the plant was used for chlorophyll fluorescence measurement at defined temperature. The leaves were collected for EL and LT50 calculation. The chosen drought–cold treatment (PEG6%+GC mode) were also applied to four other genotypes ('Centenial' 'Cardinal', 'Zhi168' and 'Beta') as former two genotypes.

Measurement of EL

EL was measured according to the method of Ma *et al.* [46] with some modifications. Briefly, the leaf samples exposed to drought, cold, and combined drought–cold stresses and their controls were collected and incubated in 6 mL of distilled water. After shaking at 0.5 *g* and 25°C for 3 h, initial conductivity (*C*1) was measured with a conductivity meter (FE30, METTLER TOLEDO, Switzerland). The samples were then autoclaved at 121°C for 20 min. After cooling to room temperature, the conductivity was re–measured as $C2$. EL was calculated using the equation EL (%)=$C1/C2$×100.

Measurement of RWC

RWC was measured using the method of Sairam et al. [47] with minor modification. For fresh weight (FW), the collected leaves were immediately weighed. The leaves were added into 100 mL of distilled water and incubated at room temperature overnight. Subsequently, the leaves were removed from the water. The liquid on the surface of the leaves was immediately dried using a filter paper and then weighed as the turgid weight (TW). The samples were oven dried at 80°C for 10 h to determine the dry weight (DW). RWC was defined as RWC (%)=(FW–DW)/(TW–DW)×100.

Measurement of chlorophyll fluorescence parameters

The third fully expanded leaf attached to the plant was subjected to a pulse–amplitude modulation fluorometer (PAM–2500, Walz, Germany) to determine chlorophyll fluorescence parameters. After 20–min dark adaptation, minimum fluorescence level (Fo) was determined with a low–intensity measuring light. Maximum fluorescence level (Fm) was measured after 0.5 s saturating pulse at 4,000 μmolm^{-2}s^{-1}. Steady–state fluorescence level (Fs) was obtained after 20–min actinic light (234 μmolm^{-2}s^{-1}) adaptation. Light–adapted maximum fluorescence level (Fm') was measured with a second saturating pulse (0.5 s, 4,000 μmolm^{-2}s^{-1}). The actinic light was then closed, and light–adapted minimum fluorescence level (Fo') was determined using a far–red light for 5s. Based on these parameters, we obtained four identification indexes: Fv/Fm=(Fm–Fo)/Fm, Fv/Fo=(Fm–Fo)/Fo, ϕ_{PSII}=(Fm'–Fs)/Fm', and qP=(Fm'–Fs)/(Fm'–Fo'). The four parameters, Fv/Fm, Fv/Fo, ϕ_{PSII}, and qP, represent the maximum photochemical quantum yield of PS II, potential activity of PS II, effective photochemical quantum yield of PS II, and coefficient of photochemical fluorescence quenching, respectively [48–50].

Model for estimation of LT50 based on leaf chlorophyll fluorescence response

Half–lethal temperature (LT50, the temperature at which the EL of leaf was reduced by 50%) was calculated by fitting the EL data to the Boltzmann 4 parameter model using R software [51].

$$y = Y_{\min} + \frac{Y_{\max} - Y_{\min}}{1 + e^{(b(x-c))}}$$

where y is the measured EL, x is the temperature, $Y_{\min}$ is the minimum value of EL, $Y_{\max}$ is the maximum value of

EL, b is the slope at inflection temperature, and c is the inflection temperature, namely, LT50.

LT50 was calculated using the same equation by replacing EL with the selected chlorophyll fluorescence parameters.

To identify a reliable and non–infusive indicator of drought–cold tolerance, we used standardized major axis linear regressions. These equations are used for quantifying the relationships between the measured chlorophyll fluorescence parameters and EL, and for comparing their slopes among different genotypes [52].

The relationship between the obtained LT50 from the newly identified chlorophyll fluorescence parameter (LT50$_{new}$) and that from the classic EL measurements (LT50$_{EL}$) was quantified using the following criteria:

Root mean squared error: $RMSE = \sqrt{\frac{1}{N}\sum_{1}^{N}(LT50_{new} - LT50_{EL})^2}$

Relative root mean squared error: $RRMSE = \frac{RMSE}{LT50_{EL}}$

Agreement index:

$$index = 1 - \frac{\sum_{1}^{N}(LT50_{new} - LT50_{EL})^2}{\sum_{1}^{N}(\mid LT50_{new} - \overline{LT50_{EL}} \mid + \mid LT50_{EL} - \overline{LT50_{EL}} \mid)^2}$$

where N is the number of genotypes used and $\overline{LT50_{EL}}$ is the average value of all LT50 obtained from EL measurements. Small RMSE and RRMSE values indicated better agreement between the two methods of LT50 estimation [53].

Statistical analysis

Data are expressed as mean±SE. T–test was used to compare EL, LT50, RWC, and chlorophyll fluorescence parameters between V. amurensis and 'Muscat Hamburg'. Paired t–test was used to compare LT50–EL and LT50–Fv/Fm, whereas the differences between the calculated and estimated LT50 among the six genotypes were analyzed through F–test by comparing the nested models [54].

Additional files

Additional file 1: Figure S1. Effect of PEG concentration levels on relative water content in V. amurensis and 'Muscat Hamburg'. The values were the mean value ± SE of results from five replicates. * and ** indicate significant differences between V. amurensis and 'Muscat Hamburg' at P< 0.05 and P< 0.01 level (t test), respectively.

Additional file 2: Figure S2. Comparison of plant growth conditions under PEG 6% treatment in V. amurensis and 'Muscat Hamburg'. Distilled water was refilled into solution to keep the initial volume every 12 hours.

Additional file 3: Figure S3. The ice frozen conditions of 1/2 Hoagland nutrient solution with different PEG concentrations at –6°C for 2 h. (a) control; (b) PEG 4%; (c) PEG 6%; (d) PEG 8%.

Additional file 4: Table S1. Summary statistics of linear regressions between the electrolyte leakage and the chlorophyll fluorescence parameter Fv/Fm under gradient cooling mode in six different grape genotypes. Slopes and intercepts are estimated by standard major axis regressions for each genotype. Their 95% confidence intervals (CI) are also provided. Different letters indicate significant differences in the intercept or slope among the genotypes at $P< 0.05$.

Additional file 5: Table S2. Correlation between electrolyte leakage (EL) and four chlorophyll fluorescence parameters (Fo, Fm, Fv/Fm and Fv/Fo) under gradient cooling mode in six different grape genotypes.

Additional file 6: Table S3. Significant analysis of LT50–EL and LT50–Fv/Fm data in *V. amurensis* and 'Muscat Hamburg' under two different cooling modes. * indicates significant differences between LT50–EL and LT50–Fv/Fm at $P<0.05$ level (paired t test).

Additional file 7: Figure S4. Responses of electrolyte leakage and Fv/Fm as a function of temperature in different grape genotypes. The open symbols are observed mean ± SE with three replicates and lines are fitted curves to the Boltzmann 4–parameter model. Filled symbols indicate where the LT50 were estimated and the corresponding genotype is indicated in each figure.

Additional file 8: Figure S5. Comparison of Fv/Fm (a), Fv/Fo (b), qP (c) and ϕ_{PSII} (d) at different time after PEG 6% treatment in *V. amurensis* and 'Muscat Hamburg'. The values were the mean value ± SE of results from five replicates. * and ** indicate significant differences between *V. amurensis* and 'Muscat Hamburg' at $P< 0.05$ and $P<0.01$ level (t test), respectively.

Abbreviations
EL: Electrolyte leakage; LT50: Half–lethal temperature; PEG: Polyethylene glycol; RWC: Relative water content.

Competing interests
The authors declare that they have no competing interests.

Authors' contributions
HPX, LYS, SHL and ZWD designed and oversaw the research. LYS performed the research. ZWD performed statistical analysis and model establishment. LYS, ZWD, SHL and HPX wrote the article. All authors read and approved the final manuscript.

Acknowledgements
This work was supported by the National Natural Science Foundation of China (NSFC Accession No.: 31130047, 31471857) and Youth Innovation Promotion Association of CAS (No.2015281).

Author details
[1]Beijing Key Laboratory of Grape Sciences and Enology and CAS Key Laboratory of Plant Resources, Institute of Botany, Chinese Academy of Sciences, Beijing 100093, China. [2]University of Chinese Academy of Sciences, Beijing 100049, China. [3]INRA, Institut des Sciences de la Vigne et du Vin, UMR 1287 Ecophysiologie et Génomique Fonctionnelle de la Vigne (EGFV), 210 Chemin de Leysotte, 33882 Villenave d' Ornon, France. [4]Key Laboratory of Plant Germplasm Enhancement and Specialty Agriculture, Wuhan Botanical Garden, Chinese Academy of Sciences, Wuhan 430074, China.

References
1. Chinnusamy V, Zhu JH, Zhu JK. Cold stress regulation of gene expression in plants. Trends Plant Sci. 2007;12(10):444–51.
2. Huang GT, Ma SL, Bai LP, Zhang L, Ma H, Jia P, et al. Signal transduction during cold, salt, and drought stresses in plants. Mol Biol Rep. 2012;39(2):969–87.
3. Ahmad P, Azooz M, Prasad M. Salt Stress in Plants. Heidelberg: Springer; 2013.
4. Mittler R. Abiotic stress, the field environment and stress combination. Trends Plant Sci. 2006;11(1):15–9.
5. Walter J, Jentsch A, Beierkuhnlein C, Kreyling J. Ecological stress memory and cross stress tolerance in plants in the face of climate extremes. Environ Exp Bot. 2013;94:3–8.
6. Rizhsky L, Liang HJ, Shuman J, Shulaev V, Davletova S, Mittler R. When Defense pathways collide. The response of Arabidopsis to a combination of drought and heat stress. Plant Physiol. 2004;134(4):1683–96.
7. de Souza CR, Maroco JP, dos Santos TP, Rodrigues ML, Lopes CM, Pereira JS, et al. Partial rootzone drying: regulation of stomatal aperture and carbon assimilation in field-grown grapevines (Vitis vinifera cv. Moscatel). Funct Plant Biol. 2003;30(6):653–62.
8. White MA, Diffenbaugh N, Jones GV, Pal J, Giorgi F. Extreme heat reduces and shifts United States premium wine production in the 21st century. Proc Natl Acad Sci. 2006;103(30):11217–22.
9. Barriopedro D, Fischer EM, Luterbacher J, Trigo RM, García-Herrera R. The hot summer of 2010: redrawing the temperature record map of Europe. Science. 2011;332(6026):220–4.
10. Rizhsky L, Liang HJ, Mittler R. The combined effect of drought stress and heat shock on gene expression in tobacco. Plant Physiol. 2002;130(3):1143–51.
11. Li JT, Wang N, Xin HP, Li SH. Overexpression of VaCBF4, a transcription factor from Vitis amurensis, improves cold tolerance accompanying increased resistance to drought and salinity in Arabidopsis. Plant Mol Biol Report. 2013;31(6):1518–28.
12. Huang HB. A preliminary evaluation of climatic regions for grape production in North China. Journal Chin Agric Univ. 1980;2:43–51.
13. Zabadal TJ, Dami IE, Goffinet MC, Martinson TE. Winter injury to grapevines and methods of protection. Chien ML: Michigan State University Extension; 2007.
14. Su HR. Apple science. Beijing: China Agricultural Press; 1999.
15. Verslues PE, Agarwal M, Katiyar-Agarwal S, Zhu J, Zhu JK. Methods and concepts in quantifying resistance to drought, salt and freezing, abiotic stresses that affect plant water status. Plant J. 2006;45(4):523–39.
16. Prášil I, Zámečnik J. The use of a conductivity measurement method for assessing freezing injury: I. Influence of leakage time, segment number, size and shape in a sample on evaluation of the degree of injury. Environ Exp Bot. 1998;40(1):1–10.
17. Bajji M, Kinet J-M, Lutts S. The use of the electrolyte leakage method for assessing cell membrane stability as a water stress tolerance test in durum wheat. Plant Growth Regul. 2002;36(1):61–70.
18. Gilmour SJ, Hajela RK, Thomashow MF. Cold acclimation in Arabidopsis thaliana. Plant Physiol. 1988;87(3):745–50.
19. Jiang H, Howell G, Flore J. Efficacy of chlorophyll fluorescence as a viability test for freeze-stressed woody grape tissues. Can J Plant Sci. 1999;79(3):401–9.
20. Ehlert B, Hincha DK. Chlorophyll fluorescence imaging accurately quantifies freezing damage and cold acclimation responses in Arabidopsis leaves. Plant Methods. 2008;4(1):12.
21. Yu X, Peng YH, Zhang MH, Shao YJ, Su WA, Tang ZC. Water relations and an expression analysis of plasma membrane intrinsic proteins in sensitive and tolerant rice during chilling and recovery. Cell Res. 2006;16(6):599–608.
22. Lanier J, Ebdon J, DaCosta M. Physiological changes associated with wilt-induced freezing tolerance among diverse turf performance perennial ryegrass cultivars. Crop Sci. 2012;52(3):1393–405.
23. Longenberger PS, Smith C, Duke S, McMichael B. Evaluation of chlorophyll fluorescence as a tool for the identification of drought tolerance in upland cotton. Euphytica. 2009;166(1):25–33.
24. Brestic M, Zivcak M. PSII fluorescence techniques for measurement of drought and high temperature stress signal in crop plants: protocols and applications. In: Molecular Stress Physiology of Plants. Dordrecht: Springer; 2013. p. 87–131.
25. Rizza F, Pagani D, Stanca A, Cattivelli L. Use of chlorophyll fluorescence to evaluate the cold acclimation and freezing tolerance of winter and spring oats. Plant Breed. 2001;120(5):389–96.
26. Maxwell K, Johnson GN. Chlorophyll fluorescence—a practical guide. J Exp Bot. 2000;51(345):659–68.
27. Christen D, Schönmann S, Jermini M, Strasser RJ, Défago G. Characterization and early detection of grapevine (Vitis vinifera) stress responses to esca disease by in situ chlorophyll fluorescence and comparison with drought stress. Environ Exp Bot. 2007;60(3):504–14.
28. Hediye SA, Rengin O, Baris U, Ismail T. Reactive oxygen species scavenging capacities of cotton (Gossypium hirsutum) cultivars under combined drought and heat induced oxidative stress. Environ Exp Bot. 2013;99:141–9.

29. Rivero RM, Mestre TC, Mittler R, Rubio F, Garcia-Sanchez F, Martinez V. The combined effect of salinity and heat reveals a specific physiological, biochemical and molecular response in tomato plants. Plant Cell Environ. 2013;37(5):1059–73.

30. Iyer NJ, Tang Y, Mahalingam R. Physiological, biochemical and molecular responses to a combination of drought and ozone in *Medicago truncatula*. Plant Cell Environ. 2013;36(3):706–20.

31. Prasch CM, Sonnewald U. Simultaneous application of heat, drought, and virus to Arabidopsis plants reveals significant shifts in signaling networks. Plant Physiol. 2013;162(4):1849–66.

32. Li XN, Cai J, Liu FL, Dai TB, Cao WX, Dong J. Physiological, proteomic and transcriptional responses of wheat to combination of drought or waterlogging with late spring low temperature. Funct Plant Biol. 2014;41(7):690–703.

33. Li F, Lei HJ, Zhao XJ, Tian RR, Li TH. Characterization of three sorbitol transporter genes in micropropagated apple plants grown under drought stress. Plant Mol Biol Report. 2012;30(1):123–30.

34. Bidabadi SS, Mahmood M, Baninasab B, Ghobadi C. Influence of salicylic acid on morphological and physiological responses of banana (Musa acuminata cv. 'Berangan', AAA) shoot tips to in vitro water stress induced by polyethylene glycol. Plant Omics J. 2012;5:33–9.

35. Sen A, Alikamanoglu S. Antioxidant enzyme activities, malondialdehyde, and total phenolic content of PEG-induced hyperhydric leaves in sugar beet tissue culture. In Vitro Cell Dev Biol Plant. 2013;49(4):396–404.

36. Gourcilleau D, Lenne C, Armenise C, Moulia B, Julien J-L, Bronner G, et al. Phylogenetic study of plant Q-type C2H2 zinc finger proteins and expression analysis of poplar genes in response to osmotic, cold and mechanical stresses. DNA Res. 2011;18(2):77–92.

37. Fennell A. Freezing tolerance and injury in grapevines. J Crop Improv. 2004;10(1–2):201–35.

38. Xin HP, Zhu W, Wang LN, Xiang Y, Fang LC, Li JT, et al. Genome wide transcriptional profile analysis of *Vitis amurensis* and *Vitis vinifera* in response to cold stress. PLoS One. 2013;8(3):e58740.

39. Wang YJ, Yang YZ, Zhang JX, Pan XJ, Wan YZ. Preliminary identification of drought resistance of Chinese wild *Vitis* species and its interspecific hybrids. Acta Horticulturae Sinica. 2004;6:1–4.

40. Rousseau C, Belin E, Bove E, Rousseau D, Fabre F, Berruyer R, et al. High throughput quantitative phenotyping of plant resistance using chlorophyll fluorescence image analysis. Plant Methods. 2013;9:17.

41. Woo NS, Badger MR, Pogson BJ. A rapid, non-invasive procedure for quantitative assessment of drought survival using chlorophyll fluorescence. Plant Methods. 2008;4:27.

42. Xu HG, Liu GJ, Liu GT, Yan BF, Duan W, Wang LJ, et al. Comparison of investigation methods of heat injury in grapevine (*Vitis*) and assessment to heat tolerance in different cultivars and species. BMC Plant Biol. 2014;14:156.

43. Wang ZX, Chen L, Ai J, Qin HY, Liu YX, Xu PL, et al. Photosynthesis and activity of photosystem II in response to drought stress in Amur Grape (*Vitis amurensis* Rupr.). Photosynthetica. 2012;50(2):189–96.

44. Jiang H, Howell GS. Applying chlorophyll fluorescence technique to cold hardiness studies of grapevines. Am J Enol Vitic. 2002;53(3):210–7.

45. Gamborg OL, Miller RA, Ojima K. Nutrient requirements of suspension cultures of soybean root cells. Exp Cell Res. 1968;50(1):151–8.

46. Ma YY, Zhang YL, Shao H, Lu J. Differential physio-biochemical responses to cold stress of cold-tolerant and non-tolerant grapes (*Vitis* L.) from China. J Agron Crop Sci. 2010;196(3):212–9.

47. Sairam RK, Rao KV, Srivastava G. Differential response of wheat genotypes to long term salinity stress in relation to oxidative stress, antioxidant activity and osmolyte concentration. Plant Sci. 2002;163(5):1037–46.

48. Kitajima M, Butler W. Quenching of chlorophyll fluorescence and primary photochemistry in chloroplasts by dibromothymoquinone. Biochim Biophys Acta Biomembr. 1975;376(1):105–15.

49. Schreiber U, Schliwa U, Bilger W. Continuous recording of photochemical and non-photochemical chlorophyll fluorescence quenching with a new type of modulation fluorometer. Photosynth Res. 1986;10(1–2):51–62.

50. Genty B, Briantais J-M, Baker NR. The relationship between the quantum yield of photosynthetic electron transport and quenching of chlorophyll fluorescence. Biochim Biophys Acta Gen Subj. 1989;990(1):87–92.

51. Team RC. R: A language and environment for statistical computing. Vienna, Austria: R Foundation for Statistical Computing; 2012.

52. Warton DI, Duursma RA, Falster DS, Taskinen S. Smatr 3–an R package for estimation and inference about allometric lines. Methods Ecol Evol. 2012;3(2):257–9.

53. Wallach D. Evaluating crop models. In: Working with dynamic crop models. Amsterdam, The Netherlands: Elsevier; 2006. p. 11–53.

54. Motulsky HJ, Christopoulos A. Fitting models to biological data using linear and nonlinear regression: A practical guide to curve fitting. San Diego: Graphpad Software Inc.; 2003.

4

Overexpression of a truncated CTF7 construct leads to pleiotropic defects in reproduction and vegetative growth in Arabidopsis

Desheng Liu and Christopher A Makaroff*

Abstract

Background: Eco1/Ctf7 is essential for the establishment of sister chromatid cohesion during S phase of the cell cycle. Inactivation of Ctf7/Eco1 leads to a lethal phenotype in most organisms. Altering Eco1/Ctf7 levels or point mutations in the gene can lead to alterations in nuclear division as well as a wide range of developmental defects. Inactivation of Arabidopsis CTF7 (AtCTF7) results in severe defects in reproduction and vegetative growth.

Results: To further investigate the function(s) of AtCTF7, a tagged version of AtCTF7 and several AtCTF7 deletion constructs were created and transformed into wild type or $ctf7^{+/-}$ plants. Transgenic plants expressing 35S:NTAP: AtCTF7$_{\Delta299-345}$ (AtCTF7ΔB) displayed a wide range of phenotypic alterations in reproduction and vegetative growth. Male meiocytes exhibited chromosome fragmentation and uneven chromosome segregation. Mutant ovules contained abnormal megasporocyte-like cells during pre-meiosis, megaspores experienced elongated meiosis and megagametogenesis, and defective megaspores/embryo sacs were produced at various stages. The transgenic plants also exhibited a broad range of vegetative defects, including meristem disruption and dwarfism that were inherited in a non-Mendelian fashion. Transcripts for epigenetically regulated transposable elements (TEs) were elevated in transgenic plants. Transgenic plants expressing 35S:AtCTF7ΔB displayed similar vegetative defects, suggesting the defects in 35S:NTAP:AtCTF7ΔB plants are caused by high-level expression of AtCTF7ΔB.

Conclusions: High level expression of AtCTF7ΔB disrupts megasporogenesis, megagametogenesis and male meiosis, as well as causing a broad range of vegetative defects, including dwarfism that are inherited in a non-Mendelian fashion.

Keywords: Meiosis, Sister chromatid cohesion, Megasporogenesis, Megagametogenesis, Megaspore mother cell, Functional megaspore-like, Epigenetic

Background

The precise establishment and release of sister chromatid cohesion are vital for cell division during mitosis and meiosis [1-3]. The cohesin complex, which forms a ring to entrap sister chromatids, is composed of four subunits: a heterodimer of Structural Maintenance of Chromosome (SMC) proteins, SMC1 and SMC3, an α-kleisin, Sister Chromatid Cohesion1 (SCC1) in somatic cells or REC8 in meiotic cells and SCC3 [2,4,5]. The core cohesin complex associates with a number of factors during the cohesion cycle [1-3,6-8] and its association with the chromosomes is controlled by

modifications, such as acetylation, phosphorylation, sumoylation and proteolysis [9-13].

Sister chromatid cohesion is established during S phase by the Establishment of Cohesion protein, ECO1/CTF7 [14-18]. In budding yeast, ECO1/CTF7 acetylates lysine residues K112, K113 of SMC3 and K84, K210 of SCC1 to stabilize the ring and keep it closed [10,19,20]. ECO1/CTF1 is an essential gene in many organisms with the complete inactivation of CTF7 resulting in lethality [1-3,14-16,21]. Inactivation or mutations in ECO1/CTF1 leads to a broad range of defects, including chromosome mis-segregation, defects in DNA double strand break repair, defects in homologous recombination and transcriptional alterations [14,15,17,21-25]. Humans contain two ECO1/CTF7 orthologs, ESCO1 and ESCO2 [24]. Mutations in ESCO2 lead to Roberts syndrome

* Correspondence: makaroca@miamioh.edu
Department of Chemistry and Biochemistry, Miami University, Oxford, OH 45056, USA

(RBS), which is associated with a variety of defects including growth retardation, limb reduction/asymmetric limb growth, cleft lip/palate and missing fingers/toes [26,27].

Arabidopsis contains a single CTF7 ortholog (AtCTF7), which can complement the temperature-sensitive yeast *ctf7-203* mutant [16]. *AtCTF7* encodes a 345 amino acid protein, containing a PIP box at residues 82 to 86, a C_2H_2 zinc finger motif at residues 92–130 and an acetyltransferase domain from residues 184 to 335. The acetyltransferase domain can be further separated into three motifs: D (amino acids 184–204), A (amino acids 266–301) and B (amino acids 311–335). Motif D provides the framework of the acetyltransferase domain. Motif A participates in acetyl-CoA binding and is critical for catalytic activity, while motif B is the most C-terminal region and participates in substrate recognition and catalytic activity regulation [28,29]. Variations in *AtCTF7* expression have been shown to result in a wide range of defects. Plants heterozygous for a T-DNA insertion in *AtCTF7* grow normally but their siliques contain approximately 25% arrested seeds, consistent with the belief that CTF7 is an essential protein [16]. However, *Atctf7* homozygous plants can be detected at very low frequencies [17]. *Atctf7* plants exhibit a wide range of developmental defects, including extreme dwarfism and sterility with *Atctf7* meiocytes exhibiting abnormal chromosome segregation and defective sister chromatid cohesion. Knockdown of *AtCTF7* mRNA levels using RNAi leads to growth retardation and defective sister chromatid cohesion [18]. Finally, overexpression of full length AtCTF7 with the CaMV 35S promoter leads to ovule abortion, but plant growth is not affected [16].

To further investigate the roles of CTF7 in Arabidopsis and to identify AtCTF7 interacting proteins, transgenic plants expressing AtCTF7 constructs that encode either TAP-tagged [30] or non-tagged versions of full length or truncated AtCTF7 were generated and analyzed. Transgenic plants that express a truncated version of AtCTF7, missing motif B, from the 35S promoter exhibited impaired reproduction and vegetative growth. Ovules in 35S:NTAP:AtCTF7ΔB plants displayed alterations in cell identity and the timing of megasporogenesis and megagametogenesis, while male meiocytes exhibited alterations during meiosis II. The transgenic plants also displayed alterations in vegetative growth that were not inherited in a Mendelian fashion.

Results

Arabidopsis plants expressing high levels of NTAP: AtCTF7ΔB display reduced fertility

An *AtCTF7* construct missing the C-terminal 46 amino acids (Δ299-345) was generated, fused with NTAP [30] and expressed from the CaMV 35S promoter in wild-type Columbia plants (35S:NTAP:AtCTF7ΔB; Additional file 1: Figure S1). Twenty out of the 36 independent lines examined exhibited reduced fertility, with fertility levels varying significantly between the lines. Plants exhibiting a weak phenotype, which accounted for two of the 20 reduced fertility lines, produced shorter siliques with reduced numbers of seeds, but the seeds appeared normal (Figure 1Aii). For example Line 19 produced 39.2 ± 3.9 seeds per silique (n = 35) compared to wild type plants that produce 54.2 ± 4.1 seeds per silique (n = 35). Anthers from Line 19 plants were smaller and contained reduced numbers of pollen (574 vs 1175 in wild type), but the pollen appeared viable (Figure 1Bii). The other 18 reduced fertility lines exhibited more severe defects. These plants produced siliques containing large numbers of unfertilized ovules and aborted seeds (Figure 1Aiii). Unfertilized ovules appeared as white dots, resembling the situation in *atctf7-1* plants [17]. Aborted seeds appeared white and plump, similar to seeds containing arrested embryos in *Atctf7-1*[+/−] plants [16]. Seed set varied considerably between the lines, ranging from 13.3 ± 5.6 seeds per silique to 34.7 ± 8.4 seeds per silique. Anthers from these plants typically contained reduced numbers of pollen, much of which was not viable. For example, anthers from Line 11 produced on average approximately 210 pollen, of which only 20% was viable (Figure 1Biii). Given that most lines exhibited severe fertility defects, one representative line (#11) was chosen and characterized in detail.

The reduced fertility phenotype of 35S:NTAP:AtCTF7ΔB plants suggested that co-suppression may be lowering *AtCTF7* transcript levels in the lines. Therefore, quantitative reverse transcription polymerase chain reaction (qRT-PCR) was conducted to examine transcript levels of native *AtCTF7* as well as the *CTF7* transgene. Surprisingly, total *AtCTF7* (35S:NTAP:AtCTF7ΔB and native *AtCTF7*) transcript levels were approximately seven-fold higher in 35S:NTAP:AtCTF7ΔB plants than *CTF7* transcript levels in wild type plants, while native *AtCTF7* transcript levels were almost three-fold higher in the transgenic plants (Figure 1C). Therefore, the transgenic lines contained high levels of *AtCTF7ΔB* transcripts along with elevated levels of native *AtCTF7* mRNA, indicating that the observed phenotypes are not the result of reduced *AtCTF7* expression. Instead these results suggest that the 35S:NTAP:AtCTF7ΔB construct may be exerting a dominant negative effect.

Reciprocal crossing experiments were carried out to determine the effect of the 35S:NTAP:AtCTF7ΔB construct on male and female gametes. The 35S:NTAP:AtCTF7ΔB construct was transmitted at reduced levels through both male and female gametes (Additional file 1: Table S1), consistent with our preliminary results that both male and female fertility were affected in 35S:NTAP:AtCTF7ΔB plants. Nonviable seeds were obtained when 35S:NTAP:AtCTF7ΔB plants were used as either the male or female parent; however the

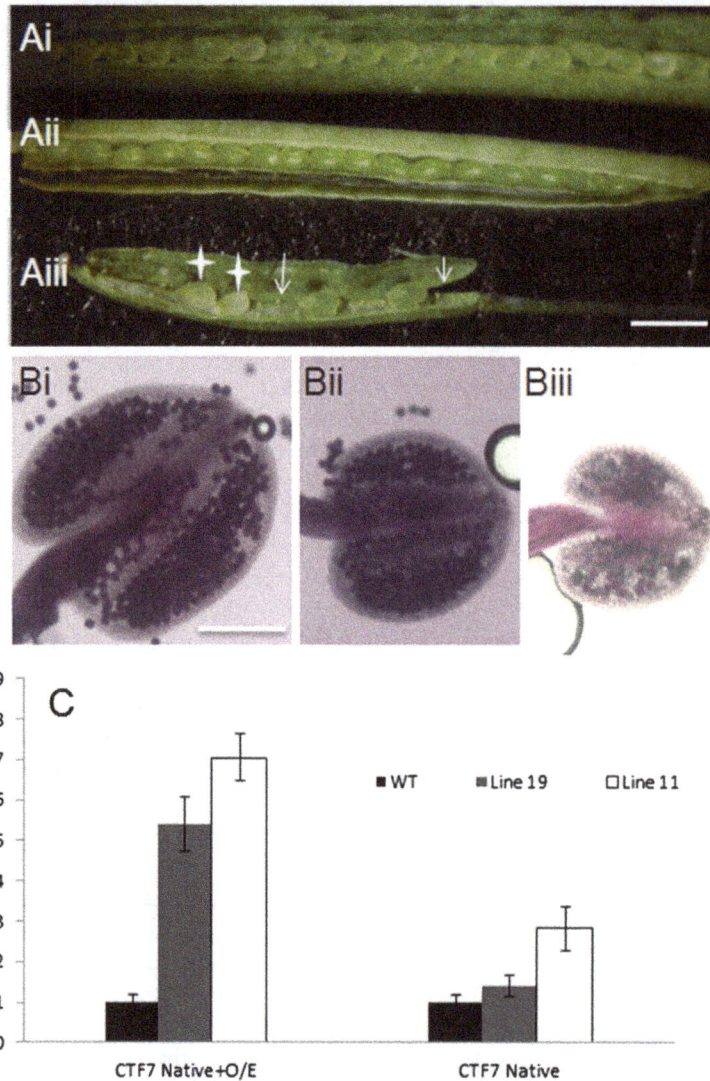

Figure 1 35S:NTAP:AtCTF7ΔB plants exhibit reduced fertility. (A) Open siliques from wild type **(Ai)** and 35S:NTAP:AtCTF7ΔB **(Aii,Aiii)** plants. **i**, Wild type silique with full seed set. **ii**, Silique from a 35S:NTAP:AtCTF7ΔB plant (Line 19) exhibiting a weak phenotype. **iii**, Silique from a 35S: NTAP:AtCTF7ΔB plant (Line 11) exhibiting a strong phenotype. Arrows indicate shriveled, unfertilized ovules. White stars show white, plump seeds, which aborted after fertilization. Scale bar = 0.5 cm. **(B)** Alexander staining of mature anthers from wild type **(i)** and 35S:NTAP:AtCTF7ΔB **(ii, iii)** plants. **i**, Wild type anther. **ii**, Anther from Line 19. The anther is smaller and contains less pollen; all the pollen is viable. **iii**, Anther from Line 11. The anther is smaller and contains low numbers of viable pollen. Scale bar = 50 μm. **(C)**. Expression analysis of *AtCTF7* in wild type and 35S:NTAP: AtCTF7ΔB plants. Transcript levels of total *AtCTF7* (35S:NTAP:AtCTF7ΔB and native AtCTF7) and native *AtCTF7* are increased in 35S:NTAP:AtCTF7ΔB plants with Line 11 plants exhibiting the highest levels. Buds of wild type, non-dwarf, 4th generation Line 11 plants and 4th generation Line19 plants were used for this experiment. Data are shown as means ± SD (n = 3).

relative proportion of nonviable seed was three times greater (9.3% verses 31.3%) when 35S:NTAP:AtCTF7ΔB plants served as the female. The number of nonviable seeds was greatest (65%) in self-pollenated plants. Therefore, while both male and female gametes are affected, the 35S:NTAP:AtCTF7ΔB construct has a greater effect on female reproduction. Inheritance of 35S:NTAP:AtCTF7ΔB-associated defects from both parents has a compounding effect on seed production.

Male meiotic chromosome segregation is altered in 35S: NTAP:AtCTF7ΔB plants

Previous studies showed that inactivation of *AtCTF7* via T-DNA insertion or reduction in *AtCTF7* mRNA levels via RNAi disrupts sister chromatid cohesion, resulting in uneven chromosome segregation during meiosis and ultimately reduced pollen viability [17,18]. The reduced fertility observed in 35S:NTAP:AtCTF7ΔB plants suggested that meiosis might also be affected in these lines. Therefore, chromosome spreading experiments were carried out to examine the effect of 35S:NTAP:AtCTF7ΔB expression on male meiosis.

Male meiocytes in 35S:NTAP:AtCTF7ΔB plants resembled wild type during early stages of meiosis, with normal chromosome morphology during pachytene (Figure 2A,E), diakinesis (Figure 2B,F) and metaphase I (Figure 2C,G). The first noticeable defect was observed at telophase I when lagging chromosomes were observed (Figure 2D,H), followed by mis-segregated chromosomes at prophase II (Figure 2M). More than twenty individual chromosomes were typically observed in meiocytes beginning at metaphase II (Figure 2N), indicating that sister chromatid cohesion was prematurely lost. Chromosomes did not segregate evenly at anaphase II (Figure 2O) resulting in the production of polyads with varying DNA contents at tetrad stage (Figure 2P). Similar to *atctf7* plants, a small number of relatively normal meiocytes were also observed throughout meiosis with fewer normal-appearing meiocytes in later stages of meiosis. For example, the percentages of defective meiocytes observed at various stages of meiosis in Line 11 were: metaphase I: 0% (0/41), telophase I: 6.8% (7/87), prophase II: 23.5% (24/102), metaphase II: 40.7% (24/59), anaphase II: 56.8% (25/44) and telophase II: 64.0% (114/178).

Immunolocalization experiments were then carried out to examine the distribution of the meiotic cohesin protein SYN1 [31] in 35S:NTAP:AtCTF7ΔB male meiocytes. Consistent with meiotic chromosome spreading experiments the overall distribution of SYN1 was not dramatically affected in meiocytes of 35S:NTAP:AtCTF7ΔB plants (Additional file 1: Figure S2). SYN1 exhibited diffuse nuclear labeling during interphase with strong signal observed on the developing chromosomal axes from early leptotene into zygotene (Additional file 1: Figures S2A,D). During late zygotene and pachytene the protein lined the chromosomes (Additional file 1: Figures S2B,E,C and F). SYN1 was released normally from the condensing chromosomes during diplotene and diakinesis and similar to the situation in wild type, SYN1 was barely detectable on chromosomes by prometaphase I. Therefore, cohesin appears to load and be removed normally from meiotic chromosomes in 35S:NTAP:AtCTF7ΔB plants.

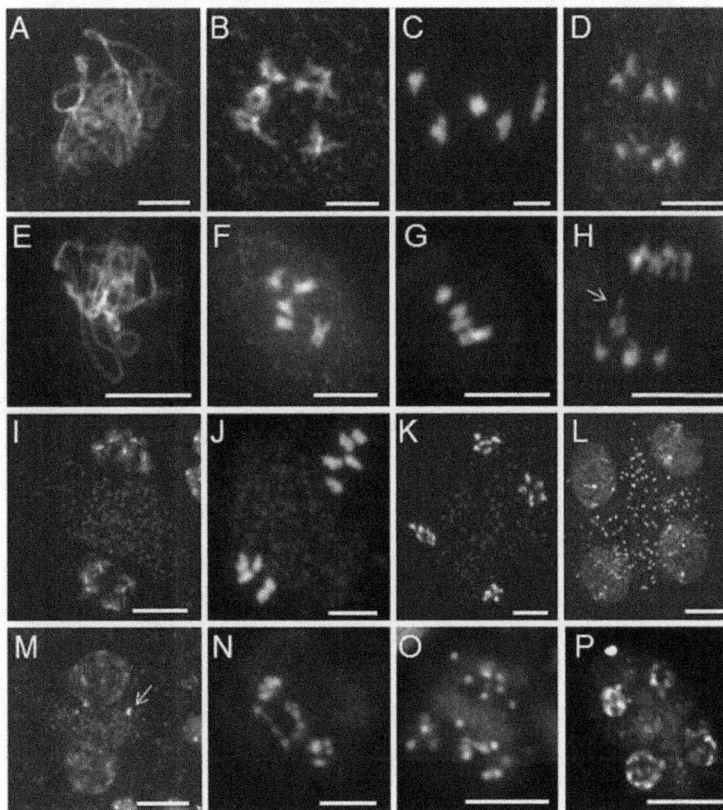

Figure 2 35S:NTAP:AtCTF7ΔB male meiocytes exhibit defective meiotic chromosome segregation. (A-D) and **(I-L)** Wild type meiocytes. **(E-H)** and **(M-P)** 35S:NTAP:AtCTF7ΔB meiocytes. **A, E** pachytene; **B, F** diakinesis; **C, G** metaphase I; **D, H** telophase I; **I, M** prophase II; **J, N** metaphase II; **K, O** anaphase II; **L, P** telophase II. Lagging chromosomes and/or chromosome fragments are denoted with arrows. Meiotic chromosomes are stained by 4', 6-diamidino-2-phenylindole (DAPI). Scale bar = 10 μm.

**Female gametophyte development is altered in 35S:
NTAP:AtCTF7ΔB plants**

Previous studies showed that ovule development is very sensitive to AtCTF7 levels [16-18]. Approximately 50% of the ovules from Atctf7-1$^{+/-}$ plants contain non-degenerated antipodal nuclei [16], while ovules from atctf7 plants degenerate early [17]. RNAi directed reduction of AtCTF7 mRNA levels, as well as 35S-mediated increases in AtCTF7 transcript levels result in ovule arrest at Female Gametophyte (FG) 1 stage [16,18].

To elucidate the function(s) of 35S:NTAP:AtCTF7ΔB in ovule development, ovules from wild type and Line 11 plants were analyzed by differential interference contrast (DIC) microscopy. In wild type siliques, archesporial cells are specified from the subepidermal cell layers and differentiate into megaspore mother cells, which are initially unpolarized and become polarized prior to the start of meiosis (Additional file 1: Figure S3A). Two rounds of meiosis produce a tetrad of four haploid megaspores (Figure 3C,D; Additional file 1: Figure S3B-E) [32]. Prior to FG1, the megaspore mother cell, dyad and tetrad are all adjacent to L1 cells (Figures 3A-D) [33]. When the ovule reaches FG1, the megaspore at the chalazal-end differentiates into the functional megaspore while the other three megaspores undergo programmed cell death (Additional file 1: Figures S3F, G; Figures S4B, C) [34,35]. In wild type siliques, adjacent ovules are similar in size and point in opposite directions (Figures 3A,B).

In mutant plants, defects were observed in ovules very early in development. Approximately 30.4% (41/135) of the ovules observed were defective at the premeiosis stage. Abnormally enlarged sub-epidermal cells were observed adjacent to normal looking megaspore mother cells in some pre-meiotic ovules (Figure 3E,F'). Adjacent ovules were often different in size and sometimes pointed in the same direction (Figure 3 F,F'). A functional megaspore was not identified in some ovules even though their shape and size was beyond FG1. Instead, large cells with prominent nuclei and megaspore-like characteristics [36] were found in the position of a normal embryo sac (Figures 3G-I). Ovules containing either one megaspore-like cell (Figure 3G,H) or two megaspore-like cells with dyad characteristics (Figure 3I) were also observed. This suggested that meiosis is delayed, altered and arrested in most megaspores. Extra cells were also observed between the megaspores and the L1 layer of cells (Figure 3G-I). In WT

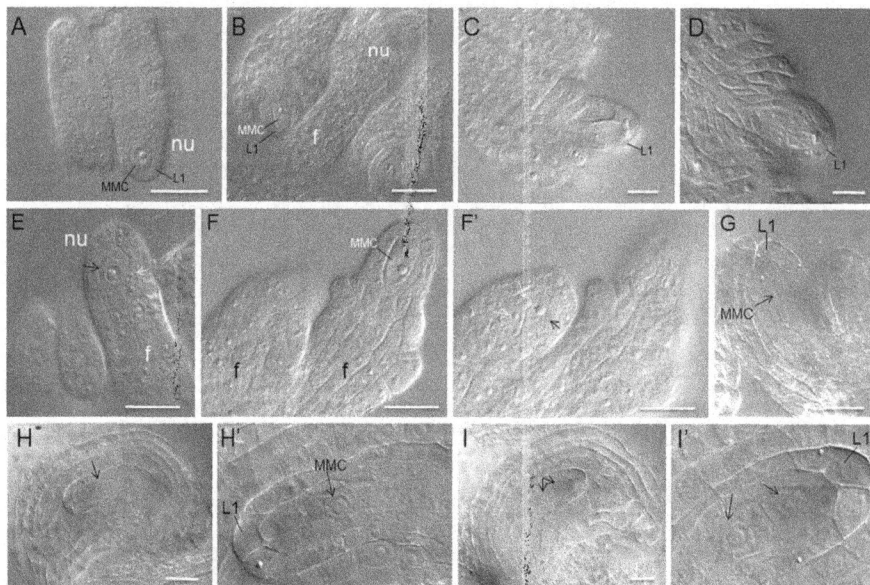

Figure 3 Early ovule development is disrupted in 35S:NTAP:AtCTF7ΔB plants. Wild type **(A-D)** and 35S:NTAP:AtCTF7ΔB **(E-I)** ovules were analyzed by differential interference contrast microscopy. **(A)** Pre-meiotic ovule containing a single megaspore mother cell (MMC; stage 1-II). **(B)** Pre-meiotic ovule at stage 2-III. Inner and outer integuments start to initiate. **(C)** Meiotic ovule containing a dyad after meiosis I (stage 2-IV). **(D)** Meiotic ovule containing a tetrad after meiosis II. **(E-F)** 35S:NTAP:AtCTF7ΔB ovules at pre-meiotic stages. **(E)** Right ovule containing an abnormal, enlarged cell (white arrow) adjacent to a MMC-like cell (black arrow). Adjacent ovules are different in size. **(F-F')** Left ovule containing two abnormal, enlarged cells (white arrows) adjacent to a MMC-like cell (black arrow). Ovules are different in size and stage (left: stage 1-II, right: stage 2-III) and point in the same direction. **(G-I)** 35S:NTAP:AtCTF7ΔB ovules at meiosis. **(G)** Ovule containing a large cell with prominent nucleus (arrow) resembling a MMC. An extra cell is present at the position of the degenerated megaspores, between the MMC and L1 cells. **(H)** Ovule containing a MMC-like cell with a prominent nucleus in the central region of the ovule. Ovule is enlarged and extra cells are between the MMC and L1 cells. **(H')** Magnified view of MMC from **H**. **(I)** Ovule containing two cells with prominent nuclei (arrows) in the central region of the ovule. The two cells are separated and resemble a dyad. Extra cells surround the dyad. **(I')** Magnified view of **I**. Size bar = 10 μm. Developmental stages are defined according to Schneitz et al. [32].

plants, a small number (6.7%, 8/120) of ovules were found to contain twin megaspore mother cells; however no defects were observed during or after meiosis. In Line 11 plants, 14.1% (19/135) of the ovules contained multiple megaspore mother-like cells at pre-meiosis. Furthermore, 72.5% (50/69) of ovules with dyads contained extra cells between the dyad and L1 layer, and 27.5% (19/69) the dyads were in elongated shape. Ultimately 50.4% (64/127) of the ovules observed were defective at FG1 stage. Therefore, alterations in archesporial cell differentiation, the onset and progression of meiosis and possibly somatic cell identity are observed in ovules of 35S:NTAP:AtCTF7ΔB plants.

The effects of 35S:NTAP:AtCTF7ΔB on megagametogenesis were investigated by confocal laser scanning microscopy (CLSM) and DIC microscopy. During wild type megagametogenesis, the functional megaspore undergoes three rounds of mitosis accompanied by nuclear migration, fusion, degeneration and cellularization to form the final embryo sac (Additional file 1: Figures S3H-L; S4D-J). At FG1, the functional megaspore undergoes mitosis to produce a two-nucleate embryo sac (FG2; Additional file 1: Figure S3H). Formation of a vacuole between the two nuclei marks stage FG3 (Figure 4,D and L; Additional file 1: Figures S3I, S4D). During FG3 the ovule becomes curved and the inner integument embraces the nucellus. A second round of mitosis produces a four-nucleate embryo sac (FG4; Figure 4E,M; Additional file 1: Figures S3J, S4E). This is followed by migration of the two chalazal nuclei from an orthogonal orientation to a chalazal-micropylar orientation. After nuclear migration, a third round of mitosis gives rise to eight nuclei in a 4n + 4n configuration (FG5; Additional file 1: Figure S3K). The two polar nuclei, one from each side, meet at the embryo sac's micropylar half and fuse to form the central cell, while the antipodal nuclei start to degenerate (Additional file 1: Figure S4H). The central cell has formed and the antipodal nuclei are completed degenerated by FG7 (Additional file 1: Figures S3K, S4I). Prior to fertilization, one synergid nucleus degenerates, such that the embryo sac consists of one egg cell, one central cell and one synergid nucleus (FG8; Additional file 1: Figures S3L, S4J).

Ovule development is typically synchronous in wild type sliques with predominately one or two developmental stages present in a given pistil (Additional file 1: Table S2). In contrast, 35S:NTAP:AtCTF7ΔB ovule development appeared slowed and asynchronous (Additional file 1: Table S3). Female gametophytes in the same pistil were often at several different stages, indicating that the synchrony of gametophyte development was disturbed and embryo sac maturation was delayed. Ovules containing a megaspore (Figures 3G,H'), FG1 embryo sacs with a functional megaspore-like cell [37] and degrading megaspores (Figures 4G,H,N), and abnormal FG2 and FG3 embryo

sacs were commonly observed in the same slique (Figure 4I,O-P). Ovules containing degraded/degrading megasporocytes, and degrading FG1 embryo sacs were also observed (Additional file 1: Figures S5A-D). Alterations in nuclear division appeared to precede arrest in some megaspores (Additional file 1: Figure S5E-H). Most embryo sacs arrest at FG2/FG3; although some terminal ovules appeared to progress beyond FG3 (Additional file 1: Table S3). Common phenotypes included degraded/degrading nuclei (Additional file 1: Figure S5J-K), degenerated embryo sacs (Additional file 1: Figure S5M), polar nuclei fusion defects (Additional file 1: Figure S5N) and vacuole development defects (Additional file 1: Figure S5N-P).

To determine if the alterations observed in Line 11 are representative of 35S:NTAP:AtCTF7ΔB plants in general, ovules from other lines exhibiting severely reduced fertility (#13 and #15) were examined. Similar to the situation in Line11, pre-meiotic ovules from Line15 contained abnormally enlarged sub-epidermal cells adjacent to megaspore mother-like cells (Figures 5A-C). During meiosis, ovules contained abnormal cells adjacent to degenerated megaspores (Figures 5D-Q). Some ovules contained functional megaspore-like cells (Figure 5F-H,Q). Mature ovules contained differentiated functional megaspore-like cells (Figures 5R,S) and female gametophytes with various defects (Figures 5U,V). Similar to Line 11, female gametophytes from Lines 13 and 15 developed slowly and asynchronously; embryo sacs arrested at FG1 and FG3 (Additional file 1: Figure S6A,C). Additional alterations not observed in Line 11 were also identified, including some ovules that appeared to contain two functional megaspore-like cells (Additional file 1: Figure S6B). In some ovules the middle megaspores appeared to differentiate into functional megaspore-like cells while the megaspores at the chalazal-end degenerated (Additional file 1: Figure S6D), suggesting that ovule polarity was disrupted. All together, 30.4% (48/158) ovules examined in Line 15 plants displayed alterations at pre-meiosis; 66.0% (68/103) of the ovules were defective at meiosis and 79.3% (73/92) of the ovules were defective at FG1. Therefore, common defects associated with NTAP:AtCTF7ΔB include a delay and alterations in both megasporogenesis and megagametogenesis with embryo sacs arresting at various stages of development.

Because the ovules of 35S:NTAP:AtCTF7ΔB plants appeared to exhibit a delay in the onset of meiosis, qRT-PCR was carried out to examine transcript levels for several genes important for meiosis and ovule development (Additional file 1: Figure S7A,B). Transcripts for WUS1 [38], MMD1 [39], SPO11-1 [40] and ZYP1a [41] were elevated between two and three fold in 35S:NTAP:AtCTF7ΔB plants relative to wild type. Transcript levels of DMC1 [42], SYN3 [43,44] and OSD1 [45] showed modest increases, while the transcript levels of other genes were unchanged (Additional file 1: Figures S7A,B).

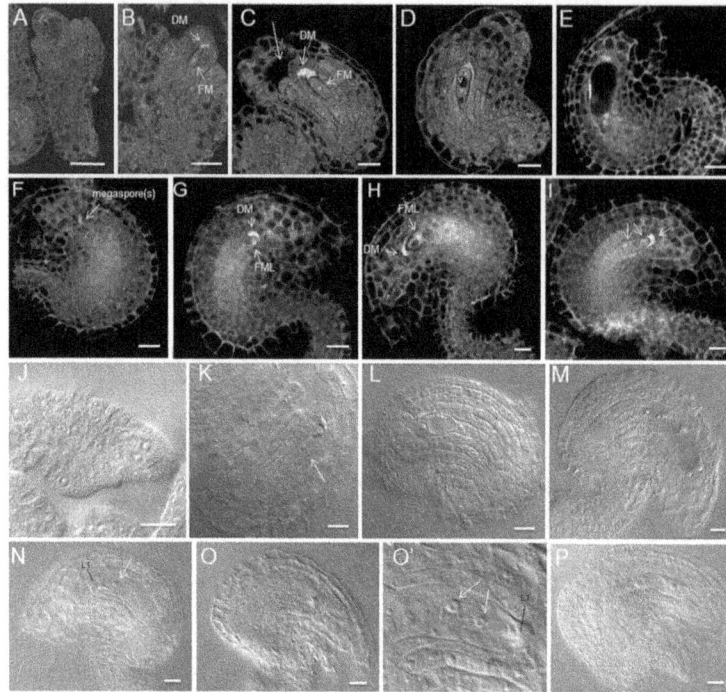

Figure 4 Embryo sac development is delayed and arrests early in 35S:NTAP:AtCTF7ΔB ovules. (A-E) and **(J-M)** Wild type ovules. **(F-I)** and **(N-P)** 35S:NTAP:AtCTF7ΔB ovules. **(A)** FG0 ovule. No FM is identified. **(B)** Early FG1 ovule showing the FM and DM. The nucellus is not surrounded by the integument. **(C)** FG1 ovule. The nucellus is surrounded by the outer integument but not the inner integument. **(D)** FG3 ovule containing a two-nucleate embryo sac. The nucellus is enclosed by the inner integument. **(E)** FG4 ovule containing a four-nucleate embryo sac. **(F-I)** Embryo sac development in 35S:NTAP:AtCTF7ΔB ovules observed by CLSM. **(F)** Ovule containing megaspore(s). **(G)** Ovule containing a FG1 embryo sac. Chalazal end megaspore becomes functional megaspore like (FML) and the other megaspores are degrading. **(H)** FG1 embryo sac. FML locates at a more chalazal position. **(I)** FG3 embryo sac containing two nuclei with a vacuole between them. **(J-M)** Embryo sac development in WT ovules visualized by DIC (also see Additional file 1: Figure S3). **(J)** Meiotic ovule containing a dyad (stage 2-IV). **(K)** FG1 ovule. FM (arrow) is uni-nucleate. **(L)** FG3 ovule, containing an embryo sac with two nuclei and a vacuole. **(M)** FG4 ovule. **(N-P)** Embryo sac development in 35S:NTAP:AtCTF7ΔB ovules observed by DIC. **(N)** FG1 embryo sac. The nucellus is surrounded by the outer integument and the inner integument. Extra cells are present between the FM (arrow) and L1 cells. **(O-O′)** Ovule containing a two-nucleate embryo sac. Non-degenerated L1 cells are present. **(P)** Ovule with a FG3 embryo sac. FMLs are identified as having distinctly bright nuclear autofluorescence and DMs contain a diffuse signal throughout the cells, but no clearly defined nucleus, defined according to Barrell and Grossniklaus [37]. Size bar = 10 μm. Developmental stages in CLSM and DIC are defined according to Christensen et al. [34,35] and Schneitz et al. [32], respectively.

Expression of 35S:NTAP:AtCTF7ΔB causes pleiotropic growth defects

During the analysis of 35S:NTAP:AtCTF7ΔB reduced fertility lines, plants displaying vegetative defects began to appear in the T2 or T3 generations. Specifically, later generations grew progressively worse in 12 of the 18 independent severely reduced fertility lines examined. The remaining six lines continued to display reduced fertility, but did not exhibit vegetative defects through the seventh generation. A wide range of morphological defects was observed in the 12 lines (Figure 6). The defects varied between lines and between progeny of the same line. The observed vegetative abnormalities included dwarf plants, fused stems and disruption of phyllotaxis (Figures 6C-E).

The proportion of plants exhibiting vegetative alterations increased in successive generations. Likewise, the severity of the vegetative alterations also became successively worse in subsequent generations. For example, the frequency of

dwarf plants increased from approximately 18% in the third generation to 82% by generation six (Figure 6; Table 1). The dwarf phenotype also became more severe with later generations containing smaller, more defective plants. Dwarf plants varied in morphology and exhibited a range of alterations, including acaulescent plants, floral abnormalities, homeotic changes and irregular leaves (Figure 6F-I). While some of the most severe 35S:NTAP:AtCTF7ΔB dwarf plants resembled *atctf7-1* plants (Figure 6J), in general the phenotype was less severe than *atctf7-1* plants. Furthermore, while most *atctf7-1* plants exhibit early senescence, 35S:NTAP:AtCTF7ΔB plants typically did not. In order to determine if the phenotypic alterations were due to increased 35S:NTAP:AtCTF7ΔB expression, native CTF7 and total CTF7 levels were measured in both dwarf and non-dwarf plants of three different generations of 35S:NTAP:AtCTF7ΔB plants, which showed progressively more severe phenotypes. Native

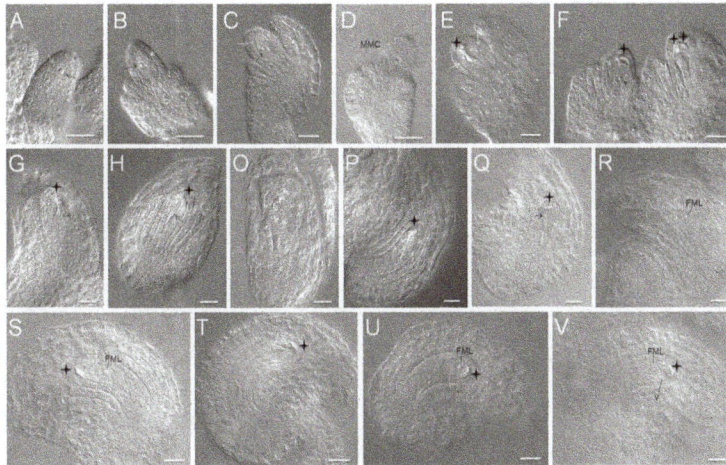

Figure 5 Phenotypes of 35S:NTAP:AtCTF7ΔB ovules from Line 15. (A-C) Pre-meiotic 35S:NTAP:AtCTF7ΔB ovules containing abnormal enlarged cells (white arrows) adjacent to megaspore mother cell-like cells (MMC like; black arrows). **(D)** Pre-meiotic ovule containing a MMC with degenerating nucleus. **(E-Q)** Meiotic ovules containing abnormal enlarged cells (white arrows) adjacent to DMs (stars). Cells with functional megaspore-like (FML) characteristics are indicated by black arrows in **F, G, H and Q**. The abnormal cell in **Q** contains two nuclei (white arrows). **(R-S)** Post-meiotic ovules containing FMLs. **(R)** The FML is associated with abnormal cells, which are not degenerating (white arrows). **(S)** Extra cells/nuclei are present between the DM (star) and FML. **(T)** Ovule containing two abnormal cells (white arrows) adjacent to the DM (star). **(U-V)** Post-meiotic ovules containing female gametophytes with one nucleus. DM, degenerated megaspore are denoted by stars; FML, functional megaspore like; MMC, megaspore mother cell. Size bars, 10 μm.

CTF7 levels ranged from 2.6-3.7 fold above wild type in the different plants, while total CTF7 transcript levels were elevated 5.4-7.6 fold relative to wild type. However, no consistent difference was observed between dwarf and non-dwarf plants or between one generation and the other.

Therefore, the phenotypic differences observed are not due to dramatic changes in AtCTF7ΔB transcript levels.

Rather, and most surprisingly, the dwarf phenotype was not inherited in a Mendelian fashion (Table 1). When dwarf plants were selfed they produced a mixture of dwarf and

Figure 6 35S:NTAP:AtCTF7ΔB plant morphology changes in later generations. (A) Wild type Columbia plant. **(B-E)** Morphological alterations get progressively worse in 35S:NTAP:AtCTF7ΔB plants through self-pollination. **(B)** Second generation plants are normal, but reduced fertile. **(C)** Both dwarf and non-dwarf, reduced fertile plants are observed in third generation plants. **(D)** Fourth generation plants. **(E)** Sixth generation plants, showing a higher frequency of dwarf plants. Defects such as reduced apical dominance (arrow) and phyllotaxis disturbances (asterisks) are observed. **(F-I)** Representative 35S:NTAP:AtCTF7ΔB dwarf plants. Inflorescence defects include aculescent **(G)**, multiple inflorescence branches at the first node **(H)** and no inflorescence **(I)**. Leave defects include aberrant rosette size and shape **(F, H and I)**. **(J)** atctf7-1 plant showing an early senescence phenotype (arrow). Plants **B** to **I** are from Line 11. All plants are grown under the same conditions. Plants in **A-H and J** are approximately 30 days old; plant in **I** is 40 days old. Scale bar = 5 cm in **A-E** and 2 cm in **F-J**.

Table 1 Non-Mendelian inheritance in 35S:NTAP: AtCTF7ΔB plants

Generation	Dwarf plants (%)	Reduced fertile, non-dwarf plants (%)
2nd		125 (100%)
3rd	24 (18.5%)	106 (81.5%)
4th		
Seeds from dwarf Parent	16 (42.1%)	22 (57.9%)
Seeds from non-dwarf Parent	25 (45.5%)	30 (54.5%)
5th		
Seeds from dwarf plants	55 (78.6%)	15 (21.4%)
Seeds from non-dwarf plants	81 (67.5%)	39 (32.5%)
6th		
Seeds from dwarf plants	55 (81.4%)	13 (18.6%)
Seeds from non-dwarf plants	66 (82.5%)	14 (17.5%)

Plants were from Line 11. Seeds from reduced fertile, non-dwarf plants and from dwarf plants were collected and sown separately.

non-dwarf, reduced fertile plants. The frequency of dwarf plants produced from selfed dwarf plants was similar to the frequency of dwarf plants resulting from selfing a non-dwarf plant. The stochastic appearance of the dwarf phenotype and variation in phenotypes suggested that the alterations could be the result of epigenetic changes. In order to investigate this possibility, qRT-PCR was carried out to measure the expression levels of several epigenetically regulated transposable elements (TEs), including *MU1*, *COPIA 28* and *solo LTR* [46], as well as several genes associated with epigenetic events [47-50]. Expression levels of *MU1*, *COPIA 28* and *solo LTR* were increased between five (*MU1*) and 24 fold (*COPIA 28*) in 35S:NTAP:AtCTF7ΔB plants (Figure 7A). Subtle changes were also observed in the transcript levels of several siRNA associated genes (Figure 7B). *ARGONAUTE1* (*AGO1*), *RDR2* and *mir156* transcript levels were reduced approximately 40-60% while *AGO4* transcripts were elevated slightly. Transcript levels of *HDA19* and *RDM4* were also decreased approximately 50% (Figure 7B), while transcript levels of the canonical DNA methylation genes, *MET1* and *DMT7*, did not vary significantly (Figure 7B).

Transcript levels of several cell cycle (*RBR*, *CYCB1.1* and *CYCA1.1*) and DNA repair (*BRCA1* and *BRCA2B*) genes have previously been shown to be elevated in *atctf7* plants [17]. These genes were tested and found to also be elevated in 35S:NTAP:AtCTF7ΔB plants (Figure 7C). The similarities in morphological defects and expression patterns observed between 35S:NTAP:AtCTF7ΔB and *atctf7-1* plants [17] are consistent with the hypothesis that the 35S:NTAP:AtCTF7ΔB construct is exerting a dominant negative effect.

Finally, experiments were carried out to determine if the 35S:NTAP:AtCTF7ΔB-associated defects are due to the presence of the N-terminal tag or the absence of the acetyltransferase B motif. A 35S:AtCTF7ΔB construct was generated and transformed into wild type Columbia plants (Additional file 1: Figure S1). Eight out of the 13 35S:AtCTF7ΔB transgenic lines examined exhibited reduced fertility, with the 35S:AtCTF7ΔB defects typically appearing more severe than those in 35S:NTAP:AtCTF7ΔB plants (Additional file 1: Figure S8). Like 35S:NTAP:AtCTF7ΔB plants, dwarf plants were not observed until the third generation in 35S: AtCTF7ΔB plants. However, dwarf plants appeared at higher frequencies and their phenotypes were more varied than 35S:NTAP:AtCTF7ΔB plants (Additional file 1: Figure S8B). For example, the viability of Line 21 decreased with successive generations, such that seeds from this line were not viable by the 4th generation. Likewise, while abnormalities associated with 35S:AtCTF7ΔB plants were similar to those observed in 35S:NTAP:AtCTF7ΔB plants, additional alterations were also present. For example, 35S:AtCTF7ΔB Line 24 segregated for two types of plants, those without an inflorescence and "normal" reduced fertility plants (Additional file 1: Figure S8C). 35S:AtCTF7ΔB Line 29 plants produced siliques that pointed downward and contained fewer seeds (46.4 ± 2.5 versus 54.2 ± 4.1 per silique in WT; n = 35) (Additional file 1: Figure S8D). This phenotype is similar to *bp/knat1* mutations [51].

Finally, in order to investigate which aspect(s) of the 35S:NTAP:AtCTF7ΔB construct was causing the fertility and growth defects, several additional constructs were generated and introduced into wild type or *Atctf7-1*$^{+/-}$ plants. Wild type plants transformed with a 35S:NTAP:AtCTF7 construct (Additional file 1: Figure S1) resembled 35S:AtCTF7 plants [16]. Specifically, the plants grew normally and produced pollen, but exhibited reduced female fertility. A CTF7:AtCTF7ΔB construct (Additional file 1: Figure S1) was also created and transformed into *Atctf7-1*$^{+/-}$ plants. The native CTF7 promoter is expressed at low levels throughout the plant [17]. Wild type plants containing the CTF7:AtCTF7ΔB construct exhibited normal growth and development and normal fertility levels (52.0 ± 2.2 seed/silique, n = 33 versus 54.2 ± 4.1 seed/silique in wild type, n = 35). No alterations were observed in six different transgenic lines over six generations. *atctf7-1* plants containing the CTF7:AtCTF7ΔB construct were obtained in T2 populations, but at frequencies (7%, 6/86) much lower than expected (25%) if the construct complemented the *atctf7-1* mutation. While these plants were dwarf, they grew better than *atctf7-1* plants (Additional file 1: Figure S8E). Somewhat similar to 35S:AtCTF7ΔB plants, plant morphology varied between plants with some plants appearing acaulescence or producing fewer rosette leaves. The plants produced approximately 40 ovules per silique; however they failed to set seed; siliques contained aborted ovules resembling *atctf7-1* plants (Additional file 1: Figure S8G). Therefore, the defects observed in 35S:NTAP:AtCTF7ΔB plants

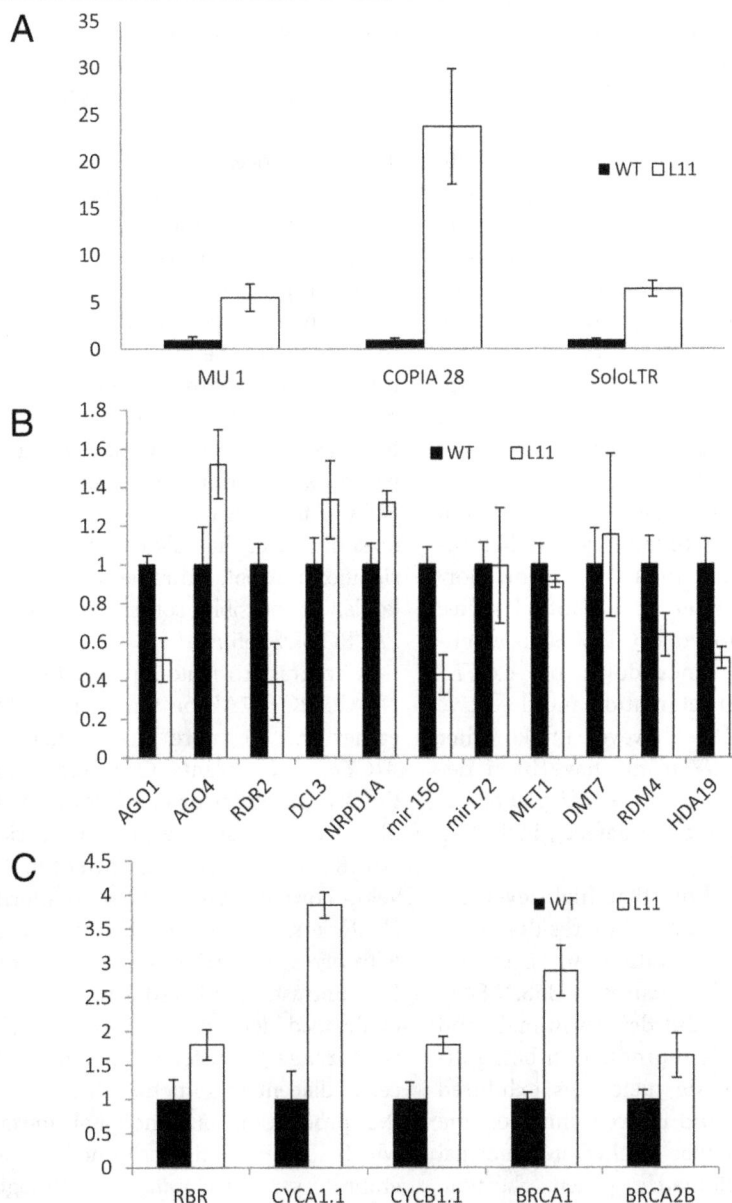

Figure 7 Transcript levels of epigenetically regulated transposable elements and other select genes in 35S:NTAP:AtCTF7ΔB plants. (A) Transcript levels of *MU1*, *COPIA28* and *soloLTR*, are increased dramatically in 35S:NTAP:AtCTF7ΔB plants. **(B)** Transcript levels of genes associated with epigenetic events are differently affected. *HDA19* and *RDM4* transcript levels are decreased, while *MET1* and *DMT7* transcripts are not altered. **(C)** Transcript levels of cell cycle genes, *CYCB1.1*, *CYCA1.1* and *RBR*, and DNA repair genes, *BRCA1* and *BRCA2B*, are increased. Buds of wild type and non-dwarf, reduced fertile 4th generation Line 11 plants were used. Data are shown as means ± SD (*n* = 3).

are caused by high-level expression of AtCTF7ΔB and not the presence of the NTAP tag. Rather, the presence of the NTAP appears to reduce the severity of the alterations, either by reducing the stability or activity of the protein, possibly by affecting its interaction with other proteins. Further, expression of a truncated version of CTF7, missing the B motif, can restore some vegetative growth to *atctf7-1* plants; however the plants are still completely sterile. Therefore, an intact acetyltransferase domain is required for full CTF7 activity.

Discussion

The acetylation of cohesin complexes at conserved lysine residues by CTF7/Eco1 plays an essential role in the establishment of cohesion during S phase and therefore nuclear division. Consistent with this, CTF7/Eco1 null mutations are typically lethal. Organisms expressing altered CTF7/Eco1 levels or point mutations in the protein typically display relatively normal levels of cohesin during nuclear division, but exhibit a range of developmental alterations. For example, mutations in the

N-terminus of the protein typically lead to defects in cohesion and often chromosome loss during mitosis [52]. In contrast certain mutations in the C-terminal acetyltransferase domain of yeast CTF7/ECO1 have little effect on S-phase cohesion and chromosome segregation, but cause an increased sensitivity to DNA-damaging agents. Likewise, Roberts Syndrome in humans has been linked to point mutations in ESCO2. Cells from patients with Roberts Syndrome are typically hypersensitive to DNA-damaging agents and show premature centromere separation; however, only 10–20% of cells show abnormal mitosis [27,53]. Finally, numerous studies have shown that cohesin mutations or reductions in cohesin levels result in transcriptional alterations that can have far-ranging developmental consequences [54].

Generally similar results have been obtained from studies on CTF7 in plants. Arabidopsis plants heterozygous for a T-DNA insertion in *AtCTF7* grow normally but produce approximately 25% aborted seeds, consistent with the conclusion that CTF7 is an essential protein [16]. Likewise knockdown of *AtCTF7* mRNA levels leads to growth retardation and defective sister chromatid cohesion [18]. However, unlike other organisms, homozygous *Atctf7* plants have been detected at very low frequencies [17]. *Atctf7* plants exhibit a wide range of developmental defects, including extreme dwarfism and sterility.

In our current study we show that high-level expression of a truncated form of AtCTF7 results in reduced fertility and dramatic alterations in vegetative growth. Specifically, plants that express a 35S:NTAP: $AtCTF7_{\Delta 299-345}$ construct exhibit defects in male and female meiocytes, with female reproduction being affected more dramatically. Male meiocytes exhibited chromosome fragmentation and uneven chromosome segregation during meiosis II that resulted in abnormal pollen development and ultimately pollen abortion. Ovules contained abnormal megasporocyte-like cells during pre-meiosis, megaspores that experienced elongated and aborted meiosis and defective megaspores and embryo sacs that arrested at various stages. A broad range of vegetative defects was also observed beginning in T_2 generations of AtCTF7ΔB transgenic plants. The appearance of these defects was stochastic and inherited in a non-Mendelian fashion. Comparison of AtCTF7ΔB transgenic plants with *AtCTF7* RNAi and *atctf7* plants and CTF7/Eco1 mutants in other organisms suggests that CTF7 may have multiple roles in the cell.

Reproductive defects in 35S:NTAP:AtCTF7ΔB plants differ from those in AtCTF7RNAi and *atctf7* plants

Inactivation of *AtCTF7* by T-DNA insertion or a reduction in *AtCTF7* levels by *AtCTF7*-RNAi lead to alterations in chromosome condensation and sister chromatid cohesion during early meiotic prophase followed by defects in homologous chromosome pairing and segregation later in meiosis [17,18]. The effect of 35S:NTAP:AtCTF7ΔB on male meiocytes was less severe and occurred later in meiosis. The first noticeable defect was the appearance of lagging and broken chromosomes during telophase I (Figure 2H). Twenty or more individual chromosomes were often observed beginning at metaphase II, suggesting that cohesion might be prematurely lost. Overexpression of AtCTF7ΔB did not have a noticeable effect on the initial establishment of cohesion, as the distribution of SYN1 on meiotic chromosomes was normal throughout prophase (Additional file 1: Figure S2). Likewise, chromosome condensation, sister chromatid cohesion and homologous chromosome pairing were normal during male meiotic prophase in 35S: NTAP:AtCTF7ΔB plants. This suggests that AtCTF7ΔB does not affect the bulk of meiotic cohesin complexes to a significant extent, but rather may alter centromeric cohesin levels, or possibly cohesin interactions with SGO1 or PATRONUS [55].

In contrast to male reproduction, the effect(s) of 35S: NTAP:AtCTF7ΔB on female reproduction are observed earlier and are more variable than those in *atctf7* and *AtCTF7* RNAi plants. Over expression of *AtCTF7* from the 35S promoter or knockdown of *AtCTF7* using RNAi blocks early ovule development, typically at FG1 or FG2 [16,18]. In contrast, *atctf7* ovules in $AtCTF7^{+/-}$ plants develop normally, but arrest soon after fertilization [16]. In all three situations the alterations are relatively uniform with arrest occurring at a specific developmental stage. In contrast, 35S:NTAP:AtCTF7ΔB causes pleiotropic ovule/seed defects. NTAP:AtCTF7ΔB lines displayed a wide range of defects, including additional abnormal cells adjacent to gametic cells, delayed/arrested meiosis, the production of functional megaspore-like cells of which some are mis-positioned, and delayed and altered embryo sac development. Although alterations were commonly first observed prior to and during meiosis, most megaspores progressed to FG2 or FG3 before arresting (Additional file 1: Table S3).

35S:NTAP:AtCTF7ΔB leads to defects consistent with epigenetic alterations

35S:NTAP:AtCTF7ΔB lines exhibited relatively normal vegetative growth and development for the first two generations. However, severe vegetative abnormalities began to appear starting in the T2 or T3 generations of different lines. The defects, which included dwarf plants, fused stems and disrupted phyllotaxis (Figure 6C-E), varied between lines and between progeny of the same line. The proportion of plants exhibiting vegetative alterations as well as the severity of the vegetative alterations increased in successive generations. It is interesting to note that phenotypic variability is very common in RBS patients [26,27].

The delayed appearance of vegetative defects and the increased frequency of defects in subsequent generations could result from the accumulation of defects in 35S: NTAP:AtCTF7ΔB plants. Consistent with this possibility is the observation that ctf7/eco1 mutations are commonly associated with sensitivity to DNA damaging agents [56]. As expected, both 35S:NTAP:AtCTF7ΔB and atctf7 plants contain elevated transcript levels for DNA repair and recombination genes (Figure 7; Additional file 1:Figure S7) [17]. While the vegetative defects in 35S:NTAP:AtCTF7 plants may result from spontaneous mutations, the situation is clearly more complex as the dwarf phenotype is not inherited in a Mendelian fashion (Table 1). When dwarf plants were selfed they produced a mixture of dwarf and non-dwarf, reduced fertile plants. Further, the frequency of dwarfs in the progeny of selfed dwarf plants was similar to the frequency of dwarf plants resulting from selfing of non-dwarf plants. This raised the possibility that epigenetic alterations may be present in 35S:NTAP:AtCTF7ΔB plants. Consistent with this possibility is our observation that transcript levels of MU1, COPIA 28 and solo LTR were increased between five (MU1) and 24 fold (COPIA 28) in 35S:NTAP:AtCTF7ΔB plants (Figure 7A). Subtle changes were observed in the transcript levels of several siRNA associated genes. AGO1, RDR2 and mir156 transcript levels were reduced approximately 40-60% while AGO4 was elevated slightly. HDA19 and RDM4 transcripts were also decreased approximately 50% (Figure 7B).

Our observation that epigenetic alterations may be present in 35S:NTAP:AtCTF7ΔB plants is consistent with the alterations we observe in female reproduction. Recent studies have shown that embryo sacs are enriched for transcripts of proteins involved in RNA metabolism and transcriptional regulation, and that they display distinct epigenetic regulatory mechanisms [57-59]. Disruption of genes in small RNA regulatory pathways, such as AGO1, AGO9, DICER-LIKE1 (DCL1) and MEIOSIS ARRESTED AT LEPTOTENE1 (MEL1), leads to multiple gametic cells at premeiosis, abnormal meiotic divisions, gametic cell fate alterations and twin female gametophytes [33,59,60]. For example, mutations in AGO9, which participates in small RNA silencing by cleaving endogenous mRNAs, results in additional gametic cells in pre-meiotic ovules, which may skip meiosis and twin female gametophytes in post-meiotic ovules [33]. Several of these defects are observed in 35S:NTAP:AtCTF7ΔB ovules before and during meiosis. Moreover, AGO9 participates in the epigenetically regulated silencing of TEs [33]. At this time it is not clear if the apparent epigenetic alterations we observed are the direct result of high-level NTAP:AtCTF7ΔB expression or a secondary effect. For example, it is possible that NTAP: AtCTF7ΔB expression directly affects the expression of

genes involved in epigenetic regulation. The involvement of cohesin complexes in transcriptional regulation is well documented in other organisms [54]. It is also possible that the changes we observe are an indirect effect of NTAP:AtCTF7ΔB expression. For example, previous studies have shown that eco1/ctf7 mutations result in defects in nucleolar integrity, rRNA production, ribosome biogenesis and protein biosynthesis in Saccharomyces cerevisiae and human [25,61]. In Arabidopsis, mutations in genes participating in mRNA production and rRNA/ribosome biogenesis slow mitotic progression in female gametophytes and result in pleiotropic defects in embryo sacs [58,59,62-66]. For example, mutations in SLOW-WALKER1 (SWA1), which participates in 18S pre-rRNA processing, results in asynchronous megagametophyte development, and embryo sac arrest over a wide range of stages [62]. Likewise, mutations in ribosomal protein genes lead to defects in inflorescence, leaf and plant stature in Arabidopsis, similar to those observed in 35S:NTAP:AtCTF7ΔB plants [66-69]. Therefore, many of the alterations we observe in 35S: NTAP:AtCTF7ΔB plants could be the result of alterations in rRNA biogenesis or ribosome biogenesis, which in turn could indirectly impact epigenetic pathways.

AtCTF7ΔB likely acts on several levels

The alterations observed in 35S:NTAP:AtCTF7ΔB plants appear to result from the presence of high levels of AtCTF7ΔB and not a reduction of native AtCTF7 levels or the presence of the NTAP. 35S:AtCTF7ΔB plants exhibit similar, if not more dramatic phenotypes than 35S: NTAP:AtCTF7ΔB plants, indicating that the NTAP is not responsible for the observed phenotypes. Likewise, expression studies show that in addition to high AtCTF7ΔB transcript levels, transgenic plants also contain elevated levels of native AtCTF7 transcripts (Figure 1C). Therefore, the 35S:AtCTF7ΔB construct does not cause co-suppression. Consistent with this are the apparently normal SYN1 cohesin patterns observed in male meiocytes (Additional file 1: Figure S2). Interestingly, the elevated levels of native AtCTF7 transcript suggest that the cellular cohesion status is altered to some extent in AtCTF7ΔB plants and that a feedback loop exists to monitor and maintain cohesion levels in plants.

High-level expression of AtCTF7ΔB appears to exert a dominant negative effect, resulting in a relatively wide range of alterations. Less clear is how high level expression of AtCTF7ΔB exerts its effect or if the alterations we observe are all related. Deletion of the last 46 amino acids of the acetyltransferase domain is expected to eliminate most of the actyltransferase activity. High-level expression of the protein may directly compete with native AtCTF7 for cohesin substrates resulting in an overall reduction or redistribution of cohesin levels throughout the genome. These changes could in turn result in a wide range of transcriptional alterations, similar to the situation observed in other organisms [54]. While this is the most-likely effect, it may not explain all of the observed alterations. For example, to our knowledge apparent epigenetic alterations have not been observed in either *AtCTF7* RNAi or *atctf7* plants [17,18]. Therefore, it is also possible that the 46 amino acid deletion alters acetyltransferase specificity such that the protein acts on off targets. For example, CTF7 has been shown to not acetylate histones; however if altered substrate specificity resulted in the acetylation of histones, then changes in chromatin structure could produce some of the alterations we observe. Finally, the possibility also exists that the deletion may alter the interaction of AtCTF7 with other proteins, either directly or indirectly involved in maintaining chromatin structure. Further experiments are required to determine how specifically AtCTF7ΔB is acting, why male and female reproduction respond differently to alterations in AtCTF7 levels and what role, if any AtCTF7 plays in epigenetic regulation.

Conclusions

Proper levels of AtCTF7 are critical for proper plant growth and development with female gametophytes being most sensitive to changes in AtCTF7 activity. High level expression of NTAP:AtCTF7ΔB results in pleiotropic defects in reproduction and vegetative growth. High levels of AtCTF7ΔB may affect small RNA processing, which in turn appears to result in epigenetic alterations. These results indicate that CTF7 may play multiple roles in plant cells.

Methods

Plant material and growth conditions

Wild type *Arabidopsis thaliana* plants (ecotype Columbia), the SALK_059500 (*ctf7-1*) insertion line and all transgenic plants described in this report were grown in Metro-Mix200 soil (Scotts-Sierra Horticultural Products; http://www.scotts.com) in a growth chamber at 22°C with a 16-h-light/8-h-dark cycle as described [16]. T-DNA insertion and transgenic plants were genotyped by PCR with primer pairs specific for the T-DNA and wild-type loci.

Cloning procedures for construction of the transgenic plants

The AtCTF7ΔB cDNA fragment (1–894 bp nucleotides, AtCTF7$_{Δ299-345}$) was digested with NdeI/HindIII and cloned into pIADL14 as described [16]. The 35S:NTAP:AtCTF7ΔB construct was generated using Gateway-compatible binary vectors containing the NTAP tag ((NTAPi) [30]; a gift from Dr. Qinn Li lab, Miami Univeristy). AtCTF7ΔB, 1–894 bp cDNA nucleotides, was first PCR-amplified with primers (1111/1201) and cloned into the pENTR vector, which was then fused with the binary vector by LR recombination reactions (Invitrogen) to make the final 35S:NTAP:AtCTF7ΔB construct. 35S:AtCTF7ΔB was generated by cloning the AtCTF7 cDNA (1–894 bp) as a NcoI/SpeI fragment into the binary vector pFGC5941. The 35S:NTAP:AtCTF7 construct was generated by the Gateway method with the corresponding primers. The AtCTF7 promoter (1.3 kb upstream of the ATG) was amplified, digested and cloned into the 35S:AtCTF7ΔB construct. All the constructs were confirmed by DNA sequencing. The primers used are listed in Additional file 1: Table S4.

Each construct was mobilized into *Agrobacterium tumefaciens* strain AGL-1 and transformed into *Arabidopsis thaliana* using the floral dip method [70]. Transgenic plants were screened by BASTA and further confirmed by PCR.

Quantitative real-time RT–PCR (qRT-PCR)

Buds of wild type and 35S:NTAP:AtCTF7ΔB Line 11 4th generation plants were harvested and pooled separately. For 35S:NTAP:AtCTF7ΔB samples, buds were only harvested from reduced fertile, non-dwarf plants or dwarf planys separately. Total RNA was extracted with the Plant RNeasy Mini kit (Qiagen, Hilden, Germany), and 10 μg of RNA was treated with Turbo DNase I (Ambion, http://www.invitrogen.com/site/us/en/home/brands/ambion.html) and used for cDNA synthesis with an oligo(dT) primer and a First Strand cDNA Synthesis Kit (Roche, http://www.roche.com). PCR was performed with the SYBR-Green PCR Mastermix (Bio-Rad, Hercules, CA, USA) and amplification was monitored on a MJR Opticon Continuous Fluorescence Detection System (Bio-Rad). Expression was normalized against β-tubulin-2. At least three biological replicates were performed, with two technical replicates for each sample. Student's *t*-test was conducted to identify transcripts that exhibit statistically significant variation at the 95% confidence level. Sequences of primers used in these studies are presented in Additional file 1: Table S4.

Chromosome analysis

Pollen morphology and viability were compared in flowers of 35S:NTAP:AtCTF7ΔB plants and wild-type plants using Alexander staining [71]. Male meiotic chromosome spreads were carried out on floral buds fixed in Carnoy's fixative (ethanol:chloroform:acetic acid: 6:3:1) and prepared as described previously [72]. Chromosomes were stained with 4,6-diamino-2-phenylindole dihydrochloride (DAPI, 1.5 μg ml^{-1}; Vector Laboratories, Inc. Burlingame, CA, USA) and observed with an Olympus BX51 epifluorescence microscope system. Images were captured with a Spot camera system (Diagnostic Instruments Inc., http://www.spotimaging.com) and processed. Meiotic stages were assigned based on chromosome structure and morphology [72].

Immunolocalization

SYN1 immunolocalization studies were carried out as previously described [31]. Primary antibodies for SYN1 were raised from rabbit and diluted 1:500. The slides were detected with Alexa Fluor 488 labeled goat anti-rabbit secondary antibody (1:2000; Molecular Probes, http://zt.invitrogen.com/). Slides were stained with DAPI and observed under an epifluorescence microscope.

Ovule analysis of 35S:NTAP:AtCTF7ΔB plants and wild type plants

Inflorescences from 35S:NTAP:AtCTF7ΔB plants and wild type plants were collected and fixed in 4% glutaraldehyde under vacuum for 2 hrs, dehydrated in a graded ethanol series (40%, 60%, 80%, 100% steps for 1 h each), and cleared in a 2:1 mixture of benzyl benzoate:benzyl alcohol. Ovules were dissected under a stereo dissecting microscope, mounted and sealed with coverslips. Ovules were observed on a Zeiss Axioskop microscope under differential interference contrast microscopy optics using a 40 objective as described [32]. Images were collected and processed. Ovules were also observed by confocal laser scanning microscopy [34,35]. Images were collected and projected with Olympus Flouview 2.0 software (http://www.olympus-global.com/) and analyzed with Image Pro Plus (Media Cybernetics; http://www.mediacy.com). All pistils from individual inflorescences were dissected and the ovules stages were recorded [34,35].

Availability of supporting data

The supporting data of this article are included with the article and its additional files.

Additional file

Additional file 1: Figure S1. Schematic diagrams of wild type AtCTF7 and AtCTF7 constructs used in this study. **Table S1.** Transfer efficiency of 35S:NTAP:AtCTF7ΔB mutants. **Figure S2.** SYN1 distribution pattern is not altered in 35S:NTAP:AtCTF7ΔB male meiocytes. **Figure S3.** Female gametophyte development in wild type plants, revealed by differential interference contrast microscopy. **Figure S4.** Female gametophyte development in wild type plants, revealed by confocal laser scanning microscopy. **Table S2.** Female gametophyte development in wild type plants. **Table S3.** Female gametophyte development in 35S:NTAP:AtCTF7ΔB plants. **Figure S5.** 35S:NTAP:AtCTF7ΔB ovules exhibit various defects during meiosis and mitosis. **Figure S6.** Female gametophytes from Line 13 and Line 15 resemble those from Line 11. **Figure S7.** Transcript levels of select genes in female gametophyte development are increased in 35S:NTAP:AtCTF7ΔB plants. **Figure S8.** Morphological alterations are associated with 35S:AtCTF7ΔB and atctf7-1 plants transformed with CTF7$_{pro}$:AtCTF7ΔB. **Table S4.** Primers used.

Abbreviations

At: Arabidopsis thaliana; ΔB: Δ299-345; SMC: Structural Maintenance of Chromosome; SCC: Sister Chromatid Cohesion; RBS: Roberts syndrome; qRT-PCR: Quantitative reverse transcription polymerase chain reaction; FG: Female Gametophyte; DIC: Differential interference contrast microscopy; MMC: Megaspore mother cell; FM: Functional megaspore; CLSM: Confocal laser scanning microscopy; EC: Egg cell; CC: Central cell; SN: Synergid nucleus; FML: Functional megaspore-like; DM: Degenerated megaspore; TEs: Transposable elements; AGO: ARGONAUTE; DAPI: 4,6-diamino-2-phenylindole dihydrochloride.

Competing interests

The authors declare that they have no competing interests.

Authors' contributions

DL and CM designed the experiments. DL performed the experiments. DL and CM prepared the manuscript. Both authors read and approved the final manuscript.

Acknowledgements

The authors thank Richard Edelmann and Matthew L. Duley for their assistance in the confocal laser scanning microscopy (CLSM) and the Makaroff lab for technical support and helpful discussions. This work was supported by a grant (MCB0718191) from the National Science Foundation to CAM.

References

1. Nasmyth K, Haering CH. Cohesin: its roles and mechanisms. Ann Rev Genet. 2009;43:525–58.
2. Brooker AS, Berkowitz KM. The roles of cohesins in mitosis, meiosis, and human health and disease. Methods Mol Biol. 2014;1170:229–66.
3. Zamariola L, Tiang CL, De Storme N, Pawlowski W, Geelen D. Chromosome segregation in plant meiosis. Front Plant Sci. 2014;5:279.
4. Watanabe Y, Nurse P. Cohesin Rec8 is required for reductional chromosome segregation at meiosis. Nature. 1999;400:461–4.
5. Haering CH, Lowe J, Hochwagen A, Nasmyth K. Molecular architecture of SMC proteins and the yeast cohesin complex. Mol Cell. 2002;9:773–88.
6. Bernard P, Schmidt CK, Vaur S, Dheur S, Drogat J, Genier S, et al. Cell-cycle regulation of cohesin stability along fission yeast chromosomes. EMBO J. 2008;27:111–21.
7. Díaz-Martínez LA, Giménez-Abián JF, Clarke DJ. Chromosome cohesion-rings, knots, orcs and fellowship. J Cell Sci. 2008;121(13):2107–14.
8. Yuan L, Yang X, Makaroff CA. Plant cohesins, common themes and unique roles. Cur Protein and Peptide Sci. 2011;12:93–104.
9. Tomonaga T, Nagao K, Kawasaki Y, Furuya K, Murakami A, Morishita J, et al. Characterization of fission yeast cohesin: essential anaphase proteolysis of Rad21 phosphorylated in the S phase. Genes Dev. 2000;14:2757–70.
10. Ben-Shahar TR, Heeger S, Lehane C, East P, Flynn H, Skehel M, et al. Eco1-dependent cohesin acetylation during establishment of sister chromatid cohesion. Science. 2008;321:563–6.
11. Huang X, Andreu-Vieyra CV, Wang M, Cooney AJ, Matzuk MM, Zhang P. Preimplantation mouse embryos depend on inhibitory phosphorylation of separase to prevent chromosome missegregation. Mol Cell Biol. 2009;29:1498–505.
12. Almedawar S, Colomina N, Bermúdez-López M, Pociño-Merino I, Torres-Rosell J. A SUMO-dependent step during establishment of sister chromatid cohesion. Curr Biol. 2012;22(17):1576–81.
13. Wu N, Kong X, Ji Z, Zeng W, Potts PR, Yokomori K, et al. Scc1 sumoylation by Mms21 promotes sister chromatid recombination through counteracting Wapl. Genes Dev. 2012;26(13):1473–85.
14. Skibbens RV, Corson LB, Koshland D, Hieter P. Ctf7p is essential for sister chromatid cohesion and links mitotic chromosome structure to the DNA replication machinery. Genes Dev. 1999;13:307–19.
15. Toth A, Ciosk R, Uhlmann F, Galova M, Schleifer A, Nasmyth K. Yeast cohesin complex requires a conserved protein, Eco1p (Ctf7), to establish cohesion between sister chromatids during DNA replication. Genes Dev. 1999;13:320–33.
16. Jiang L, Yuan L, Xia M, Makaroff CA. Proper levels of the Arabidopsis cohesion establishment factor CTF7 are essential for embryo and megagametophyte, but not endosperm, development. Plant Physiol. 2010;154(2):820–32.

17. Bolanos-Villegas P, Yang X, Wang H, Juan C, Chuang M, Makaroff CA, et al. Arabidopsis CHROMOSOME TRANSMISSION FIDELITY 7 (AtCTF7/ECO1) is required for DNA repair, mitosis and meiosis. Plant J. 2013;75:927–40.

18. Singh DK, Andreuzza S, Panoli AP, Siddiqi I. AtCTF7 is required for establishment of sister chromatid cohesion and association of cohesin with chromatin during meiosis in Arabidopsis. BMC Plant Biol. 2013;13:117–24.

19. Heidinger-Pauli JM, Unal E, Guacci V, Koshland D. The kleisin subunit of cohesin dictates damage-induced cohesion. Mol Cell. 2008;31:47–56.

20. Unal E, Heidinger-Pauli JM, Kim W, Guacci V, Onn I, Gygi SP, et al. A molecular determinant for the establishment of sister chromatid cohesion. Science. 2008;321:566–9.

21. Whelan G, Kreidl E, Wutz G, Egner A, Peters JM, Eichele G. Cohesin acetyltransferase Esco2 is a cell viability factor and is required for cohesion in pericentric heterochromatin. EMBO J. 2012;31:71–82.

22. Tanaka K, Yonekawa T, Kawasaki Y, Kai M, Furuya K, Iwasaki M, et al. Fission yeast eso1p is required for establishing sister chromatid cohesion during S phase. Mol Cell Biol. 2000;20:3459–69.

23. Williams BC, Garrett-Engele CM, Li Z, Williams EV, Rosenman ED, Goldberg ML. Two putative acetyltransferases, San and Deco, are required for establishing sister chromatid cohesion in Drosophila. Curr Biol. 2003;13:2025–36.

24. Hou F, Zou H. Two human orthologues of Eco1/Ctf7 acetyltransferases are both required for proper sister-chromatid cohesion. Mol Biol Cell. 2005;16(8):3908–18.

25. Gard S, Light W, Xiong B, Bose T, McNairn AJ, Harris B, et al. Cohesinopathy mutations disrupt the subnuclear organization of chromatin. J Cell Biol. 2009;187:455–62.

26. Vega H, Waisfisz Q, Gordillo M, Sakai N, Yanagihara I, Yamada M, et al. Roberts syndrome is caused by mutations in Esco2, a human homolog of yeast Eco1 that is essential for the establishment of sister chromatid cohesion. Nat Genet. 2005;37:468–70.

27. Vega H, Trainer AH, Gordillo M, Crosier M, Kayserili H, Skovby F, et al. Phenotypic variability in 49 cases of ESCO2 mutations, including novel missense and codon deletion in the acetyltransferase domain, correlates with ESCO2 expression and establishes the clinical criteria for Roberts syndrome. J Med Genet. 2010;47(1):30–7.

28. Dyda F, Klein DC, Hickman AB. GCN5-related N-acetyltransferases: a structural overview. Annu Rev Biophys Biomol Struct. 2000;29:81–103.

29. Ivanov D, Schleiffer A, Eisenhaber F, Mechtler K, Christian H, Nasmyth K. Eco1 is a novel acetyltransferase that can acetylate proteins involved in cohesion. Curr Biol. 2002;12:323–8.

30. Rohila JS, Chen M, Cerny R, Fromm ME. Improved tandem affinity purification tag and methods for isolation of protein heterocomplexes from plants. Plant J. 2004;38:172–81.

31. Cai X, Dong FG, Edelmann RE, Makaroff CA. The Arabidopsis SYN1 cohesin protein is required for sister chromatid arm cohesion and homologous chromosome pairing. J Cell Sci. 2003;116:2999–3007.

32. Schneitz K, Hulskamp M, Pruitt RE. Wild-type ovule development in Arabidopsis thaliana-a light microscope study of cleared whole-mount tissue. Plant J. 1995;7:731–49.

33. Olmedo-Monfil V, Durán-Figueroa N, Arteaga-Vázquez M, Demesa-Arévalo E, Autran D, Grimanelli D, et al. Control of female gamete formation by a small RNA pathway in Arabidopsis. Nature. 2010;464:628–32.

34. Christensen CA, King EJ, Jordan JR, Drews GN. Megagametogenesis in Arabidopsis wild type and the Gf mutant. Sex Plant Reprod. 1997;10:49–64.

35. Christensen CA, Subramanian S, Drews GN. Identification of gametophytic mutations affecting female gametophyte development in Arabidopsis. Dev Biol. 1998;202:136–51.

36. Siddiqi I, Ganesh G, Grossniklaus U, Subbiah V. The dyad gene is required for progression through female meiosis in Arabidopsis. Development. 2000;127:197–207.

37. Barrell PJ, Grossniklaus U. Confocal microscopy of whole ovules for analysis of reproductive development: the elongate1 mutant affects meiosis II. Plant J. 2005;43:309–20.

38. Gross-Hardt R, Lenhard M, Laux T. WUSCHEL signaling functions in interregional communication during Arabidopsis ovule development. Genes Dev. 2002;16:1129–38.

39. Yang X, Makaroff CA, Ma H. The Arabidopsis MALE MEIOCYTE DEATH1 gene encodes a PHD-finger protein that is required for male meiosis. Plant Cell. 2003;15(6):1281–95.

40. Grelon M, Vezon D, Gendrot G, Pelletier G. AtSPO11-1 is necessary for efficient meiotic recombination in plants. EMBO J. 2001;20(3):589–600.

41. Higgins JD, Sanchez-Moran E, Armstrong SJ, Jones GH, Franklin FCH. The Arabidopsis synaptonemal complex protein ZYP1 is required for chromosome synapsis and normal fidelity of crossing over. Genes Dev. 2005;19:2488–500.

42. Couteau F, Belzile F, Horlow C, Grandjean O, Vezon D. Random chromosome segregation without meiotic arrest in both male and female meiocytes of a dmc1 mutant of Arabidopsis. Plant Cell. 1999;11:1623–34.

43. Jiang L, Xia M, Strittmatter LI, Makaroff CA. The Arabidopsis cohesin protein SYN3 localizes to the nucleolus and is essential for gametogenesis. Plant J. 2007;50:1020–34.

44. Yuan L, Yang X, Ellis JL, Fisher NM, Makaroff CA. The Arabidopsis SYN3 cohesin protein is important for early meiotic events. Plant J. 2012;71:147–60.

45. d'Erfurth I, Jolivet S, Froger N, Catrice O, Novatchkova M. Turning meiosis into mitosis. PLoS Biol. 2009;7:e1000124.

46. Moissiard G, Cokus S, Cary J, Feng S, Billi AC, Stroud H, et al. MORC Family ATPases Required for Heterochromatin Condensation and Gene Silencing. Science. 2012;336(6087):1448–51.

47. Tian L, Chen ZJ. Blocking histone deacetylation in Arabidopsis induces pleiotropic effects on plant gene regulation and development. Proc Natl Acad Sci U S A. 2001;98:200–5.

48. Chen X. Small RNAs and their roles in plant development. Annu Rev Cell Dev Biol. 2009;35:21–44.

49. Matzke M, Kanno T, Daxinger L, Huettel B, Matzke AJ. RNA-mediated chromatin-based silencing in plants. Curr Opin Cell Biol. 2009;21:367–76.

50. Law JA, Jacobsen SE. Establishing, maintaining and modifying DNA methylation patterns in plants and animals. Nat Rev Genet. 2010;11:204–20.

51. Shi CL, Stenvik GE, Vie AK, Bones AM, Pautot V, Proveniers M, et al. Arabidopsis class I KNOTTED-like homeobox proteins act downstream in the IDA-HAE/HSL2 floral abscission signaling pathway. Plant Cell. 2011;23:2553–67.

52. Brands A, Skibbens RV. Ctf7p/Eco1p exhibits acetyltransferase activity–but does it matter? Curr Biol. 2005;15(2):R50–51.

53. Van Den Berg DJ, Francke U. Roberts syndrome, a review of 100 cases and a new rating system for severity. Am J Med Genet. 1993;47:1104–23.

54. Dorsett D, Merkenschlager M. Cohesin at active genes: a unifying theme for cohesin and gene expression from model organisms to humans. Curr Opin Cell Bio. 2013;25(3):327–33.

55. Cromer L, Jolivet S, Horlow C, Chelysheva L, Heyman J, De Jaeger G, et al. Centromeric cohesion is protected twice at meiosis, by SHUGOSHINs at anaphase I and by PATRONUS at interkinesis. Curr Biol. 2013;23(21):2090–99.

56. Lu S, Goering M, Gard S, Xiong B, McNairn AJ, Jaspersen SL, et al. Eco1 is important for DNA damage repair in S. cerevisiae. Cell Cycle. 2010;9:3315–27.

57. Wuest SE, Vijverberg K, Schmidt A, Weiss M, Gheyselinck J, Lohr M, et al. Arabidopsis female gametophyte gene expression map reveals similarities between plant and animal gametes. Curr Biol. 2010;20:506–12.

58. Schmidt A, Wuest SE, Vijverberg K, Baroux C, Kleen D, Grossniklaus U. Transcriptome analysis of the Arabidopsis megaspore mother cell uncovers the importance of RNA helicases for plant germline development. PLoS Biol. 2011;9(9):e1001155.

59. Shi DQ, Yang WC. Ovule development in Arabidopsis: progress and challenge. Curr Opin Plant Biol. 2011;14(1):74–80.

60. Nonomura K, Morohoshi A, Nakano M, Eiguchi M, Miyao A, Hirochika H, et al. A germ cell–specific gene of the ARGONAUTE family is essential for the progression of premeiotic mitosis and meiosis during sporogenesis in rice. Plant Cell. 2007;19:2583–94.

61. Bose T, Lee KK, Lu S, Xu B, Harris B. Cohesin proteins promote ribosomal RNA production and protein translation in yeast and human cells. PLoS Genet. 2012;8:e1002749.

62. Shi DQ, Liu J, Xiang YH, Ye D, Sundaresan V, Yang WC. SLOW WALKER1, essential for gametogenesis in Arabidopsis, encodes a WD40 protein involved in 18S ribosomal RNA biogenesis. Plant Cell. 2005;17:2340–54.

63. Coury D, Zhang C, Ko A, Skaggs M, Christensen C, Drews GN, et al. Segregation distortion in Arabidopsis gametophytic factor 1 (gfa1) mutants is caused by a deficiency of an essential RNA splicing factor. Sex Plant Reprod. 2007;20:87–97.

64. Groß-Hardt R, Kagi C, Baumann N, Moore JM, Baskar R, Gagliano WB, et al. LACHESIS restricts gametic cell fate in the female gametophyte of Arabidopsis. PLoS Biol. 2007;5:494–500.

65. Huang CK, Huang LF, Huang JJ, Wu SJ, Yeh CH, Lu CA. A DEAD-Box protein, AtRH36, is essential for female gametophyte development and is involved in rRNA biogenesis in Arabidopsis. Plant Cell Physiol. 2010;51:694–706.

66. Zsögön A, Szakonyi D, Shi X, Byrne ME. Ribosomal protein RPL27a promotes female gametophyte development in a dose-dependent manner. Plant Physio. 2014;165:1133–43.

67. Van Lijsebettens M, Vanderhaeghen R, De Block M, Bauw G, Villarroel R, Van Montagu M. An S18 ribosomal protein gene copy at the Arabidopsis PFL locus affects plant development by its specific expression in meristems. EMBO J. 1994;13:3378–88.

68. Byrne ME, Simorowski J, Martienssen RA. ASYMMETRIC LEAVES1 reveals knox gene redundancy in Arabidopsis. Development. 2002;129:1957–65.

69. Stirnberg P, Liu JP, Ward S, Kendall SL, Leyser O. Mutation of the cytosolic ribosomal protein-encoding RPS10B gene affects shoot meristematic function in Arabidopsis. BMC Plant Biol. 2012;12:160.

70. Clough SJ, Bent AF. Floral dip: a simplified method for Agrobacterium-mediated transformation of *Arabidopsis thaliana*. Plant J. 1998;16:735–43.

71. Alexander P. Differential staining of aborted and nonaborted pollen. Stain Technol. 1969;44:117–22.

72. Ross KJ, Fransz P, Jones GH. A light microscopic atlas of meiosis in *Arabidopsis thaliana*. Chromosome Res. 1996;4:507–16.

5

Mineral nitrogen sources differently affect root glutamine synthetase isoforms and amino acid balance among organs in maize

Bhakti Prinsi[*] and Luca Espen

Abstract

Background: Glutamine synthetase (GS) catalyzes the first step of nitrogen assimilation in plant cell. The main GS are classified as cytosolic GS1 and plastidial GS2, of which the functionality is variable according to the nitrogen sources, organs and developmental stages. In maize (*Zea mays* L.) one gene for GS2 and five genes for GS1 subunits are known, but their roles in root metabolism are not yet well defined. In this work, proteomic and biochemical approaches have been used to study root GS enzymes and nitrogen assimilation in maize plants re-supplied with nitrate, ammonium or both.

Results: The plant metabolic status highlighted the relevance of root system in maize nitrogen assimilation during both nitrate and ammonium nutrition. The analysis of root proteomes allowed a study to be made of the accumulation and phosphorylation of six GS proteins. Three forms of GS2 were identified, among which only the phosphorylated one showed an accumulation trend consistent with plastidial GS activity. Nitrogen availabilities enabled increments in root total GS synthetase activity, associated with different GS1 isoforms according to the nitrogen sources. Nitrate nutrition induced the specific accumulation of GS1-5 while ammonium led to up-accumulation of both GS1-1 and GS1-5, highlighting co-participation. Moreover, the changes in thermal sensitivity of root GS transferase activity suggested differential rearrangements of the native enzyme. The amino acid accumulation and composition in roots, xylem sap and leaves deeply changed in response to mineral sources. Glutamine showed the prevalent changes in all nitrogen nutritions. Besides, the ammonium nutrition was associated with an accumulation of asparagine and reducing sugars and a drop in glutamic acid level, significantly alleviated by the co-provision with nitrate.

Conclusion: This work provides new information about the multifaceted regulation of the GS enzyme in maize roots, indicating the involvement of specific isoenzymes/isoforms, post-translational events and biochemical factors. For the first time, the proteomic approach allowed to discriminate the individual contribution of the GS1 isoforms, highlighting the participation of GS1-5 in nitrate metabolism. Moreover, the results give new insights about the influence of amino acid metabolism in plant C/N balance.

Keywords: Amino acids, Ammonium, Glutamine synthetase, Maize, Nitrate, Roots

Background

Nitrogen (N) represents one of the main minerals required throughout plant development. In agronomic terms, this results in a worldwide ever-increasing use of fertilizers and its consequent environmental and socioeconomic costs [1]. This N requirement is emphasized with regard to cereal crops [2], for which maize (*Zea mays* L.) is a model species because of its economic importance and high metabolic capacity [3]. In agricultural soils the main mineral N sources are nitrate (NO_3^-) and ammonium (NH_4^+). In order to balance their N nutritional requirements with environmental availability, plants have to modulate the individual steps of N metabolism such as up-take, reduction of NO_3^- to NH_4^+, NH_4^+ assimilation and N recycling. The contribution of root and leaf systems depends on species, developmental stage and environmental conditions [4,5], and it is also deeply influenced by C metabolism [6].

* Correspondence: bhakti.prinsi@unimi.it
Dipartimento di Scienze Agrarie e Ambientali - Produzione, Territorio, Agroenergia (DISAA), Università degli Studi di Milano, Via Celoria, 2, 20133 Milano, Italy

All the NH_4^+ in the cell, derived from soil, from NO_3^- reduction or from other metabolic processes, is channelled through the glutamine synthetase (GS, EC6.3.1.2) reaction. The GS catalyzes the fixation of NH_4^+ on glutamic acid (Glu) to form glutamine (Gln), and in the assimilation process it is generally coupled with plastidial glutamate synthase (GOGAT, EC1.4.1.13/14) that incorporates C skeletons. Gln and Glu can be recruited as amino group donors as well as main N transport molecules [7]. Several evidence indicate that GS activity is deeply influenced by metabolic and environmental factors mainly linked to the balance between C and N metabolism [8]. For instance, Glu level seems to be fundamental in sensing plant nutritional status and in joining C and N metabolisms [9]. Moreover, the inter-conversion with other amino acids greatly influences N plant economy, especially regarding asparagine (Asn) and alanine (Ala) [10].

Plant responses are deeply affected by the proportion of mineral N sources [11]. While NH_4^+ as sole nutrient can induce toxicity symptoms, its co-provision with NO_3^- generally promotes a synergistic effect leading to growth enhancement [12]. It is noteworthy that NH_4^+ tolerance was related to high root N metabolism sustained by high GS activities [13], which in maize appear to be associated with the capacity to cope with the C skeleton demands [14].

The main GS are decameric enzymes [15] classified on the basis of subcellular localization in cytosolic GS1 and plastidial GS2. In plants, multigenic families encode several GS1 isoforms while the plastidial GS2 derives from one or few nuclear genes. In general, GS2 is associated with the leaf NH_4^+ (re)assimilation while GS1 is associated with plant N recycling. But the relative activity of GS1 and GS2 is variable according to the species, organs, N sources, developmental stages and environmental conditions, suggesting a multifaceted participation of isozymes [16]. Moreover, recent studies conducted both in dicotyledonous [17] and in monocotyledonous crops [18,19] showed non-overlapping functions for the GS1 isoforms. Besides, distinct post-translational modifications were described for both isoenzymes [20,21].

In maize, one gene for GS2 [SwissProt:P25462] and 5 genes codifying for different GS1 subunits were identified, named from GS1-1 to GS1-5 according to the reviewed UniProtKB/Swiss-Prot database [22] [Swiss-Prot:P38559; Swiss-Prot:P38560; Swiss-Prot:P38561; Swiss-Prot:P38562; Swiss-Prot:P38563]. GS1 and GS2 are differentially regulated in roots and leaves in response to growing conditions. The cytosolic isoforms also have different kinetic properties, stabilities and tissue localizations [23-25]. By means of Quantitative Trait Loci analyses and characterization of maize mutants, Hirel and co-workers indicated the key roles of GS1-3 and GS1-4 both in grain yield and germination [19,26]. GS1-3 and GS1-4 represent the major leaf

isoforms [19] and in maize mutants the deficiency of these enzymes affects leaf gene transcripts, proteins and metabolite accumulations [27]. Moreover, the transcript localizations confirmed the involvement of GS1-1 in root metabolism and suggested that GS1-2 acts in N phloem translocation [19]. It is worth noting that in mutants deficient for GS1-3 and GS1-4 the dry weight and total N content in the shoot vegetative parts were unaltered, providing evidence of how such parameters are prevalently determined by root metabolism [19]. This observation, together with the finding that the N stored before silking supplies up to 70% of the grain N content [28], draws attention to the need to study the root system during the early phases of maize development. After a first localization of GS1-1 and GS1-5 in tip and/or cortex tissues [25], root GS arrangement had scarcely been investigated, probably due to technical limitations. N availability is associated with the accumulation of a specific root isoform (GSr in [24]) that is theoretically assigned to GS1-1, but had not been precisely characterized. Similarly, the responses to different N sources were not fully elucidated and information about GS1-5 was still lacking.

By the means of Two-Dimensional Western Blotting (2D-WB) and Liquid Chromatography-nanoElectroSpray Ionization-Tandem Mass Spectrometry (LC-nESI-MS/MS) techniques, this work profiles the GS patterns in maize roots in response to NO_3^-, NH_4^+ or both, during vegetative growth, describing for the first time the differential modulations of cytosolic and plastidial forms and the active involvement of GS1-5. Moreover, the determination of amino acid composition in roots, xylem sap and leaves provides new information about the roles of Gln, Glu, Asn, and Ala metabolisms in plant C/N balance.

Results and discussion
Effects of the nutritional treatments on leaf and root metabolic status

The aim of this work was to investigate the responses of the different root GS isoforms in maize plants exposed to different inorganic N sources, during early vegetative growth. To better appreciate the effects at metabolic level, the changes in plant amino acid balance were also evaluated. Plantlets of the T250 inbred line were grown in a hydroponic system in the absence of N for 10 days to reach a developmental stage corresponding to the third-leaf expansion (Additional file 1: Figure A1 and Table A1). Since in field conditions maize N fertilization consists of a single application at sowing [29], the third-leaf stage corresponds to a vegetative phase in which plants are exposed to a high level of inorganic N and that is indicated as one of the more susceptible to NH_4^+ toxicity [14]. Moreover, it is important to note that the optimal dose of NO_3^- fertilization also depends on maize varieties [30]. In order to better appreciate the short-term responses and compare our proteomic results with previous works, the plants were exposed to a total N availability of 10 mM. In details, plants were exposed for 30 h to four nutritional treatments: N absence (c), 10 mM NO_3^- (n), 10 mM NH_4^+ (a), 5 mM $NO_3^- + 5$ mM NH_4^+ (na).

The concentration of NO_3^- and NH_4^+ in roots, xylem saps and leaves were measured (Figure 1A and B), together with the content of the reducing sugars and sucrose in root and leaf systems (Figure 1C and D). Moreover, the accumulation of the main N assimilative enzymes such as Nitrate Reductase (NR, EC 1.6.6.1; [31]) and GS in roots and leaves was estimated by One-Dimensional Western Blotting (1D-WB) (Figure 2). The analysis conducted against NR in roots detected the expected single band at 99 kDa (Figure 2A) while in leaf profiles two bands were visible (Figure 2C). The lower band at about 94 kDa, corresponding to pyruvate phosphate dikinase (EC2.7.9.1) that is the most abundant enzyme in maize leaves (Additional file 1: Figure A2), was considered as an unspecific signal. On the base of molecular masses, the three bands in GS profiles (Figure 2B and D) were

assigned to GS2 (44 kDa), to GS1 (40 kDa) and to the root isoform GS1r (39 kDa), as described by Sakakibara and co-workers [23]. According to previous works [19,23,32], higher levels of GS1 were detected in root profiles (Figure 2B) while GS2 was predominant in leaves (Figure 2D).

It appeared evident that the exposure of the plants to NO_3^- induced strong accumulation of the anion in all organs and especially in roots, reaching the value of 39.08 ± 0.62 µmol g^{-1}FW in (n) plants (Figure 1A). This trend was similar to that observed in T250 by previous time-course experiments [32]. The increment of NO_3^- in xylem sap attested the concomitant induction of anion translocation. In particular, in (n) and (na) xylem saps the values of 27.70 ± 1.02 and 16.10 ± 0.71 µmol g^{-1}FW, respectively (Figure 1A), quite a lot higher than the 10.5 mM value observed in maize crops [33], suggested an extensive plant response to restoring tissue N levels. The N management in plants exposed to NH_4^+ was different. Both (a) and (na) plants did not show any toxicity symptoms and the NH_4^+ root accumulation and translocation incremented proportionally to the medium concentrations (Figure 1B). On the contrary, the NH_4^+ levels in leaf systems were

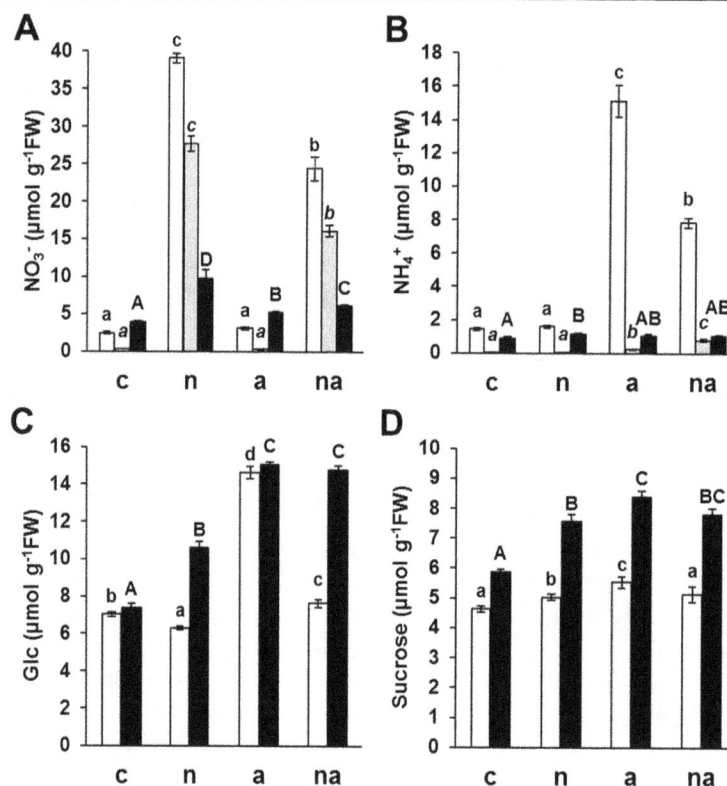

Figure 1 Nitrogen and carbon metabolites in maize in response to different inorganic N sources. Concentration of NO_3^- **(A)**, NH_4^+ **(B)**, reducing sugars **(C)** and sucrose **(D)** in roots (white bars), xylem sap (grey bars) and leaves (black bars) in maize plants grown for 10 days without N sources and then exposed for the last 30 h to absence of N (c), to 10 mM NO_3^- (n), to 10 mM NH_4^+ (a) or to 5 mM $NO_3^- + 5$ mM NH_4^+ (na). Graphs show average values ± SE (n = 6). The upper letters indicate differences among the four treatments within each organ according to Student's t-test (p < 0.05).

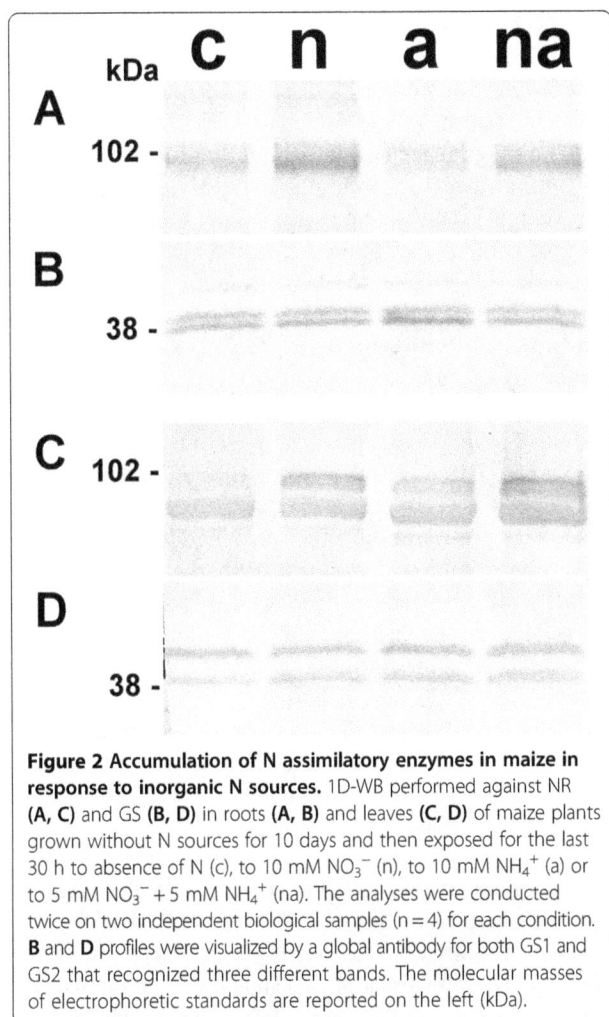

Figure 2 Accumulation of N assimilatory enzymes in maize in response to inorganic N sources. 1D-WB performed against NR **(A, C)** and GS **(B, D)** in roots **(A, B)** and leaves **(C, D)** of maize plants grown without N sources for 10 days and then exposed for the last 30 h to absence of N (c), to 10 mM NO_3^- (n), to 10 mM NH_4^+ (a) or to 5 mM NO_3^- + 5 mM NH_4^+ (na). The analyses were conducted twice on two independent biological samples (n = 4) for each condition. **B** and **D** profiles were visualized by a global antibody for both GS1 and GS2 that recognized three different bands. The molecular masses of electrophoretic standards are reported on the left (kDa).

carbohydrates towards the root system in order to sustain NH_4^+ assimilation [14]. Hence, the results confirmed that the T250 line shows some physiological traits typical of tolerant maize cultivars, especially those deriving from a high capability for NH_4^+ assimilation by the root system. At the same time, the reduction of reducing sugars in (n) roots (6.29 ± 0.09 µmol g^{-1}FW) indicated that the roots were also involved in NO_3^- assimilation, as proposed by Prinsi and co-workers [32].

The 1D-WB supported the induction of the assimilatory pathways (Figure 2). The NR accumulated proportionally to the NO_3^- concentration in the external medium both in roots and leaves (Figure 2A and C). Similarly, the GS2 levels in roots (Figure 2B) responded to NO_3^- availability. This response was scarcely detected in leaves, where the enzyme level was quite constant (Figure 2D). This behaviour is similar to that observed in young vegetative maize plants grown at high and low N fertilization in the field [35]. Interestingly, in comparison to (c) plants, the exposure to NH_4^+ led to a reduction of NR level in roots and to a slight increment in leaves. Together with the increment of the NO_3^- accumulated in the shoot (5.16 ± 0.12 µmol g^{-1}FW, Figure 1A), these results suggest that the NH_4^+ availability induced a shift in plant N economy, promoting the usage of NO_3^- reserves by the shoot. Finally, the GS1 enzyme accumulated in response to the increasing cell NH_4^+ contents, by the specific induction of the GSr isoform (Figure 2B and D).

In summary, the plant metabolic status results confirmed that the four experimental treatments induced the assimilation and translocation of the different N sources, highlighting the fundamental role of the root system in the plant physiological adaptation to both NO_3^- and NH_4^+ nutrition.

Characterization of the GS isoforms in maize roots

In order to distinguish at protein level the different GS isoforms accumulated in the maize root system, the WB with the GS global antibody (GS1 + GS2) was applied on the 2D-electrophoretic map of root soluble proteins. A preliminary investigation conducted in a wide pH range of 3–10 on 10% acrylamide Sodium Dodecyl Sulphate-PolyAcrylamide Gel Electrophoresis (SDS-PAGE) indicated that all recognizable isoforms were embraced in the acidic portion of the profile and in a molecular range of 35–50 kDa (*i.e.* 2D-WB profiles similar to Figure 3). To improve the analytical resolution, the next protein separations were performed in the narrower 4–7 pH range. This approach allowed a good reproducibility among the four nutritional treatments (Additional file 1: Figure A3), providing an overall proteomic map composed of about 1210 ± 26 spots (Figure 3A). Considering the electrophoretic adjustments, this map compared well with the one proposed by Prinsi and co-workers [32].

scarcely affected especially in comparison with roots in which NH_4^+ reached the highest value of 15.13 ± 0.93 µmol g^{-1}FW in (a) condition (Figure 1B). In several plant species, a similar limitation of NH_4^+ translocation to the shoot, considered more sensitive to NH_4^+ than roots [12], was associated to a high rate of metabolisation by GS in roots, which was proposed as one of the main traits of tolerance to high NH_4^+ inputs [13,34]. Finally, no synergistic effect on the ions accumulation emerged (Figure 1).

The availability of N led to an increase of both reducing sugars and sucrose levels in leaves, especially in response to NH_4^+ exposure (Figure 1C and D). However, only in the roots of (a) plants a peculiar doubling of reducing sugars (14.63 ± 0.32 µmol g^{-1}FW) was observed (Figure 1C). Considering that the sucrose slightly increased, it is possible that this increment originated from the delivery of photoassimilates from the shoot. Interestingly, this trait was almost completely alleviated by the co-provision with NO_3^- (Figure 1C). In a comparative study among maize genotypes, the physiological traits of NH_4^+ tolerance seemed to rely on the plant's capability to shift the partitioning of

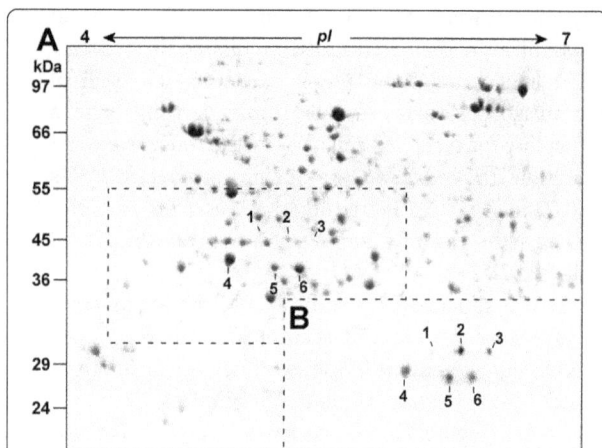

Figure 3 **Localization of the GS isoforms in 2D profile of maize roots by Western Blotting.** Representative 2D-Electrophoretic map of soluble protein fraction from roots of maize plants. Proteins were analyzed by isoelectric focusing at pH 4–7 and 10% SDS-PAGE and visualized by cCBB staining **(A)**. For each nutritional treatments, the gel portion comprised in the broken-line was analyzed by WB against all GS (**B**, GS1 + GS2 global antibody). The analyses were conducted twice on two independent biological samples (n = 4) showing the same profile among all the experimental conditions. The visualized spots are numbered (1–6) and traced on the 2D-Electrophoretic map. The LC-nESI-MS/MS characterizations of the six spots are reported in Table 1. The molecular masses of electrophoretic standards are reported on the left (kDa).

The 2D-WB showed a unique pattern (Figure 3B), recurrent in all experimental conditions, consisting of six stains that were numbered from 1 to 6 and assigned to six spots in the 2D-electrophoretic maps (Figure 3A). These 24 spots (*i.e.* six per treatment) were separately analysed by LC-nESI-MS/MS. Each spot was assigned to a specific GS isoform according to the discriminating peptides sequenced. Since for every reference spot (Figure 3) the assignment was independently confirmed among the four 2D profiles, the data were then pooled to get the overall spot characterization (Table 1, Additional file 1: Table A2-A7). The spots 1, 2 and 3 were identified as

different GS2 forms from the same gene, named GS2a, GS2b and GS2c (Table 1). This observation was in agreement with the fact that two distinct GS2 forms were highlighted in the maize leaf proteome [19]. Here, as well as in several other plant species, the GS2 protein forms exceed the number of encoding genes ([21,36] and references therein). As a whole, this trait supports the presence of post-transcriptional/translational modifications (PTMs, [37]). The similar incidence of specific peptides for both the two isoforms revealed that spot 4 derived from an overlapping of GS1-3 and GS1-4, explainable by the almost identical isoelectric point that made the two proteins inseparable by denaturing electrophoresis (Table 1), as previously noted by Martin and co-workers [19]. Conversely, spots 5 and 6 were distinctly identified as GS1-5 and GS1-1 isoforms, respectively (Table 1). This proteomic investigation revealed a pattern of GS accumulation very consistent with the expression profile proposed by Li and co-workers [25]. These authors, by means of a comprehensive analysis of the transcript levels of the six GS genes in maize, provided evidence that all isoforms were expressed in the roots, with the prevalence of GS1-3, GS1-4 and GS1-1 mRNAs. At the same time, the fact that GS1-2 was characterized as a low abundance and vascular-specific isoform [19,25] might explain why it was not detectable.

Considering the electrophoretic positions of the GS proteins in roots, it was possible to conclude that the middle band observed in the 1D-WB (Figure 2B) included GS1-3 and GS1-4, while the GSr form (Figure 2B), accumulated in response to external N availability, was composed by GS1-1 with GS1-5. This conclusion was previously suggested [24], but to our knowledge, this proteomic approach allowed the first discrimination between the individual responses of GS1-1 and GS1-5. The accumulation levels of the six GS proteins (%*Vol*) in the 2D profiles showed different and specific changes in responses to inorganic N sources (Figure 4). Moreover, the 2D profiles were staining with the Pro-Q® diamond

Table 1 Spot assignments to the specific isoform of glutamine synthetase (GS) by LC-nESI-MS/MS

N	Acr	Protein AN	pI/MW(t)	pI/MW(e)	u.p. s/t	%Cov.	Avg n. p.
1	GS2a	Glutamine synthetase chloroplastic P25462	6.42/46.0	5.36/47.6	9/9	8.0 ± 2.4	3.2 ± 0.9
2	GS2b	Glutamine synthetase chloroplastic P25462	6.42/46.0	5.44/45.5	10/10	16.9 ± 1.3	6.7 ± 0.4
3	GS2c	Glutamine synthetase chloroplastic P25462	6.42/46.0	5.52/45.2	4/4	7.2 ± 0.6	2.7 ± 0.2
4*	GS1-3/4	Glutamine synthetase root isozyme 3 P38561	5.24/39.2	5.22/41.6	9/10	35.9 ± 1.8	8.4 ± 0.4
		Glutamine synthetase root isozyme 4 P38562	5.23/39.0		9/12		
5	GS1-5	Glutamine synthetase root isozyme 5 P38563	5.52/39.3	5.40/39.4	6/14	44.3 ± 1.3	11.2 ± 0.2
6	GS1-1	Glutamine synthetase root isozyme 1 P38559	5.60/39.2	5.48/39.1	3/17	40.5 ± 1.8	11.2 ± 0.6

The different spots indicated in Figure 3 were separately collected from the root 2D profile of each nutritional treatment. Values are the mean ± SE of four independent biological samples (one per treatment) analysed in triplicate (n = 12). 4*: the spot was assigned as a co-migration of the GS1-3 and GS1-4 isoforms. **N**: spot number on Figure 3. **Acr**: acronym reported in Figure 4. **AN**: accession number. **pI/MW** are expressed in kDa and compare the theoretical values (**t**) with the experimental ones (**e**); **u.p.**: number of unique peptides identified, **s**: specific to the isoform, **t**: total. **%Cov**: amino acid coverage. **Avg n. p.**: number of distinct peptides. Detailed information on LC-nESI-MS/MS sequencing is reported in Additional file 1.

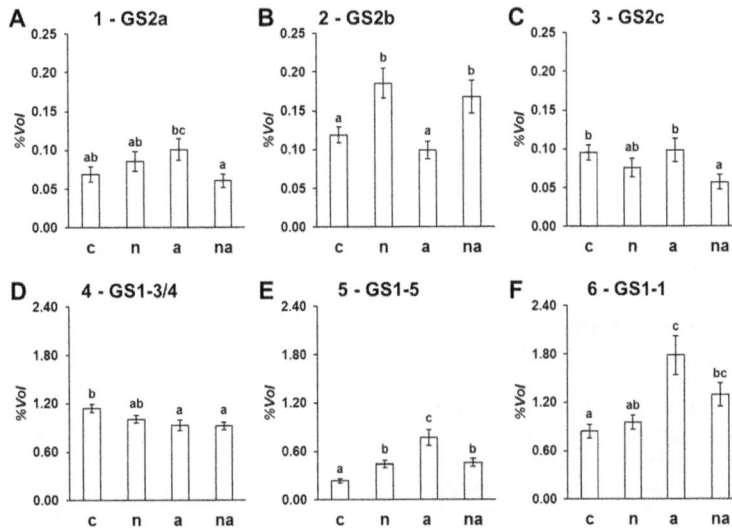

Figure 4 Accumulation of the GS isoforms in maize roots in response to N sources. The graphs report the levels *(%Vol)* of the different isoforms evaluated in the root 2D-electrophoretic profiles stained with cCBB (Figure 3) of maize plants grown without N sources for 10 days and then exposed for the last 30 h to absence of N (c), to 10 mM NO_3^- (n), to 10 mM NH_4^+ (a) or to 5 mM NO_3^- + 5 mM NH_4^+ (na). The graph number and isoform acronyms refer to Figure 3 and Table 1, respectively. **A**. Spot 1 (GS2a). **B**. Spot 2 (GS2b). **C**. Spot 3 (GS2c). **D**. Spot 4 (GS1-3/4). **E**. Spot 5 (GS1-5). **F**. Spot 6 (GS1-1). Graphs show average values ± SE (n = 6). The upper letters indicate differences among the four treatments according to Student's t-test (p < 0.05).

that is specific for phosphoproteins [38], applying this approach to investigate the phosphorylation state of GS isoforms, to our knowledge for the first time. The results showed that four of the six spots were phosphorylated and that the spot phosphorylation state was constant among all conditions (Figure 5). The results, together with the evaluations of root GS activities (Figure 6), contributed to improve the characterization of the six proteins.

The responses to inorganic N sources of the GS2 proteins in maize roots

GS2 was a very faint protein in the maize roots, with a maximum value of only 0.19 ± 0.02 *%Vol* for GSb in (n)

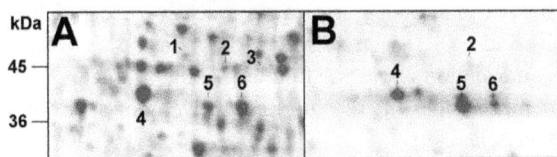

Figure 5 Analysis of phosphorylation state of root GS isoforms. The 2D-electrophoretic profiles in 4–7 p*I* range of soluble proteins from root of maize plants grown in the different experimental conditions were stained with sequential fluorescence staining procedures. **A.** Representative magnification of the profile containing the GS isoforms stained with Sypro Ruby®, showing all the proteins. **B.** Representative magnification of the same gel portion after the Pro-Q® Diamond phosphoprotein gel staining that points out only the phosphorylated proteins. The experiment was conducted for each nutritional condition showing a similar profile for all the samples. The spots are numbered (1–6) according to Figure 3 and Table 1. The molecular masses of electrophoretic standards are reported on the left (kDa).

plants (Figure 4). Similarly, the highest GS synthetase activity of the plastidial fraction was detected in (n) plants, where it represented about the 1.4% of the total enzymatic activity (Figure 6A and B). This proportion appeared similar to that measured in rice (*Oryza sativa* L.) roots, in which the contribution of GS2 was less than 4% of the total GS activity [39]. In detail, while in (c) and (a) plants the plastidial activity was almost undetectable, it increased proportionally to NO_3^- availability in (n) and (na) roots, reaching the values of 0.0017 ± 0.0001 and 0.0007 ± 0.0001 μmol mg_{prot}^{-1} min^{-1}, respectively (Figure 6A). These results strongly support the idea that, in maize roots, GS2 is involved only in NO_3^- assimilation and it is not influenced by NH_4^+ nutrition. Among the three identified GS2 forms, the GS2b is the only one of which the accumulation level was significantly increased in plants exposed to NO_3^-. This observation confirms our former proteomic analyses [32]. The comparison of GS2 accumulation levels (Figure 4A, B and C) and GS2 activity (Figure 6A) suggests the hypothesis that GS2b represented the form really participating in catalytic activity. Interestingly, this spot represented also the only phosphorylated GS2 (Figure 5).

The reciprocal position of the GS2 spots was comparable with those observed in the leaf maize proteome. Indeed, in leaves two GS2 spots were identified of which the accumulation levels were differently related to the enzymatic activity. In particular, the more acidic spot, matching with GS2b, was accumulated proportionally, while the basic one disappeared with the increment of the enzymatic activity [19]. The GS2c electrophoretic

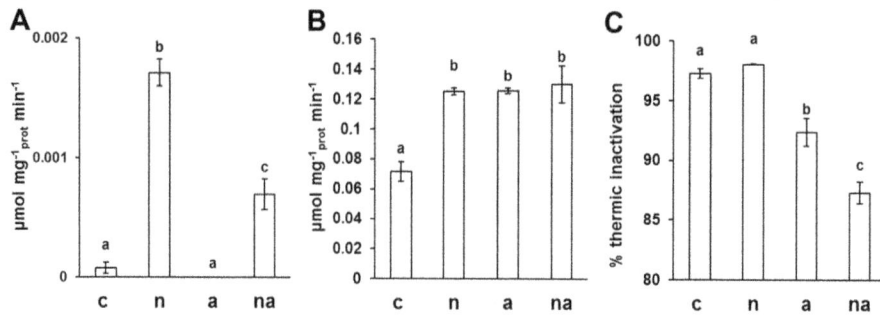

Figure 6 Effects of the different inorganic N sources on GS activity in maize root. Maize plants were grown without N sources for 10 days and then exposed for the last 30 h to absence of N (c), to 10 mM NO_3^- (n), to 10 mM NH_4^+ (a) or to 5 mM NO_3^- + 5 mM NH_4^+ (na). Both the GS synthetase **(A, B)** and the transferase activities **(C)** were evaluated and expressed as specific activity. The synthetase activity was measured in the enriched fraction of plastidial proteins **(A)** and in total protein fraction **(B)** representing the GS2 activity and the total activity, respectively. The transferase activity was measured in total protein fraction prior to and after the exposure of the samples to 45°C for 10 min. The % thermal inactivation **(C)** represents the portion of transferase activity lost after this thermic treatment. Graphs show average values ± SE (n = 3). The upper letters indicate differences among the four treatments according to Student's t-test (p < 0.05).

position (Figure 3) as well as its trend of accumulation (Figure 4) showed a high degree of similarity with this last observation. As a whole, the results suggest that GS2a and GS2c might be transitional forms of the enzyme, probably originated by PTMs. The chemical and/or immunological detection of sugar and nitric oxide moieties did not lead to any positive results (*i.e.* no signal on PVDF membranes corresponding to GS isoforms), but it is not possible to exclude the occurrence of several other PTMs.

The responses to inorganic N sources of the GS1 isoforms in maize roots

According to the proposed model, the presence of phosphorylation on GS1 is associated with active enzymatic forms, where it promotes protection from degradation and the interaction with activating 14-3-3 proteins [20]. The evaluation that the spots 4, 5 and 6 were phosphorylated in all conditions tested (Figure 5) supports the hypothesis that they represented active subunits of the root GS1 enzyme.

Firstly, it is possible to note that all three GS1 spots were detected in (c) plants, in which spot 4 was predominant (1.14 ± 0.05 %*Vol*, Figure 4D, E and F). Their presence in roots of starved plants confirmed a GS1 involvement in the use and/or recycling of endogenous N reserves. Spot 4 was the least influenced by the plant nutritional status, as it only decreased by about 20% in response to NH_4^+ (Figure 4D). This trend was in agreement with previous transcriptional analyses highlighting that GS1-3 and GS1-4 mRNAs slightly decreased in response to N as well as with the unchanged intensity of GS1 band in SDS-PAGE (Figure 2B; [23,24]). Together with the fact that in maize *gln1-3* and *gln1-4* mutants the vegetative biomass is not affected [19], these results reinforce the hypothesis that other root isoforms are able to sustain N assimilation during vegetative growth.

Interestingly, spots 5 and 6 showed marked and different changes (Figure 4E and F). The GS1-1 isoform specifically increased in response to NH_4^+ nutrition, becoming the most abundant one in (a) roots with the highest measured value (1.78 ± 0.24 %*Vol*). It is worth noting that the changes in GS1-1 accumulation reflected the NH_4^+ availability, reaching an increment of about +112% and +54% in (a) and (na) plants, respectively. The responses of GS1-5 isoform attested an even higher increase of about +222% in (a) condition. Likewise, in (a) plants the GS synthetase activity significantly increased (Figure 6B). These results confirm that NH_4^+ induces GSr (Figure 2B, [24]) and, for the first time, they allowed us to discern the differential contribution of its components. Moreover, GS1-5 showed a peculiar doubling (+86%) in (n) plants. The fact that GS1-5 is metabolically active in (n) roots is reinforced by the estimation that its change is the only one to be associated with the increase in total GS synthetase activity in the (n) condition (Figure 6B). Interestingly, these observations confirm at protein and enzymatic order the induction of GS1-5 transcript by NO_3^- exposure recently observed in the T250 line [40]. Taken together, these results provide the first information about the functional role of the GS1-5 isoform, providing evidence of its involvement in root NO_3^- metabolisation.

The GS synthetase activity in the total protein fraction, starting from 0.071 ± 0.007 $\mu mol\ mg^{-1}_{prot}\ min^{-1}$ in starved plants (c), increased to a similar extent in N treated ones (Figure 6B). This feature suggests that GS1 followed a saturation kinetic, probably because the elevated concentrations of NO_3^- and/or NH_4^+ overfilled the metabolic capability of the maize root organ. However, the root metabolic capability seemed to be sustained by different GS1 isoforms, according to different mineral N sources. This was particular evident in (na) roots where GS1-1, GS1-5 and GS2 were accumulated to an intermediate level (Figure 4). In order to get better information about this aspect, the percentage of thermal inactivation of total GS transferase activity, induced by 45°C for 10 min, was evaluated (Figure 6C). According to the physicochemical characterization, the Isoleucine-161 in the GS1-4 sequence confers thermal stability, while the substitution with Ala-161 clearly renders the GS1-1 more heat-labile than GS1-4 [15,24]. Considering that GS1-4 and GS1-1 shared this feature with GS1-3 and GS1-5 respectively, it is possible to assume that the thermal inactivation measured reflected the proportion of GS activity ascribable to the GS1-1 with GS1-5 subunits. Such evaluation allowed us to confirm that GS1-1 and GS1-5 were fundamental for the assembling of GS active enzyme in (c) and (n) roots, where the thermal treatment provoked a loss in activity of about 97% and 98%, respectively. It is also worth noting how the contribution of GS1-3 and GS1-4 gained in importance in roots of plants exposed to NH_4^+, especially if in co-provision with NO_3^- (Figure 6C).

Overall, this proteomic investigation confirmed the GS1 involvement in N recycling as well as in the root assimilation of NO_3^- and NH_4^+, and at the same time, it allowed us to propose that specific and combined isoforms sustain these different metabolic tasks.

The inorganic N sources differently affected the amino acid accumulation and composition in root, leaf and xylem sap

In order to appreciate the extent by which the GS activation induced by the experimental treatments affected the plant N metabolism, the composition of amino acids in different tissues was measured. Figure 7 points out the changes related to the mostly more abundant amino acids, gathering the others in a single group, detected in roots (Figure 7A), in xylem sap (Figure 7B) and in leaves (Figure 7C) of plants exposed to the four N conditions. The comprehensive amino acid compositions of the three tissues are detailed in the Additional file 1: Tables A8, A9 and A10, respectively.

The provision of N provoked a significant increase in amino acid level at whole plant scale, but the extent of these increments varied in relation to the inorganic N source and/or organ. The total amino acids concentration in roots and leaves of (a) plants was almost double the amounts in (n) plants, confirming that NH_4^+ nutrition promoted a more intensive N assimilation than NO_3^- (Figure 7; [4]). This aspect was mirrored by the total protein amounts in the root systems (Additional file 1: Figure A4). Because the total amino acid concentration reached similar values both in roots and leaves during the (a) and (na) treatments (Additional file 1: Tables A8 and A10), the T250 line did not show a marked synergistic response to the co-provision of NO_3^- with NH_4^+.

Gln was the amino acid subjected to the greatest and most prevalent changes after the exposure of the plants to N as well as being the main compound for N translocation in the xylem sap, in which it reached the maximum value of 6.55 ± 0.28 mM in (a) plants (Figure 7B). Moreover, all three organs showed high amounts of Ala, which represented up to 40%, 44% and 55% of the total amino acids in

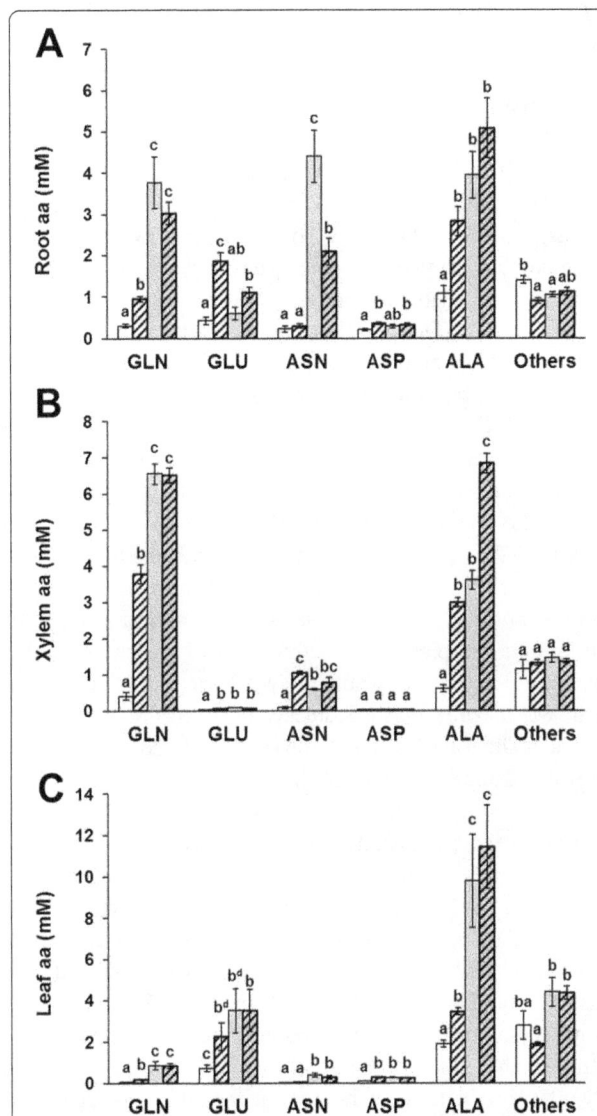

Figure 7 Levels of amino acids in maize plants in response to inorganic N sources. Concentration of the main amino acids in roots **(A)**, xylem sap **(B)** and leaves **(C)** in maize plants grown for 10 days without N sources and then exposed for the last 30 h to absence of N (c, white bars), to 10 mM NO_3^- (n, crossed white bars), to 10 mM NH_4^+ (a, grey bars) or to 5 mM NO_3^- + 5 mM NH_4^+ (na, crossed and grey bars). Graphs show the amino acid concentrations (mM) as average values ± SE (roots and leaves n = 8, xylem sap n = 6). The letters above are assigned according to Student's t-test (p ≤ 0.01; [d] p ≤ 0.05). Detailed quantification of individual amino acids in root, xylem and leaf is reported in Additional file 1: Table A8, A9 and A10, respectively.

roots, xylem sap and leaves, respectively (Additional file 1: Tables A8, A9 and A10). Considering that previous studies on maize plants grown in high N reported Ala percentages in xylem and leaves ranging from 5% to 20-29%, respectively [19,35], the higher percentage of Ala may be a peculiarity of the T250 line. Taking into account the involvement of Ala and aspartic acid (Asp) in the C4 photosynthesis, it is also possible that this variability derives from differences in the times of the day when leaves were sampled as well as from differences in leaf developmental stage and in plant N regime [35,41].

Looking at the root system, the Gln level starting from 0.30 ± 0.04 mM in (c) condition increased to 0.94 ± 0.06 mM in (n) plants but it reached the higher values of 3.77 ± 0.62 mM and 3.02 ± 0.27 mM in (a) and (na) plants, respectively (Figure 7A). This trend was associated with comparable upsurges in the xylem sap (Figure 7B). Since the root GS synthetase activity was very similar in all three N treatments (Figure 6B), it is reasonable to suppose that the lowest level of Gln in (n) roots did not result from an enzymatic control on GS but rather from a metabolic regulation on NO_3^- reduction steps, limiting the free NH_4^+ in the cell. The slower N assimilation in (n) roots was associated with an accumulation of Glu and Asp (Figure 7A), supporting the idea that the GS/GOGAT and the Tricarboxylic Acid (TCA) cycles were reciprocally balanced, sufficiently to sustain the storage of intermediates. In addition, the similar Gln concentrations observed in roots and xylem saps of (a) and (na) plants (Figure 7A and B) indicated that the maximum capacity of plants to synthesize and translocate Gln was already reached during the exposure to the lowest availability of NH_4^+ (i.e. 5 mM). This is also consistent with the saturation kinetic of the GS synthetase activity described above (Figure 6B).

Otherwise, Asn significantly incremented only in roots exposed to NH_4^+, reaching the values of 4.40 ± 0.63 mM and 2.09 ± 0.32 mM in (a) and (na) conditions, respectively (Figure 7A). Asn represents one of the main compounds for N storage and transport due to its high N/C ratio and stability. It is synthesized by Asn synthetase (AS, EC6.3.5.4) by the amidation of Asp using Gln as amino donor, but several studies have indicated an NH_4^+-dependent synthetase activity in plants ([42] and references therein). The Asn changes are in agreement with the observation that in maize roots the AS gene expression is influenced by C/N ratio since it is induced by carbohydrate limitation and by supplies of NH_4^+, Gln, Asn, Asp but not of Glu [43]. Moreover, in maize mutants deficient for GS1-3 and/or GS1-4 a higher leaf content of Asn and Ala compared with than of the wild-type was reported, suggesting compensatory involvement in NH_4^+ (re)assimilation [44]. In this work, the levels of Asn in root tissues were quite proportional to the root NH_4^+ concentrations, but not to the Gln ones, suggesting that the Asn accumulation could be involved in a

mechanism of cell protection from high NH_4^+, which appeared specifically induced by the cation and not by Gln. This induction was not associated with a comparable upsurge of Asn translocation. On the contrary, the Ala concentration strongly increased in all three organs in response of N availability (Figure 7). The highest increment of Ala translocation was observed during the (na) treatment, showing a synergistic trait related to co-provision of NO_3^- with NH_4^+. Recently it was proposed that Ala and pyruvate translocations might have important roles for the maintenance of C/N balance throughout plants [45]. Considering that in leaves the Asn re-assimilation necessarily releases free NH_4^+ [46], the sequestration of Asn in roots and the preferential translocation of Gln and Ala might participate in the mechanism of NH_4^+ tolerance observed in T250.

The Glu levels showed marked differences between root and leaf systems under the same nutritional treatment. In roots, Glu accumulation was increased in the presence of NO_3^-, reaching the values of 1.10 ± 0.13 mM and 1.86 ± 0.21 mM in (na) and (n) plants, respectively. However, the provision of NH_4^+ (a) was associated with a very low level of 0.60 ± 0.16 mM, comparable with the (c) plants (Figure 7A). Glu was scarcely translocated in the xylem sap, while in leaves the N availability sustained a generalized increment of Glu that mirrored the total amino acid concentrations (Figure 7B and C). These observations suggest that the balancing of the GS/GOGAT cycle distinctly diverged between the two organs. In particular, in leaf the high amount of Gln received from the xylem sap appeared re-assimilated to restore Glu. On the contrary, in the root system the NH_4^+ exposure seemed to hamper the accumulation of Glu, resulting in a large prevalence of Gln. Several studies have given evidence of the existence of a Glu homeostasis in plants, probably involved in plant C/N perception, which is perturbed by NADH (reduced Nicotinamide Adenine Dinucleotide) and 2OG (2-oxoglutarate) availability [6,9,47]. Hence, it is possible to suppose that the peculiar shortage of Glu in (a) roots was associated with a scarce provision of 2OG by the TCA cycle to the GS/GOGAT system. The involvement of the TCA cycle also seems to be supported by the concerted changes of Asp (Figure 7). Considering that the (a) and (na) roots showed similar extent of N assimilation as well as that the (a) roots were characterized by the highest content of reducing sugars, it is unlikely that this imbalance derived from a lack of C skeletons. Instead, it is more conceivable that the lack of Glu was related to an excess of reducing power. In fact, the high N assimilation in roots exposed to NH_4^+ could be associated with a strong activation of anaplerotic reactions for C skeletons, leading to a production of NADH exceeding the metabolic requests. Considering that this excess could cause a feedback inhibition on the TCA cycle [48] it is possible that the outcome could be a very low availability of 2OG in the cell. This consideration is consistent with the evidence that in roots of *Arabidopsis thaliana* the supply of NH_4^+ compared to NO_3^- promotes a higher capacity of respiratory bypass pathways involved in the dissipation of excess of redox equivalents [49]. In addition, the co-provision of NO_3^- with NH_4^+ sustained a higher accumulation of Glu in (na) roots (Figure 7A). This could be a synergistic effect by which, even if the N assimilation in (a) and (na) roots was similar, the consumption of reducing power by NO_3^- reducing steps could be associated with a minor inhibition of the TCA cycle. It is interesting to note that, because the Asn and Ala are synthesized by the transfer of the amino group of Gln on C skeletons (*i.e.* pyruvate and oxaloacetate) available out of the TCA cycle [6], their synthesis could contribute towards regenerating Glu with a NADH production lower than the GS/GOGAT route.

Overall, the analysis of the amino acid composition confirms the activation of GS observed by the proteomic and enzymatic approaches, highlighting the relevance of root responses in the N economy. Moreover, the results provide new information about the metabolic regulation of the GS/GOGAT cycle that seem to be deeply influenced by several aspects, such substrates and coenzymes, as well as by the biosynthetic pathways of other amino acids.

Conclusion

Taken together, the results give novel insights about the multiplicity of factors involved in GS regulation. Firstly, the work provides new evidence that in maize different GS isoenzymes/isoforms have distinct metabolic functions, diverging between root and leaf system. Interestingly, the proteomic discrimination of the GS1 proteins revealed that in roots the cytosolic enzyme also contributes in NO_3^- assimilation by GS1-5 activity, providing first indications about the role of this isoform. At the same time, the changes in enzymatic properties as well as the presence of phosphorylation confirmed the involvement of PTMs. It is conceivable that these observations may be useful for future studies aimed to investigate the rearrangement of GS native enzyme and its interaction with regulatory proteins. Furthermore, the analyses of amino acid composition in roots, xylem sap and leaves provides novel information about the fact that in roots the GS/GOGAT cycle was not only regulated at molecular level but it was also deeply influenced by biochemical factors, like substrates and cell redox status. Finally, from a physiological point of view, it is interesting to note that the work gives new insights about the relevance of Glu, Asn and Ala in plant C/N balance in response to nitrate and/or ammonia nutritions.

Methods

Plant materials

Maize seeds of the T250 inbred line, kindly provided by Prof. Zeno Varanini of the University of Verona, Italy, were germinated in the dark at 26°C for 72 h. The seedlings were transferred to a hydroponic system in a growth chamber with a photoperiod of 16/8 h at 26/22°C, assuring PPFD of 200 μmol m^{-2} s^{-1} and at constant relative humidity of 65%. After incubation in 4 mM $CaSO_4$ for 48 h, the plants were grown for the following eight days in a solution of 400 μM $CaSO_4$, 200 μM K_2SO_4, 175 μM KH_2PO_4, 100 μM $MgSO_4$, 20 μM Fe-EDTA, 5 μM KCl, 2.5 μM H_3BO_3, 0.2 μM $CuSO_4$, 0.2 μM $ZnSO_4$, 0.2 μM $MnSO_4$, 0.05 μM Na_2MoO_4, pH = 6.1. All hydroponic solutions were continuously aerated and renewed every three days. After this period of N starvation, at the beginning of the light period plants were transferred for 30 h into fresh growing solutions of the following four treatments, balanced with K_2SO_4: i) N absence (c); ii) 10 mM NO_3^- (n); iii) 10 mM NH_4^+ (a); iv) 5 mM NO_3^- + 5 mM NH_4^+ (na). (For details, see Additional file 1: Figure A1 and Table A1). At the time of sampling, roots and leaves were separately collected, frozen in liquid N_2 and stored at –80°C. The root systems of plants destined for NO_3^- and NH_4^+determination were rinsed in aerated ice-cold solution (5 mM K_2SO_4, 0.4 mM $CaSO_4$) in the growth chamber for 15 min before sampling. For xylem sap collection, the plants were maintained in the hydroponic solution and de-topped by cutting the stem with a razor blade just above the first internode. The cut surface was rinsed twice with distilled water and blotted with paper. Then the stem was encircled with a silicon tube and the liquid drawn in the first 5 min was discarded. Finally, the xylem saps collected from 5 to 25 min from six plants were pooled into a biological sample, weighed and stored at –80°C.

Determination of nitrate, ammonium, reducing sugars and sucrose

For NO_3^- and sugar content determination, organ samples were treated as described by Prinsi and co-workers [32]. NO_3^- and sugars were quantified according to Cataldo *et al.* [50] and Nelson [51], respectively. NH_4^+ was extracted from roots and leaves by adding 3% (w/w) of polyvinylpolypyrrolidone (PVPP), homogenizing in 4 vol of 50 mM Tris–HCl pH 7.4, 10 mM imidazole, 10 mM ascorbic acid, 0.5% (v/v) β-mercaptoethanol in ice and then centrifuging at 10,000 *g* for 20 min at 4°C. NH_4^+ concentration was determined by the Ammonia Assay Kit (Sigma-Aldrich) according to manufacturer's instructions. For NO_3^- and NH_4^+ detection leaf and root samples were filtered by Millipore Millex HV cartridges (0.45 μm) while the xylem saps were directly analysed. All of the three analyses were conducted on three biological samples, each composed by three plants (and six for xylem sap), analysed in duplicate (n = 6).

Extraction of soluble protein fraction and electrophoretic analyses

The leaves or roots collected from 18 plants were pooled into one sample used for the further analysis as one independent biological replicate, powdered in liquid N_2 and stored at –80°C. The soluble protein fraction (*i.e.* whole proteome depleted of membrane proteins) was separated by centrifugation at 100,000 *g* at 4°C for 38 min and then the protein components were purified by consecutive precipitations in 0.1 M ammonium acetate in methanol and acetone as described by Prinsi and co-workers [32], optimizing the procedure by the addition of phosphatase inhibitors in the extraction buffer (10 mM NaF, 1 mM Na_3VO_4). 1D-PAGE was conducted on 10% acrylamide gel [52]. The 2D-PAGE were done according to Prinsi and co-workers [32] but adapting the isoelectric focusing on pH 4–7, 13 cm IPG strips (GE Healthcare) for a total of 25 kV and the SDS-PAGE into 10% acrylamide gels. The qualitative investigations by Western Blot (WB) was conducted as described by Bernardo and co-workers [53] both on 1D and 2D profiles on two biological replicates analyzed in duplicate (n = 4), using separately two primary antibodies: against nitrate reductase (NR, EC1.7.99.4, 1:1000, Agrisera AS08310, polyclonal [54]) and against all GS (1:10,000, GS1 + GS2 global antibody, Agrisera AS08295, polyclonal with reactivity in maize proved in producer's technical sheet and in Arabidopsis by [55]). The WB were visualized by anti-rabbit IgG conjugated with alkaline phosphatase. For the 2D approach, the map sector corresponding to p*I* 4–6.5 and MW 32–55 kDa was blotted and, before the blocking procedure, the filters were reversibly stained with Ponceau S 0.5% (w/v) in 1% (v/v) acetic acid in order to orientate the pattern. For the qualitative determination of GS phosphorylation, we employed Sypro Ruby® coupled with Pro-Q® Diamond phosphoprotein gel stain (Molecular Probes) according to the manufacturer's instructions (n = 4). For the quantitative analysis, three biological replicates in duplicate (n = 6) were stained with colloidal Coomassie Brilliant Blue G-250 (cCBB; [56]) and analyzed with ImageMaster 2-D Platinum Software (GE Healthcare) in order to quantify the spots (%*Vol*: percentage of the total spot volume) assigned to GS by WB.

Protein identification by LC-nESI-MS/MS

The six spots visualized by WB against GS were excised from cCBB 2D gels independently for each experimental condition, obtaining 24 samples. After trypsin digestion [32], the samples were analyzed by a 6520 Q-TOF mass spectrometer with HPLC Chip Cube source driven by 1200 series nano/capillary LC system (Agilent Technologies). The nLC separation was done on 75 μm x 43-mm column

(Zorbax SB, C18, 300 Å), applying a 13-min Acetonitrile (ACN) gradient (from 5% to 60% v/v) in 0.1% (v/v) formic acid at 0.4 μl min^{-1}. The mass spectrometer ran in positive ion mode acquiring 4 MS spectra s^{-1} from 300 to 3000 m/z. The auto-MS/MS mode was applied from 50 to 3000 m/z with a maximum of 4 precursors per cycle and an active exclusion of 2 spectra for 0.1 min. Peptide identification was performed by Spectrum Mill MS Proteomics Workbench (Rev B.04.00.127; Agilent Technologies). Cysteine carbamidomethylation and methionine oxidation were set as fixed and variable modifications, respectively, accepting two missed cleavages per peptide. The search was conducted against the subset of *Zea mays* protein sequences (Oct 2013, 172261 *entries*) downloaded from the National Center for Biotechnology Information [57] and concatenated with the reverse one. The threshold used for peptide identification was Spectrum Mill score ≥ 9, Score Peak Intensity $\geq 70\%$, mass MH^+ Error ≤ 10 ppm, Local False Discovery Rate $\leq 0.1\%$ and Database Fwd-Rev Score ≥ 2. Each sample was analyzed in triplicate and independently assigned to a GS isoform, according to the discriminating peptides. For each spot position (n. 1 to 6 in Figure 3A), the isoform assignment was independently confirmed among all the four 2D profiles corresponding to the four nutritional conditions. Then, to provide the overall information obtained about protein sequencing, MS data regarding each isoform were summed. Physical properties of the isoforms were predicted by *in silico* tools at ExPASy [58].

Determination of GS activity

For the determination of total GS activity, frozen root systems were powdered in liquid N_2 to which was added 0.5% (w/w) PVPP, and extracted in 3 volumes of 50 mM Tris–HCl pH = 7.8, 10 mM $MgSO_4$, 1 mM dithiothreitol, 10% (v/v) ethylene glycol. The samples were centrifuged at 12,000 *g* for 20 min at 4°C and filtered on G-25 columns (PD-10, GE Healthcare), eluting in 25 mM Tris–HCl pH = 7.8, 10 mM $MgSO_4$. Root plastids were isolated according to Redinbaugh and Campbell [59]. The GS synthetase and transferase activities were measured by the spectrophotometric determination of γ-glutamylhydroxamate, as described by Lea *et al.* [60] and Cullimore and Sims [61], respectively. The transferase activities were measured prior to and after thermic treatments at 45°C for 10 min. Protein contents were quantified by the BioRad protein assay. All experiments were replicated on three independent biological samples, each derived from six plants (n = 3).

Determination of amino acid composition by LC-ESI-MS analysis

The frozen root and leaf samples, each collected from six plants, were homogenized in 3 volumes of 1 mM tridecafluoroheptanoic acid (TDFHA), 50% (v/v) methanol. Samples were shaken for 10 min at 4°C and then centrifuged twice at 14,000 g for 20 min at 4°C. The underivatized supernatants and the collected xylem saps were finally diluted to 0.5 mM TDFHA, 25% (v/v) methanol. The LC-ESI-MS analyses were conducted by an Agilent Technologies 1200 Series capillary pump coupled with dual ESI source on 6520 Q-TOF mass spectrometer according to Armstrong et al. [62]. Briefly, LC runs were done on an XDB-C18 column (2.1 x 50 mm, 1.8 μm, Agilent Technologies) applying a 30 min non-linear gradient of 0.5 mM TDFHA/ACN with a flow rate of 200 μl min^{-1}. The ESI source was set at 350°C, 3500 V and the fragmentor at 100 V. The data acquisition range was 50–350 m/z at 0.93 scans s^{-1}. The quantitation was conducted on EIC for single MH$^+$ in ±0.02 m/z window, accepting a mass error of ±5 mDa in ion identification and referring to calibration curves (Additional file 1: Table A11). The analyses were conducted on four root and leaf samples and three xylem sap samples analysed in duplicate (n = 8, n = 6).

Additional file

Additional file 1: A single .pdf file containing:
Figure A1. Experimental design. **Figure A2.** 1D-electrophoretic profiles of leaf protein samples and mass spectrometry characterization of the most prominent band. **Figure A3.** 2D-electrophoretic reproducibility of maize root protein samples. **Figure A4.** Soluble protein contents in maize roots. **Table A1.** Composition of the hydroponic solutions used for the 30 h nutritional treatments. **Table A2, A3, A4, A5, A6** and **A7.** LC-nESI-MS/MS characterization of spots n.1, 2, 3, 4, 5 and 6. **Table A8.** Levels of amino acids in roots. **Table A9.** Levels of amino acids in xylem sap. **Table A10.** Levels of amino acids in leaves. **Table A11.** Experimental parameters used for amino acid quantitation.

Abbreviations
1D: One-dimensional; 2D: Two-dimensional; 2OG: 2-oxoglutarate; (a): Plants exposed to 10 mM NH$_4^+$; ACN: Acetonitrile; Ala: Alanine; AS: Asparagine synthetase; Asn: Asparagine; Asp: Aspartic acid; C: carbon; (c): control plants exposed to N absence; cCBB: colloidal Coomassie Brilliant Blue G-250; Glc: Glucose; Gln: Glutamine; Glu: Glutamic acid; GOGAT: Glutamate synthase; GS: Glutamine synthetase; LC-nESI-MS/MS: Liquid Chromatography-nanoElectroSpray Ionization-Tandem Mass Spectrometry; LC-ESI-MS: Liquid Chromatography-ElectroSpray Ionization-Mass Spectrometry; N: Nitrogen; (n): plants exposed to 10 mM NO$_3^-$; (na): Plants exposed to 5 mM NO$_3^-$ + 5 mM NH$_4^+$; NADH: Reduced Nicotinamide Adenine Dinucleotide; NH$_4^+$: Ammonium; NO$_3^-$: Nitrate; NR: Nitrate Reductase; PTMs: Post transcriptional/translational modifications; PVPP: Polyvinylpolypyrrolidone; SDS-PAGE: Sodium Dodecyl Sulphate-PolyAcrylamide Gel Electrophoresis; TCA cycle: Tricarboxylic acid cycle; TDFHA: Tridecafluoroheptanoic acid; WB: Western Blotting.

Competing interests
The authors declare that they have no competing interests.

Authors' contributions
BP contributed to the conception of the study and of the experimental design, participated to the determination of biochemical parameters and enzyme assays, carried out the protein extraction, 2D-electrophoresis, western blot analyses, protein characterization by LC-nESI-MS/MS and determination of amino acid composition by LC-ESI-MS, contributed to the interpretation of the results, wrote and edited the manuscript. LE contributed to the conception of

the study and of the experimental design, participated to the determination of biochemical parameters and enzymatic assays, contributed to the interpretation of the results and took part in the critical revision of the manuscript. All authors read and approved the final manuscript.

Acknowledgements
This work was supported by the Italian Ministry of Education, University and Research [MIUR-PRIN 2009].

References
1. Galloway JN, Townsend AR, Erisman JW, Bekunda M, Cai Z, Freney JR, et al. Transformation of the nitrogen cycle: recent trends, questions, and potential solutions. Science. 2008;320:889–92.
2. Raun WR, Johnson GV. Improving nitrogen use efficiency for cereal production. Agron J. 1999;91:357–63.
3. Hirel B, Le Gouis J, Ney B, Gallais A. The challenge of improving nitrogen use efficiency in crop plants: towards a more central role for genetic variability and quantitative genetics within integrated approaches. J Exp Bot. 2007;58:2369–87.
4. Miller AJ, Cramer MD. Root nitrogen acquisition and assimilation. Plant Soil. 2004;274:1–36.
5. Masclaux-Daubresse C, Daniel-Vedele F, Dechorgnat J, Chardon F, Gaufichon L, Suzuki A. Nitrogen uptake, assimilation and remobilization in plants: challenges for sustainable and productive agriculture. Ann Bot. 2010;105:1141–57.
6. Nunes-Nesi A, Fernie AR, Stitt M. Metabolic and signaling aspects underpinning the regulation of plant carbon nitrogen interactions. Mol Plant. 2010;3:973–96.
7. Hirel B, Lea PJ. Ammonia assimilation. In: Lea PJ, Morot-Gaudry JF, editors. Plant nitrogen. Hidelberg: Springer-Verlag Berlin Hidelberg; 2001. p. 79–99.
8. Thomsen HC, Eriksson D, Møller IS, Schjoerring JK. Cytosolic glutamine synthetase: a target for improvement of crop nitrogen use efficiency? Trends Plant Sci. 2014;19:656–63.
9. Forde BG, Lea PJ. Glutamate in plants: metabolism, regulation, and signalling. J Exp Bot. 2007;58:2339–58.
10. Lea PJ, Azevedo RA. Nitrogen use efficiency. 2. Amino acid metabolism. Ann Appl Biol. 2007;151:269–75.
11. Andrews M, Raven JA, Lea PJ. Do plants need nitrate? The mechanism by which nitrogen form affects plants. Ann Appl Biol. 2013;163:174–99.
12. Britto DT, Kronzucker HJ. NH$_4^+$ toxicity in higher plants: a critical review. J Plant Physiol. 2002;159:567–84.
13. Cruz C, Bio AFM, Domínguez-Valdivia MD, Aparicio-Tejo PM, Lamsfus C, Martins-Loução MA. How does glutamine synthetase activity determine plant tolerance to ammonium? Planta. 2006;223:1068–80.
14. Schortemeyer M, Stamp P, Feil B. Ammonium tolerance and carbohydrate status in maize cultivars. Ann Bot. 1997;79:25–30.
15. Unno H, Uchida T, Sugawara H, Kurisu G, Sugiyama T, Yamaya T, et al. Atomic structure of plant glutamine synthetase: a key enzyme for plant productivity. J Biol Chem. 2006;281:29287–96.
16. Cren M, Hirel B. Glutamine synthetase in higher plants: regulation of gene and protein expression from the organ to the cell. Plant Cell Physiol. 1999;40:1187–93.
17. Orsel M, Moison M, Clouet V, Thomas J, Leprince F, Canoy A-S, et al. Sixteen cytosolic glutamine synthetase genes identified in the Brassica napus L. genome are differentially regulated depending on nitrogen regimes and leaf senescence. J Exp Bot. 2014;65:3927–47.
18. Yamaya T, Kusano M. Evidence supporting distinct functions of three cytosolic glutamine synthetase and two NADH-glutamate synthases in rice. J Exp Bot. 2014;65:5519–25.
19. Martin A, Lee J, Kichey T, Gerentes D, Zivy M, Tatout C, et al. Two cytosolic glutamine synthetase isoforms of maize are specifically involved in the control of grain production. Plant Cell. 2006;18:3252–74.
20. Finnemann J, Schjoerring JK. Post-translational regulation of cytosolic glutamine synthetase by reversible phosphorylation and 14-3-3 protein interaction. Plant J. 2000;24:171–81.
21. Lima L, Seabra A, Melo P, Cullimore J, Carvalho H. Phosphorylation and subsequent interaction with 14-3-3 proteins regulate plastid glutamine synthetase in Medicago truncatula. Planta. 2006;223:558–67.

22. UniProtKB/Swiss-Prot (reviewed entries) [http://www.uniprot.org/uniprot/?query=reviewed%3Ayes]

23. Sakakibara H, Kawabata S, Hase T, Sugiyama T. Differential effects of nitrate and light on the expression of glutamine synthetase and ferredoxin-dependent glutamate synthase in maize. Plant Cell Physiol. 1992;33:1193–8.

24. Sakakibara H, Shimizu H, Hase T, Yamazaki Y, Takao T, Shimonishi Y, et al. Molecular identification and characterization of cytosolic isoforms of glutamine synthetase in maize roots. J Biol Chem. 1996;271:29561–8.

25. Li M, Villemur R, Hussey PJ, Silflow CD, Gantt JS, Snustad DP. Differential expression of six glutamine synthetase genes in Zea mays. Plant Mol Biol. 1993;23:401–7.

26. Limami AM, Rouillon C, Glevarec G, Gallais A, Hirel B. Genetic and physiological analysis of germination efficiency in maize in relation to nitrogen metabolism reveals the importance of cytosolic glutamine synthetase. Plant Physiol. 2002;130:1860–70.

27. Amiour N, Imbaud S, Clément G, Agier N, Zivy M, Valot B, et al. An integrated "omics" approach to the characterization of maize (Zea mays L.) mutants deficient in the expression of two genes encoding cytosolic glutamine synthetase. BMC Genomics. 2014;15:1005. doi:10.1186/1471-2164-15-1005.

28. Gallais A, Coque M. Genetic variation and selection for nitrogen use efficiency in maize: a synthesis. Maydica. 2005;50:531–47.

29. Plénet D, Lemaire G. Relationships between dynamics of nitrogen uptake and dry matter accumulation in maize crops. Determination of critical N concentration. Plant Soil. 2000;216:65–82.

30. Saiz-Fernández I, De Diego N, Sampedro MC, Mena-Petite A, Ortiz-Barredo A, Lacuesta M. High nitrate supply reduces growth in maize, from cell to whole plant. J Plant Physiol. 2015;173:120–9.

31. Sivasankar S, Rothstein S, Oaks A. Regulation of the accumulation and reduction of nitrate by nitrogen and carbon metabolites in maize seedlings. Plant Physiol. 1997;114:583–9.

32. Prinsi B, Negri AS, Pesaresi P, Cocucci M, Espen L. Evaluation of protein pattern changes in roots and leaves of Zea mays plants in response to nitrate availability by two-dimensional gel electrophoresis analysis. BMC Plant Biol. 2009;9:113. doi:10.1186/1471-2229-9-113.

33. Oaks A. Biochemical Aspects of Nitrogen Metabolism in a Whole Plant Context. In: Lambers H, Neeteson JJ, Stulen I, editors. Fundamental, Ecological and Agricultural Aspects of Nitrogen Metabolism In Higher Plants. Doordrecht, Boston, Lancaster: Martinus Nijhoff Publishers; 1986. p. 133–51.

34. Schjoerring JK, Husted S, Mäck G, Mattsson M. The regulation of ammonium translocation in plants. J Exp Bot. 2002;53:883–90.

35. Hirel B, Martin A, Tercé-Laforgue T, Gonzalez-Moro M-B, Estavillo J-M. Physiology of maize I: A comprehensive and integrated view of nitrogen metabolism in a C4 plant. Physiol Plantarum. 2005;124:167–77.

36. Riedel J, Tischner R, Mäck G. The chloroplastic glutamine synthetase (GS-2) of tobacco is phosphorylated and associated with 14-3-3 proteins inside the chloroplast. Planta. 2001;213:396–401.

37. Rabilloud T. Two-dimensional gel electrophoresis in proteomics: old, old fashioned, but it still climbs up the mountains. Proteomics. 2002;2:3–10.

38. Steinberg TH, Agnew BJ, Gee KR, Leung W-Y, Goodman T, Schulenberg B, et al. Global quantitative phosphoprotein analysis using Multiplexed Proteomics technology. Proteomics. 2003;3:1128–44.

39. Ishiyama K, Inoue E, Tabuchi M, Yamaya T, Takahashi H. Biochemical background and compartmentalized functions of cytosolic glutamine synthetase for active ammonium assimilation in rice roots. Plant Cell Physiol. 2004;45:1640–7.

40. Zamboni A, Astolfi S, Zuchi S, Pii Y, Guardini K, Tononi P, Varanini Z: Nitrate induction triggers different transcriptional changes in a high and a low nitrogen use efficiency maize inbred line. J Integr Plant Biol. 2014, doi: 10.1111/jipb.12214.

41. Khamis S, Lamaze T, Lemoine Y, Foyer C. Adaptation of the photosynthetic apparatus in maize leaves as a result of nitrogen limitation. Plant Physiol. 1990;94:1436–43.

42. Gaufichon L, Reisdorf-Cren M, Rothstein SJ, Chardon F, Suzuki A. Biological functions of asparagine synthetase in plants. Plant Sci. 2010;179:141–53.

43. Chevalier C, Bourgeois E, Just D, Raymond P. Metabolic regulation of asparagine synthetase gene expression in maize (Zea mays L.) root tips. Plant J. 1996;9:1–11.

44. Broyart C, Fontaine J-X, Molinié R, Cailleu D, Tercé-Laforgue T, Dubois F, et al. Metabolic profiling of maize mutants deficient for two glutamine synthetase isoenzymes using ^{1}H-NMR-based metabolomics. Phytochem Analysis. 2010;21:102–9.

45. Miyashita Y, Dolferus R, Ismond KP, Good AG. Alanine aminotransferase catalyses the breakdown of alanine after hypoxia in Arabidopsis thaliana. Plant J. 2007;49:1108–21.

46. Lea PJ, Sodek L, Parry MAJ, Shewry PR, Halford NG. Asparagine in plants. Ann Appl Biol. 2007;150:1–26.

47. Dutilleul C, Lelarge C, Prioul J-L, De Paepe R, Foyer CH, Noctor G. Mitochondria-driven changes in leaf NAD status exert a crucial influence on the control of nitrate assimilation and the integration of Carbon and Nitrogen metabolism. Plant Physiol. 2005;139:64–78.

48. Fernie AR, Carrari F, Sweetlove LJ. Respiratory metabolism: glycolysis, the TCA cycle and mitochondrial electron transport. Curr Opin Plant Biol. 2004;7:254–61.

49. Escobar MA, Geisler DA, Rasmusson AG. Reorganization of the alternative pathways of the Arabidopsis respiratory chain by nitrogen supply: opposing effects of ammonium and nitrate. Plant J. 2006;45:775–88.

50. Cataldo DA, Maroon M, Schrader LE, Youngs VL. Rapid colorimetric determination of nitrate in plant tissue by nitration of salicylic acid. Commun Soil Sci Plant Anal. 1975;6:71–80.

51. Nelson N. A photometric adaptation of the Somogy method for the determination of glucose. J Biol Chem. 1944;153:375–80.

52. Laemmli UK. Cleavage of structural proteins during the assembly of the head of bacteriophage T4. Nature. 1970;227:680–5.

53. Bernardo L, Prinsi B, Negri AS, Cattivelli L, Espen L, Valè G. Proteomic characterization of the Rph15 barley resistance gene-mediated defence responses to leaf rust. BMC Genomics. 2012;13:642. doi:10.1186/1471-2164-13-642.

54. Beyzaei Z, Sherbakov RA, Averina NG. Response of nitrate reductase to exogenous application of 5-aminolevulinic acid in barley plants. J Plant Growth Regul. 2014;33:745–50.

55. Podgórska A, Gieczewska K, Łukawska-Kuźma K, Rasmusson AG, Gardeström P, Szal B. Long-term ammonium nutrition of Arabidopsis increases the extrachloroplastic NAD(P)H/NAD(P)$^+$ ratio and mitochondrial reactive oxygen species level in leaves but does not impair photosynthetic capacity. Plant Cell Environ. 2013;36:2034–45.

56. Neuhoff V, Arold N, Taube D, Ehrhardt W. Improved staining of proteins in polyacrylamide gels including isoelectric focusing gels with clear background at nanogram sensitivity using Coomassie Brilliant Blue G-250 and R-250. Electrophoresis. 1988;9:255–62.

57. National Center for Biotechnology Information [http://www.ncbi.nlm.nih.gov/]

58. ExPASy Proteomics Server [http://web.expasy.org/compute_pi/]

59. Redinbaugh MG, Campbell WH. Nitrate regulation of the oxidative pentose phosphate pathway in maize (Zea mays L.) root plastids: induction of 6-phosphogluconate dehydrogenase activity, protein and transcript levels. Plant Sci. 1998;134:129–40.

60. Lea PJ, Blackwell RD, Chen FL, Hecht U. Enzymes of ammonia assimilation. In: Lea PJ, editor. Methods in Plant Biochemistry. San Diego: Academic Press; 1990. p. 257–76.

61. Cullimore JV, Sims AP. An association between photorespiration and protein catabolism: studies with Chlamydomonas. Planta. 1980;150:392–6.

62. Armstrong M, Jonscher K, Reisdorph NA. Analysis of 25 underivatized amino acids in human plasma using ion-pairing reversed-phase liquid chromatography/time-of-flight mass spectrometry. Rapid Commun Mass Sp. 2007;21:2717–26.

6

New insights into the evolutionary history of plant sorbitol dehydrogenase

Yong Jia[1], Darren CJ Wong[1,2], Crystal Sweetman[1,3], John B Bruning[4] and Christopher M Ford[1*]

Abstract

Background: Sorbitol dehydrogenase (SDH, EC 1.1.1.14) is the key enzyme involved in sorbitol metabolism in higher plants. SDH genes in some *Rosaceae* species could be divided into two groups. L-idonate-5-dehydrogenase (LIDH, EC 1.1.1.264) is involved in tartaric acid (TA) synthesis in *Vitis vinifera* and is highly homologous to plant SDHs. Despite efforts to understand the biological functions of plant SDH, the evolutionary history of plant SDH genes and their phylogenetic relationship with the *V. vinifera* LIDH gene have not been characterized.

Results: A total of 92 SDH genes were identified from 42 angiosperm species. SDH genes have been highly duplicated within the *Rosaceae* family while monocot, *Brassicaceae* and most *Asterid* species exhibit singleton SDH genes. Core Eudicot SDHs have diverged into two phylogenetic lineages, now classified as SDH Class I and SDH Class II. *V. vinifera* LIDH was identified as a Class II SDH. Tandem duplication played a dominant role in the expansion of plant SDH family and Class II SDH genes were positioned in tandem with Class I SDH genes in several plant genomes. Protein modelling analyses of *V. vinifera* SDHs revealed 19 putative active site residues, three of which exhibited amino acid substitutions between Class I and Class II SDHs and were influenced by positive natural selection in the SDH Class II lineage. Gene expression analyses also demonstrated a clear transcriptional divergence between Class I and Class II SDH genes in *V. vinifera* and *Citrus sinensis* (orange).

Conclusions: Phylogenetic, natural selection and synteny analyses provided strong support for the emergence of SDH Class II by positive natural selection after tandem duplication in the common ancestor of core Eudicot plants. The substitutions of three putative active site residues might be responsible for the unique enzyme activity of *V. vinifera* LIDH, which belongs to SDH Class II and represents a novel function of SDH in *V. vinifera* that may be true also of other Class II SDHs. Gene expression analyses also supported the divergence of SDH Class II at the expression level. This study will facilitate future research into understanding the biological functions of plant SDHs.

Keywords: Sorbitol dehydrogenase, L-idonate-5-dehydrogenase, Gene duplication, Functional divergence, Tartaric acid, Ascorbic acid, Grapevine

Background

Sorbitol dehydrogenase (SDH, EC 1.1.1.14) is commonly found in all kinds of life forms, including animals [1-4], yeasts [5], bacteria [6] and plants [7-13]. It represents the early divergence within the NAD (H)-dependent medium-chain dehydrogenase/reductase (MDR) super-family (with a typical ~350-residue subunit), sharing a distant homology with alcohol dehydrogenase (ADH, EC 1.1.1.1) [14-17]. SDH catalyses the reversible oxidation of a range of related sugar alcohols into their corresponding

ketoses [7,13,18-21], preferring polyols with a d-*cis*-2,4-dihydroxyl (2S,4R) configuration and a C1 hydroxyl group next to the oxidation site at C2, such as sorbitol, xylitol and ribitol (Additional file 1). It exhibits the highest activity on sorbitol while also being able to oxidize the other polyols at lower reaction rates [6,13,18,20]. The process of sorbitol oxidation by human SDH requires a catalytic zinc atom which is coordinated by the side chains of three amino acids (44C, 69H, 70E, numbering in human SDH) and one water molecular. NAD^+ binds to the protein first, followed by sorbitol. The backbone of sorbitol stacks against the nicotinamide ring while the C1 and C2 oxygen atoms are coordinated to the zinc. The water molecule co-ordinating the zinc atom acts a general base and abstracts

* Correspondence: christopher.ford@adelaide.edu.au
[1]School of Agriculture, Food and Wine, University of Adelaide, Adelaide 5005, Australia
Full list of author information is available at the end of the article

the proton of the C2 hydroxyl, which creates an electron flow to NAD^+, leading to the oxidation of sorbitol at C2 and the final production of NADH [22].

Plant SDH is the key enzyme in the sorbitol metabolism pathway [7,13,20,21,23] and has been associated with resistance to abiotic stresses such as drought and salinity. SDH activity regulates the levels of polyols [13,23], which act as important osmolytes during drought stress and recovery processes [24]. In Rosaceae species sorbitol occurs as the major photosynthate and phloem transported carbohydrate [25]. In these plants, which include apple [26-31], pear [32,33] and loquat [34,35], SDH plays a crucial role in the oxidation of sorbitol and its translocation to sink tissues such as developing fruits and young leaves. Gene transcript level and enzyme activity remain high during fruit development and maturation, dropping gradually in later stages, and contributing to the sugar accumulation in the ripening fruits [27-30,34-36]. The role of sink strength regulation for SDH is of particular research interest given the economic importance of these fruit species. Additionally, SDH has been shown to be involved in the sugar metabolism process during seed germination of some herbaceous plants including soybean [37] and maize [8,38].

Despite efforts to understand the physiological role of SDH in plants, little attention has been paid toward the evolutionary history of the plant SDH gene family. The distribution of the SDH genes in higher plants appears to be species-dependant. In particular, 9 paralogous SDH genes have been reported in apple [27] and 5 in Japanese pear [39]. In contrast, other plant genomes such as A. thaliana [23], tomato [11] and strawberry [12] contain only one SDH gene. Recent studies have indicated that there are two groups of SDH present in some Rosaceae plants. Park et al. [10] isolated four SDH isoforms (MdSDH1-4) from Fuji apple and found that MdSDH2-4 could be clearly distinguished from MdSDH1 based on the deduced amino acid sequence, showing 69–71% identity with MdSDH1 and 90–92% identity with each other. In addition, MdSDH2-4 were expressed only in sink tissues such as young leaves, stems, roots and maturing fruits while MdSDH1 was highly expressed in both sink and source organs [10]. Nosarzewski et al. [27] identified nine SDHs (SDH1-9) from the Borkh apple genome and showed that all isoforms except SDH1 (71–73% identity with SDH2-9) were highly homologous with an identity of 91–97%. Similar observations have been made with the SDH isoforms (PpySDH1-5) identified in pear whereby PpySDH5 differed from PpySDH1-4 at both the primary structure level and the gene transcriptional level [39]. Preliminary phylogenetic analyses have classified these homologous SDHs into two groups based on primary protein structures [10,29,33,40]. However, these studies focused on only one or just a few related Rosaceae species. No comprehensive phylogenetic analysis has been performed on SDH across a broad range of angiosperm species.

Gene duplication is widespread in plant genomes. Functional divergence after gene duplication is the major mechanism by which genes with novel function evolve; this phenomenon plays a key role in the evolution of phenotypic diversity [41-44]. The current understanding of gene evolution via duplication suggests that duplicated genes could arise through different mechanisms including unequal crossing over (resulting in tandem duplication), retrotransposition, segmental duplication and chromosomal (or whole genome) duplication [42,45]. Most duplicated genes are lost due to the accumulation of mutations that render them non-functional (pseudogenization) [42]. However, they can be retained under certain circumstances whereby the acquisition of beneficial mutations leads to novel function (neofunctionalization), which requires positive natural selection, or through adoption of part of the functions of the ancestral gene (sub-functionalization), which could occur by expression divergence or functional specialization of protein [41,42,46,47]. The latter usually involves a shift in the enzyme substrate specificity.

Protein structural analyses have shown that the LIDH of V. vinifera, which catalyses the inter-conversion of L-idonate and 5-keto-D-gluconate (5KGA) in the tartaric acid (TA) synthesis pathway [48], is highly homologous to plant SDHs, sharing ~77% amino acid sequence similarity with SDH from tomato (Gene ID: 778312) and A. thaliana (Gene ID: AT5G51970) [48]. The 366 amino acid LIDH (UniProt ID: Q1PSI9) contains an N-terminal GroES-like fold and a C-terminal Rossmann fold [48], characteristics of the ADH family [49], which has a distant homology to SDH [14-17]. However, unlike other plant SDHs, LIDH displays principal activity against L-idonate and has a low reaction rate with sorbitol [48]. The unique substrate specificity of LIDH was suggested to be due to small changes in amino acid sequence encoded by paralogous genes [48].

In this study, a comprehensive phylogenetic analysis of angiosperm SDHs was conducted using currently available genomic data. A computational approach was employed to characterise the natural selection pressure on plant SDH. The protein structures of the SDH homologues in V. vinifera were modelled based on human SDH (PDB:1PL8) to identify the putative active site residues of plant SDHs. Transcription and co-expression data of SDH genes were also extracted from recent publicly available microarray and co-expression databases and analysed. New insights into the evolution history of the plant SDH family and the evolutionary origin of V. vinifera LIDH will be discussed.

Results and discussion

Identification of sorbitol dehydrogenase (SDH) homologous genes in higher plants

A database homology search identified 92 SDH homologous genes from 42 species (Figure 1; See Additional file 2: Table S1 for identified gene IDs and Additional file 3 for gene sequences in corresponding species). At least one putative SDH gene was present in each plant genome studied, consistent with previous studies [17] that suggested the ubiquity of SDH and its functional importance across all life forms. However, the distribution of SDH homologous genes varied dramatically across species. Monocot species (n = 8) uniformly presented a single SDH gene, and this same observation was made with *Brassicaceae* plants (n = 7) from the Eudicot group. It was recently reported that there are

2 SDH genes in both rice (monocot) and *A.thaliana* (*Brassicaceae*) [50], however, in both cases these SDH genes were found to be alternative transcripts of a single gene. All except one species from the *Asterid* clade and the *Leguminosae* family had one SDH gene, the exceptions being *Solanum tuberosum* (potato) and *Glycine max* (soybean), respectively, which both had two copies. By contrast, numerous copies of SDH genes were found in *Rosaceae* species, which employ sorbitol as the major transported carbohydrate [25]. *Malus × domestica* (apple) contained 16 putative SDH genes, the highest number among all species investigated. A previous study [50] identified 17 SDH genes in the apple genome, however, the extra putative SDH (MDP0000506359) was only a partial gene (177 residues) and was excluded from the present study. In addition to apple, other *Rosaceae*

			Species Name	Class I	Class II	Total
Monocots			*Brachypodium distachyon*	0	0	1
			Oryza sativa	0	0	1
			Panicum virgatum	0	0	1
			Setaria italica	0	0	1
			Zea mays	0	0	1
			Sorghum bicolor	0	0	1
			Aegilops tauschii	0	0	1
			Hordeum vulgare	0	0	1
Lower Eudicots			*Aquilegia coerulea*	0	0	7
	Asterids		*Solanum lycopersicum*	1	0	1
			Solanum tuberosum	1	1	2
			Capsicum annuum cv. CM334	1	0	1
		Vitaceae	**Vitis vinifera**	1	2	3
			Eucalyptus grandis	1	1	2
			Citrus sinensis	2	1	3
			Theobroma cacao	1	1	2
			Cucumis sativus	2	0	2
			Gossypium raimondii	3	0	3
			Carica papaya	1	0	1
		Brassicaceae	*Thellungiella halophila*	1	0	1
			Brassica rapa Chiifu-401	1	0	1
			Brassica oleracea	1	0	1
Core Eudicots			*Capsella rubella*	1	0	1
			Arabidopsis lyrata	1	0	1
			Arabidopsis thaliana	1	0	1
			Eutrema salsugineum	1	0	1
		Geraniaceae	*Pelargonium x hortorum*	1	1	2
	Rosids		*Fragaria vesca*	1	0	1
			Malus domestica	15	1	16
		Rosaceae	*Eriobotrya japonica*	0	1	1
			Prunus persica	1	3	4
			Prunus mume	1	2	3
			Pyrus bretschneideri	4	1	5
			Glycine max	2	0	2
		Leguminosae	*Cajanus cajan*	1	0	1
			Phaseolus vulgaris	1	0	1
			Medicago truncatula	1	0	1
			Populus trichocarpa	1	1	2
			Linum usitatissimum	2	2	4
		Euphorbiaceae	*Ricinus communis*	1	0	1
			Jatropha curcas	2	1	3
			Manihot esculenta	1	1	2

Figure 1 Distribution of SDH homologous genes in higher plants. Closely related species were specified accordingly. The gene abundance heat map was based on the total copy number of SDH genes in each species. SDHs of *P. bretschneideri* [39] and *E. japonica* (loquat) [35] were obtained from literature; additional SDHs may be identified in these two species when complete genome information becomes available. The classification of SDH Class I and SDH Class II was based on the phylogenetic analysis carried out in the present study.

species such as *Prunus persica* (peach), *Prunus mume* (Chinese plum), *Eriobotrya japonica* (loquat) and *Pyrus bretschneideri* (pear) had 4, 3, 1 and 5 putative SDH genes respectively. It should be noted that the information of SDH numbers in loquat [35] and pear [39] was retrieved from earlier reports, and that more SDH genes may be found when complete genome data for these species become available. Although *Fragaria vesca* (strawberry) belongs to the *Rosaceae* family, only one SDH gene was present in this species. Unlike other *Rosaceae* fruit species, *F. vesca* utilizes sucrose instead of sorbitol as the main translocated carbohydrate [51]. According to a recent development in the evolution by duplication theory, a proper gene dosage should be kept to maintain a stoichiometric balance in macromolecular complexes such as functional proteins, thereby ensuring the normal functioning of a particular biological process [41,52]. Transportation and assimilation of sorbitol is a *Rosaceae*-specific metabolism. The retention of highly duplicated SDH genes in *Rosaceae* species suggests that a higher dosage of SDH transcription or enzyme activity is needed to facilitate sorbitol metabolism in these species.

Three putative SDH genes were identified in the *V. vinifera* genome. One (GSVIVT01010646001) corresponded to the previously characterized LIDH (Uniprot No. Q1PSI9) [48] while the other two shared 99% (GSVIVT01010644001) and 77% (GSVIVT01010642001) amino acid sequence identity with *V. vinifera* LIDH (Additional file 2: Table S4). Other important crops such as *C. sinensis* (orange), *Theobroma cacao* (cocoa), and *Pelargonium hortorum* (a geranium species) had 3, 2 and 2 SDH genes respectively. *P. hortorum* and *S. tuberosum* are of particular interest in this study because they have also been shown to accumulate significant levels of TA, like *V. vinifera* [53,54]. Another species that should be noted is *Aquilegia coerulea* (a flower native to the Rocky Mountains), which belongs to the Eudicot family but has been recognized as an evolutionary intermediate [55] between monocot and core Eudicot plants, and contained 7 SDH paralogues.

Phylogenetic analysis of plant sorbitol dehydrogenase families

To determine the evolutionary history of plant SDH family and the phylogenetic relationship between LIDH and SDH, a phylogeny of the SDH family was reconstructed. Consistent results were obtained using both Neighbour Joining (Figure 2A; Additional file 4) and Maximum Likelihood (Figure 2B) methods. As can be seen in the Maximum Likelihood tree (Figure 2B), the target proteins divided at the basal nodes into three major clusters, corresponding to the three life kingdoms: fungi, animal and plant (Bootstrap supports at 0.98, 1 and 1 respectively). The overall topology of the plant

SDH clade was in agreement with the Phytozome species tree (http://www.phytozome.net/), indicating that the phylogeny results were reliable. Specifically, monocot plants (n = 8) formed a single clade with strong support (0.91), corresponding to the early split between monocot and dicot lineages. *A. coerulea* SDHs separated into a single group (0.91) which positioned itself between monocot and core Eudicot plants. The *Aquilegia* genus belongs to the Eudicot order *Ranunculales* which has been established as a sister clade to the rest of the core Eudicot [56-58] and agrees with the present phylogenetic analysis.

The core Eudicot SDHs split into two distinct lineages in the Maximum Likelihood tree (Figure 2B). The first lineage (classified as Class I) covered all core Eudicot species included in this study while the second (Class II) had a narrower coverage and was less expanded compared to SDH Class I. The divergence of core Eudicot SDHs into two lineages was in agreement with previous reports that SDHs from some *Rosaceae* species could be separated into two groups [10,29,33]. All *Rosaceae* plants (n = 5) investigated in this study except *F. vesca* (strawberry) had multiple copies of SDH genes that covered both SDH Class I and SDH Class II. However, within these species, the distribution of SDHs among the two SDH classes varied greatly. In particular, 15 out of the 16 SDHs from *M. domestica* and 4 out of the 5 SDHs from *P. bretschneideri* fell into SDH Class I while 3 out of the 4 SDHs from *P. persica* and 2 out of the 3 SDHs from *P. mume* belonged to SDH Class II. Other species retaining two classes of SDHs included *S. tuberosum*, *V. vinifera*, *Eucalyptus grandis*, *C. sinensis*, *T. cacao*, *P. hortorum*, *Populus trichocarpa*, *Linum usitatissimum*, *Jatropha curcas* and *Manihot esculenta*, from different orders or families. In contrast, *Brassicaceae* plants (n = 7), *Leguminosae* plants (n = 4) and *Asterid* plants (n = 2) except *S. tuberosum* contained either a single SDH or two SDHs that could only be classified into SDH Class I. Within both SDH Class I and Class II clades, *Rosaceae* SDHs (except *F. vesca*) formed separate phylogeny groups (Figure 2B), implying divergent molecular characteristics for SDHs from this family. Most recent phylogenetic analyses [59,60] have placed *Vitaceae* as a sister clade to the *Rosid* plants in the core Eudicot group. The presence of two classes of SDHs in both *V. vinifera* and *S. tuberosum* (*Asterids*) indicated that the divergence between SDH Class I and Class II occurred before the species radiation of the core Eudicot plants. Moreover, although 7 SDH genes were retained in the genome of the evolutionarily intermediate species *A. coerulea*, none of them could be classified into SDH Class I or SDH Class II. Taken together, our results suggested that SDH Class I and Class II might have diverged during the common ancestor of core Eudicot plants

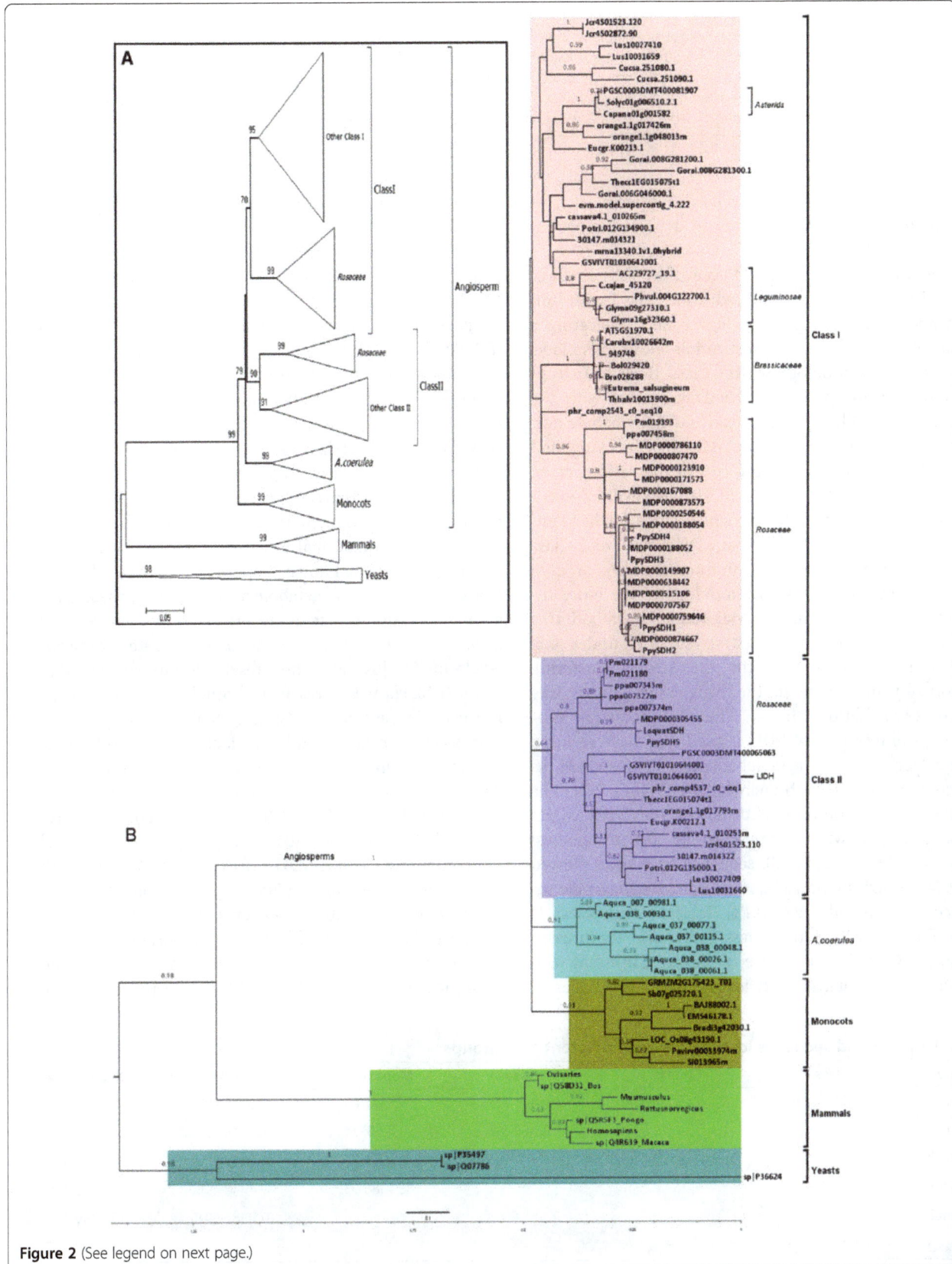

Figure 2 (See legend on next page.)

(See figure on previous page.)
Figure 2 Phylogenetic tree showing the evolutionary history of the angiosperm SDH family. **A**: A simplified schematic phylogeny of the SDH family inferred by MEGA 6.0 [97] software using the Neighbour Joining method. Values (as percentage, cutoff value 50) of Internal branch test (1000 replicates) supports are indicated above the corresponding branches. **B**: The Maximum Likelihood phylogeny of the SDH family developed by MEGA 6.0 [97] software using the selected best-fitting substitution model JTT + G [99]. 1000 times Bootstraping supports (cut off at 0.5) are displayed above corresponding branch. Closely related species are annotated accordingly. The *V. vinifera* LIDH (GSVIVT01010646001) is also marked.

but after the branching of the basal Eudicots such as *Ranunculales.* This corresponds to a period of about 125Mya ~ 115Mya [55,58].

In the Maximum Likelihood tree, the Class II clade was well-supported and separated from Class I with longer branch length in general (Figure 2B), suggesting a higher level of amino acid substitution within this clade. In addition, the topology of the Class II clade (except the *Rosaceae* group) was in good agreement with the species tree at Phytozome (http://www.phytozome.net/search.php), with *S. tuberosum* (*Asterids*) diverging first followed by *V. vinifera* and the rest of the rosid species. This indicates that the Class II SDHs have evolved vertically within respective species, which lends further support to the suggestion above that SDH Class I and Class II have existed during the common ancestry of core Eudicot plants. The backbone topology of the more inclusive Class I clade in the Maximum Likelihood tree was weakly supported (Bootstrap support under 0.5; Figure 2B), in contrast with the strong clustering support for this clade in the Neighbour Joining tree (Figure 2A; Additional file 4). The weak bootstrap support for the topology of SDH Class I may have resulted from a lack of amino acid substitution in this clade, as reflected by the short branch length (Figure 2B). The calculation of evolutionary distances for plant SDHs revealed a pair-wise distance under 0.3 in general (Additional file 2: Table S2), sequence alignment showed that Class I SDHs tend to be more conserved (average sequence pair-wise identity 83.4%; Table 1) than Class II (79%; Table 1), which means less amino acid substitution within the Class I clade. These results are consistent with the strong clustering support for the major sub-clades of

the Class I branch in the Neighbour Joining tree (Figure 2A; Additional file 4).

In contrast to the ubiquity of Class I SDHs, the absence of Class II SDHs in some species may be due to gene loss after duplication, a common mechanism in gene evolution via duplication [42,61]. This also indicated that SDH Class II members may not be essential for the normal growth of plants, suggesting a divergent function for this class of SDH genes. Interestingly, the previously characterized *V. vinifera* LIDH (GSVIVT01010646001) [48] was grouped into SDH Class II, providing direct support that in at least one case SDH Class II may have acquired a novel function, in this instance its involvement in the synthesis of TA. While the identity of additional functions for Class II SDHs in other species is unknown, support for a role of some Class II SDHs in TA metabolism may be proposed. Only a few plant families, including *Vitaceae*, *Geraniaceae* and *Leguminosae* have been shown to accumulate significant levels of TA [54] and the present results showed that Class II SDHs were present in both *Vitaceae* and *Geraniaceae*. The absence of Class II SDHs in *Leguminosae* plants could be explained by the fact that the synthesis of TA in *Leguminosae* proceeds via a different pathway, which bypasses the interconversion of L-idonate and 5KGA (catalysed by LIDH) [62]. Recent studies have revealed that potato [53], citrus fruits [63] and pear [64,65] (all containing Class II SDHs) also produce TA, although to a lesser degree than *V. vinifera*. This is consistent with the potential correlation between Class II SDHs and TA synthesis. However, it has also been reported that TA is absent or found only in trace amount in apple [66], and no information is available about the occurrence of TA in

Table 1 Amino acid sequence identity between different SDH groups

Identity	Class I	Class II	A. coerulea	Monocot	Mammal	Yeast
Class I	83.4 (71-99.7)	75.2 (67-83)	78.5 (71-86)	77.5 (71-83)	48.0 (44-50)	40.9 (38-43)
Class II		79.0 (71-99)	73.2 (68-80)	71.0 (67-74)	46.4 (43-49)	39.3 (37-42)
A. coerulea			86.7 (83-99.7)	75.7 (72-79)	48.0 (47-50)	41.4 (40-43)
Monocot				88.4 (86-93)	47.4 (46-49)	41.5 (40-45)
Mammal					87.8 (82-99.8)	42.3 (39-44)
Yeast						65.5 (48-99.7)

SDH sequences were divided into six groups (Class I, Class II, *A. coerulea*, Monocot, Mammal and Yeast SDHs) according to the phylogenetic analysis carried out in the present study (Figure 2). The amino acid sequence identity (as percentage) was obtained using all-vs-all BLAST tool. The average pair-wise identity between each group is presented, followed by the identity range (in bracket).

peach even though three copies of Class II SDH genes were identified in this species (Figure 1). It is possible that Class II SDHs have evolved varied functions to meet the different environmental challenges faced by respective plants. In this context, it would also be valuable for future work to investigate the in-planta function of SDH and the occurrence of TA in the evolutionarily intermediate plant *A.coerulea*, for which 7 SDH paralogues were identified.

Sequence alignment and protein subdomain analysis

Sequence alignment and protein subdomain analyses were performed to investigate the molecular characteristics of plant SDHs. Results showed that plant SDHs shared an overall identity above 67% (Table 1), while having ca 48% and ca 41% identities with mammal and yeast SDHs respectively (Additional file 2: Table S4). Plant SDHs were clustered into four groups in the present phylogenetic analysis: monocot SDH, *A. coerulea* SDH, core Eudicot SDH Class I and SDH Class II. Protein BLAST results showed that Class I and Class II SDHs within the same species generally had an inter-class identity of around 70% and an intra-class identity above 90% (Additional file 2: Table S4). When compared with monocot and *A. coerulea* SDHs, Class I SDHs always demonstrated a significantly higher similarity than Class II SDHs (77.5% vs 71.0% and 78.5% vs 73.2% respectively; Table 1), suggesting that core Eudicot Class I SDHs have a closer distance to monocot and *A. coerulea* SDHs and that SDH Class II may have diverged from SDH Class I. In addition, Class I SDHs tend to be more homologous than Class II SDHs (83.4% vs 79.0%; Table 1). No significant difference between the two SDH classes was observed when compared to mammal or yeast SDHs (48.0% vs 46.4% and 40.9% vs 39.3% respectively; Table 1). Protein functional domain prediction identified two functional domains for plant SDHs: an N-terminal GroES-like fold and a C-terminal Rossmann fold (Figure 3; See Additional file 5 for the complete sequence alignment). Secondary structure analysis showed that these two domains tended to be highly conserved among all plant SDHs, and amino acid substitutions mainly occurred at boundary regions linking secondary structural elements such as alpha-helices and beta-sheets (Figure 3).

Gene duplication pattern characterization and synteny analysis

To characterise the expansion patterns of plant SDH gene family, nine species that were from different families and contained both classes of SDHs were selected for gene duplication and synteny analyses (*C. sinensis*, *E. grandis*, *P. mume*, *P. persica*, *Populus trichocarpa*, *M. domestica*, *S. tuberosum*, *T. cacao* and *V. vinifera*). As shown in Table 2 (See Additional file 6 for the original output data), tandem duplication contributed the most to the expansion of the core Eudicot SDH family, followed by WGD/Segmental duplication. Dispersed SDHs (MDP0000305455, MDP0000759646 and PGSC0003DMC400055323) and a single proximal SDH (MDP0000188054) were identified only in *M. domestica* and *S. tuberosum*. Based on phylogenetic classification in the present study, Class I and Class II SDH genes from *E. grandis*, *P. trichocarpa*, *T. cacao* and *V. vinifera* are located in a tandem manner in their corresponding chromosomes, which provides strong support that SDH Class I and SDH Class II are tandem duplications. A similar pattern was observed with *C. sinensis* whereby Cs9g16660.1 (SDH Class II) is separated by a single-gene insertion with the two Class I SDH genes (Cs9g16680.1, Cs9g16690.1; data not shown). This may be caused by gene insertion after tandem duplication. Class I and Class II SDH genes in the three *Rosaceae* species (*M. domestica*, *P. mume*, *P. persica*) and in *S. tuberosum* are separated either on the one chromosome or on separate chromosomes altogether, indicating a divergent evolutionary history for SDH genes in the *Rosaceae* family and in *S. tuberosum* compared to other plants. SDH genes on chromosome 1 (md1) and chromosome 7 (md7) in *M. domestica* were highly duplicated by tandem duplication (Table 2), in contrast to the other *Rosaceae* species (*P. mume*, *P. persica*). Notably, the Class I SDH gene from *S. tuberosum* (PGSC0003DMC400055323) and the Class II SDH gene from *M. domestica* (MDP0000305455) were identified as dispersed duplicates, which may underpin the divergent sorbitol metabolism profiles across these species.

To investigate the conservation of SDH genes across species, collinear SDH gene pairs were identified within and across species. SDH genes from the nine above-mentioned species were analysed. The single SDH gene (AT5G51970) from the model plant *A. thaliana* was also used as a reference for collinear block identification. As shown in Figure 4, all target plant genomes contained at least one SDH gene (corresponding to chromosome positions A, B, C, D, E, H, J, L, N, P and Q in Figure 4) with collinear SDH genes in all other nine species studied, indicating a conserved collinear SDH block. SDH genes at gene positions F, G, I, K and O, concerning only the *Rosaceae* species investigated, were collinear with SDH genes in only some of the species included in the present analysis. In particular, position F at chromosome 8 (pp8) of *P. persica* paired only with position I at chromosome 6 (Pm6) of *P. mume*. While position F was found collinear only with position I, position I had another collinear region at position O from *E. grandis*. Position G at chromosome 4 (pp4) of *P. persica* was

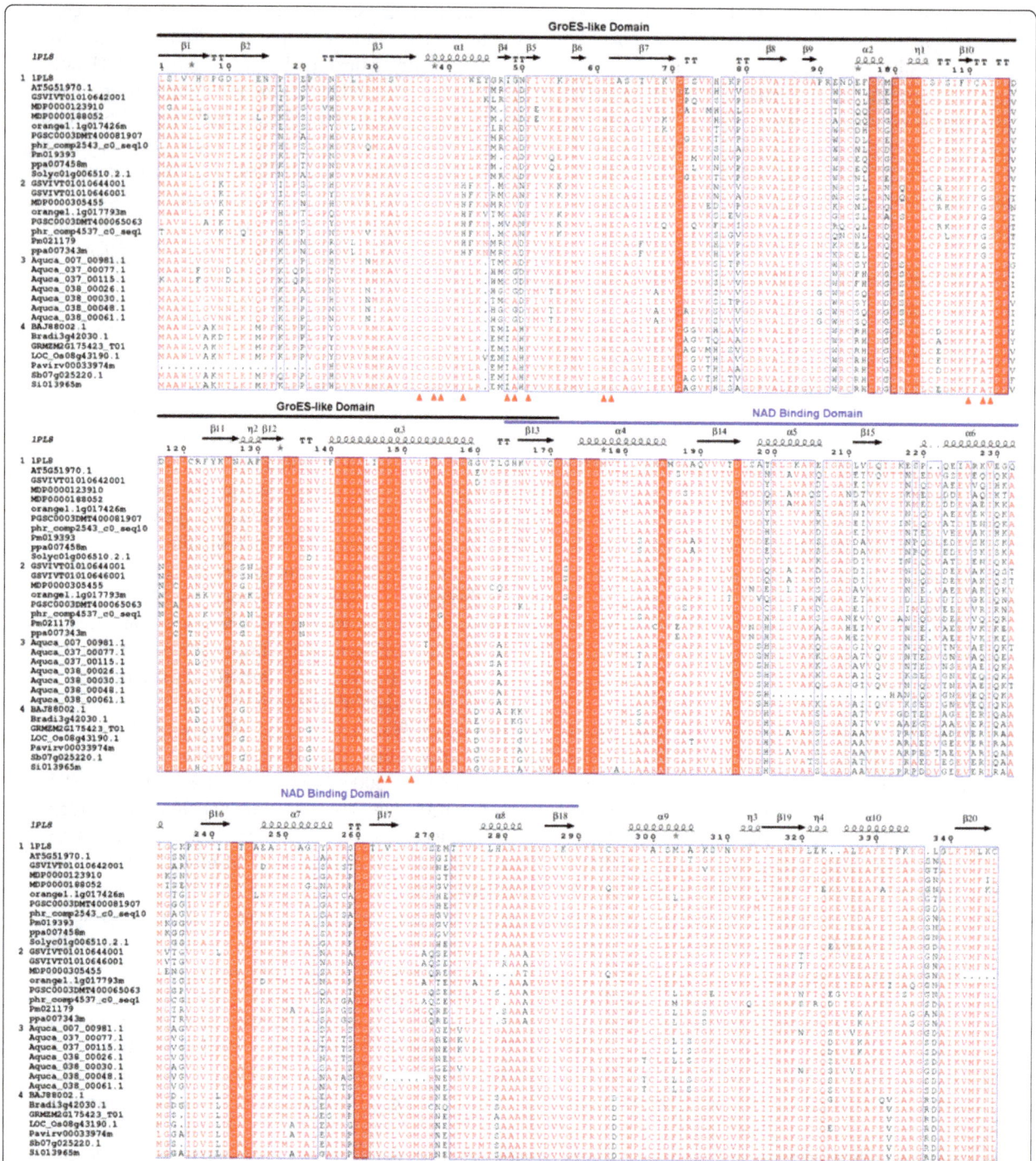

Figure 3 Multiple sequence alignment of plant SDH family. ESPript output was obtained with the sequence alignment of plant SDHs and human SDH. Secondary structures were inferred using human SDH (PDB: 1PL8) as a template, with springs representing helices and arrows representing beta-strands. Sequences were grouped into 1 (1PL8 and core Eudicot SDH Class I), 2 (core Eudicot SDH Class II), 3 (*A.coerlea* SDH) and 4 (monocot SDH). Amino acid site numbering above the alignment is according to LlDH (Q1PSI9) without the first 20 amino acids. Adjacent similarity amino acid sites were boxed in blue frame. Similarity calculations were based on the complete SDH alignments but only partial sequences for SDH Class I and SDH Class II were displayed. The active site residues identified in this study are marked with red triangles. Conserved domains are indicated above the alignment.

Table 2 Gene duplication patterns of plant SDH

Species	Chromosome ID	SDH gene ID	SDH class	Duplication pattern	Start position	End position
C. sinensis	cs9	Cs9g16680.1 (orange1.1g017426m)	I	Tandem	16143063	16147624
	cs9	Cs9g16690.1 (orange1.1g048013m)	I	Tandem	16150122	16154404
	cs9	Cs9g16660.1 (orange1.1g017793m)	II	WGD or Sgm	16135216	16138066
E. grandis	eg11	Eucgr. K00213.1	I	Tandem	2624187	2627945
	eg11	Eucgr.K00212.1	II	Tandem	2615486	2618589
M. domestica	md1	MDP0000786110	I	Tandem	25191824	25193641
	md1	MDP0000873573	I	Tandem	25182502	25183812
	md1	MDP0000707567	I	Tandem	25180931	25182241
	md1	MDP0000515106	I	Tandem	25177288	25178612
	md1	MDP0000250546	I	Tandem	25173127	25174375
	md1	MDP0000874667	I	Tandem	25157544	25158783
	md1	MDP0000638442	I	WGD or Sgm	25149134	25150444
	md1	MDP0000123910	I	WGD or Sgm	25087036	25088743
	md1	MDP0000305455	II	Dispersed	14150327	14159200
	md7	MDP0000188052	I	Tandem	23301490	23302735
	md7	MDP0000171573	I	WGD or Sgm	23281847	23283529
	md7	MDP0000188054	I	Proximal	23310942	23312187
	md7	MDP0000167088	I	Tandem	23405354	23406795
	md7	MDP0000807470	I	WGD or Sgm	23390960	23392683
	md14	MDP0000759646	I	Dispersed	24043122	24044360
P. mume	Pm5	Pm019393	I	WGD or Sgm	23673441	23675177
	Pm6	Pm021180	II	Tandem	7217228	7219256
	Pm6	Pm021179	II	Tandem	7217228	7225304
P. persica	pp2	ppa007458m|PACid:17644502	I	WGD or Sgm	24766424	24768515
	pp4	ppa007327m|PACid:17655491	II	WGD or Sgm	17729024	17731238
	pp8	ppa007343m|PACid:17644328	II	Tandem	15254677	15256888
	pp8	ppa007374m|PACid:17655656	II	Tandem	15249947	15251989
P .trichocarpa	pt12	POPTR_0012s13780	II	WGD or Sgm	13789342	13787442
	pt12	POPTR_0012s13790	I	WGD or Sgm	13790093	13792804
S. tuberosum	st01	PGSC0003DMC400055323	I	Dispersed	1594220	1598967
	st06	PGSC0003DMC400043871	II	WGD or Sgm	24156879	24158593
T. cacao	tc03	Tc03_g019280	I	WGD or Sgm	18300080	18303115
	tc03	Tc03_g019270	II	WGD or Sgm	18298897	18296706
V. vinifera	vv16	GSVIVT01010642001	I	WGD or Sgm	15653874	15651701
	vv16	GSVIVT01010646001	II	Tandem	15675560	15678887
	vv16	GSVIVT01010644001	II	Tandem	15666264	15664425

SDH gene duplication patterns were characterized by the *duplicate_gene_classifier* program in the MCScanX package. "WGD or Sgm" refers to Whole Genome Duplication or segmental duplication. "SDH Class" is defined according to the present phylogenetic analysis. Notably, MDP0000149907 from *M. domestica* could not be anchored in any chromosome and was therefore absent in this table.

only paired with positions A, E and K from *A. thaliana*, *P. trichocarpa* and *M. domestica* respectively. Some collinear SDH gene pairs, such as F-I, G-K and K-O, were restricted to *Rosaceae* species only, reflecting genetic features shared only by these plants. Notably, intra-species collinear SDH pairs were identified only within *M. domestica* but not in *P. mume*, *P. persica* and *S. tuberosum* although all of these species have SDH genes located on multiple chromosomes (Figure 4; See Additional file 2: Table S5 for identified collinear SDH gene pairs). This observation could be explained by the fact that the apple genome

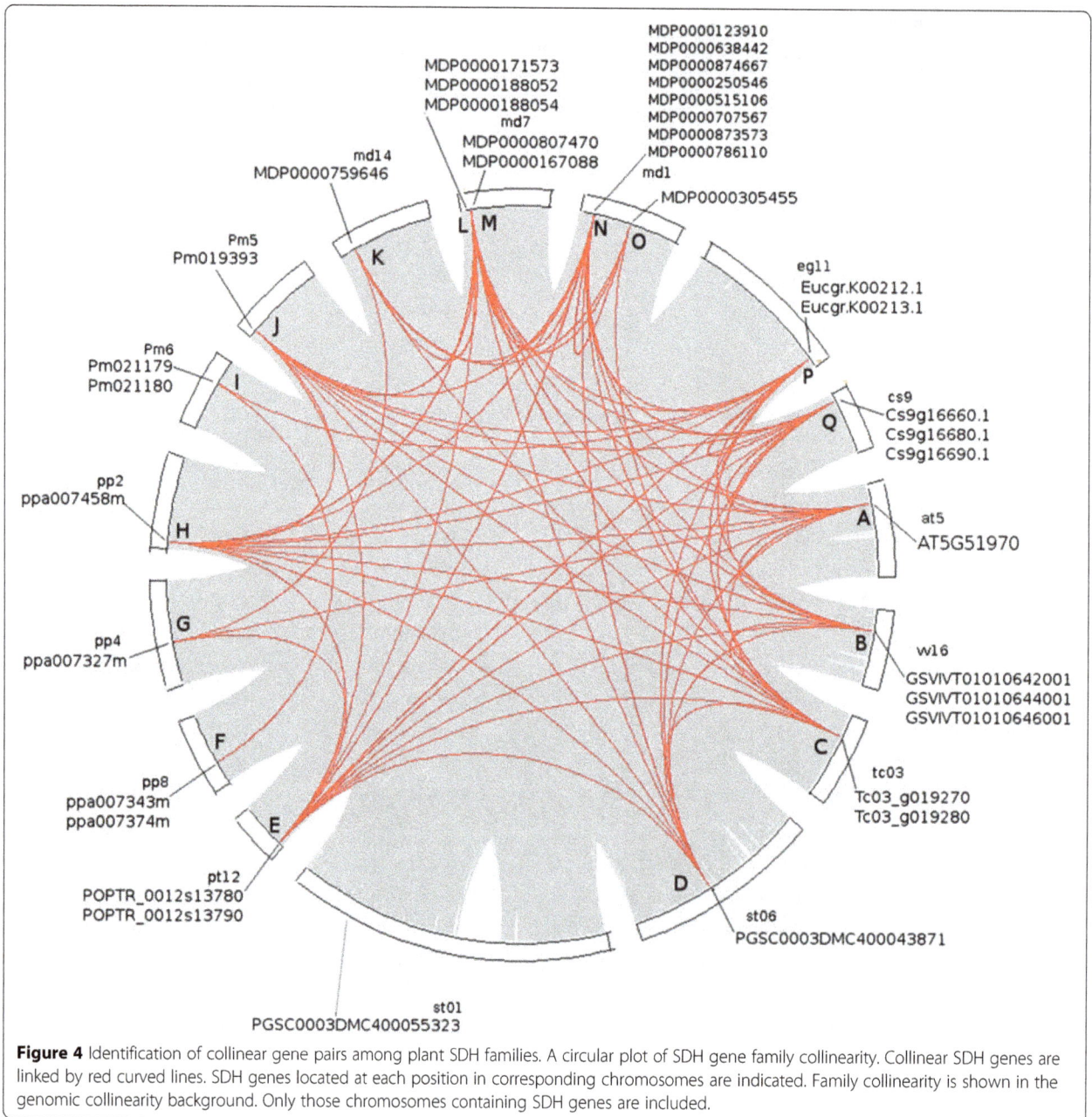

Figure 4 Identification of collinear gene pairs among plant SDH families. A circular plot of SDH gene family collinearity. Collinear SDH genes are linked by red curved lines. SDH genes located at each position in corresponding chromosomes are indicated. Family collinearity is shown in the genomic collinearity background. Only those chromosomes containing SDH genes are included.

underwent a recent (>50Mya) WGD, which doubled the chromosome number from nine to 17 in the *Pyreae* [50] while most other *Rosaceae* plants have a haploid chromosome number of 7, 8 or 9. *S. tuberosum* was unique among the species investigated in that it had a Class II SDH gene (PGSC0003DMC400043871) but no Class I SDH gene preserved in the collinear region (Figure 4). The Class I SDH gene (PGSC0003DMC400055323), which was identified as a dispersed duplication (Table 2), was the only SDH gene for which no collinear gene was identified in the present analysis. Since the Class II SDH homologue

(LIDH) in *V. vinifera* has been shown to be involved in TA synthesis [48], it would be of great interest to investigate the potential role of SDHs in *S. tuberosum*, which has also been shown to accumulate a significant amount of TA [53]. Noteworthy, *S. lycopersicum*, another species from the *Solanale* order, accumulates no TA [67] and contains only a single SDH, which belongs to Class I (Figure 2B).

Natural selection analysis

Assessment of synonymous and non-synonymous substitution ratios is important to understand molecular evolution at the amino acid level [68,69]. To examine the intensity of natural selection acting on the specific clade, the ratio (w) of non-synonymous substitution to synonymous substitution in the developed plant SDH phylogeny was investigated, whereby w<1, w=1 and w>1 indicated purifying selection, neutral evolution and positive selection respectively. Based on our phylogeny results, four branches ("monocot SDH", "*A. coerulea* SDH", "core Eudicot SDH Class I" and "core Eudicot SDH Class II") were specified for w assessments (w [mono], w [Aer], w [sdhC1] and w [sdhC2] respectively). Firstly, the branch-specific likelihood model [70] was applied to the SDH data. As can be seen in Table 3, Likelihood-ratio tests (LRT) showed that the two-ratio model and the four-ratio model fit the dataset significantly better ($2\Delta l = 12.6$ with p = 0.0004, df = 1 and $2\Delta l = 13.2$ with p = 0.0042, df = 3 respectively) than the one-ratio model. In contrast, the three-ratio model assumption lacked statistical support ($2\Delta l = 0.2$ with p = 0.9048, df = 2). Given that the two-ratio and four-ratio models assume unequal w ratios for the Class I and Class II branches while the three-ratio model specifies w(sdhC1)=w (sdhC2) (Table 3), the above calculation suggested that the w ratio for the core Eudicot SDH Class II was significantly different from that of Class I. Moreover, the

four-ratio model, which assumes unequal w ratios for the monocot, *A.coerulea* and Class I branches (Table 3), was not significantly better ($2\Delta l = 0.6$ with p = 0.7408, df = 2) than the two-ratio model (assuming uniform ratio for these branches; Table 3). This indicated that the w ratios for monocot, *A. coerulea* and core Eudicot Class I branches had no significant difference. Notably, all branch-specific models tested demonstrated a low w value for the monocot, *A. coerulea* and Class I branches (w[mono]=w[Aer]=w[sdhC1]=0.10415 with the two-ratio model and w[mono]=0.10428, w[Aer]=0.09731, w [sdhC1]=0.0001with the four-ratio model), suggesting that plant SDHs have been under strong purifying selection. This agrees well with the suggestion that functional proteins are usually under strong structural and functional constraints [71]. It should be noted that w[sdhC2] were infinite in both multi-ratio models (w[sdhC2]=859 and 999 respectively). This is because an extremely low level of synonymous substitution or no synonymous substitution was detected in the SDH Class II clade. On the other hand, the number of non-synonymous substitutions in the core SDH Class II clade was estimated to be 12.7 and 12.8 respectively for the two-ratio model and the four-ratio model. In contrast, only 0.4 non-synonymous substitution was detected for the SDH Class I clade with the two-ratio model

Table 3 Natural selection tests of plant SDH

Model	np	$l = \ln L$	Estimates of parameters	Positively selected sites
M0: one-ratio				
w(mono)=w (Aer)=w(sdhC1)=w(sdhC2)	1	-30147.4	w(mono)=w(Aer)=w(sdhC1)=w(sdhC2)=0.10492	Not Allowed (NA)
Branch-specific models				
w(mono)=w(Aer)=w(sdhC1)≠w(sdhC2) (two ratios)	2	-30141.1	w(mono)=w(Aer)=w(sdhC1)=0.10415, w(sdhC2)=859.33956	NA
w(mono)≠w(Aer)≠w(sdhC1)=w(sdhC2) (three ratios)	3	-30147.3	w(mono)=0.10510, w(Aer)=0.10821, w(sdhC1)=w(sdhC2)=0.06935	NA
w(mono)≠w(Aer)≠w(sdhC1)≠w(sdhC2) (four ratios)	4	-30140.8	w(mono)=0.10428, w(Aer)=0.09731, w(sdhC1)=0.0001, w(sdhC2)=999	NA
w(mono)=w(Aer)=w(sdhC1)≠w(sdhC2) (two ratios with w(sdhC2) fixed to 1)	1	-30141.4	w(mono)=w(Aer)=w(sdhC1)=0.10424 (w(sdhC2)=1)	NA
Site-specific models				
M1:Neutral (2 site classes)	2	-29650.0	p0=0.87775 (p1=1-p0=0.12225); w0=0.07628 (w1=1)	NA
M2:Selection (3 site classes)	3	-29650.0	p0=0.87775, p1=0.07499 (p2=1-p0-p1=0.04726); w0=0.07628 (w1=1), w2=1	None
Branch-site models (SDH Class II as foreground lineage)				
Model A Null (4 site classes)	3	-29643.2	p0=0.33951, p1=0.04783 (p2+p3=0.61266); w0=0.07544	NA
Model A (4 site classes)	4	-29640.9	p0=0.82864, p1=0.11666 (p2+p3=0.0547); w0=0.07544 (w1=1), w2=132.6226	Sites for foreground lineage: 42H,43F,112G, 113S,116T, 270Q (p > 0.99);

All calculations were implemented using codeml at PAML4.7. Different models were specified according to the software instruction. "np" refers to the number of parameters, "l = (ln L)" refers to the log value of the likelihood. The estimated parameters w and p refer to the K_a/K_s ratio and the percentage of the corresponding site classes respectively. In the one-ratio model M0 and the Branch-specific models, w(mono), w(Aer), w(sdhC1) and w(sdhC2) stand for the w ratios for the monocot, *A. coerulea*, SDH Class I and SDH Class II branches respectively. In the Site-specific models and the Branch-site models, w0, w1 and w2 represent the w ratios for the specific site classes in respective models (see the Methods section for more details). For the Branch-site models, the SDH Class II branch was specified as the foreground branch. Amino acid site numbering is according to LIDH (Uniprot No: Q1PSI9) without the first 20 amino acids.

(Additional file 7: branch-specific-two-ratio-output) and no non-synonymous substitution was detected with the four-ratio model (Additional file 7: branch-specific-four-ratio-output). These results provided clear evidence that positive selection had occurred in the lineage leading to core Eudicot SDH Class II. To test whether w[sdhC2] is significantly higher than 1, the log likelihood value (Table 3; Additional file 7: branch-specific-two-ratio-null-output) was calculated for the two-ratio model with w[sdhC2]=1 fixed. Results showed that this model was not significantly worse than the two-ratio model without the "w[sdhC2]=1" constraint ($2\Delta l = 0.6$ with p = 0.4386, df = 1), suggesting that w[sdhC2] was not significantly greater than 1 at the 5% significance level. This leads to the hypothesis that positive selection in SDH Class II might have only affected particular amino acid residues in the protein sequence, which is possible for a functional protein under strong structural and functional constraints [72]. To test this, Site-specific likelihood analysis was performed on the same data, which assumes variable selection pressures among amino acid sites but no variation among branches in the phylogeny. Results (Table 3: model M2) showed that the selection model (M2) fitted the dataset significantly better ($2\Delta l = 994.8$ with p = 0.0001, df = 2) than the one-ratio model but was not better ($2\Delta l = 0$ with p = 1, df = 1) than the neutral model (M1). These results indicated a significant variation of selection pressure among amino acid sites of plant SDH. However, the Selection model failed to detect any positively selected amino acid site at a significant level (Table 3; Additional file 7: site-specific-output), which suggested that no positively selected amino acid site could be identified across all branches. Therefore, we speculate that the positive selection might have only acted on a few amino acid sites in the core Eudicot SDH Class II clade.

In this context, a Branch-site model [73] that permits variable w ratios among both amino acid sites and branches was applied. Model A successfully identified the potential amino acid sites under positive selection in the SDH Class II branch (Table 3; Additional file 7: branch-site-modelA-output). Specifically, 42H, 43F, 112G, 113S, 116T and 270Q (numbering in LIDH (Q1PSI9) without the first 20 amino acids) were identified with Model A (Bayes Empirical Bayes analysis possibility >0.99; Additional file 7: branch-site-modelA-output). LRTs test showed that Model A fit the data significantly better ($2\Delta l = 18.2$ with p = 0.0001, df = 2) than the neutral model M1. The comparison ($2\Delta l = 4.6$ with p = 0.0320, df = 1) of Model A with its null hypothesis which assumes w2=1 (Additional file 7: branch-site-modelA-null-output) indicated that these amino acid sites had undergone positive selection in SDH Class II but not in the background branches. In addition, the Model A test

demonstrated that 82.90% (model A: p0 = 0.82864; Table 3) of the amino acids of SDH were under strong purifying selection (model A: w0=0.07544; Table 3) and 11.7% were under neutral selection (model A: p1=0.11666, w1=1; Table 3) in all branches. No positive selection could be detected in the background branches (Additional file 7: branch-site-modelA-output). Taken together, these calculations demonstrated that plant SDHs were under strong purifying selection pressure and were highly conserved across all the plant species, and more importantly, that positive natural selection had occurred in the SDH Class II clade, affecting specific amino acids, namely 42H, 43F, 112G, 113S, 116T and 270Q.

Ancestral sequence reconstruction and evolution rate analysis

To characterize the evolutionary rates for different groups of plant SDHs, ancestral amino acid sequences for the developed SDH phylogeny were reconstructed. Results (Additional file 8: ancestral-sequence-construction-output) showed that 9 potential amino acid substitutions (Y42H, L43F, A112G, T113S, V116T, Q228K, H270Q, N271S, R283A; numbering in LIDH (Q1PSI9) without the first 20 amino acids) occurred in the branch leading to SDH Class II from the common ancestor of core Eudicot SDH. This finding corresponded well with the natural selection analysis, whereby six out of the nine amino acid sites were identified to be under positive selection (42H, 43F, 112G, 113S, 116T and 270Q; Table 3). In contrast, no substitution was detected in the branch leading to core Eudicot SDH Class I (Additional file 8: ancestral-sequence-construction-output and interpreted-ancestral-sequences.fasta). Relative rate tests (RRT) [74] using monocot SDH as the out-group showed that core Eudicot SDH Class II evolved significantly faster than core Eudicot SDH Class I (Additional file 9: ClassI-vs-ClassII.txt), indicating a relaxed selection pressure on SDH Class II. In contrast, *A. coerulea* SDH and core Eudicot Class I SDH demonstrated no significant difference (Additional file 9: Aer-vs-ClassI.txt).

Protein structure modelling analysis

To deduce the reaction mechanism and identify the potential active sites of plant SDHs, protein structure models of *V. vinifera* Class I SDH (Vv_SDH, UniProt No: D7TMY3) and Class II SDH (Vv_LIDH, UniProt No: Q1PSI9) were created based on human SDH (PDB: 1PL8; 46 ~ 47% identity with Vv_SDH and Vv_LIDH). Ligands including zinc, NAD^+, D-sorbitol and L-idonate were docked into the models (Additional file 10). Our models contain one zinc binding site, located in the active site. Some published SDH crystal structures (eg. PDB: 1E3J) contain a second, structural zinc-binding site distant from the active site catalytic zinc atom; this is not however a universal feature of these enzymes. No function has been correlated with the second, structural zinc-binding site. The sequence of our homology models does not support a second, structural zinc-binding site, as the necessary side chains required for zinc coordination are absent. A ribbons diagram of the overall structure of the homology models can be seen in Figure 5A, with Vv_SDH and Vv_LIDH adopting a typical dehydrogenase fold with an NAD$^+$ binding site conforming to a Rossmann fold. The catalytic zinc ion in the active site was modelled coordinating to 36C, 61H and 62E (Figure 5C; numbering in LIDH (Q1PSI9) without the first 20 amino acids). All three of these residues together with 147E (corresponding to 155E in human SDH, mediating the water molecule coordinating the zinc atom [22]) are strictly conserved in plant SDHs (Figure 3). The 2′ and 3′ hydroxyls of the NAD$^+$ ribose in our model were poised to 195D (203D in human SDH), potentially forming hydrogen bonds (Additional file 10: Asp195-NAD.png). The preservation of 195D instead of 195A at this amino acid site has been shown to be the structural basis for the selection of NAD (H) over NADP (H) as co-enzyme [75]. This amino acid site is strictly conserved in all plant SDHs (Figure 3), implying that plant SDHs preferably utilize NAD (H). This suggestion is consistent with the lack of NADP-SDH activity for plant SDHs [7,10,11,13]. Previous characterizations of SDHs from Arabidopsis [13], tomato [11], apple [7,76] and pear [20]

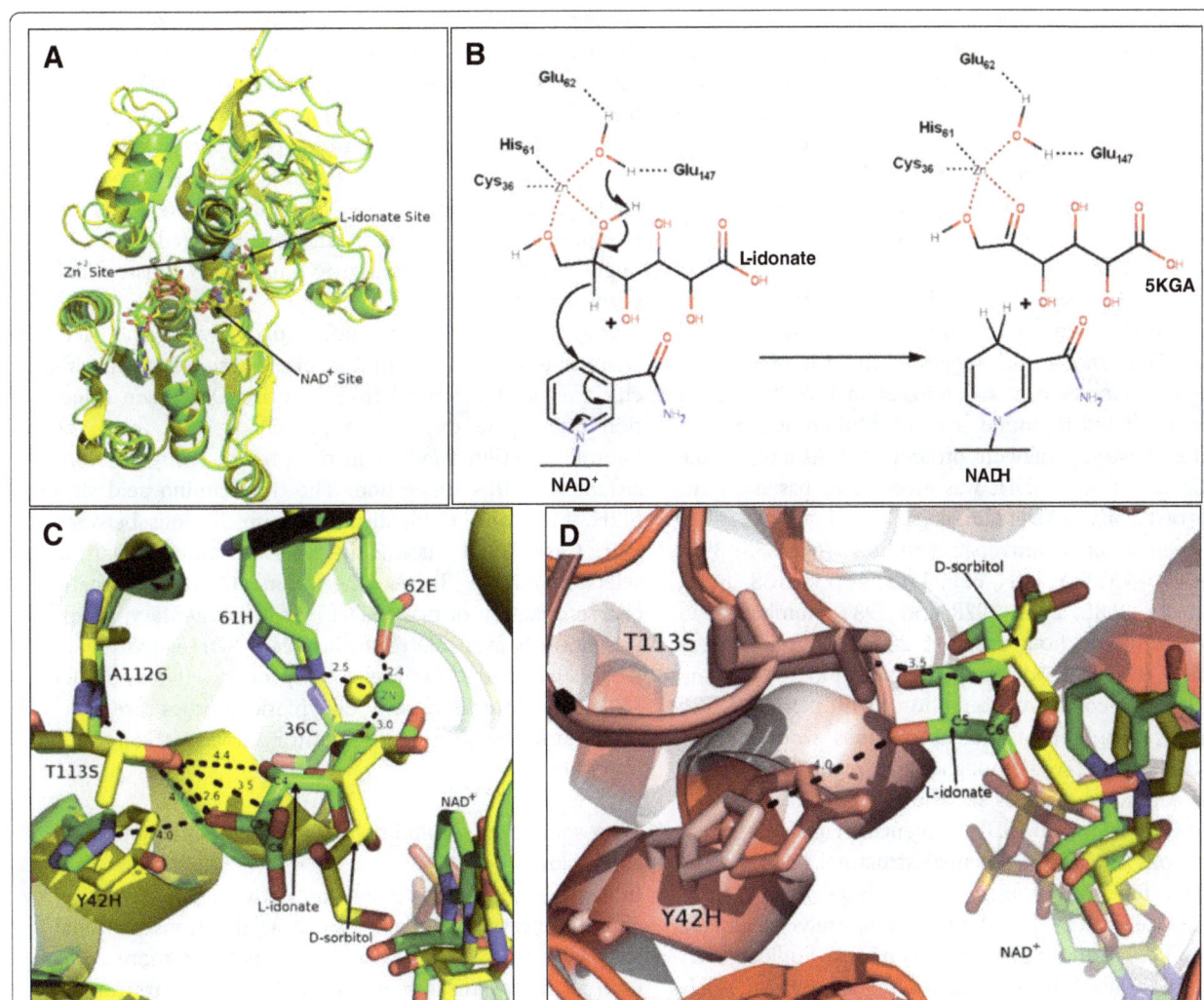

Figure 5 Homology models of Vv_LIDH and Vv_SDH and proposed reaction mechanisms. **A.** Structure superimposition of Vv_LIDH_idonate (green) and Vv_SDH_sorbitol (yellow) in Ribbon forms. **B.** The proposed reaction mechanism for Vv_LIDH on the oxidation of L-idonate into 5-keto-D-gluconate (5KGA). **C.** Superimposition of the active site residues of Vv_LIDH (green) and Vv_SDH (yellow). The distances (Å) between corresponding atoms are labelled. Target active site residues are shown in stick forms and labelled correspondingly. **D.** Hydrophobicity variance at Y42H between Vv_LIDH (green) and Vv_SDH (yellow) with red and white colours representing the highest hydrophobicity and the lowest hydrophobicity respectively. (All amino acid site numbering is according to LIDH (UniProt No: Q1PSI9) without the first 20 amino acids).

have suggested that plant SDHs exhibit highest activity for the oxidation of sorbitol, while also being able to oxidize other polyols such as xylitol and ribitol at lower reaction rates. However, the characterization of *V. vinifera* LIDH showed that this enzyme demonstrated the highest reaction rate on L-idonate but had a low reaction rate with sorbitol [48]. Upon docking of L-idonate, we found overall similar hydrogen bonding patterns with sorbitol as those proposed by Pauly et al. [22] and Yennawar et al. [77]. Earlier studies on enzyme substrate specificity also indicated that SDHs preferentially use substrates with a d-cis-2,4-dihydroxyl (2S,4R) configuration [6,13,18,20] (Additional file 1). L-idonate and D-sorbitol have the same molecular configuration from C1 to C4 and differ only at C5 (D and L chirality) and C6 (a hydroxyl group in sorbitol is replaced by a carboxyl group in L-idonic acid) (Additional file 1). Protein modelling analyses showed that L-idonate occupied a comparable position in the active site to sorbitol (Figure 5C). Therefore a similar reaction mechanism for L-idonate oxidation by *V. vinifera* LIDH is possible with D-sorbitol oxidation by human SDH [22]. The hydroxyl groups at C1 and C2 of L-idonate were modelled within interacting distance of the zinc atom in *V. vinifera* LIDH (Additional file 10: C1-C2-Zn.png), which may facilitate the proton transfer from C2 hydroxyl to NAD^+, ultimately resulting in an oxidized C2 with ketone and the production of NADH (Figure 5B). Previous work suggested that the preferential binding of L-idonate over sorbitol seen in *V.vinifera* LIDH may be attributed to amino acid substitution at the catalytic sites between paralogous proteins [48]. As a result, the catalytic site of plant SDHs was investigated based on our models of *V.vinifera* SDH homologs.

Nineteen putative active site residues (36C, 38S, 39D, 42H, 48C, 49A, 51F, 61H, 62E, 110F, 112G, 113S, 147E, 148P, 151V, 268L, 291F, 292R and 293Y; numbering in LIDH(Q1PSI9) without the first 20 amino acids) were identified either coordinating the zinc ion or forming potential non-covalent interactions with NAD(H) and L-idonate. Ten out of the 19 residues were considered strictly conserved throughout all plant SDH forms, and six additional residues are also largely conserved with variations in only a few SDH sequences (Figure 3). These observations revealed a potential structural basis for the preserved function of plant SDHs. Interestingly, three other residues were found to be uniformly exchanged (Y42H, A112G and T113S) between core Eudicot SDH Class I and Class II while monocot and *A. coerulea* SDHs resemble SDH Class I at these amino acid sites (Figure 3). A closer inspection of these residues showed that the oxygen atom of C5 hydroxyl of L-idonate was poised to potentially interact with both 42H and 113S within distances of 4 Å and 2.6 Å respectively (Figure 5C). Additionally, the oxygen atom of the C6 ketone group of L-idonate was

within non-covalent interaction distance to 113S (3.5 Å; Figure 5C). Notably, the replacement of 42Y (hydrophobic aromatic side chain) with 42H (charged side chain) in LIDH has the potential to change the hydrophobicity in the substrate-binding pocket (Figure 5D), which may lead to the preferential binding of L-idonate over D-sorbitol. These observations potentially provided a structural explanation for the unique activity of *V. vinifera* LIDH compared to other plant SDHs. Previous studies have indicated that the chiral configuration at C5 is not a determining factor for SDH substrate specificity [18,20], however, our analysis suggested that the C5 hydroxyl group and the C6 ketone group of L-idonate potentially affect substrate binding affinity due to amino acid substitutions at 42H, 112G and 113S in Class II SDHs. A previously identified SDH from apple fruit [9] was found to be the single Class II SDH (MDP0000305455) in *M. domestica* in the present study. This SDH has a much lower affinity for sorbitol (K_m 247 mM [9]) compared to other SDHs purified (K_m 40.3 mM [76], 86.0 mM [7]) or cloned (K_m 83.0 mM [10]; SDH Class I) from apple species. While the kinetic differences were suggested to be due to protein configuration changes between the fusion protein and native protein [9], the present analysis indicated that they might have been be due also to amino acid substitutions at the catalytic site.

From an evolutionary point of view, amino acid changes leading to the shift of enzyme substrate specificity are usually derived from positive Darwinian selection after gene duplication [41,43]. Results from the natural selection analyses in the present study are consistent with this suggestion. The three amino acid sites (42H, 112G and 113S) displaying substitutions between SDH Class I and Class II are all under positive natural selection (Table 3). At the moment, the enzymatic characterization of plant SDH is still fragmentary; no information is available regarding plant SDH activity with L-idonate, except for the activity of *V. vinifera* LIDH [48]. Site mutation and enzymatic studies are currently underway in our laboratory to investigate this hypothesis.

Meta-analysis of sorbitol dehydrogenase related gene expression

In addition to changes in enzyme activity, gene evolution after duplication can also occur at the transcriptional level [42]. Expression division appears to be more common than structural evolution and often occurs rapidly after gene duplication [42,78,79]. To further characterize the evolutionary pattern of plant SDH genes and also to explore the role of SDH related genes during plant development, a survey of transcriptional data was undertaken. Based on the availability of microarray and RNA sequencing data and the presence of both classes of

SDH in the genome, grapevine and citrus species were selected. In addition, the expression profile of the single Class I SDH (AT5G51970, Figure 2) in *A. thaliana* was used as a model reference [80]. This gene was highly expressed in cotyledons, leaves and late stages of seed development compared to organs such as flowers (stamen, petal, carpel) and shoots (inflorescence, vegetative, transition), where it was marginally expressed (data not shown).The results support a potential role for SDH Class I during seed germination in *A. thaliana* [23], soybean [37] and maize [8,38]. In grapevines, transcriptional patterns of VIT_16s0100g00290 (SDH Class II, LIDH) and VIT_16s0100g00300 (SDH Class I, SDH) were analysed using the normalised grapevine gene expression atlas of the 'Corvina' cultivar [81]. Notable differences in gene expression intensities and dynamics were observed between SDH Class I and Class II (Figure 6A; Additional file 11: Table S1). The transcript abundance of grapevine SDH Class I was highest in the ripening stages of berries (measured in pericarp, pulp, seeds and skins), resembling the expression profiles reported for Class I SDHs in apple [10,27,29]. In most cases, transcript abundance was lowest in young berry growth stages and increased gradually until harvest in berry tissues. Developmental up-regulation of SDH Class I transcripts in other cultivars such as 'Shiraz' [82] and 'Tempranillo' [83] during berry development under normal conditions was also evident. In addition, the latter work showed sorbitol is present in leaves and berries, and that the biochemical activity of SDH Class I, involving sorbitol oxidation, coincided with SDH class I transcripts levels in these berries during development [83]. Similarly, developmental increases of the grapevine SDH Class I transcript were observed in leaf, rachis, seed and tendrils. Interestingly, gene expression of grapevine SDH Class I was highly induced in winter buds and followed a gradual down-regulation during dormancy release. A similar gene expression and protein activity pattern reported in raspberry [84] and pear [39] respectively may reflect a response to the environment where dormancy periods encompasses dehydration and temperature (cold) stress, although developmental processes could take place concurrently. Taken together, this suggests an active role for SDH Class I in developmental processes through the coordinated regulation of transcript and protein activities in controlling the flux of sorbitol (and related polyols) in grapevines which may be critical in maintaining cell and tissue homeostasis in the mature tissues [83] where oxidative stress is inherent [85,86].

Expression profiles of SDH Class II were well represented in most grapevine organs with the highest expression in berries at fruit-set and in flower carpels. A striking developmental down-regulation of grapevine SDH Class II genes was evident in most grapevine organs, where expression levels in young tissues of berries (pericarp, flesh, skin and seed), buds, leaves, stems and tendrils were high and gradually decreased during development (Figure 6A). We have previously demonstrated in a cross-comparison study involving RNA-seq, microarray and qRT-PCR in young, early veraison, late veraison and ripening berries of grapevine [82] that SDH Class II genes were developmentally down-regulated consistently in all profiling platforms. This distinct expression coincides with the accumulation of TA biosynthesis in young/immature tissues [48,87].

In citrus, SDH Class I and SDH Class II genes were represented by probesets "Cit.9778.1.S1_s_at" and "Cit.9780.1.S1_s_at" respectively. Although gene expression studies encompassing developmental series in citrus are not as comprehensive compared to *A. thaliana* and grapevine, several striking observations could be inferred (Figure 6B; Additional file 11: Table S1). The citrus SDH Class I gene was highly expressed regardless of organ and tissue, including stems, roots, leaves, ovules and fruit tissues (albedo, flavedo, juice sacs), similar to that of grapevine SDH Class I. Interestingly, SDH Class II genes were expressed to a very low level (possibly in fact not at all) in the majority of organs, including fruit tissues, except for the root where expression was highest. It is speculated that this may reflect the trace amount of TA detected in fruits of sweet oranges and other citrus species [63]. Until now, no information, to our knowledge, has been reported on the function of citrus SDHs. Given the novel transcription profiles of one the two citrus Class II SDHs (specifically expressed in root tissues), and the presence of an additional Class II SDH (albeit this sequence was not represented in the array from which these data were analysed), these features may indicate a novel function of SDHs specific to root tissues of sweet oranges and therefore, deserve more attention in future research. In addition to *V. vinifera* and citrus, divergent transcription profiles have also been reported for SDHs from apple [10] and pear [39] where the single copy Class II SDH genes were shown to be under independent transcriptional regulation from other SDH genes. Taken together, divergent expression profiles for SDH Class I and SDH Class II appear to be true to all species where two classes are present, supporting a gene functional divergence at the expression level.

Gene co-expression mining in various plant species

Gene co-expression network analysis (GCA) is based on the principle that genes involved in similar and/or related biological processes may be expressed in a proportional manner, thereby providing a unique tool to understand gene function. Based on information availability, co-expressed gene lists of SDHs from *A. thaliana*,

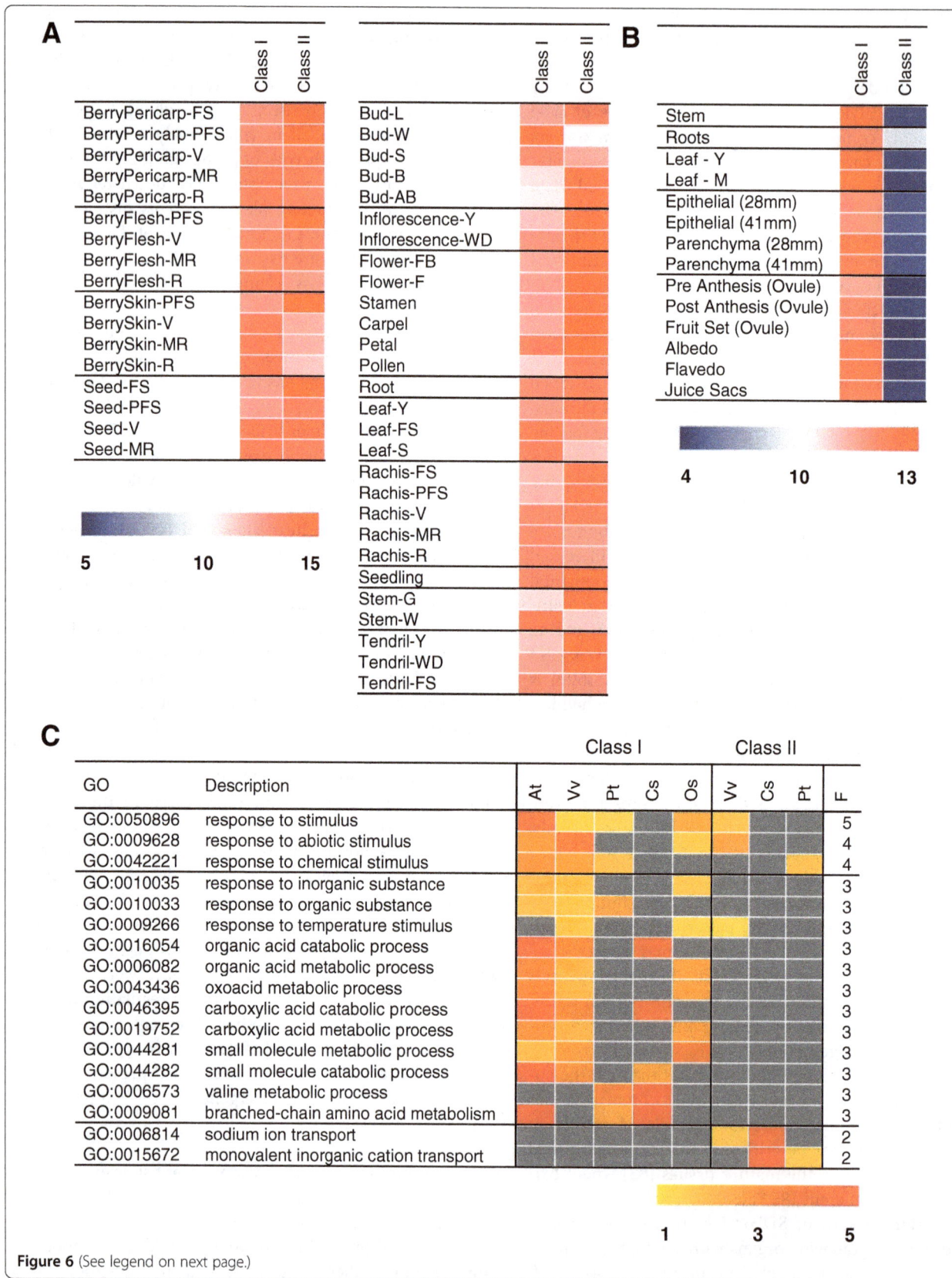

Figure 6 (See legend on next page.)

(See figure on previous page.)
Figure 6 Transcript and gene co-expression profiles of SDH in different plants. **A**. Expression profiles for Class I and Class II SDH genes in various tissues and developmental stages of *V. vinifera*. Class I and II SDH genes were moderately to highly expressed in most tissues (Log2 intensity > 10; 50th percentile of all gene expression values, see Methods). The heatmap was adjusted to colour ranges between log2 intensity of 5 (blue), 10 (white) and 15 (red) to illustrate low, moderate and high expression when compared to all other genes respectively. **B**. Expression profiles for Class I and Class II SDH gene in citrus. The heatmap was adjusted to colour ranges between log2 intensity of 4 (blue), 10 (white) and 14 (red) to illustrate low, moderate and high expression when compared to all other genes respectively. **C**. Heatmap of selected enriched GO terms (−log10 (adj. *p*-value) for genes co-expressed with SDHs from *A. thaliana* (At), *V. vinifera* (Vv), *C. sinensis* (Cs), *P .trichocarpa* [84], *O. sativa* (Os) and associated frequencies in the plants tested. Light and dark orange denote enrichment scores between 1 and 3 respectively. Highly enriched scores (>5) are coloured in red. Grey colour denotes no significant enrichment.

rice, poplar, grapevine and citrus (Additional file 11: Table S2-S9) were retrieved from publicly available co-expression databases [88-90]. In *A. thaliana*, the SDH Class I homologue (At5g51970) was significantly co-expressed with 67 genes (33% of total genes in the list) involved in branched chain amino acid metabolism, 72 genes (36%) involved in response to various stimuli, 37 genes (19%) involved in protein import in the peroxisome and 17 genes (9%) involved in auxin metabolism (Additional file 11: Table S2). In grapevines, the SDH Class I homologue (VIT_16s0100g00300) was significantly co-expressed with genes involved in abiotic stress (21%), peptide metabolism (13%) and lipid metabolism (13%) (Additional file 11: Table S3; Additional file 12: Table S2–S3). The co-expression results presented here corroborated with recent findings that the importance of SDH Class I lies in regulating sorbitol levels via its biochemical activity and gene expression during various abiotic stresses [83]. More importantly, intracellular accumulation of sorbitol to high levels, accentuated under salt and osmotic stress, significantly reduced stress-induced biomass loss of grapevine berry cell suspensions which were likely the results of the polyol utilisation as an effective osmoprotectant and cellular homeostasis buffer [83]. Similar to its Arabidopsis counterpart (At5g51970), it is therefore likely that grapevine SDH Class I plays an important role in abiotic stress tolerance via the synergistic regulation of polyol transport and metabolism. The SDH Class II homologue (LIDH, VIT_16s0100g00290) was also significantly co-expressed with genes related to abiotic stress response (35%). Other genes related to hexose biosynthetic pathways and carbohydrate metabolism (25%), protein biogenesis and catabolism (8%) and malic acid transport (6%) were also evident in the list of co-expressed genes (Additional file 11: Table S4). GO terms associated with these genes were also enriched within the gene lists (FDR < 0.05). Interestingly, GO enrichment analysis of co-expressed genes showed that terms associated with "malate trans-membrane transport" and "response to abiotic stimulus" were highly enriched (FDR < 1.51E-04 and 3.5E−03 respectively) (Additional file 12: Table S2). Similarly to the grapevine SDH Class I

gene, SDH Class II transcription was also stress responsive, being down-regulated during the heat stress recovery of grapevine leaves and up-regulated during exposure to UV-C light irradiation (Additional file 12: Table S3). Based on our coexpression analysis, we speculate that the involvement of Class II SDHs in abiotic stress responses is likely to occur via a separate mechanism from that of sorbitol metabolism, namely the ascorbate-glutathione cycle [91] and specifically in regulating the balance between the biosynthesis of ascorbate by the L-galactose pathway [92] and its catabolism. This is supported in part in grapevines in which a marked down-regulation of SDH Class II (LIDH) protein (impeding TA formation) and the up-regulation of proteins involved in L-galactose pathway (favouring Asc formation) in shoots of grapevines during drought stress were observed [93]. Therefore, the stress responsive nature of SDH Class II gene and enzyme could potentially function as an extra level of control (preventing loss of Asc to TA). The *C. sinensis* SDH Class II gene (Cit.9780.1.S1_at) was significantly co-expressed with genes involved in ion transport (11%), ubiquinone biosynthesis/oxidative phosphorylation (20%) and ribosome biogenesis (9%) (Additional file 11: Table S6). GO terms associated with these genes were highly enriched within the co-expressed gene lists (Additional file 12: Table S5). Unlike Class I SDHs, enriched GO terms associated with Class II SDH co-expressed genes were more specialised to each corresponding plant but shared a common set of co-expressed genes related to transporters (Additional file 11: Table S7; Additional file 12: Table S6). In rice, the top 200 genes co-expressed with SDH (Os08g0545200) were primarily enriched for genes involved in stress response (31%), carboxylic acid biosynthesis (16%), plastid organisation (11%), protein transport (10%) and starch metabolism (5%) (Additional file 11: Table S5; Additional file 12: Table S4).

Enriched GO parent terms such as "response to stimulus" and descendent terms "response to abiotic stimulus", were frequently enriched in SDH Class I co-expressed lists and slightly in SDH Class II containing plant species (Figure 6C; Additional file 12: Table S1-S9). These observations agreed with previous reports that SDHs (Class I) in *A. thaliana* [13,23] and grapevine [83] play an active role during drought stress and recovery processes and also suggest some shared functions related to stress tolerance between the two classes of SDH, even though to a conservative degree and potentially involving a separate mechanistic route. Therefore, enriched GO parent terms associated with "organic acid metabolic process" and "branched-chain amino acid metabolism" were demonstrated to be more relevant to SDH Class I co-expressed genes but not to SDH Class II (Figure 6C). This is not surprising as response to various stresses involves the coordinated regulation of amino acid and polyol accumulation [94]. On the other hand, co-expression analysis showed that plant SDH Class II could be tightly linked to mechanisms related to transport and compartmentation of cations and solutes (Figure 6C). In membrane transport and compartmentation systems involving pumps, carriers and ion channels are also pivotal for ion homeostasis and equivocally involved in a wide range of stress conditions [95]. In addition, divergent co-expression profiles across species have also been observed for both classes of SDH. In general, monocot rice SDH-related genes have more common co-expression responses with core Eudicot SDH Class I than with SDH Class II, corresponding with the finding that monocot SDH has a closer relationship with core Eudicot SDH Class I than SDH Class II at the enzyme structural level.

Conclusions

SDH is the key enzyme involved in sorbitol metabolism in higher plants. The results of the present study demonstrated that core Eudicot SDHs have evolved into two distinct lineages: SDH Class I and SDH Class II. Class I SDH genes were present in all core Eudicot species investigated in this study and appear to be essential for the normal growth of plants. Class II SDH genes were found to be absent in *Brassicaceae*, *Leguminosae*, most *Asterid*s (except *S. tuberosum*) and some other plants. The previously characterized LIDH involved in TA synthesis in *V. vinifera* has now been identified as a Class II SDH and represents a novel function of SDH genes in *V. vinifera*. The role of LIDH in TA synthesis may be relevant to the function of Class II SDHs in other species. Phylogeny, natural selection and genomic structure analyses supported the emergence of SDH Class II as a result of positive natural selection after tandem duplication, which might occur in the common ancestor of core Eudicot plants. Furthermore, positive natural selection has only acted on specific amino acid sites in the SDH Class II lineage. Protein modelling analyses revealed substitutions of three putative active site residues for Class I and Class II SDHs, which may be responsible for the unique enzyme activity of *V. vinifera* LIDH. Gene expression analysis demonstrated a clear transcriptional

divergence between SDH Class I and Class II in several plants and supports the divergence of Class II SDHs at the expression level as well. Future work should be dedicated to uncovering the enzymatic activities and roles of Class II SDH gene products in plant metabolism.

Methods
Identification of sorbitol dehydrogenase homologous genes in higher plants

To identify homologous SDHs in angiosperm plants, the amino acid sequence of *A. thaliana* SDH (accession no. At5g51970) was used as a query to BLAST against the genomes of angiosperm species at Phytozome (http://www.phytozome.net/), with the exception of *M. domestica* for which genome dataset at Plant Genome Duplication Database (PGDD, http://chibba.agtec.uga.edu/duplication/) was used instead. To increase dataset coverage, the genomes of 8 recently sequenced species including *Cajanus cajan*, *Jatropha curcas*, *Capsicum annuum*, *Brassica oleracea*, *Eutrema saisugineum*, *P. mume*, *Hordeum vulgare* and *Aegilops tauschii* were also queried using the corresponding genome databases. BLAST hits with an expectancy value (E value) of zero were selected as SDH homologs were subjected to another round of BLAST searches within the genomes from which they were identified. Only the primary transcript was chosen when alternative transcripts occurred. In addition, five partial SDH protein sequences of *P. bretschneideri* [39] and one SDH sequence of *Eriobotrya japonica* [35] were obtained from literature searches. Homologous SDHs of *P. hortorum* were provided by the *P. hortorum* genome sequencing project author (Prof. Robert K. Jansen, The University of Texas at Austin).

Phylogenetic analysis of sorbitol dehydrogenase

The Uniprot database was queried for previously identified MDR mammal SDHs and yeast SDHs. Only reviewed entries were selected and used as the outgroup in this phylogenetic analysis. Multiple sequence alignments of 102 sequences (92 plant SDHs, 7 mammal SDHs and 3 yeast SDHs) were carried out using ClustalW2 [96]. The evolutionary distances of target SDHs (pairwise p-distance) were estimated using MEGA6 software [97]. The Neighbour Joining tree was inferred by MEGA6 software [97] using the p-distance [98] substitution model, the certainty at each node was assessed by the Interior-branch Test method (1000 times iteration). Maximum likelihood trees were estimated by MEGA6 software [97] using the JTT+GAMMA substitution model [99], the best fitting model as determined by the "Find Best DNA/Protein Models" function in MEGA6. Bootstrap supports for Maximum likelihood trees were calculated from 1000 replicates. For both Neighbour Joining and Maximum likelihood methods, the Gaps/Missing Data Treatment parameter was set as Complete-Deletion to eliminate the effects of gaps and insertions. The developed phylogenetic trees were rooted on the yeast SDHs and annotated using the FigTree version 1.4.2 software (http://tree.bio.ed.ac.uk/software/figtree/).

Sequence alignment and protein subdomain analysis

Preliminary sequence identity of SDHs was obtained by local all-vs-all BLAST using NCBI-BLAST-2.2.29 tool [100] downloaded from ftp://ftp.ncbi.nlm.nih.gov/blast/executables/blast+/LATEST/. The BLAST results were sorted according to respective phylogeny groups. Average pair-wise sequence identities were calculated using Microsoft Excel software based on the BLAST results. Protein functional domains were predicted using InterPro (http://www.ebi.ac.uk/interpro/). Secondary structure analysis was implemented with ESPript3.0 tool (http://espript.ibcp.fr/ESPript/ESPript/) using human SDH (PDB: 1PL8) as a template. All residue numberings in the present study are according to LIDH (Q1PSI9) without the first 20 amino acids (unless otherwise declared) which was predicted to be a mitochondria-targeting signal sequence (data not shown; alignment corresponding to this region was highly divergent).

Gene duplication pattern characterization and synteny analysis

The MCScanX package [101] from http://chibba.pgml.uga.edu/mcscan2/ was employed to investigate gene duplication patterns of plant SDHs. In order to elaborate on the origin of the core Eudicot Class II SDHs, plant genomes containing SDHs from both Class I and Class II were selected. These were further refined to genomes for which predicted genes have been mapped into corresponding chromosome locations. *A.thaliana* was included as a reference for inter-species collinear block analysis. Amino acid sequence files and gene position files were downloaded either from PGDD or from Phytozome databases and were further modified to suit the requirements of the MCScanX software. BLAST tool NCBI-BLAST-2.2.29 [100] was used for intra and inter species genome comparisons. The E-value threshold was set at 10^{-5} for all analyses. For gene duplication pattern identification, self-genome all-vs-all BLAST was performed. The *duplicate_gene_class ifier* program from the MCScanX package was applied to each dataset. For collinear SDH gene pair identification, amino acid sequences and genetic position information of chromosomes containing SDHs were extracted from each species, then combined to perform the multi-species MCScanX analysis. The SDH gene family file was created manually by including all the SDHs identified from the selected species. The *family_circle_plotter.*

java tool at MCScanX package was used to display the results.

Natural selection analysis

Natural selective pressure on plant SDH was examined by measuring the ratio of non-synonymous to synonymous substitutions (dN/dS=w). Codon-based maximum-likelihood estimates of w was performed using codeml in PAML4.7 [73]. Multiple-alignment of conserved domain sequences (CDS) for those identified plant SDHs was carried out using ClustalW2 [96]. Significant insertions and gaps were removed manually. To facilitate the input data requirements of codeml, an additional Maximum Likelihood tree was constructed using a smaller dataset where SDHs with no CDS sequence available were removed. The sub-tree covering the plant SDHs was used in codeml. Branch pattern specification was implemented using Treeview1.6.6 (http://taxonomy.zoology.gla.ac.uk/rod/treeview.html). Four target clades were specified based on the present phylogenetic analysis: monocot SDH, *A. coerulea* SDH, core Eudicot SDH Class I and core Eudicot SDH Class II. The w values for these clades were represented as w[mono], w[Aer], w[sdhC1] and w[sdhC2] respectively. Nested likelihood ratio tests(LRTs) were performed to assess the significance of the model under different hypothesises: (w[mono]≠w[Aer]≠w[sdhC1]=w[sdhC2], w[mono]=w[Aer]≠w[sdhC1]≠w[sdhC2], w[mono]≠w[Aer]≠w[sdhC1]≠w[sdhC2], w[mono]=w[Aer]=w[sdhC1]≠w[sdhC2], w[mono]=w[Aer]=w[sdhC1]≠w[sdhC2] with w[sdhC2]=1). The corresponding p values were calculated using the online tool at http://graphpad.com/quickcalcs/PValue1.cfm. In the Site-specific model M1, two site classes were specified: highly conserved sites (w0) and neutral sites (w1=1). For the Site-specific model M2, there were three site classes: highly conserved sites (w0), neutral sites (w1=1) and positively selected sites (w2). For w assessments with the Branch-site models, core Eudicot SDH Class II was specified as the foreground group. In the Branch-site model A, four site classes were specified. The first two classes have w ratios of w0 and w1 respectively, corresponding to highly conserved sites and neutral sites across all lineages. In the other two site classes, the background lineages have w0 or w1 while the foreground lineages have w2.

Ancestral sequence reconstruction and evolution rate analyses

The ancestral sequence (amino acid) reconstruction for the internal nodes of the obtained plant SDH phylogeny was carried out using codeml in PAML4.7 [73]. The Empirical_Frequency model, which allowed the estimates of the stationary frequencies based on user dataset, was performed on the plant SDHs. Ancestral amino acid sequences for nodes representing monocot SDH, *A. coerulea* SDH, core Eudicot SDH Class I and core Eudicot SDH Class II were used for Tajima's RRT analysis [74] using MEGA6.0 software [97].

Protein structure modelling analysis

SDH homology modelling was carried out using ICM Pro (Molsoft LLC, La Jolla, CA, USA). Models of *V. vinifera* LIDH (Uniprot ID: Q1PSI9; accession no: GSVIVT01010646001) and *V. vinifera* SDH (Uniprot ID: D7TMY3; accession no: GSVIVT01010642001) structures were generated with the human SDH (PDB:1PL8) as a template. Given that no plant SDH structures exist in the protein data bank we chose the model with the highest identity as performed within the Molsoft software package. Ligands including the zinc atom, NAD^+, D-sorbitol and L-idonate were docked into the models using the Molsoft Monte Carlo method [102]. Residues within 5 Å to the ligands were inspected for enzyme-ligand interaction potential. All molecular visualizations were obtained using the PyMOL graphic tool (The PyMOL molecular graphics system, Version 1.3r1. Schrodinger, LLC). The deduced reaction mechanism of *V. vinifera* LIDH on the oxidation of L-idonate was created using the Marvin online tool (http://www.chemaxon.com/marvin/sketch/index.php). Protein hydrophobicity profiles were implemented in PyMOL using the Color_h script (http://www.pymolwiki.org/index.php/Color_h), based on the hydrophobicity scale defined at http://us.expasy.org/tools/pscale/Hphob.Eisenberg.html. All residue numberings are according to LIDH (Q1PSI9) without the first 20 amino acids.

Meta-analysis of developmental gene expression

Identification of corresponding probesets in the microarray platforms of *A. thaliana*, rice, poplar, grapevine and citrus were performed using the BLAST software (NCBI-BLAST-2.2.29+) [100], and grapevine Class I (VIT_16s0100g00290) and Class II (VIT_16s0100g00290) SDH sequences with default settings. The top hits for each corresponding probeset in the microarray platform of each species were selected for downstream analysis (Additional file 11). Normalised gene expression atlases encompassing transcriptional data during growth and development of *A. thaliana*, grapevine and citrus were retrieved from the Botany Array Resource (BAR) [80], *Vitis* co-expression database (VTCdb) [88] and Network inference of citrus co-expression (NiCCE) [89] webservers, respectively. Only experimental conditions relating to tissue/organ development and probesets intensities (normalised) corresponding to Class I and Class II SDHs were retained. Normalised log2 intensities were deemed highly, well and lowly/not expressed when the intensities of total background

distribution > 95th, at the 50th and < 20th percentile respectively.

Gene co-expression mining in various plant species

Information on co-expressed genes with Class I and Class II SDHs in plants such as *A. thaliana*, poplar and rice (version 7.1) [90], grapevine (version 2.1) [88] and citrus [89] were retrieved from the various plant gene co-expression webservers. The top 200 co-expressed genes (unless otherwise specified) for each SDH class in each species were empirically chosen as a cut-off for significant co-expression, and to provide comparisons of enriched gene ontology (GO) terms within the co-expressed gene lists from each species. Enrichment of GO terms (i.e. biological processes, BP; molecular function, MF; cellular component, CC) were evaluated by hypergeometric distribution, adjusted by false discovery rate (FDR) for multiple hypothesis correction and using the 'gProfileR' package [103] in R (http://www.r-project.org) which interfaces g:profiler webserver (http://gprofiler.at.mt.ut.ee/gprofiler/). The 'ordered query' option was enabled to perform incremental enrichment analysis, which prioritises highly co-expressed genes and results in better functional GO term associations. GO terms were considered to be significantly enriched when FDR < 0.05 and > 2 genes were annotated with the same GO term. Enriched GO terms from the SDH co-expressed gene lists across tested plants (*A. thaliana*, poplar, rice, grapevine and citrus), were considered 'commonly occurring' when more than 3 counts were present for each enriched GO term.

Availability of supporting data

All relevant supporting data can be found within the additional files accompanying this article. Phylogenetic data supporting the results of this article are available in the TreeBASE repository at http://purl.org/phylo/treebase/phylows/study/TB2:S17300.

Additional files

Additional file 1: Displays the molecular structures of SDH substrates.

Additional file 2: Table S1. Contains SDH gene IDs from corresponding species and organisms. **Table S2.** Contains pairwise p-distance values of SDH sequences. **Table S3.** Contains information on sequence renaming. **Table S4.** Contains the all-vs-all BLAST results of SDH amino acid sequences. Table S5 contain the identified collinear SDH gene pairs.

Additional file 3: Contains the original amino acid sequences of the identified plant, mammal and yeast SDHs.

Additional file 4: Displays the complete Neighbour Joining tree for Figure 2A.

Additional file 5: Displays complete sequence alignment for Figure 3.

Additional file 6: Contains gene duplication pattern information.
Tables "cs", "eg", "md", "pm", "pp", "pt", "st", "tc", "vv" refer to *C. sinensis*, *E. grandis*, *M. domestica*, *P. mume*, *P. persica*, *P. trichocarpa*, *S. tuberosum*, *T. cacao* and *V. vinifera* respectively.

Additional file 7: Contains input and output data for natural selection modelling analyses. "-output" files are codeml outputs and are recommended to be viewed using Microsoft WordPad. ".phy" is phylogenetic tree file and can be viewed using Treeview software. ".ctl" is a control file and can be viewed using any text viewer. "sdh-pep2.fas" sequence file was produced from Additional file 3 by manually removing the significant gaps, insertions; sequences with no CDS sequence available were also removed. "sdh-cds2.fas" is the corresponding CDS sequences for "sdh-pep2.fas". "sdh-pep2.nwk" is the phylogenetic tree produced from "sdh-cds2.fas" and can be viewed using any phylogenetic tree viewer software. Sequence IDs are represented by numbers for software input convenience (see Additional file 2: Table S3 for sequence ID renaming information). Amino acid site numbering is according to LIDH (Uniprot No: Q1PSI9) without the first 20 amino acids.

Additional file 8: Contains input and output data for the reconstruction of ancestral SDH sequences. "sdh-pep.fas" contains amino acid sequences for the plant SDH sub-branch. The "ancestral-sequence-construction_output" file is codeml output and can be viewed using any text viewer. Ancestral sequences for corresponding branches were extracted and put in the "interpreted-ancestral-sequence.fas" file for readers' convenience.

Additional file 9: Contains the Tajima's RRT test outputs.

Additional file 10: Contains the modelled structures files of Vv_LIDH and Vv_SDH and additional illustration figures. "Asp195_NAD.png" displays the interaction of Asp195 with the hydroxyl groups at C1 and C2 of L-idonate. "LIDH-hydrophobicity.png" and "SDH-hydrophobicity.png" display the overall hydrophobicity profiles of Vv_LIDH and Vv_SDH respectively. Amino acid site numbering is according to LIDH (Uniprot No: Q1PSI9) without the first 20 amino acids.

Additional file 11: Contains a Microsoft Excel spread sheet with detailed results of transcript and gene co-expression analysis of Class I and Class II SDH in plants. Table S1 contains gene expression profile of Class I and Class II SDH profile in various tissues of (A) grapevine and (B) sweet oranges. Table S2 – S9 contains lists of all significantly co-expressed genes and respective rank, function description, and co-expression metric with class I and II SDH in *A. thaliana* (Table S2), grapevine (Table S3 and S4), rice (Table S5), sweet orange (Table S6 and S7) and poplar (Table S8 and S9).

Additional file 12: Contains a Microsoft Excel spread sheet with detailed results of functional (GO) enrichment analysis of significantly co-expressesed genes of class I and II SDH in plants. Table S1 – S8 contains outputs of GO enrichment analysis containing enriched GO ID, description, adjusted p-value, and lists of genes having the enriched GO term for A. thaliana (Table S1), grapevine (Table S2 and S3), rice (Table S4), sweet orange (Table S5 and S6) and poplar (Table S7 and S8). Table S9 contains a summary of common enriched GO ID/term identified among the co-expressed genes with SDHs in the aforementioned plants tested.

Abbreviations
SDH: Sorbitol dehydrogenase; LIDH: L-idonate-5-dehydrogenase; TA: Tartaric acid; MDR: Medium-chain dehydrogenase/reductase; ADH: Alcohol dehydrogenase; 5KGA: 5-keto-D-gluconate; Mya: Million years ago; WGD: Whole genome duplication; RRT: Relative rate tests; GCA: Gene co-expression network analysis; PGDD: Plant Genome Duplication Database; CDS: Conserved domain sequences; LRTs: Likelihood ratio tests; BAR: Botany Array Resource; VTCdb: *Vitis* co-expression database; NiCCE: Network inference of citrus co-expression; GO: Gene ontology; BP: Biological processes; Asc: Ascorbate; MF: Molecular function; CC: Cellular component; FDR: False discovery rate.

Competing interests
The authors declare that they have no competing interests.

Authors' contributions
YJ conceived the research. YJ and DCJW did sequence retrieval, curation and gene duplication characterization. YJ performed phylogenetic, synteny, natural selection modeling and ancestral sequence analyses and drafted the manuscript. JBB and YJ carried out protein modeling analyses. DCJW performed the transcript expression and gene co-expression analysis. CS

and DCJW assisted with the drafting of the manuscript. CMF and JBB supervised the project. All authors have read and approved the final manuscript.

Acknowledgements
We acknowledge the related research groups for making the genomic information and microarray data available to the public. We are very grateful to Dr Anthony Borneman and Dr Julian Schwerdt for their valuable suggestions about the phylogenetic analyses. We thank the anonymous referees for their constructive comments and suggestions. This work was part-supported by Australia's grape growers and winemakers through the Grape and Wine Research and Development Corporation with matching funds from the Australian Government (project UA 10/01). YJ is supported by a postgraduate scholarship from China Scholarship Council.

Author details
[1]School of Agriculture, Food and Wine, University of Adelaide, Adelaide 5005, Australia. [2]Present address: Wine Research Center, Faculty of Land and Food Systems, University of British Columbia, Vancouver V6T 1Z4BC, Canada. [3]Present address: School of Biological Sciences, Flinders University, GPO Box 2100, Adelaide 5001, Australia. [4]School of Biological Sciences, University of Adelaide, Adelaide 5005, Australia.

References
1. Iwata T, Hoog JO, Reddy VN, Carper D. Cloning of the human Sorbitol Dehydrogenase gene. Invest Ophth Vis Sci. 1993;34(4):712–2.
2. Karlsson C, Jornvall H, Hoog JO. Sorbitol Dehydrogenase - cDNA coding for the rat enzyme - variations within the Alcohol-Dehydrogenase family independent of quaternary structure and metal content. Eur J Biochem. 1991;198(3):761–5.
3. Wang T, Hou M, Zhao N, Chen Y, Lv Y, Li Z, et al. Cloning and expression of the sorbitol dehydrogenase gene during embryonic development and temperature stress in *Artemia sinica*. Gene. 2013;521(2):296–302.
4. Niimi T, Yamashita O, Yaginuma T. A cold-inducible Bombyx gene encoding a protein similar to mammalian Sorbitol Dehydrogenase - yolk nuclei-dependent gene-expression in diapause eggs. Eur J Biochem. 1993;213(3):1125–31.
5. Sarthy AV, Schopp C, Idler KB. Cloning and sequence determination of the gene encoding Sorbitol Dehydrogenase from *Saccharomyces cerevisiae*. Gene. 1994;140(1):121–6.
6. Ng K, Ye RQ, Wu XC, Wong SL. Sorbitol Dehydrogenase from *Bacillus subtilis* - purification, characterization, and gene cloning. J Biol Chem. 1992;267(35):24989–94.
7. Negm FB, Loescher WH. Detection and characterization of Sorbitol Dehydrogenase from apple callus-tissue. Plant Physiol. 1979;64(1):69–73.
8. Doehlert DC. Ketose reductase-activity in developing maize endosperm. Plant Physiol. 1987;84(3):830–4.
9. Yamada K, Oura Y, Mori H, Yamaki S. Cloning of NAD-dependent sorbitol dehydrogenase from apple fruit and gene expression. Plant Cell Physiol. 1998;39(12):1375–9.
10. Park SW, Song KJ, Kim MY, Hwang JH, Shin YU, Kim WC, et al. Molecular cloning and characterization of four cDNAs encoding the isoforms of NAD-dependent sorbitol dehydrogenase from the Fuji apple. Plant Sci. 2002;162(4):513–9.
11. Ohta K, Moriguchi R, Kanahama K, Yamaki S, Kanayama Y. Molecular evidence of sorbitol dehydrogenase in tomato, a non-Rosaceae plant. Phytochemistry. 2005;66(24):2822–8.
12. Sutsawat D, Yamada K, Shiratake K, Kanayama Y, Yamaki S. Properties of sorbitol dehydrogenase in strawberry fruit and enhancement of the activity by fructose and auxin. J Jpn Soc Hortic Sci. 2008;77(3):318–23.
13. Aquayo MF, Ampuero D, Mandujano P, Parada R, Muñoz R, Gallart M, et al. Sorbitol dehydrogenase is a cytosolic protein required for sorbitol metabolism in *Arabidopsis thaliana*. Plant Sci. 2013;205–206(1):63–75.
14. Persson B, Zigler JS, Jornvall H. A super-family of medium-chain dehydrogenases/reductases (MDR) - Sub-lines including zeta-crystallin, alcohol and polyol dehydrogenases, quinone oxidoreductases, enoyl reductases, Vat-1 and other proteins. Eur J Biochem. 1994;226(1):15–22.

15. Persson B, Hedlund J, Jornvall H. The MDR superfamily. Cell Mol Life Sci. 2008;65(24):3879–94.

16. Jornvall H, Persson M, Jeffery J. Alcohol and polyol dehydrogenases are both divided into 2 protein types, and structural-properties cross-relate the different enzyme-activities within each type. P Natl Acad Sci-Biol. 1981;78(7):4226–30.

17. Nordling E, Jornvall H, Persson B. Medium-chain dehydrogenases/reductases (MDR) - family characterizations including genome comparisons and active site modelling. Eur J Biochem. 2002;269(17):4267–76.

18. Lindstad RI, Koll P, McKinley-McKee JS. Substrate specificity of sheep liver sorbitol dehydrogenase. J Biochemical. 1998;330(Pt 1):479–87.

19. Lindstad RI, Hermansen LF, McKinley-Mckee JS. The kinetic mechanism of sheep liver sorbitol dehydrogenase. Eur J Biochem. 1992;210(2):641–7.

20. Oura Y, Yamada K, Shiratake K, Yamaki S. Purification and characterization of a NAD(+)-dependent sorbitol dehydrogenase from Japanese pear fruit. Phytochemistry. 2000;54(6):567–72.

21. Guo ZX, Pan TF, Li KT, Zhong FL, Lin L, Pan DM, et al. Cloning of NAD-SDH cDNA from plum fruit and its expression and characterization. Plant Physiol Biochem. 2012;57:175–80.

22. Pauly TA, Ekstrom JL, Beebe DA, Chrunyk B, Cunningham D, Griffor M, et al. X-ray crystallographic and kinetic studies of human sorbitol dehydrogenase. Structure. 2003;11(9):1071–85.

23. Nosarzewski M, Downie AB, Wu B, Archbold DD. The role of sorbitol dehydrogenase in *Arabidopsis thaliana*. Funct Plant Biol. 2012;39(6):462–70.

24. Yancey PH, Clark ME, Hand SC, Bowlus RD, Somero GN. Living with water-stress - evolution of osmolyte systems. Science. 1982;217(4566):1214–22.

25. Loescher WH. Physiology and metabolism of sugar alcohols in higher-plants. Physiol Plantarum. 1987;70(3):553–7.

26. Loescher WH, Marlow GC, Kennedy RA. Sorbitol metabolism and sink-source interconversions in developing apple leaves. Plant Physiol. 1982;70(2):335–9.

27. Nosarszewski M, Clements AM, Downie AB, Archbold DD. Sorbitol dehydrogenase expression and activity during apple fruit set and early development. Physiol Plantarum. 2004;121(3):391–8.

28. Nosarzewski M, Archbold DD. Tissue-specific expression of sorbitol dehydrogenase in apple fruit during early development. J Exp Bot. 2007;58(7):1863–72.

29. Wang XL, Xu YH, Peng CC, Fan RC, Gao XQ. Ubiquitous distribution and different subcellular localization of sorbitol dehydrogenase in fruit and leaf of apple. J Exp Bot. 2009;60(3):1025–34.

30. Yamaguchi H, Kanayama Y, Soejima J, Yamaki S. Changes in the amounts of the NAD-dependent sorbitol dehydrogenase and its involvement in the development of apple fruit. J Am Soc Hortic Sci. 1996;121(5):848–52.

31. Wu BH, Li SH, Nosarzewski M, Archbold DD. Sorbitol dehydrogenase gene expression and enzyme activity in apple: tissue specificity during bud development and response to rootstock vigor and growth manipulation. J Am Soc Hortic Sci. 2010;135(5):379–87.

32. Iida M, Bantog NA, Yamada K, Shiratake K, Yamaki S. Sorbitol- and other sugar-induced expressions of the NAD+–dependent sorbitol dehydrogenase gene in Japanese pear fruit. J Am Soc Hortic Sci. 2004;129(6):870–5.

33. Kim HY, Ahn JC, Choi JH, Hwang B, Choi DW. Expression and cloning of the full-length cDNA for sorbitol-6-phosphate dehydrogenase and NAD-dependent sorbitol dehydrogenase from pear (*Pyrus pyrifolia N.*). Sci Hortic. 2007;11(5):406–12.

34. Bantog NA, Shiratake K, Yamaki S. Changes in sugar content and sorbitol- and sucrose-related enzyme activities during development of loquat (*Eriobotrya japonica Lindl. cv. Mogi*) fruit. J Jpn Soc Hortic Sci. 1999;68(5):942–8.

35. Bantog NA, Yamada K, Niwa N, Shiratake K, Yamaki S. Gene expression of NAD(+)-dependent sorbitol dehydrogenase and NADP(+)-dependent sorbitol-6-phosphate dehydrogenase during development of loquat (*Eriobotrya japonica Lindl.*) fruit. J Jpn Soc Hortic Sci. 2000;69(3):231–6.

36. Beruter J. Sugar accumulation and changes in the activities of related enzymes during development of the apple fruit. J Plant Physiol. 1985;121(4):331–41.

37. Kuo TM, Doehlert DC, Crawford CG. Sugar metabolism in germinating soybean seeds - evidence for the sorbitol pathway in soybean axes. Plant Physiol. 1990;93(4):1514–20.

38. de Sousa SM, Paniago MD, Arruda P, Yunes JA. Sugar levels modulate sorbitol dehydrogenase expression in maize. Plant Mol Biol. 2008;68(3):203–13.

39. Ito A, Hayama H, Kashimura Y. Partial cloning and expression analysis of genes encoding NAD(+)-dependent sorbitol dehydrogenase in pear bud during flower bud formation. Sci Hortic. 2005;103(4):413–20.

40. Hartman MD, Figueroa CM, Piattoni CV, Iglesias AA. Glucitol Dehydrogenase from peach (*Prunus persica*) fruits is regulated by thioredoxin h. Plant Cell Physiol. 2014;55(6):1157–68.

41. Flagel LE, Wendel JF. Gene duplication and evolutionary novelty in plants. New Phytol. 2009;183(3):557–64.

42. Zhang JZ. Evolution by gene duplication: an update. Trends Ecol Evol. 2003;18(6):292–8.

43. Hughes AL. The evolution of functionally novel proteins after gene duplication. P Roy Soc B-Biol Sci. 1994;256(1346):119–24.

44. Hughes AL. Adaptive evolution after gene duplication. Trends Genet. 2002;18(9):433–4.

45. Hurles M. Gene duplication: the genomic trade in spare parts. Plos Biol. 2004;2(7):900–4.

46. Force A, Lynch M, Pickett FB, Amores A, Yan YL, Postlethwait J. Preservation of duplicate genes by complementary, degenerative mutations. Genetics. 1999;151(4):1531–45.

47. Conant GC, Wolfe KH. Turning a hobby into a job: how duplicated genes find new functions. Nat Rev Genet. 2008;9(12):938–50.

48. DeBolt S, Cook DR, Ford CM. L-Tartaric acid synthesis from vitamin C in higher plants. P Natl Acad Sci USA. 2006;103(14):5608–13.

49. Strommer J. The plant ADH gene family. Plant J. 2011;66(1):128–42.

50. Velasco R, Zharkikh A, Affourtit J, Dhingra A, Cestaro A, Kalyanaraman A, et al. The genome of the domesticated apple (*Malus x domestica Borkh.*). Nat Genet. 2010;42(10):833–9.

51. Forney CF, Breen PJ. Growth of strawberry fruit and sugar uptake of fruit disks at different inflorescence positions. Sci Hortic. 1985;27(1–2):55–62.

52. Veitia RA, Bottani S, Birchler JA. Cellular reactions to gene dosage imbalance: genomic, transcriptomic and proteomic effects. Trends Genet. 2008;24(8):390–7.

53. Galdon BR, Mesa DR, Rodriguez EMR, Romero CD. Influence of the cultivar on the organic acid and sugar composition of potatoes. J Sci Food Agric. 2010;90(13):2301–9.

54. Stafford HA. Distribution of tartaric acid in the leaves of certain angiosperms. Am J Bot. 1959;46(5):347–52.

55. Kramer EM. *Aquilegia*: a new model for plant development, ecology, and evolution. Annu Rev Plant Biol. 2009;60:261–77.

56. Worberg A, Quandt D, Barniske AM, Lohne C, Hilu KW, Borsch T. Phylogeny of basal eudicots: insights from non-coding and rapidly evolving DNA. Org Divers Evol. 2007;7(1):55–77.

57. Hoot SB, Magallon S, Crane PR. Phylogeny of basal eudicots based on three molecular data sets: atpB, rbcL, and 18S nuclear ribosomal DNA sequences. Ann Mo Bot Gard. 1999;86(1):1–32.

58. Moore MJ, Bell CD, Soltis PS, Soltis DE. Using plastid genome-scale data to resolve enigmatic relationships among basal angiosperms. P Natl Acad Sci USA. 2007;104(49):19363–8.

59. Wang HC, Moore MJ, Soltis PS, Bell CD, Brockington SF, Alexandre R, et al. Rosid radiation and the rapid rise of angiosperm-dominated forests. P Natl Acad Sci USA. 2009;106(10):3853–8.

60. Bremer B, Bremer K, Chase MW, Fay MF, Reveal JL, Soltis DE, et al. An update of the angiosperm phylogeny group classification for the orders and families of flowering plants: APG III. Bot J Linn Soc. 2009;161(2):105–21.

61. Roulin A, Auer PL, Libault M, Schlueter J, Farmer A, May G, et al. The fate of duplicated genes in a polyploid plant genome. Plant J. 2013;73(1):143–53.

62. Saito K, Loewus FA. Formation of tartaric acid in Vitaceous plants - relative contributions of L-ascorbic acid-inclusive and acid-noninclusive pathways. Plant Cell Physiol. 1989;30(6):905–10.

63. Nour V, Trandafir I, Ionica ME. HPLC organic acid analysis in different citrus juices under reversed phase conditions. Not Bot Horti Agrobo. 2010;38(1):44–8.

64. Hudina M, Stampar F. Sugars and organic acids contents of European (*Pyrus communis L.*) and Asian (*Pyrus serotina Rehd.*) pear cultivars. Acta Aliment Hung. 2000;29(3):217–30.

65. Sha SF, Li JC, Wu J, Zhang SL. Characteristics of organic acids in the fruit of different pear species. Afr J Agr Res. 2011;6(10):2403–10.

66. Fuleki T, Pelayo E, Palabay RB. Carboxylic-acid composition of varietal juices produced from fresh and stored apples. J Agr Food Chem. 1995;43(3):598–607.

67. Suarez MH, Rodriguez ER, Romero CD. Analysis of organic acid content in cultivars of tomato harvested in Tenerife. Eur Food Res Technol. 2008;226(3):423–35.

68. Ina Y. Pattern of synonymous and nonsynonymous substitutions: an indicator of mechanisms of molecular evolution. J Genet. 1996;75(1):91–115.

69. Kimura M. Preponderance of synonymous changes as evidence for the neutral theory of molecular evolution. Letters to Nature. 1977;267(5608):275–6.

70. Yang ZH. PAML: a program package for phylogenetic analysis by maximum likelihood. Comput Appl Biosci. 1997;13(5):555–6.

71. Worth CL, Gong S, Blundell TL. Structural and functional constraints in the evolution of protein families. Nat Rev Mol Cell Bio. 2009;10(10):709–20.

72. Yang ZH, Nielsen R. Codon-substitution models for detecting molecular adaptation at individual sites along specific lineages. Mol Biol Evol. 2002;19(6):908–17.

73. Yang ZH. PAML 4: phylogenetic analysis by maximum likelihood. Mol Biol Evol. 2007;24(8):1586–91.

74. Tajima F. Simple methods for testing the molecular evolutionary clock hypothesis. Genetics. 1993;135(2):599–607.

75. Baker PJ, Britton KL, Rice DW, Rob A, Stillman TJ. Structural consequences of sequence patterns in the fingerprint region of the nucleotide binding fold - implications for nucleotide specificity. J Mol Biol. 1992;228(2):662–71.

76. Yamaguchi H, Kanayama Y, Yamaki S. Purification and properties of NAD-dependent Sorbitol Dehydrogenase from apple fruit. Plant Cell Physiol. 1994;35(6):887–92.

77. Yennawar H, Moller M, Gillilan R, Yennawar N. X-ray crystal structure and small-angle X-ray scattering of sheep liver sorbitol dehydrogenase. Acta Crystallogr D. 2011;67(Pt5):440–6.

78. Gu ZL, Nicolae D, Lu HHS, Li WH. Rapid divergence in expression between duplicate genes inferred from microarray data. Trends Genet. 2002;18(12):609–13.

79. Wagner A. Decoupled evolution of coding region and mRNA expression patterns after gene duplication: implications for the neutralist-selectionist debate. P Natl Acad Sci USA. 2000;97(12):6579–84.

80. Toufighi K, Brady SM, Austin R, Ly E, Provart NJ. The botany array resource: e-northerns, expression angling, and promoter analyses. Plant J. 2005;43(1):153–63.

81. Fasoli M, Dal Santo S, Zenoni S, Tornielli GB, Farina L, Zamboni A, et al. The grapevine expression atlas reveals a deep transcriptome shift driving the entire plant into a maturation program. Plant Cell. 2012;24(9):3489–505.

82. Sweetman C, Wong DCJ, Ford CM, Drew DP. Transcriptome analysis at four developmental stages of grape berry (*Vitis vinifera cv. Shiraz*) provides insights into regulated and coordinated gene expression. BMC Genomics. 2012;13(1):691.

83. Conde A, Regalado A, Rodrigues D, Costa JM, Blumwald E, Chaves MM, et al. Polyols in grape berry: transport and metabolic adjustments as a physiological strategy for water-deficit stress tolerance in grapevine. J Exp Bot. 2015;66(3):889–906.

84. Mazzitelli L, Hancock RD, Haupt S, Walker PG, Pont SDA, McNicol J, et al. Co-ordinated gene expression during phases of dormancy release in raspberry (*Rubus idaeus L.*) buds. J Exp Bot. 2007;58(5):1035–45.

85. Pilati S, Perazzolli M, Malossini A, Cestaro A, Dematte L, Fontana P, et al. Genome-wide transcriptional analysis of grapevine berry ripening reveals a set of genes similarly modulated during three seasons and the occurrence of an oxidative burst at veraison. BMC Genomics. 2007;8(1):428.

86. Fortes AM, Agudelo-Romero P, Silva MS, Ali K, Sousa L, Maltese F, et al. Transcript and metabolite analysis in *Trincadeira* cultivar reveals novel information regarding the dynamics of grape ripening. BMC Plant Biol. 2011;11(1):149.

87. Melino VJ, Soole KL, Ford CM. A method for determination of fruit-derived ascorbic, tartaric, oxalic and malic acids, and its application to the study of ascorbic acid catabolism in grapevines. Aust J Grape Wine Res. 2009;15(3):293–302.

88. Wong DCJ, Sweetman C, Drew DP, Ford CM. VTCdb: a gene co-expression database for the crop species *Vitis vinifera* (grapevine). BMC Genomics. 2013;14(1):17.

89. Wong DCJ, Sweetman C, Ford CM. Annotation of gene function in citrus using gene expression information and co-expression networks. BMC Plant Biol. 2014;14(1):17.

90. Obayashi T, Okamura Y, Ito S, Tadaka S, Aoki Y, Shirota M, et al. ATTED-II in 2014: evaluation of gene coexpression in agriculturally important plants. Plant Cell Physiol. 2014;55(1):e6.

91. Foyer CH, Noctor G. Ascorbate and glutathione: the heart of the redox hub. Plant Physiol. 2011;155(1):2–18.

92. Wheeler GL, Jones MA, Smirnoff N. The biosynthetic pathway of vitamin C in higher plants. Nature. 1998;393(6683):365–9.

93. Cramer GR, Van Sluyter SC, Hopper DW, Pascovici D, Keighley T, Haynes PA. Proteomic analysis indicates massive changes in metabolism prior to the inhibition of growth and photosynthesis of grapevine (*Vitis vinifera L.*) in response to water deficit. BMC Plant Biol. 2013;13(1):49.

94. Krasensky J, Jonak C. Drought, salt, and temperature stress-induced metabolic rearrangements and regulatory networks. J Exp Bot. 2012;63(4):1593–608.

95. Conde A, Chaves MM, Geros H. Membrane transport, sensing and signaling in plant adaptation to environmental stress. Plant Cell Physiol. 2011;52(9):1583–602.

96. Larkin MA, Blackshields G, Brown NP, Chenna R, McGettigan PA, McWilliam H, et al. Clustal W and Clustal X version 2.0. Bioinformatics. 2007;23(21):2947–8.

97. Tamura K, Stecher G, Peterson D, Filipski A, Kumar S. MEGA6: molecular evolutionary genetics analysis Version 6.0. Mol Biol Evol. 2013;30(12):2725–9.

98. Nei M, Kumar S. Molecular Evolution and Phylogenetics. New York: Oxford University Press; 2000.

99. Jones DT, Taylor WR, Thornton JM. The rapid generation of mutation data matrices from protein sequences. Comput Appl Biosci. 1992;8(3):275–82.

100. Altschul SF, Gish W, Miller W, Myers EW, Lipman DJ. Basic local alignment search tool. J Mol Biol. 1990;215(3):403–10.

101. Wang YP, Tang HB, DeBarry JD, Tan X, Li JP, Wang XY, et al. MCScanX: a toolkit for detection and evolutionary analysis of gene synteny and collinearity. Nucleic Acids Res. 2012;40(7):e49.

102. Abagyan R, Totrov M. Biased probability Monte-Carlo conformational searches and electrostatic calculations for peptides and proteins. J Mol Biol. 1994;235(3):983–1002.

103. Reimand J, Arak T, Vilo J. g: profiler-a web server for functional interpretation of gene lists (2011 update). Nucleic Acids Res. 2011;39(suppl2):W307–15.

7

Keeping the rhythm: light/dark cycles during postharvest storage preserve the tissue integrity and nutritional content of leafy plants

John D Liu[1], Danielle Goodspeed[1,4], Zhengji Sheng[1], Baohua Li[2], Yiran Yang[1], Daniel J Kliebenstein[2,3] and Janet Braam[1*]

Abstract

Background: The modular body structure of plants enables detached plant organs, such as postharvest fruits and vegetables, to maintain active responsiveness to environmental stimuli, including daily cycles of light and darkness. Twenty-four hour light/darkness cycles entrain plant circadian clock rhythms, which provide advantage to plants. Here, we tested whether green leafy vegetables gain longevity advantage by being stored under light/dark cycles designed to maintain biological rhythms.

Results: Light/dark cycles during postharvest storage improved several aspects of plant tissue performance comparable to that provided by refrigeration. Tissue integrity, green coloration, and chlorophyll content were generally enhanced by cycling of light and darkness compared to constant light or darkness during storage. In addition, the levels of the phytonutrient glucosinolates in kale and cabbage remained at higher levels over time when the leaf tissue was stored under light/dark cycles.

Conclusions: Maintenance of the daily cycling of light and dark periods during postharvest storage may slow the decline of plant tissues, such as green leafy vegetables, improving not only appearance but also the health value of the crops through the maintenance of chlorophyll and phytochemical content after harvest.

Keywords: Biological clock, Chlorophyll, Circadian clock, Circadian rhythms, Vegetable and fruit preservation, Diurnal, Glucosinolates, Nutritional value, Vegetable and fruit shelf life

Background

Approximately one-third of food produced globally is lost or wasted [1], yet fewer resources are devoted to postharvest research and development than to efforts for improving productivity [2]. The modular design of plants [3] allows plant tissues and organs to remain biologically active even after harvest [4,5]. Therefore, capitalizing on the ability of harvested vegetables and fruits to continue to sense and respond to diverse stimuli, similarly to intact plants, may be a powerful approach to promote postharvest quality.

Research demonstrating the biological advantage of a functional circadian clock in plants led us to investigate whether maintaining diurnal (24-hour light/dark) cycles

may promote longevity and therefore reduced yield loss during postharvest storage of vegetables. The circadian clock enables plants to anticipate and prepare for the daily environmental changes that occur as a consequence of the rotation of the earth. Coordination of plant circadian rhythms with the external environment provides growth and reproductive advantages to plants [6], as well as enhanced resistance to insects [7] and pathogens [8,9]. The circadian clock also regulates aspects of plant biology that may have human health impact, such as levels of carbohydrates [10], ascorbic acid [11], chlorophyll [12], and glucosinolates [5] in edible plant species.

Plants exhibit exquisite sensitivity to light stimuli, and isolated plant leaves maintain responsiveness to light after harvest and can continue light-dependent biological processes, such as photosynthesis [13]. Additionally, the clocks of postharvest fruit and vegetable tissues can be

* Correspondence: braam@rice.edu
[1]Department of BioSciences, Rice University, Houston, TX 77005, USA
Full list of author information is available at the end of the article

entrained with 12-hour light/12-hour darkness cycles producing rhythmic behaviors not observed in tissues stored in constant light or constant dark [5]. A few studies have examined the effects of light on performance and longevity during postharvest storage [14]. For example, light exposure delays broccoli senescence and yellowing [13,15] but accelerates browning in cauliflower, a close relative of broccoli [16,17]. Other studies report that light exposure to broccoli during postharvest storage either provides no additional benefits [17] or decreases performance [18]. Postharvest light exposure improves chlorophyll content in cabbage [19], but leads to increased browning of romaine lettuce leaves [20]. Although exposure of spinach to light during postharvest storage can improve nutritional value [21,22], light can also accelerate spinach water loss, leading to wilting [22]. Together, these findings are inconclusive as to whether light exposure during postharvest storage can be generally beneficial, and the variation of the results may be attributable to differences in the plant species examined and the specific conditions used during postharvest storage, such as lighting intensities, temperature, humidity or packaging. Alternatively, light may be advantageous but only if present in its natural context with 24-hour periodicity because of such timing on circadian clock function.

This study aimed to examine whether mimicking aspects of the natural environment predicted to maintain circadian biological rhythms during postharvest storage of green leafy vegetables improves performance and longevity compared to postharvest storage under constant light or constant darkness. We focused this work on several popular and nutritionally valuable species, including kale (*Brassica oleracea* cv. acephala group) and cabbage (*Brassica oleracea*), members of the Brassicaceae family with worldwide production of approximately 70 million tons [23]. In addition, we analyzed green leaf lettuce (*Lactuca sativa*) and spinach (*Spinacia oleracea*), which have worldwide production of approximately 25 and 22 million tons, respectively [23]. Here, we report on the promotion of postharvest longevity, including tissue integrity and nutritional value, of green leafy vegetables by provision of 24-hour light/dark cycles during storage compared to storage under constant light or constant darkness.

Methods
Plant materials and storage conditions
Kale (*Brassica oleracea* cv. acephala group), cabbage (*Brassica oleracea*), green leaf lettuce (*Lactuca sativa*) and spinach (*Spinacia oleracea*) were purchased from a local organic grocer. Leaf tissue was cut into 2 cm disks and placed on 0.5% agar as described previously [5]. Light-treated leaf disks were stored under $120 \pm 10 \, \mu E \cdot m^{-2} \cdot s^{-1}$ (E = Einstein, defined as one mole of photons) at 22°C for

either cycles of 12 hours of light followed by 12 hours of darkness or constant 24-hour light. Leaf disks stored in constant darkness were either maintained at 22°C or refrigerated at 4°C. Samples for analysis were collected 6 hours after initiation of the light period for samples stored under 12-hour light/12-hour dark cycles or at comparable times for samples stored in constant light, in constant darkness, or under refrigeration to avoid time-of-day dependent differences in measured values.

Chlorophyll measurements
Frozen leaf disks were ground using a mortar and pestle, and approximately 50 mg of the resulting powder was mixed with 0.5 mL of 80% (v/v) acetone and incubated overnight at 4°C. Samples were centrifuged at 14,000 x g for 10 minutes at 4°C. The absorbance of the supernatant was measured spectrophotometrically at 645 and 663 nm using a Tecan Infinite M200 PRO (Tecan, Morrisville, NC). Chlorophyll concentrations relative to fresh weight were determined using the formula: total chlorophyll (μg/g fresh weight) = $(17.76A_{645} + 7.34A_{663})$/g plant tissue, as described [24]. Chlorophyll concentrations per gram dry weight were determined after plant materials were freeze dried at –80°C at 0.04 mBar overnight using a FreeZone 4.5 liter benchtop freeze dry system (Labconco, Kansas City, M.O.).

Electrolyte leakage measurements
For electrolyte leakage measurements, 2 cm leaf disks were placed into 50 ml of deionized room temperature water. After 30 minutes, electrical conductivity was measured using a Horiba B-173 Twin Cond Conductivity Meter (Horiba Instruments Inc., Kyoto, Japan). To avoid complications associated with electrolyte leakage due to initially cutting out the leaf disks, the first electrolyte leakage measurements were delayed as specified in the text.

Glucosinolate measurements
Tissue frozen with liquid nitrogen was ground using a mortar and pestle and then submerged in 90% methanol and extracted. Glucosinolates were identified and quantified in comparison to reference standards using HPLC-DAD, as previously described [25]. Total glucosinolate levels were determined by totaling the concentrations of individually quantified glucosinolate species (Additional file 1: Figure S1 and S2) as described previously [5].

Statistical analyses
Data were subjected to analysis of variance (ANOVA) with Bonferroni Post Hoc analysis to examine differences between leaf tissue stored under light/dark cycles to those stored under alternative condition such as constant light, constant dark or refrigerated in constant dark at each time point using SPSS Statistics software. Comparisons of

means between the beginning and end of the experiment were analyzed by Student's *t*-test using SPSS Statistics software.

Results

Kale, cabbage, lettuce and spinach leaf tissues show improved appearance when stored under 12-hour light/12-hour dark cycles

Fruits and vegetables after harvest can respond to repeated cycles of 12-hour light/12-hour dark, resulting in circadian clock function and rhythmic behaviors [5]. Because a functional plant circadian clock is physiologically advantageous [7,26,27] we sought to address whether postharvest storage under conditions that simulate day/night cycles, thereby potentially maintaining biological rhythms, would affect postharvest longevity. We chose to address this question using green leafy vegetables, including commonly consumed kale (*Brassica oleracea* cv. acephala group), cabbage (*Brassica oleracea*), green leaf lettuce (*Lactuca sativa*) and spinach (*Spinacia oleracea*), because we anticipated that the leaf organ would likely maintain light sensitivity and responsiveness even after harvest. To begin to determine whether daily light/dark cycles during postharvest storage affects leaf longevity, we compared the overall appearance of leaf disks that were stored at 22°C under cycles of

12-hour light/12-hour darkness (LD) versus leaf disks stored under constant light (LL) or constant darkness (DD) for various lengths of time (Figure 1).

Under cycles of 12-hour light/12-hour darkness, kale leaf disks were dark green after 3 days of storage (Figure 1A; LD). After 6 days and 15 days of storage, the kale disks showed lighter green coloration than the kale disks stored for 3 days (Figure 1A; LD). However, the kale leaf disks stored under constant light were lighter green than the kale disks stored under light/dark cycles and showed some brown or yellow discoloration after 3 and 6 days (Figure 1A; LL). By 15 days, the kale leaf disks stored under constant light lost nearly all green coloration and showed light and dark shades of browning with shape changes resulting from leaf folding and shrinkage (Figure 1A; LL). The kale leaf disks stored under constant darkness resembled those stored under constant light, except that the 3-day kale samples were darker green than the 3-day constant light-stored kale leaf disks (Figure 1A; DD), suggesting that the constant light may have constituted a greater stress on the kale leaves than constant darkness. These results indicate that postharvest storage with daily cycling of light and darkness improved the appearance of the kale leaf tissue compared to storage under either constant light or constant darkness. However, the

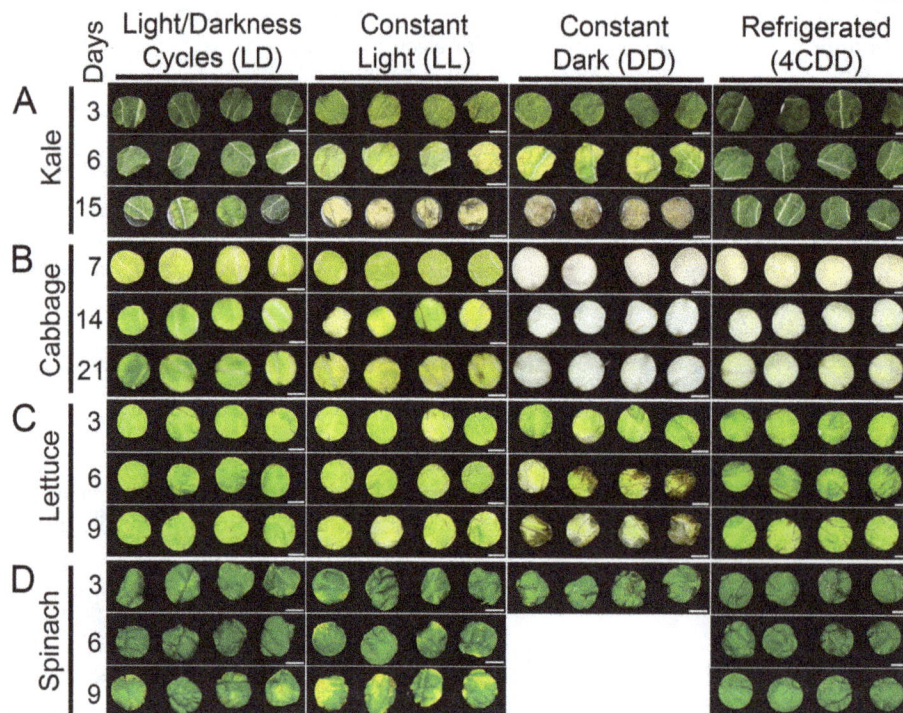

Figure 1 Leaf disk appearance depends upon light exposure during post harvest storage. Representative photographic images of kale (**A**), cabbage (**B**), lettuce (**C**), and spinach (**D**) leaf disks after varying number of days (number of days indicated at left) of storage at 22°C under 12-hours light/12-hours darkness (Light/Dark Cycles, LD), constant light (LL), or constant dark (DD), or at 4°C under constant dark (Refrigerated, 4CDD). Spinach disks disintegrated after 3 days when stored under constant darkness and therefore images were not available. (Scale bars, 1 cm).

preservation benefit obtained from postharvest storage under light/dark cycles at 22°C appeared to be less than that provided by refrigeration; kale leaf disks stored at 4°C with constant darkness, were comparable in their dark green coloration whether stored for 3, 6 or 15 days (Figure 1A; 4CDD).

Cabbage leaf disks stored under cycles of 12-hour light/12-hour darkness showed brown spots along the disk edges that increased in intensity over the storage period of 7, 14, and 21 days (Figure 1B; LD). However, although the 7-day cabbage leaf samples were light green in coloration, the 14- and 21-day cabbage leaf disks stored under light/dark cycles had darker green coloration (Figure 1B; LD), suggesting increased photosynthetic activity over storage time. In contrast, although the cabbage leaf disks stored under constant light were also light green after 7 days of storage, the 14- and 21-day cabbage leaf disks were more yellow and included more brown discolorations (Figure 1B; LL). Remarkably, the absence of light exposure during post-harvest storage had a dramatic effect on the cabbage leaf disk coloration. Cabbage leaf disks stored under constant darkness at either 22°C or 4°C were pale tan or yellow after 3 days of storage (Figure 1B; DD and 4CDD). The constant darkness-exposed cabbage leaf disks stored at 22°C appeared nearly white in color by 14 and 21 days; those at 4°C had a yellowish appearance after 2 or 3 weeks of storage (Figure 1B; DD and 4CDD).

Lettuce and spinach leaf disks tissue were nearly uniformly green, with little difference in color intensity between 3 and 6 days of storage under cycles of 12-hour light/12-hour darkness (Figure 1C,D; LD). By 9 days of storage under light/dark cycles, however, both lettuce and spinach leaf disks looked slightly less green, and most of the spinach leaf disks had distinct patches of yellow (Figure 1C,D; LD). In contrast, the loss of green coloration and increased yellowing over time was much more apparent in the lettuce and spinach leaf disks stored under constant light; the lettuce leaf disks were pale green by 9 days (Figure 1C; LL), and all the spinach disks had large yellow patches (Figure 1D; LL). Lettuce and spinach leaf disks stored under constant darkness displayed small brown patches by 3 days (Figure 1C,D; DD). After 6 and 9 days of storage under constant darkness, the lettuce disks had large wet patches of darkened tissue (Figure 1C; DD). However, the spinach leaf disks stored under constant darkness at 22°C for 6 days completely disintegrated and therefore could not be moved for photographic imaging. Lettuce and spinach leaf disks stored in constant darkness at 4°C largely maintained dark green coloration at 6 days and were lighter green at 9 days, similar to that of the disks stored under light/dark cycles at 22°C (Figure 1C,D; 4CDD). However, after 6 days of storage at 4°C, the lettuce leaf disks also displayed browning around the vascular tissues (Figure 1C; 4CDD).

Overall, the image analysis shown in Figure 1 suggests that postharvest storage under cycles of 12-hour light and 12-hour darkness may enable kale, cabbage, lettuce, and spinach leaf tissues to maintain physiological functioning for longer durations after harvest. The reduction in green color and appearance of brown discoloration suggests that postharvest storage in constant light or constant darkness may accelerate loss of tissue viability.

Chlorophyll content retention is higher in leaf tissue stored under cycles of 12-hour light/12-hour darkness

To further characterize kale, cabbage, lettuce and spinach leaf health and viability during postharvest storage, we quantified chlorophyll content in leaf samples after storage under cycles of 12-hour light/12-hour darkness to leaf tissues stored under constant light or constant darkness. Three sets of comparative data are shown (Figure 2, Additional file 1: Figure S3 and Figure S4). Because our primary focus was to determine whether light/dark cycles were advantageous relative to constant light or constant dark storage conditions, we first conducted two-way comparative statistical analyses between data derived from the samples stored under light/dark cycles relative to comparable samples stored under the alternative condition

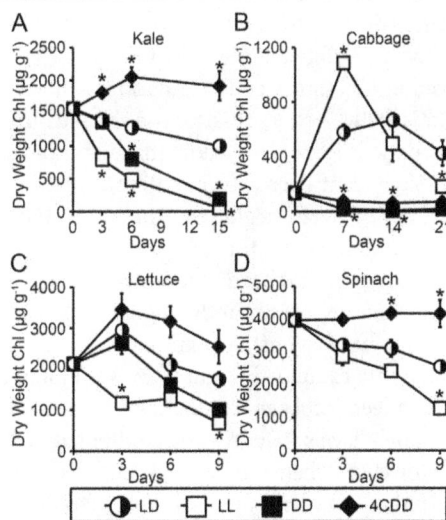

Figure 2 Chlorophyll content retention was higher in kale, cabbage, lettuce and spinach leaves stored in light/dark cycles compared to constant light or constant darkness at 22°C.
Chlorophyll content relative to dry weight was quantified from leaf tissue disks of kale (**A**), cabbage (**B**), lettuce (**C**) and spinach (**D**) stored under cycles of 12-hours light/12-hours darkness (LD, half-filled circles), constant light (LL, open squares), or constant darkness (DD, filled squares) at 22°C, or under constant darkness at 4°C (4CDD, filled diamonds). Mean ± SE; n = 4. Asterisks indicate significant differences (p < 0.05. ANOVA Test with Bonferroni Post Hoc analysis) between data derived from leaf disks stored under light/dark cycles (22°C) and that derived from leaf disks stored under other conditions at each time point. Statistical analyses of differences over time for each plant type are shown in Additional file 1: Figure S4.

(that is, constant light, constant dark or refrigerated/dark). Figure 2 presents statistical analysis of storage-dependent differences in chlorophyll levels relative to dry weight at each time point. Additional file 1: Figure S3 shows similar analyses but of storage-dependent differences in chlorophyll levels relative to fresh weight. Finally, to evaluate whether there were significant changes in chlorophyll levels of each plant type over time, statistical analyses of differences in chlorophyll content at the beginning and end of the experiments for kale, cabbage, lettuce, and spinach are shown in Additional file 1: Figure S4.

Consistent with the loss of green coloration in the representative leaf disk samples shown in Figure 1A, postharvest storage of kale leaf disks in either constant light or constant dark led to significantly greater losses in kale chlorophyll content within 3 or 6 days of postharvest storage compared to storage under 12-hour light/12-hour dark cycles (Figure 2A; Additional file 1: Figure S3A). Kale leaf disks stored for 15 days under constant light lost 97% and 93% of their original chlorophyll content relative to dry and fresh weight, respectively (Additional file 1: Figure S4A, E); kale leaf disks stored under constant darkness lost 88% and 89% of their chlorophyll content relative to total dry and fresh weight, respectively (Additional file 1: Figure S4A, E). In contrast, 15 days of storage under cycles of 12-hour light/12-hour darkness led to loss of only 36% and 9% of the kale leaf disk chlorophyll relative to dry and fresh weight, respectively (Additional file 1: Figure S4A, E). Kale leaf disks stored at 4°C under constant darkness, however, performed statistically better than those stored under cycles of light/dark (Figure 2A; Additional file 1: Figure S3A), with no significant decreases in chlorophyll content relative to total dry or fresh weight over the full 15 days of the experiment (Additional file 1: Figure S4A, E).

Postharvest storage of cabbage leaf disks under light/dark cycles resulted in significantly higher chlorophyll levels than storage under constant dark either at 22°C or 4°C at all time points examined (Figure 2B; Additional file 1: Figure S3B). Indeed, cabbage leaf disks began with only modest chlorophyll levels (Figure 2B; Additional file 1: Figure S3B). However, when the cabbage leaf disks were stored under either constant light or light/dark cycles, chlorophyll content increased over time (Figure 2B, Additional file 1: Figure S3B) with significantly higher levels remaining even after three weeks of storage (Additional file 1: Figure S4B, F). Light-induced synthesis is likely responsible for the elevated chlorophyll levels (Figure 2B; Additional file 1: Figure S3B, S4B, F) and the enhanced green coloration observed in the cabbage leaf images (Figure 1B) of the samples stored under light/dark cycles or constant light but absent in the cabbage leaf disks stored under constant darkness either at 22°C or 4°C (Figure 1B, 2B; Additional file 1: Figure S3B). Storage under light/dark cycles was also more successful than

constant light exposure in maintaining higher chlorophyll levels after long-term storage of 3 weeks (Figure 2B; Additional file 1: Figure S3B). Over time, storage under constant light may be counterproductive; whereas cabbage leaf disks stored for 7 days under constant light had significantly higher chlorophyll content than leaf disks stored under light/dark cycles (Figure 2B; Additional file 1: Figure S3B), by three weeks of storage, the leaf disks stored under light/dark cycles retained at least 2-fold more chlorophyll than samples stored under constant light (Figure 2B; Additional file 1: Figure S3B). These results indicate that light during postharvest storage can have a profound effect on chlorophyll levels in cabbage, consistent with previous reports [19], and that diurnal cycling of light and darkness prolongs this benefit during longer term storage.

Storage under cycles of 12-hour light/12-hour darkness also promoted chlorophyll retention (relative to both dry and fresh weight) in lettuce leaf disks, comparable to that of lettuce leaf disks stored under refrigeration; chlorophyll levels were statistically indistinguishable between lettuce leaf disks stored under light/dark cycles versus those refrigerated under constant darkness conditions after 3, 6 or 9 days of storage (Figure 2C; Additional file 1: Figure S3C). Postharvest storage of lettuce leaf disks either at 22°C under light/dark cycles or under refrigeration resulted in no significant change in chlorophyll content over the course of the 9-day experiment, whereas the lettuce leaf disks stored under either constant light or constant darkness, lost more than 50% of their starting chlorophyll content (Additional file 1: Figure S4C, G).

Chlorophyll content of spinach leaf disks was not significantly affected by treatment conditions for the first 3 days of postharvest storage (Figure 2D; Additional file 1: Figure S3D, S4D, H). However, the spinach leaf disks stored at 22°C in constant darkness disintegrated by 6 days and were therefore unable to be further analyzed. Spinach leaf disks stored under light/dark cycles had similar chlorophyll content to those stored under constant light with relatively stable chlorophyll retention until day 9 when chlorophyll levels in both samples decreased significantly from initial levels (Figure 2D; Additional file 1: Figure S3D, S4D, H). In contrast, refrigeration led to stable chlorophyll levels in the spinach leaf disks over the course of the experiment (Additional file 1: Figure S4D, H).

These results indicate that chlorophyll content of postharvest green leafy vegetables varies depending upon the storage conditions and suggests that storage under 12-hour cycles of light and darkness, known to maintain the plant circadian clock [5], can improve kale, cabbage and lettuce chlorophyll content maintenance relative to storage in constant light or constant dark. Perhaps surprisingly light/dark cycles during postharvest storage may be at least as beneficial as refrigeration with respect to chlorophyll content for cabbage and lettuce.

Tissue integrity improved by postharvest storage under light/dark cycles

Over time during postharvest storage, plant tissues typically show visible signs of tissue disintegration (e.g., Figure 1). To determine if maintaining light/dark cycles during storage of post harvest leafy vegetables could prolong tissue integrity, we compared electrolyte leakage from kale, cabbage, lettuce, and spinach leaf disks stored over time under cycles of 12-hour light/12-hour darkness to leaf disks stored under constant light or constant darkness at 22°C or constant darkness at 4°C. Figure 3 shows that postharvest storage under light/dark cycles and refrigeration were comparable, with respect to leaf tissue integrity maintenance of kale, cabbage and lettuce, as measured by electrolyte leakage, (Figure 3A-C; LD and 4CDD). When directly comparing light/dark storage to other conditions, a statistically significant benefit to diurnal stimuli during storage was apparent relative to constant light for kale (Figure 3A; LL), constant darkness for cabbage and lettuce (Figure 3B,C; DD), and constant darkness (at 3 days) and constant light (at 6 days) for spinach (Figure 3D; LL, and Figure 3E; DD).

Postharvest storage under constant dark was detrimental to kale, cabbage, and lettuce tissue integrity, with at least 4-fold increases in electrolyte leakage, whereas storage under light/dark cycles at 22°C or refrigeration resulted in no significant increase in electrolyte leakage over

the course of the experiment (Additional file 1: Figure S5). Constant light treatment also led to significant increases in electrolyte leakage from kale and lettuce leaf disks, but not cabbage leaf disks, over the storage periods examined (Additional file 1: Figure S5).

Overall the results shown in Figure 3 and Additional file 1: Figure S5 provide evidence that daily cycles of light and darkness during postharvest storage resulted in superior leaf tissue integrity maintenance largely comparable to refrigeration, whereas either constant light or constant dark storage conditions were detrimental.

Total glucosinolate levels are maintained when kale and cabbage are stored in light/dark cycles

Our results indicate that storage of kale, cabbage, lettuce, and spinach in light/dark cycles can improve the postharvest longevity of chlorophyll levels and tissue integrity. Next we were interested in determining whether plant maintenance under daily cycles of light and darkness affects human-health relevant metabolite content. In particular, we sought to examine whether kale and cabbage stored under light/dark cycles maintain their glucosinolate content longer than when stored under constant light, constant darkness, or refrigeration.

Figure 4A shows total glucosinolate levels in kale leaf disks after 0, 3, 6 and 15 days of postharvest storage under different conditions. Individual glucosinolate levels are shown in Additional file 1: Figure S1. Total glucosinolate levels were comparable between kale leaf disks stored at 22°C under light/dark cycles and leaf disks stored at 4°C in the dark (Figure 4A); after 15 days of postharvest storage under these conditions, total glucosinolate levels decreased by less than 35% (Additional file 1: Figure S6A). In comparison to light/dark storage conditions, constant light or constant darkness exposure during storage resulted in significantly reduced glucosinolate content in the kale leaf disks (Figure 4A). By 15 days of postharvest storage under constant light or constant darkness at 22°C, the kale leaf disks lost over 80% and 99% of initial levels, respectively.

Daily cycles of light and darkness also promoted maintenance of glucosinolate content during postharvest storage of cabbage (Figure 4B; Additional file 1: Figure S2). Total glucosinolate content in the cabbage leaf disks stored under light/dark cycles remained stable with no significant fluctuation in levels over the 21 days of analysis (Figure 4B; Additional file 1: Figure S6B). In comparison to total glucosinolate levels in light/dark-stored cabbage, the glucosinolate levels were significantly lower by 7 days when cabbage leaf disks were stored under constant darkness (Figure 4B; DD) and by 21 days when stored under constant light (Figure 4B; LL). Total glucosinolate levels declined by 70% and 88%, respectively, in cabbage disks stored at 22°C under constant light or constant darkness

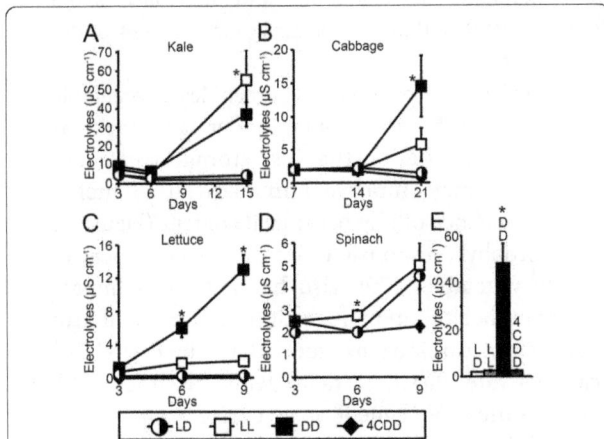

Figure 3 Electrolyte leakage from kale, cabbage, lettuce, and spinach leaf disks was affected by light exposure during storage. Electrolytes released from leaves of kale (**A**), cabbage (**B**), lettuce (**C**) or spinach (**D & E**) were measured after storage under cycles of 12-hour light/12-hour darkness (LD, half-filled circles), constant light (LL, open squares), or constant darkness (DD, filled squares) at 22°C or under constant darkness at 4°C (4CDD, filled diamonds). Mean ± SE; n = 4. Asterisks indicate significant differences (p < 0.05, ANOVA Test with Bonferroni Post Hoc analysis) between data derived from leaf disks stored under light/dark cycles (22°C) and that derived from leaf disks stored under other conditions at each time point. Statistical analyses of differences over time for each plant type are shown in Additional file 1: Figure S5.

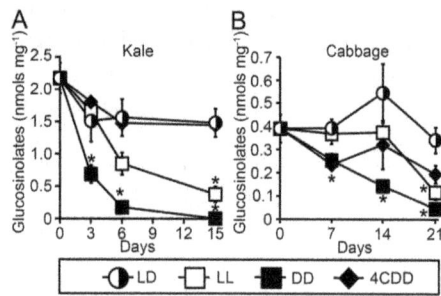

Figure 4 Maintenance of total glucosinolate levels in kale and cabbage leaves when stored under light/dark cycles. Glucosinolate species was quantified from leaf disks of kale **(A)** and cabbage **(B)** stored under cycles of 12-hour light/12-hour darkness (LD, half-filled circles), constant light (LL, open squares), or constant darkness (DD, filled squares) at 22°C or under constant darkness at 4°C (4CDD, filled diamonds). Mean ± SE; n = 4. Asterisks indicate significant differences (p < 0.05. ANOVA Test with Bonferroni Post Hoc analysis) between samples kept in light/darkness cycles (22°C) and those stored in constant light (22°C), constant dark (22°C) and under constant darkness (4°C) for a specified time point. Statistical analyses of differences over time for each plant type are shown in Additional file 1: Figure S6.

(Additional file 1: Figure S6B). Remarkably, glucosinolate levels of the cabbage leaf disks stored at 4°C also showed a significant decrease by 21 days, with a loss of 50% of the initial glucosinolate levels (Additional file 1: Figure S6B), indicating that storage under cycle of light/darkness led to enhanced retention of this valuable phytochemical even relative to refrigeration.

Discussion

In this work, we examined whether kale, cabbage, lettuce and spinach leaf tissue maintain the ability to respond to light/dark cycles during postharvest storage and whether under these conditions that better mimic the natural light cycles of the environment tissue deterioration would be reduced. Our goal was to expose plant tissues to diurnal conditions known to maintain the functioning of the circadian clock [5] and thereby capitalize on physiological enhancements conferred by robust circadian rhythms. Plants grown under light/dark cycles that match the endogenous cycling of their internal circadian clock have a growth and reproductive advantage over plants exposed to light/dark cycles that do not match their internal oscillator [26,28]. Furthermore, phasing of circadian rhythms so as to be synchronized with the external environment promotes biotic stress resistance [5,7,27].

We found that storing green leafy vegetables in cycles of 12 hours of light followed by 12 hours of darkness improved several postharvest performance markers compared to postharvest storage of the leaf tissues under constant light or constant darkness. Similarly, a modest reduction in senescence was noted for post-harvest broccoli stored under natural light/dark cycles [15]. Perhaps

surprisingly, we found that storage in light/dark cycles resulted in several aspects of postharvest performance being comparable to storage under refrigeration, a commonly practiced method of postharvest storage thought to slow down cellular breakdown [2]. The longevity of kale and lettuce leaf color, chlorophyll levels, and tissue integrity, which are important contributors to the appeal of green leafy vegetables to consumers [29], were largely indistinguishable whether the kale and lettuce leaf samples were stored at 22°C under light/dark cycles or were stored under refrigeration in constant darkness (Figures 1 and 3). Spinach leaf samples also maintained green coloration and chlorophyll levels under light/dark cycles at 22°C as well as when refrigerated, but refrigeration was more successful at preventing spinach leaf tissue breakdown. Significant improvement of green coloration and chlorophyll content was seen when cabbage leaves were stored under light/dark cycles at 22°C compared to refrigeration, demonstrating that light may not only be important for clock entrainment but also can provide the additional benefit of promoting continued photosynthesis during postharvest storage. Promotion of photosynthesis and/or chlorophyll levels was previously observed in post-harvest crops stored under light [15,17,30]. However, constant light during postharvest storage can also cause detrimental physiological activity, such as respiration leading to browning [30] and transpiration contributing to weight loss [16,18,31]. Therefore, cycling of light treatment with darkness periods may not only maintain clock function but may also avoid physiological damage that may occur in plant tissues under too much light.

In addition to improvement of green leafy vegetable appearance by postharvest storage under light/dark cycles, we found that this postharvest storage treatment of plant crops may improve human health benefits through maintenance of phytochemical content (Figures 2 and 4). Chlorophyll, responsible for the visual appeal of green leafy vegetables [29], also has beneficial impact upon human health upon ingestion. Chlorophyll can limit efficacy of carcinogens, such aflatoxin B_1 [32-36] and can activate Phase II detoxifying enzymes [37]. Additional anticancer benefit may derive from glucosinolates in kale and cabbage. Glucosinolates, sulfur-containing compounds that play a major role in *Brassicaceae* plant herbivore defense [38], also underlie the human health benefits attributed to *Brassicaceae* (cruciferous) vegetable consumption [39,40]. For example, the glucosinolate glucoraphanin (4-methylsulfinylbutyl) has potent anticancer activity [37,41]. Previous studies have shown that glucosinolate levels can be maintained by refrigeration [42] or exposure to radiation [43]; here we find that post-harvest storage under light/dark cycles can also lead to sustained glucosinolate levels (Figure 4).

Light/dark cycles also maintain the circadian clock function of other edible crops after harvest, including zucchini, carrots, sweet potatoes, and blueberries [5]. These fruits and vegetables displayed time-dependent differences in insect resistance strongly suggesting temporal fluctuations in diverse metabolites, some of which may have important human health impact. Whether continued promotion of circadian periodicity postharvest can also improve longevity of tissue integrity and phytochemical content in diverse vegetables and fruits, as we have shown with kale, cabbage, lettuce, and spinach, remains to be investigated.

Conclusions

Here we show that detached kale, cabbage, lettuce and spinach leaves show enhanced tissue longevity through continued exposure to diurnal light/dark cycles during storage. In addition, human-health relevant metabolites, such as glucosinolates and chlorophyll, are also retained at higher levels under diurnal storage conditions, suggesting that postharvest vegetables that retain natural rhythms during storage may be of greater nutritional value. These results provide additional evidence that, even postharvest, plant tissues retain the ability to sense external stimuli and respond in ways that affect tissue integrity and cellular metabolite levels. Translation of our understanding of plant physiology to postharvest crops, including the profound effects of the circadian clock on plant performance, may help to improve postharvest crop performance and reduce loss.

Additional file

Additional file 1: Figure S1. Accumulation of individual glucosinolate species in kale disks. Figure S2. Accumulation of individual glucosinolate species in cabbage disks. Figure S3. Chlorophyll content in fresh weight leaf tissue was maintained at higher levels in light/darkness stored vegetables. Figure S4. Chlorophyll content in fresh and dry weight leaf tissue was maintained at higher levels over time when stored in light/darkness cycles or under refrigeration. Figure S5. Electrolyte leakage from kale, cabbage, lettuce, and spinach leaf disks is increased when stored under constant light or constant dark in kale, cabbage and spinach. Figure S6. Maintenance of total glucosinolate levels in kale and cabbage leaves when stored under light/dark cycles.

Abbreviations

LD: 12-hour light/12-hour darkness cycles at 22°C; LL: 24-hour constant light at 22°C; DD: 24-hour constant dark at 22°C; 4CDD: 24-hour constant dark under refrigeration at 4°C.

Competing interests

The authors declare that they have no competing interests.

Authors' contributions

JDL participated in study design, data collection and statistical analysis and drafted the manuscript. DG participated in study design and statistical analysis. ZS participated in chlorophyll and electrolyte leakage measurements of kale and cabbage samples. YY participated in chlorophyll and electrolyte leakage measurements of green leaf lettuce and spinach samples. BL and DJK participated in glucosinolate measurements in kale and cabbage

samples. JB conceived of the study, participated in study design and drafted the manuscript. All authors read and approved the final manuscript.

Acknowledgements

This work was supported by the Agriculture and Food Research Initiative competitive grant 2015-67013-22813 of the USDA National Institute of Food and Agriculture, the National Science Foundation [grant number MCB 0817976], and a Rice University Institute of Biosciences and Bioengineering Medical Innovations Award to JB, and the National Science Foundation [grant number DBI 0820580] to D.J.K.

Author details

[1]Department of BioSciences, Rice University, Houston, TX 77005, USA. [2]Department of Plant Sciences, University of California, Davis, CA 95616, USA. [3]DynaMo Centre of Excellence, Department of Plant and Environmental Sciences, Faculty of Science, University of Copenhagen, Thorvaldsensvej 40, 1871 Frederiksberg C, Denmark. [4]Current Address: Department of Obstetrics and Gynecology, Baylor College of Medicine, Houston, TX 77030, USA.

References

1.　FAO. Global Food Losses and Food Waste: Extent, Causes and Prevention. Rome: Food and Agriculture Organization of the United Nations; 2011.
2.　Kader AA. A Perspective on Postharvest Horticulture (1978–2003). HortSci. 2003;38:1004–8.
3.　Watkinson AR, White J. Some life-history consequences of modular construction in plants. Philos Trans R Soc Lond B Biol Sci. 1986;313:31–51.
4.　Burton WG. Post-Harvest Physiology of Food Crops. London, New York: Longman; 1982.
5.　Goodspeed D, Liu JD, Chehab EW, Sheng Z, Francisco M, Kliebenstein DJ, et al. Postharvest circadian entrainment enhances crop pest resistance and phytochemical cycling. Curr Biol. 2013;23:1235–41.
6.　Harmer SL. The circadian system in higher plants. Annu Rev Plant Biol. 2009;60:357–77.
7.　Goodspeed D, Chehab EW, Min-Venditti A, Braam J, Covington MF. Arabidopsis synchronizes jasmonate-mediated defense with insect circadian behavior. Proc Natl Acad Sci. 2012;109:4674–7.
8.　Wang W, Barnaby JY, Tada Y, Li H, Tör M, Caldelari D, et al. Timing of plant immune responses by a central circadian regulator. Nature. 2011;470:110–4.
9.　Bhardwaj V, Meier S, Petersen LN, Ingle RA, Roden LC. Defence responses of Arabidopsis thaliana to infection by pseudomonas syringae are regulated by the Circadian clock. PLoS One. 2011;6:e26968.
10.　Sicher RC, Kremer DF, Harris WG. Diurnal carbohydrate metabolism of barley primary leaves. Plant Physiol. 1984;76:165–9.
11.　Kiyota M, Numayama N, Goto K. Circadian rhythms of the l-ascorbic acid level in Euglena and spinach. J Photochem Photobiol B. 2006;84:197–203.
12.　Hasperué JH, Chaves AR, Martínez GA. End of day harvest delays postharvest senescence of broccoli florets. Postharvest Biol Technol. 2011;59:64–70.
13.　Costa L, Millan Montano Y, Carrión C, Rolny N, Guiamet JJ. Application of low intensity light pulses to delay postharvest senescence of Ocimum basilicum leaves. Postharvest Biol Technol. 2013;86:181–91.
14.　Nilsson T. Postharvest handling and storage of vegetables. In: Shewfelt RL, Bruckner B, editors. Fruit and Vegetable Quality: An Integrated View. Pennsylvania: CRC Press; 2000. p. 96–121.
15.　Büchert AM, Gómez Lobato ME, Villarreal NM, Civello PM, Martínez GA. Effect of visible light treatments on postharvest senescence of broccoli (Brassica oleracea L.). J Sci Food Agric. 2011;91:355–61.
16.　Sanz Cervera S, Olarte C, Echávarri JF, Ayala F. Influence of exposure to light on the sensorial quality of minimally processed cauliflower. J Food Sci. 2007;72:S012–8.
17.　Olarte C, Sanz S, Federico Echávarri J, Ayala F. Effect of plastic permeability and exposure to light during storage on the quality of minimally processed broccoli and cauliflower. LWT - Food Sci Technol. 2009;42:402–11.
18.　Kasim R, Kasim MU. Inhibition of yellowing in Brussels sprouts (B. oleraceae var. gemmifera) and broccoli (B. oleraceae var. italica) using light during storage. J Food Agric Environ. 2007;5:126.
19.　Perrin PW. Poststorage effect of light, temperature and nutrient spray treatments on chlorophyll development in cabbage. Can J Plant Sci. 1982;62:1023–6.

20. Martínez-Sánchez A, Tudela JA, Luna C, Allende A, Gil MI. Low oxygen levels and light exposure affect quality of fresh-cut Romaine lettuce. Postharvest Biol Technol. 2011;59:34–42.

21. Toledo MEA, Ueda Y, Imahori Y, Ayaki M. L-ascorbic acid metabolism in spinach (Spinacia oleracea L.) during postharvest storage in light and dark. Postharvest Biol Technol. 2003;28:47–57.

22. Lester GE, Makus DJ, Hodges DM. Relationship between fresh-packaged spinach leaves exposed to continuous light or dark and bioactive contents: effects of cultivar, leaf size, and storage duration. J Agric Food Chem. 2010;58:2980–7.

23. FAO.ORG. [http://faostat3.fao.org/].

24. Porra RJ. The chequered history of the development and use of simultaneous equations for the accurate determination of chlorophylls a and b. In: Discoveries in Photosynthesis. Netherlands: Springer; 2005. p. 633–40.

25. Kliebenstein DJ, Kroymann J, Brown P, Figuth A, Pedersen D, Gershenzon J, et al. Genetic control of natural variation in Arabidopsis Glucosinolate accumulation. Plant Physiol. 2001;126:811–25.

26. Dodd AN, Salathia N, Hall A, Kévei E, Tóth R, Nagy F, et al. Plant Circadian clocks increase photosynthesis, growth, survival, and competitive advantage. Science. 2005;309:630–3.

27. Goodspeed D, Chehab EW, Covington MF, Braam J. Circadian control of jasmonates and salicylates: the clock role in plant defense. Plant Signal Behav. 2013;8:e23123.

28. Green RM, Tingay S, Wang Z-Y, Tobin EM. Circadian rhythms confer a higher level of fitness to Arabidopsis plants. Plant Physiol. 2002;129:576–84.

29. Hutchings JB. Food Color and Appearance. Gaithersburg, Maryland: Aspen Publishers, Inc.; 1999.

30. Ayala F, Echávarri JF, Olarte C, Sanz S. Quality characteristics of minimally processed leek packaged using different films and stored in lighting conditions. Int J Food Sci Technol. 2009;44:1333–43.

31. Barbieri G, Bottino A, Orsini F, De Pascale S. Sulfur fertilization and light exposure during storage are critical determinants of the nutritional value of ready-to-eat friariello campano (Brassica rapa L. subsp. sylvestris). J Sci Food Agric. 2009;89:2261–6.

32. Egner PA, Muñoz A, Kensler TW. Chemoprevention with chlorophyllin in individuals exposed to dietary aflatoxin. Mutat Res Mol Mech Mutagen. 2003;523–524:209–16 [Dietary and Medicinal Antimutagens and Anticarcinogens: Molecular Mechanisms and Chemopreventive Potential].

33. Simonich MT, Egner PA, Roebuck BD, Orner GA, Jubert C, Pereira C, et al. Natural chlorophyll inhibits aflatoxin B1-induced multi-organ carcinogenesis in the rat. Carcinogenesis. 2007;28:1294–302.

34. Simonich MT, McQuistan T, Jubert C, Pereira C, Hendricks JD, Schimerlik M, et al. Low-dose dietary chlorophyll inhibits multi-organ carcinogenesis in the rainbow trout. Food Chem Toxicol Int J Publ Br Ind Biol Res Assoc. 2008;46:1014–24.

35. Jubert C, Mata J, Bench G, Dashwood R, Pereira C, Tracewell W, et al. Effects of chlorophyll and chlorophyllin on low-dose aflatoxin B1 pharmacokinetics in human volunteers. Cancer Prev Res (Phila Pa). 2009;2:1015–22.

36. McQuistan TJ, Simonich MT, Pratt MM, Pereira CB, Hendricks JD, Dashwood RH, et al. Cancer chemoprevention by dietary chlorophylls: a 12,000-animal dose–dose matrix biomarker and tumor study. Food Chem Toxicol. 2012;50:341–52.

37. Fahey JW, Stephenson KK, Dinkova-Kostova AT, Egner PA, Kensler TW, Talalay P. Chlorophyll, chlorophyllin and related tetrapyrroles are significant inducers of mammalian phase 2 cytoprotective genes. Carcinogenesis. 2005;26:1247–55.

38. Hopkins RJ, van Dam NM, van Loon JJA. Role of glucosinolates in insect-plant relationships and multitrophic interactions. Annu Rev Entomol. 2009;54:57–83.

39. Higdon JV, Delage B, Williams DE, Dashwood RH. Cruciferous vegetables and human cancer risk: epidemiologic evidence and mechanistic basis. Pharmacol Res. 2007;55:224–36 [Nutritional Pharmacology].

40. Hayes JD, Kelleher MO, Eggleston IM. The cancer chemopreventive actions of phytochemicals derived from glucosinolates. Eur J Nutr. 2008;47:73–88.

41. Zhang Y, Talalay P, Cho CG, Posner GH. A major inducer of anticarcinogenic protective enzymes from broccoli: isolation and elucidation of structure. Proc Natl Acad Sci. 1992;89:2399–403.

42. Rangkadilok N, Tomkins B, Nicolas ME, Premier RR, Bennett RN, Eagling DR, et al. The effect of post-harvest and packaging treatments on glucoraphanin concentration in broccoli (Brassica oleracea var. Italica). J Agric Food Chem. 2002;50:7386–91.

43. Banerjee A, Variyar PS, Chatterjee S, Sharma A. Effect of post harvest radiation processing and storage on the volatile oil composition and glucosinolate profile of cabbage. Food Chem. 2014;151:22–30.

8

Impacts of nucleotide fixation during soybean domestication and improvement

Shancen Zhao[1,2*†], Fengya Zheng[1†], Weiming He[2], Haiyang Wu[2], Shengkai Pan[1] and Hon-Ming Lam[1*]

Abstract

Background: Plant domestication involves complex morphological and physiological modification of wild species to meet human needs. Artificial selection during soybean domestication and improvement results in substantial phenotypic divergence between wild and cultivated soybeans. Strong selective pressure on beneficial phenotypes could cause nucleotide fixations in the founder population of soybean cultivars in quite a short time.

Results: Analysis of available sequencing accessions estimates that ~5.3 million single nucleotide variations reach saturation in cultivars, and then ~9.8 million in soybean germplasm. Selective sweeps defined by loss of genetic diversity reveal 2,255 and 1,051 genes were involved in domestication and subsequent improvement, respectively. Both processes introduced ~0.1 million nucleotide fixations, which contributed to the divergence of wild and cultivated soybeans. Meta-analysis of reported quantitative trait loci (QTL) and selective signals with nucleotide fixation identifies a series of putative candidate genes responsible for 13 agronomically important traits. Nucleotide fixation mediated by artificial selection affected diverse molecular functions and biological reactions that associated with soybean morphological and physiological changes. Of them, plant-pathogen interactions are of particular relevance as selective nucleotide fixations happened in disease resistance genes, cyclic nucleotide-gated ion channels and terpene synthases.

Conclusions: Our analysis provides insights into the impacts of nucleotide fixation during soybean domestication and improvement, which would facilitate future QTL mapping and molecular breeding practice.

Keywords: Soybean domestication, Genetic improvement, Artificial selection, Nucleotide fixation, Plant-pathogen interaction

Background

The cultivated soybean [*Glycine max* (L.) Merr.] is an economically important crop that grown all over the world. With an average of ~38% protein and ~18% oil content in seeds, soybean provides 69% of dietary protein and 30% of vegetable oil consumption worldwide (www.usda.gov). Modern soybean cultivars were originally domesticated from its wild progenitor (*Glycine soja* Sieb. & Zucc.) more than 3000 years ago, which was an endemic species in China [1]. Since then, a variety of morphological and physiological changes except for reproductive isolation have occurred that distinguish soybean

cultivars from their wild ancestor. Wild soybeans possess much higher adaptability to various natural environments such as drought and salt stress, whereas cultivated soybeans exhibit a bush-type growth habit with large seeds, variable seed coat colors and a stout primary stem. Wild soybeans also differ in the extent of photosynthesis capacity, pod dehiscence and number from cultivated soybeans [2-4].

Heritable changes occurred during plant domestication are being revealed by gene mapping and genomic analyses [5]. The availability of soybean genome and high throughput sequencing technologies provides excellent opportunity to excavate the domestication events and phenotypic diversification at the genome level [6]. Re-sequenced soybeans representing wild and cultivated accessions revealed the nature and extent of genetic diversity in both populations [7-9]. Another research reported a reservoir of genes that were affected by early domestication

* Correspondence: zhaoshancen@genomics.cn; honming@cuhk.edu.hk
†Equal contributors
[1]Centre for Soybean Research, Partner State Key Laboratory of Agrobiotechnology, The Chinese University of Hong Kong, Shatin, New Territories, Hong Kong
[2]BGI-Shenzhen, Main Building, Beishan Industrial Zone, Yantian District, Shenzhen 518083, China

and modern genetic improvement [10]. Besides, several domestication-related traits have been studied and proposed to be controlled by a small number of genes or several major QTLs [11,12]. However, more analyses are needed to delimit the regions of these QTLs and the footprints of domestication for further gene mapping.

From an evolutionary perspective, if a mutation happens to be beneficial to the species, it will spread to the population immediately by selection [13]. During crop domestication, strong selective pressure caused traits of interests to be fixed in a founder population in quite a short time [14]. Probably, advantageous mutations underlying traits of interests will be subject to fixation in the population. These fixation events differ from those in natural populations, because artificial selection usually acted on alleles that were likely neutral or nearly neutral before domestication. Thus, understanding nucleotide fixation driven by artificial selection is indispensable to complete the picture of soybean evolution. In this research, the published soybean sequencing data were collected to identify single nucleotide variations (SNVs), based on which we detected the genomic regions affected by artificial selection during domestication and improvement. In these footprints, nucleotide fixations that happened in all cultivars were potentially caused by artificial selection, and the genes with these nucleotides were further analyzed, and some of these genes were associated with agronomic traits through functional annotation and QTL meta-analysis. This kind of investigation will provide clues to understand the differentiation of wild and cultivated soybeans. Besides, fundamental practical information will be obtained for future enhancement of cultivars through traditional breeding and transgenic methods.

Results

Estimation of single nucleotide variations among soybean populations

Recently, a set of diverse soybean individuals was sequenced and reported based on NGS platforms [7,8,10]. These soybeans, representing wild and cultivars that mainly consist of landrace and modern elite accessions in East Asia, were selected based on intensive molecular and phenotypic analysis to maximally reflect the genetic diversity of soybeans (Additional file 1: Table S1). It provides us an important resource to depict the genetic diversity of wild and cultivated populations, and to detect the footprints of domestication events. Thus, we downloaded all the short reads of sequencing soybeans from NCBI Short Read Archive under accession numbers SRA020131, SRA009252, SRP015830, and ERP002622. These reads were aligned to the soybean reference genome *Glycine max* (Phytozome v9) with SOAP2 [15], and were subsequently used to detect SNVs with SOAPsnp pipeline [16]. A total of 9,820,934 SNVs were identified across all accessions, of which 8,168,883 and 5,201,747 appear in wild and cultivars, respectively. Previous reports with the same pipeline have shown that the SNV calling accuracy is 95-99%, with false-positive and false-negative rates to be ~2% and ~3%, respectively [17-19].

To estimate the coverage of these SNVs in the whole soybean germplasm, we employed a random sampling approach to investigate the accumulation of SNVs detected in different accessions (Figure 1A). The end of the SNV curve tends to be flat, which indicates that the SNVs identified here probably reach saturation in soybean germplasm. It is sufficient for as few as 48 accessions to detect 95% of all SNVs in different populations. For

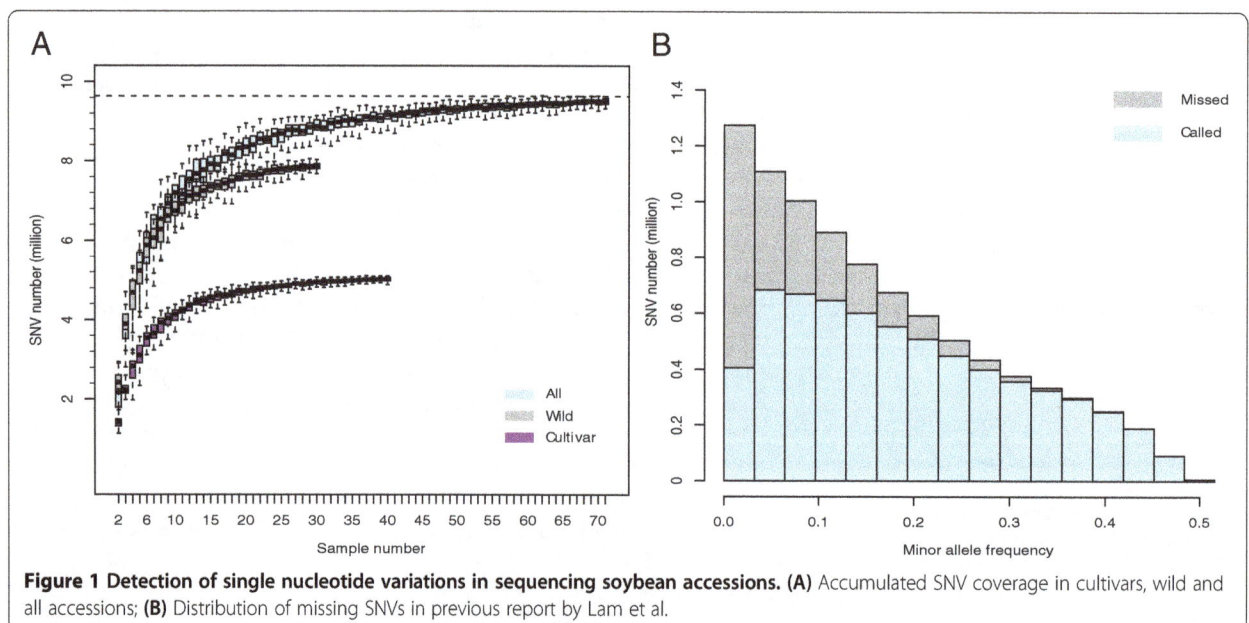

Figure 1 Detection of single nucleotide variations in sequencing soybean accessions. (A) Accumulated SNV coverage in cultivars, wild and all accessions; **(B)** Distribution of missing SNVs in previous report by Lam et al.

cultivated soybeans, only 30 individuals can achieve 95% of SNVs. Approximately 5.2 million SNVs would reach saturation in cultivars, which are far less than those in wild soybeans. In previous work [7], Lam *et al* reported 6.3 million SNVs in 31 soybeans, while we discovered 2,481,645 more in the same individuals by a larger population. A large number of rare SNVs and those with low allele frequency were missed in former analysis due to strict filtering conditions and a small number of individuals (Figure 1B). Although some very rare SNVs still remain to be discovered, we have identified a substantial majority of the common SNVs in soybeans.

Soybean has suffered several genetic bottlenecks, such as early domestication producing lots of Asian landrace, the introduction of few landraces to North America, and modern extensive breeding activities [20]. Subsequently, different level of genetic diversity was reduced during these human-mediated events. More SNVs were identified in wild than in cultivated accessions. Two common statistics used to measure nucleotide diversity are the pairwise divergence per nucleotide θ_π [21] and Watterson estimator θ_w [22] that corrected for sample size. Whole-genome analysis using these parameters shows a higher level of genetic diversity in wild populations (Figure 2A). Estimated by θ_π, the average diversity within wild, landrace and elite cultivars are 3.84×10^{-3}, 2.40×10^{-3}, and 2.08×10^{-3} per nucleotide, respectively. Considering the cultivars consist of landrace and elites, the average θ_π is 2.25×10^{-3} in cultivated population. It is notable that the cultivars have retained only 58.6% of the sequence diversity present in wild soybeans, which is lower than previous estimation [7,20]. The genetic diversity was reduced by 37.5% in early domestication and further reduced by 8.3% in genetic improvement.

The reduction of genetic diversity eroded by artificial selection could also be reflected by phylogenetic tree (Figure 2B) and principle component analysis (PCA, Figure 2C). The wild soybeans shattered in a loose 3-dimension space, while cultivated soybeans formed a relatively tight cluster distinct from the wild individuals. Within the cluster, however, the landraces were not clearly separated from elite cultivars. Some landraces mixed with wild group in our analysis, indicating the early domestication process probably accompanied with considerable gene flow with the wild ancestors. In addition to artificial selection, the genetic erosion can also reflect the narrow genetic base of cultivated soybeans [23]. Analysis of representative wild and cultivated soybeans provides us a comprehensive insight into such evolutionary events that affected population dynamics of soybeans.

Detecting artificial selection and nucleotide fixation in soybeans

The signal of artificial selection could be detected by the loss of genetic diversity, which shaped selective sweeps around beneficial alleles on the genomes [24-26]. To further elucidate the effects of domestication, we detected the genomic regions showing artificial selection signals by genetic bottleneck model [18,19] and population branch statistics [27]. The sequenced accessions except C12 and C16 were grouped into wild and cultivated population to detect selection signals in early domestication process. Using a sliding window approach, we calculated the distribution of θ_π and Tajima's D [28] in wild and cultivated populations along the genome. Regions with significantly lower θ_π (Z test, $P < 0.05$) and lower Tajima's D (Z test, $P < 0.05$) in cultivars than that in wild accessions were treated as putative candidates that were affected by early domestication (Figure 3A). However, signals of very recent natural selection could be easily omitted using the above bottleneck model. To detect signatures that shaped in modern crop improvement, we employed an effective method known as population branch statistics. Taking wild soybeans as control, we recalculated the divergence index F_{st} [29] in a sliding window along the genome, based on which we detected significant signals ($P < 0.001$ after Bonferroni correction) to infer selective footprints from landraces to elite cultivars (Figure 3B). This approach had been shown to be effective in identifying recent artificial selection considering the very short time of modern breeding practice [18]. A total of 598 regions comprising 27.9 Mb genome sequences and 286 regions with a length of 12.7 Mb were affected by early domestication and genetic improvement, respectively. Based on the latest annotation, 2,255 genes with 3,100 transcripts were involved in early domestication, whereas 1,051 genes with 1,462 transcripts were affected in subsequent improvement.

During the human-mediated breeding process, the strongly selected advantageous mutations could become fixed as these mutations increase in frequency in a population [11,13]. A selective sweep is shaped when a selected mutation goes to fixation, because it reduces variability in the neighboring region where neutral variants are segregating [30,31]. A nucleotide fixation locus was defined when a SNV has a unique genotype in one population while it exhibits polymorphic genotypes in the others. To better understand how genes were affected by domestication events, we primarily focused on those with nucleotide fixation in the selective footprints. We calculated the likelihood of genotypes of each individual and then we allocated the allele type with the maximum likelihood back to each individual as the consensus genotype. After calibration, 101,292 nucleotide fixations were identified in the selective regions in cultivars, which could be potentially caused by artificial selection.

Compared with the genome-wide distribution, nucleotide fixations happened more frequently in the candidate regions of artificial selection (Figure 4). Nucleotide fixation

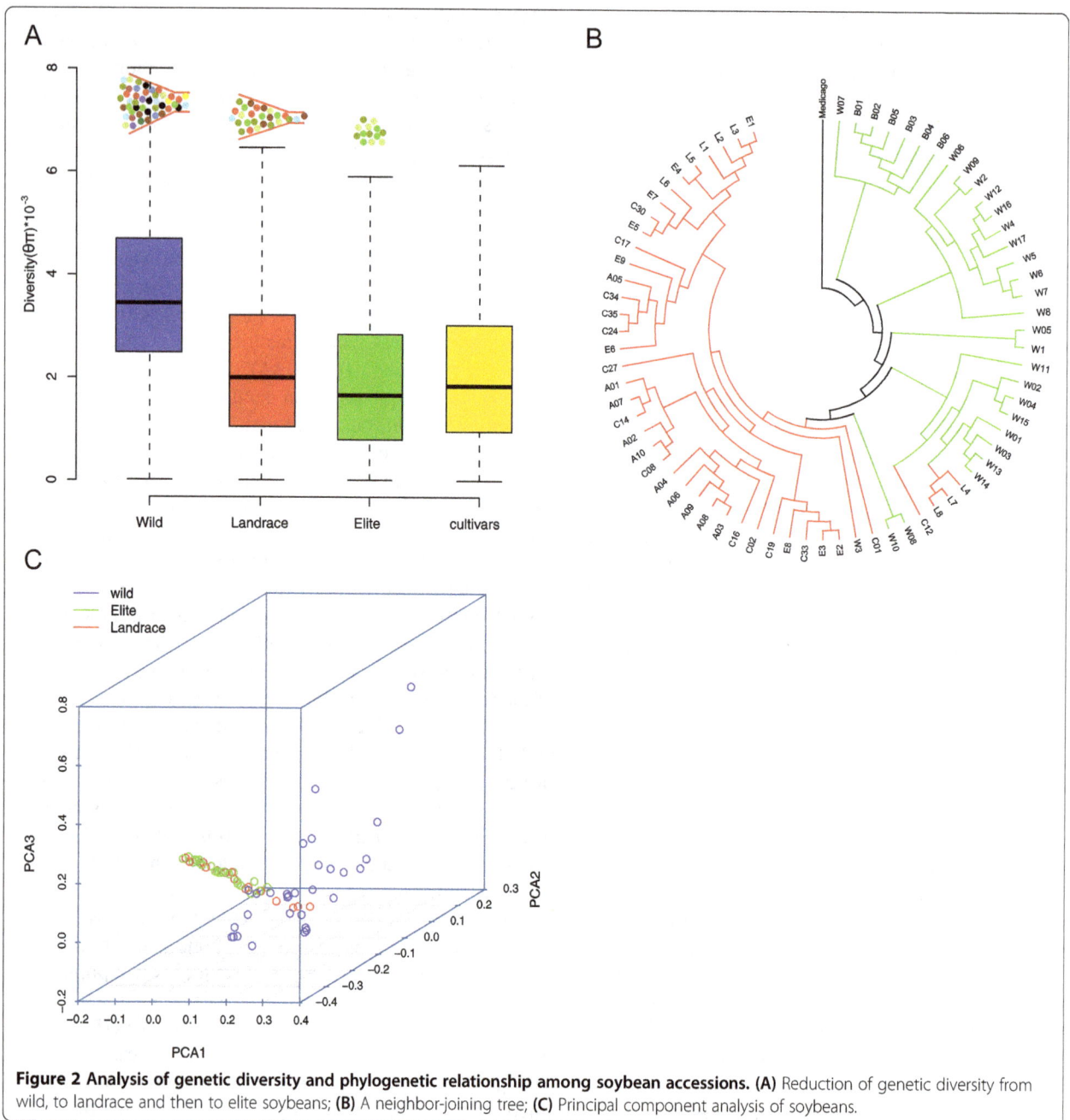

Figure 2 Analysis of genetic diversity and phylogenetic relationship among soybean accessions. (A) Reduction of genetic diversity from wild, to landrace and then to elite soybeans; **(B)** A neighbor-joining tree; **(C)** Principal component analysis of soybeans.

accumulated substantially in cultivars and happened unevenly along chromosomes (Additional file 2: Figure S1), indicating that some chromosomes were more susceptible to be affected by artificial selection. Nucleotide fixation also explains the reduction of genetic diversity in cultivated crops compared with their wild ancestors. We analyzed the allele frequency of SNVs in wild soybeans that were fixed in cultivars, as it represents the initial status of these nucleotide fixations before domestication. The frequency spectrum shows that these SNVs were almost neutral at the beginning of domestication (Additional file 3: Figure S2). Since non-synonymous substitutions may

result in a change in functions, they are subject to natural or artificial selection [32]. Of the nucleotide fixation happened in early domestication, 24,316 located in coding sequences and 2,162 of them caused non-synonymous substitutions in 1,188 genes, which altered the amino acid sequences of the proteins. For those loci fixed in modern improvement, 8,065 located in coding sequences with 756 non-synonymous in 489 genes. Apparently, more nucleotide fixations were introduced to cultivars during domestication than those during improvement.

A central question in analyzing the genetic variations in a given population is to explore whether the population

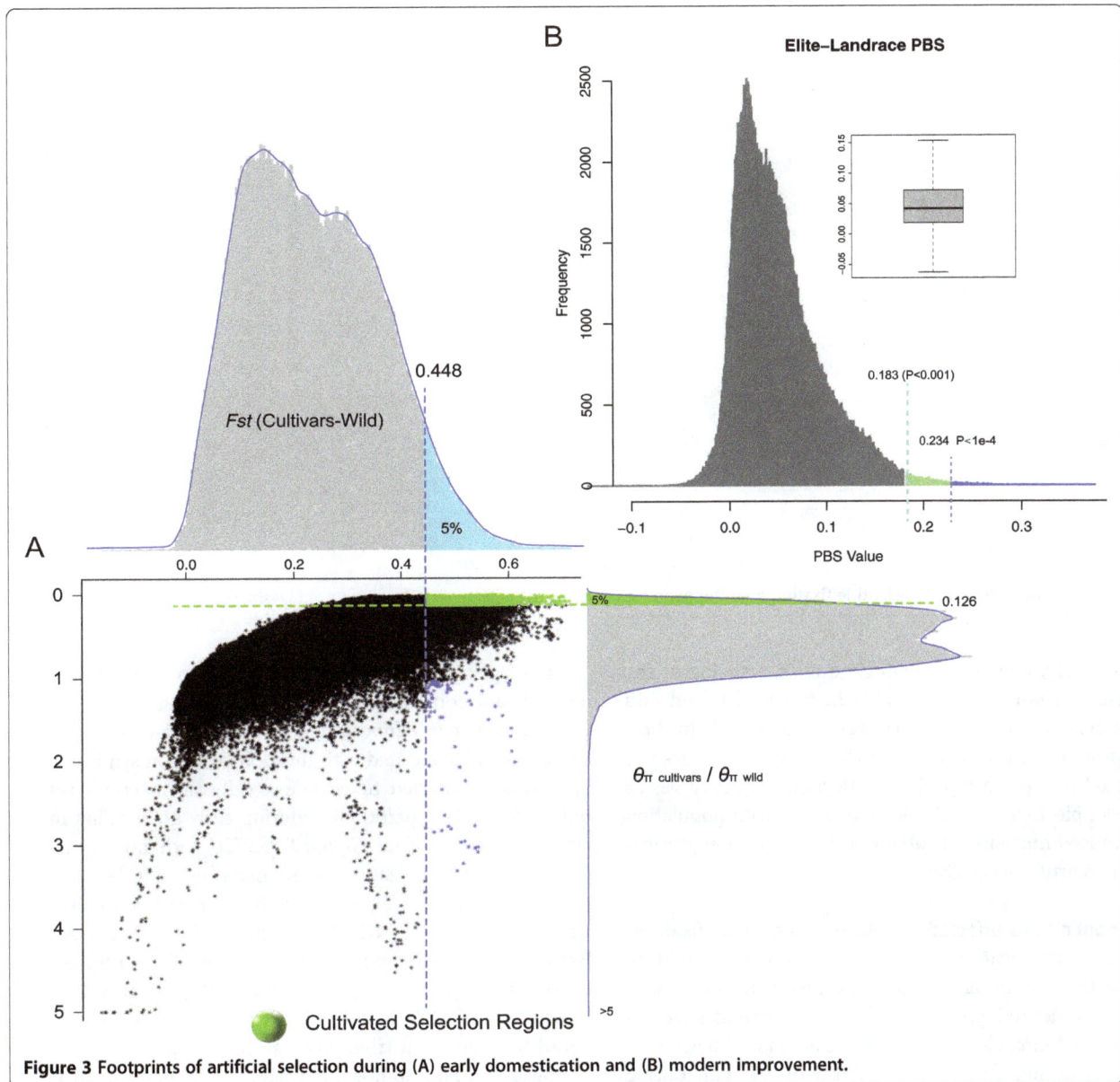

Figure 3 Footprints of artificial selection during (A) early domestication and (B) modern improvement.

has different substructures [29,33]. When analyzing the nucleotide fixations by PCA and phylogenetic tree, two distinct clusters shaped between the cultivars and wild soybeans (Additional file 4: Figure S3). Some noise always exists in inferring phylogenetic relationships among individuals, especially when they are subject to introgressive hybridization [34,35]. Cultivars tightly joined together without noise, supporting the hypothesis of a single rather than multiple evolutionary origins in soybean domestication [36,37].

Nucleotide fixation in wild soybeans

In the process of nucleotide substitution, the fixation of a mutation could spread through the population by random genetic drift or extreme natural selection [38]. In the regions affected by artificial selection, 4,111

nucleotide fixations happened in wild soybeans, which located in 875 transcripts corresponding to 723 genes. Nucleotide fixation happened more frequently in cultivars compared with wild soybeans. To some degree, artificial selection could have promoted the occurrence of fixation events. However, genetic bottlenecks caused by domestication often results in a smaller effective population size of cultivars than that of wild soybeans, which would also contribute to an elevated level of nucleotide fixation. Genes affected by nucleotide fixations were involved in kinds of biological activities as described in the Kyoto Encyclopedia of Genes and Genomes (KEGG) database (Additional file 5: Figure S4).

The ability of resistance to pathogen in wild soybeans is much broader than that in cultivated soybeans [23,39].

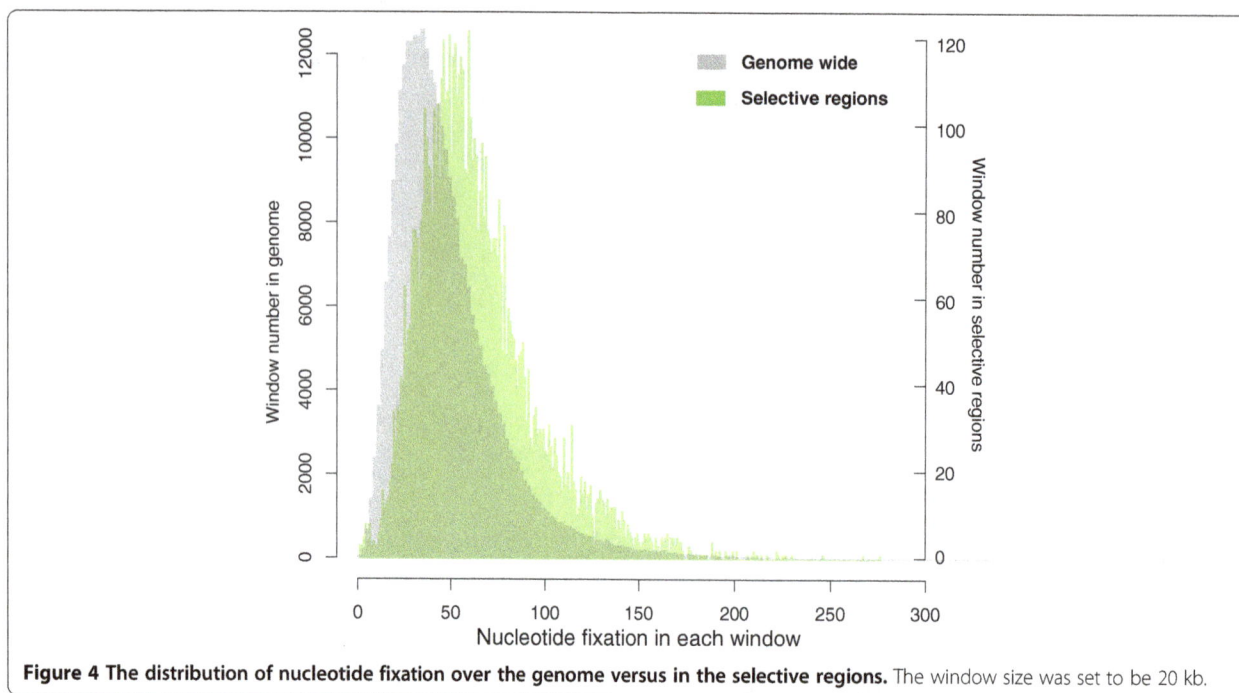

Figure 4 The distribution of nucleotide fixation over the genome versus in the selective regions. The window size was set to be 20 kb.

Interestingly, *Glyma20g08290* gene is an ortholog of the disease resistance gene *RPM1*, which was detected and characterized using molecular genetic approach in Arabidopsis [40]. In soybeans, the *RPM1* gene was recently reported being under purifying selection [41]. It serves as an example that natural selection in the wild population also caused nucleotide fixations, although its strength was less than artificial selection.

Agronomic traits affected by selective nucleotide fixation
During domestication, artificial selection is thought to have extremely strong selective pressure on ancestral population for desired phenotypes [42]. The strong selection exerted by human led to an excessive amount of nucleotide fixations during domestication. Artificial selection during soybean domestication has modified a number of traits including seed size, seed color, plant height and prostrate habitat, shaping the domestication syndrome [11,43]. To analyze the effects of nucleotide fixation during artificial selection, we focused on genes within QTLs responsible for domestication-related traits (www.soybase.org), such as oil content, pod number, lodging, plant height, *etc.* Meta-analysis of these QTLs revealed that 51 of them responsible for 13 traits and 33 for 11 traits were affected by nucleotide fixation in early domestication and modern improvement, respectively (Additional file 1: Table S2, S3). Total QTL regions were narrowed down from 214.9 Mb to 8.1 Mb assisted by selective signals. Analysis of related genes, as well as their orthologs through comparative genomics, could provide information on their potential functions under artificial selection.

As an agriculturally important trait, grain filling makes a significant contribution to grain weight [44]. The gene *Grain Incomplete Filling 1* (*GIF1*) was detected to be responsible and associated with this domestication syndrome [45]. It was reported to encode a cell-wall invertase required for carbon partitioning during early grain filling in rice. A selective gene *Glyma03g35520* with nucleotide fixation in domestication is an ortholog of *GIF1* and this gene was involved in the carbohydrate metabolism pathway by searching KEGG (Additional file 1: Table S4). Besides, this gene was covered by the QTLs responsible for lodging and pod number. It indicates that *Glyma03g35520* is a potential candidate gene, which could be used for further soybean breeding.

Flower and pod numbers per plant are important agronomic traits for grain yield in soybean. To detect the genes involved in flower and pod numbers will help to understand the genetic basis of soybean yield [46]. Two genes, *Glyma07g05470* and *Glyma07g05480*, with nucleotide fixation introduced in improvement, are orthologs of *COMT2* gene encoding caffeic acid 3-O-methyltransferase (Additional file 1: Table S5). It differentially expressed in hair cells of growing pod, the possible location of vanillin biosynthesis [47]. Another five selective genes with nucleotide fixation mediated by domestication and improvement encode a kind of protein responsible for the transportation of inositol. These genes were covered by QTLs responsible for seed-coat color, protein and pod number. Previous study showed that the total number of mature pods considerably higher due to the application of inositol, indicating the positive effect in pod number [48]. It suggested that deficiency of lignin biosynthesis resulted in growth reduction and dwarfing [49]. The gene *Glyma13g21010* is linked to marker Sat103 that associate with seed weight. As an orthologs of *NifU* gene, it is required for full activation of nitrogenase catalytic components [50]. *NifU* protein has been suggested to either mobilize the Fe necessary for nitrogenase Fe-S cluster formation or provide an intermediate Fe-S cluster assembly site [51]. In addition, the gene was reported to be related to seed weight [52]. As nitrogen fixation is imperative in soybean growth, *Glyma13g21010* gene could also be a putative candidate gene responsible for seed weight through activating biological nitrogen fixation.

The flowering of soybean represents the transition from a vegetative state to a reproductive state, making a contribution to the yield. Meta-analysis of QTLs identified 14 selective genes with non-synonymous nucleotide fixation responsible for flower number in soybean. Carbon fixation in the process of photosynthesis is pivotal to soybean production. Seven selective genes with nucleotide fixation were involved in photosynthesis or photosystem. Besides, two selective genes *Glyma03g36970* and *Glyma19g39620* with nucleotide fixation were identified as orthologs of *Luminidependens*, which is involved in the timing of flowering in Arabidopsis [53].

Interestingly, 63 and 27 selective genes with nucleotide fixation in domestication and improvement, respectively, were annotated to be, or related with transcription factors. Analysis of all the genes subject to artificial

selection with agriGO [54] also told an accumulation of transcription factors by Fisher's exact test and the permutation test (Additional file 1: Table S6). Most of the genes cloned to date that responsible for domestication related traits in crops were proved to be transcription factors, such as *teosinte branched 1* (*tb1*) [55], shattering (*sh4*) [56], six-rowed spike (*vsr1*) [57,58], *etc.* It is probably because the human mediated domestication history was momentary compared with the long natural evolution; changing the transcription factors probably the easiest way happened to affect the agricultural traits of interest. However, putative candidate genes underlying these domestication-targeted phenotypes have diverse functions, which need to be validated by further experiments.

Plant-pathogen interaction affected by artificial selection

Domestication caused complex morphological and physiological changes in soybeans. Annotated by the KEGG and agriGO database, selective genes were associated with different biological functions, among which, plant-pathogen interaction, sequence-specific DNA binding, phenylpropanoid biosynthesis, starch and sucrose metabolism are over-represented categories (Figure 5; Additional file 6: Figure S5). Plant-pathogen interactions are conducted between a pathogen and the host plant. In nature, plants are generally resistant to most invading pathogens due to innate ability to recognize them through successful defenses. When an exception happens, a pathogen would

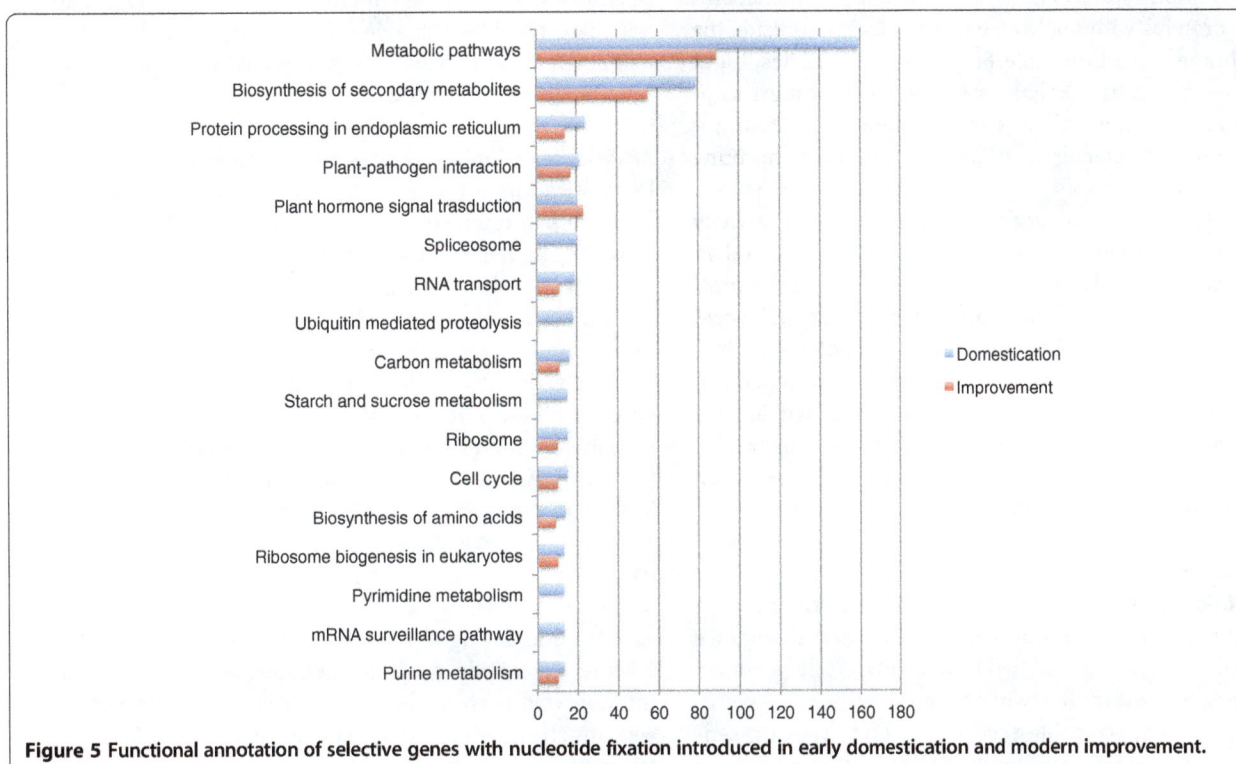

Figure 5 Functional annotation of selective genes with nucleotide fixation introduced in early domestication and modern improvement.

cause diseases in its host [59]. However, pathogens could also cause diseases if they have evolved to evade detection or suppress host defense mechanisms, or both. The effects of plant-pathogen interactions are of particular relevance during early domestication events on agricultural systems [60]. Thus, understanding the genetic basis of why a certain pathogen causes disease in its host plant instead of others has long intrigued and motivated plant pathologists.

A total of 37 selective genes with nucleotide fixation were involved in plant-pathogen interactions (Additional file 7: Figure S6). Of them, two selective genes Glyma14g36511 and Glyma08g12560 with nucleotide fixation are orthologs of RPS2 gene. The disease resistance gene RPS2 was isolated using positional cloning and further screen for susceptible mutant [61,62]. The RPS2 protein contains two characteristics of a large family of plant R genes: a nucleotide-binding site and a leucine-rich repeat region [63]. It is consistent with previous report that RPS2 locus exhibit selection signals by examining a worldwide sample of 27 Arabidopsis accessions, and the N-terminal part of the leucine-rich repeat region was a probable target of selection [64].

Cyclic nucleotide-gated ion (CNG) channels are ion channels that function in the pathogen signaling cascade by facilitating Ca^{2+} uptake into the cytosol [65]. Two selective genes with nucleotide fixation were detected to encode CNG channels. The topology of their proteins was predicted using TMHMM, which is based on a hidden Markov model [66]. The two genes encode transmembrane proteins with nucleotide fixation located outside the membrane (Additional file 8: Figure S7). Besides, eight selective genes are orthologs of transmembrane receptor kinase FLS2, which acts as pathogen-associated molecular pattern signals in triggering the innate immune response [67].

In addition, the category of terpene synthase activity was also enriched with six selective genes involved in (Additional file 1: Table S6). Terpenes are one of the most important defensive plant compounds against herbivores and pathogens [68]. Recently, a new monoterpene synthase gene GmNES was identified and characterized in soybean [69]. Its transcription was up-regulated in soybean when infested with cotton leafworm. Our analysis indicates the gene was possibly affected by artificial selection during soybean domestication.

Discussion
Nucleotide fixation was crucial in soybean divergence
Domestication led to significant morphological divergence between cultivated and wild soybeans. Wild soybean exhibits, for example, twining and vine stem, severer shattering, impermeable seed coats, pod cracking sensitivity, small seeds, and low oleic acid, all of which are seldom observed in cultivars [70]. Deciphering how cultivated soybean have been transformed from its wild ancestor is advantageous both from genetic and evolutionary perspectives. With the available sequencing data, we comprehensively estimated the saturation number of SNVs in soybean germplasm and detected a set of candidate genes showing artificial selection signals. To some degree, analysis of artificial selection and nucleotide fixation unravels the mystery of soybean domestication and subsequent improvement. Based on nucleotide fixation, our analysis supports a single evolutionary origin of domesticated soybean. During domestication, only lines with certain agriculturally important traits were selected, resulting in a genome-wide reduction of genetic diversity or so-called selective sweep in cultivated crops [42,71,72]. One possible explanation for the reduction is that an excess of nucleotide fixation happened in cultivars compared to wild soybeans.

Meta-analysis of QTLs responsible for domestication related traits and the selective genes provided insights into the role of nucleotide fixation played in morphological differentiation between wild and cultivated soybeans. Using comparative genomics, an amount of genes was found to be orthologs of those whose function was validated and responsible for corresponding traits in other plants. Nucleotide fixation happened in those genes responsible for agronomically important traits. Although traditional linkage and association mapping were used to dissect these traits, they failed to detect genetic changes caused by domestication and improvement [73]. Our analysis here provides valuable information for further QTL mapping and will facilitate molecular assisted selection in soybean breeding practice.

Artificial selection accelerates nucleotide fixation
Domestication was an evolutionary process where the characters of interests were selected, such as loss of seed dispersal, higher yield and increasing abiotic resistance. The detection of selective loci during crop domestication contributes to modern breeding efforts and the opportunity to improve genomic selection models [74]. Recently, genome-wide scans based on genetic bottlenecks have been successfully applied to detect footprints of selection in plants by surveying both natural and cultivated species [19,75,76]. Artificial selection of a beneficial mutation will lead to an elevated frequency in a population. Eventually, allele frequencies will be skewed and nucleotide fixation happened after plant domestication. Our analysis focused on to what degree nucleotide fixation was caused by artificial selection during soybean domestication.

More nucleotide fixation happened in cultivars than those in wild soybeans, indicating that artificial selection was much stronger than natural selection. However, the effective population size of cultivated soybeans was substantially reduced during domestication [77], which could make a nucleotide seem to be fixed in cultivars. That mainly explains why nucleotide fixations were observed in cultivars across the soybean genome. Considering nucleotide fixation accumulated in footprints of domestication and improvement, artificial selection probably accelerated the occurrence of fixation in soybean breeding activities. Even thought, some of them could be also caused by the shrinking population size, especially when different haplotypes shaped in those selective sweeps. These fixations are extremely hard to be distinguished in current samples.

Morphological transition can be achieved by a mutation at a single locus [78,79], and artificial selection can rapidly change domestication targeted phenotypes within 20 generations [31,80]. Domestication could be a rapid instead of a slow or gradual process, given strong selective pressures and a suitable genetic architecture. This was supported by the severe reduction of genetic diversity and large selective sweeps. In the process of domestication, any mutations detrimental to the traits of interests were eliminated immediately, whereas those advantageous ones were strongly selected, diffused and eventually fixed in a population. The environments wild soybeans grow in are various and usually harsh, resulting in diversifying selection instead of strong directional selection. What's more, selection intensity imposed by natural selection was disparate in diverse habitats. These reasons also explain why artificial selection was much stronger than natural selection in crop domestication.

Evolutionary perspective of nucleotide fixation
A long-term goal of crop genomics is to determine to what extent artificial selection impacts genomic variation patterns within and between populations. There are both genetic and statistical approaches to detect signals of hitchhiking caused selective sweeps [13]. The hitchhiking effect is contingent on the nature of genetic variations and how selection acts on them. Generally, there are at least three evolutionary routes by which a novel mutation may fix: drift to fixation for nearly neutral mutation; rapidly sweep to fixation, so-called hard sweep for beneficial mutation; and soft sweep to fixation for those initially neutral but later become beneficial for some reason. Affected by artificial selection, a pre-exist mutation which became beneficial during domestication rapidly increased in frequency toward nucleotide fixation, as what we found in our analysis. When traits of interests during domestication were determined by multiple adaptive mutations at the same locus, artificial selection usually generates soft rather hard selective sweeps. Many studies focus on hard sweeps in which only a single adaptive haplotype was skewed to fixation in the population [81], whereas multiple adaptive haplotypes formed simultaneously in a

soft sweep. Lots of nucleotide fixations happened within quantitative traits, indicating the corresponding traits of interests were incrementally changed at various causal loci. As a consequence, these sweeps related with artificial selection are likely to be both soft and incomplete. In soybean, some traits related to yield were selected, such as seed weight, seed blooming and prostrate habit, for which these are usually major QTLs responsible. Nevertheless, during intensive breeding human pursuits quality related traits such as protein content and lipid content, for which there are lots of small effect QTLs responsible. Analysis of nucleotide fixation indicates that more soft selective sweeps happened in extensive breeding than in early domestication in soybean, which still needs further investigation.

Conclusion
We integrated the available sequencing accessions to describe a whole picture of soybean genetic diversity, artificial selection and concomitant nucleotide fixation. There are approximately 9.8 million SNVs in soybean germplasm, of which about 5.3 million reserved in cultivars. The genetic diversity was reduced by 37.5% in early domestication and subsequently reduced by 8.3% in genetic improvement. A total of 2,255 and 1,051 genes were involved in early domestication and subsequent improvement, respectively. Both processes introduced about 0.1 million nucleotide fixations, which contributed to the divergence of wild and cultivated soybeans. Artificial selection probably accelerated the occurrence of nucleotide fixation, which affected some agronomic traits, as well as related biological pathways such as plant-pathogen interaction.

Methods
Data collection and SNP detection
The sequenced soybean accessions representing 31 wild, 15 landrace, and 24 elites were described in several published papers [7-10]. These accessions originate from large ecological regions in China and South Korea. All sequence reads were downloaded in Sequence Read Archive (SRA) under accession number SRP015830, SRA020131 SRA009252, and ERP002622. These reads were then mapped to the soybean reference (*Glycine max* var. Williams 82, Phytozome v9.0) with SOAP2 software [15]. PCR duplication in each sequencing library was removed before SNV calling.

In the SNV calling process, genotype likelihood of each genomic locus was first calculated with Bayesian theory implemented in SOAPsnp [16]. The genotype with the highest probability at each site was selected with a quality value to create a consensus sequence for each individual. High quality SNVs were obtained with certain criteria such as sequencing depth, copy number (<=1.5), quality value (>20) and the rank sum test.

Detection of artificial selection signals

As described in previous report [10], we used two outlier approaches to detect signals of artificial selection. Using a 20 kb sliding window with a 2 kb step-size, we calculated θ_π and Tajima's D between wild and cultivated groups. Those regions showing significantly low $\theta_{\pi.cultivated}/\theta_{\pi.wild}$ and low D values (Z test, $P < 0.05$ for both) in cultivars were treated as putative selection signals. Besides, we chose the population branch statistic [27] on the basis of F_{st} to infer the selective footprints from landrace to elite cultivar, considering the very short divergence time between them.

Identification of nucleotide fixation

We screened the SNVs located in the regions showing signals of artificial selection. Short reads of each individual were re-aligned to the reference for individual genotyping at each SNV. The likelihood of individual genotypes was calculated and then the allele type with the maximum likelihood was allocated back to each individual. If a SNV has a unique genotype in all wild soybeans or in cultivars, it will be identified as a nucleotide fixation locus.

PCA and phylogenetic analysis

Using the principal component analysis (PCA), the population subdivision pattern was then inferred [82]. We constructed a phylogenetic tree by a neighbor joining method in the software PHYLIP (version 3.68) [83]. A total of 1,000 replicates generated the bootstrap values.

Enrichment of selective genes

The functions of selective genes were analyzed with KEGG (www.genome.jp/kegg/) and agriGO (http://bioinfo.cau.edu.cn/agriGO/), and the results were displayed using a Cytoscape plugin BiNGO [84]. For enrichment P value (<0.05) was calculated using Fisher's exact test and Permutation test. For multiple hypotheses testing, false discovery rate correction of Benjamini and Hochberg method was used to reduce false negatives.

Inferring protein topology

We predicted transmembrane protein topology with a hidden Markov model (TMHMM) to infer the protein topology with default parameters [66] (http://www.cbs.dtu.dk/services/TMHMM/).

Additional files

Additional file 1: Table S1. Summary of sequencing soybean accessions collected from publications. **Table S2.** Meta-analysis of the published QTLs responsible for agriculturally important traits and selective genes with nucleotide fixation during early domestication. **Table S3.** Meta-analysis of the published QTLs responsible for agriculturally important traits and selective genes with nucleotide fixation during modern improvement. **Table S4.** Functional analysis of the selective genes with nucleotide fixation during early domestication. **Table S5.** Functional analysis of the selective genes with nucleotide fixation during modern improvement. **Table S6.** Functional analysis of the selective genes with nucleotide fixation based on agriGO.

Additional file 2: Figure S1. Fixed SNP distribution on each chromosome.

Additional file 3: Figure S2. The allele frequency of SNVs in wild soybeans that were fixed in cultivars. The allele frequency < 0.1 was underestimated in SNV calling to improve accuracy.

Additional file 4: Figure S3. (A) PCA and (B) phylogenetic tree among soybean accessions based on nucleotide fixation.

Additional file 5: Figure S4. The accumulated KEGG pathway in the genes with nucleotide fixation in wild soybeans.

Additional file 6: Figure S5. Over-represented GO categories in the selective genes with nucleotide fixation (Fisher's exact test < 0.05 and false discovery rate (FDR) < 0.05).

Additional file 7: Figure S6. Selective genes with nucleotide fixation involved in plant hormone signal transduction pathway. Red: affected by early domestication; Green: affected both by domestication and improvement.

Additional file 8: Figure S7. The protein topology CNG channels involved in plant-pathogen interaction pathway. The stars denote nucleotide fixation in the protein.

Abbreviations

QTL: Quantitative Trait Loci; SNV: Single nucleotide variation; PCA: Principle component analysis; KEGG: Kyoto Encyclopedia of Genes and Genomes; GIF1: Grain Incomplete Filling 1; CNG: Cyclic Nucleotide-Gated ion; tb1: teosinte branched 1.

Competing interests

The authors declare that they have no competing interests.

Authors' contributions

H-ML and SCZ conceived and designed the research; SCZ, FYZ, HWM, WHY, and PSK performed the research and analyzed the data; SCZ wrote and H-ML revised the manuscript. All authors read and approved the final manuscript.

Acknowledgements

This work was financially supported by the Hong Kong RGC Collaborative Research Fund (CUHK3/ CRF/11G), the Hong Kong RGC General Research Fund (468610), and the Lo Kwee-Seong Biomedical Research Fund and Lee Hysan Foundation.

References

1. Hymowiltz T. Speciation and cytogenetics. In: Boerma HR, Specht JE, editors. Soybeans: Improvement, production, and uses. 3rd ed. Madison (WI): ASA, CSSA, SSSA; 2004. p. 97–136.
2. Kato S, Sayama T, Fujii K, Yumoto S, Kono Y, Hwang T-Y, et al. A major and stable QTL associated with seed weight in soybean across multiple environments and genetic backgrounds. Theor Appl Genet. 2014;127:1365–74.
3. Saitoh K, Nishimura K, Kuroda T. Comparisons of growth and photosynthetic characteristics between wild and cultivated types of soybeans. Brisbane, Australia: 4th International Crop Science Congress; 2008.
4. Liu B, Fujita T, Yan ZH, Sakamoto S, Xu D, Abe J. QTL Mapping of Domestication-related Traits in Soybean (Glycine max). Ann Bot. 2007;100:1027–38.
5. Vaughan DA, Balazs E, Heslop-Harrison JS. From crop domestication to super-domestication. Ann Bot. 2007;100:893–901.
6. Schmutz J, Cannon SB, Schlueter J, Ma J, Mitros T, Nelson W, et al. Genome sequence of the palaeopolyploid soybean. Nature. 2010;463:178–83.
7. Lam H-M, Xu X, Liu X, Chen W, Yang G, Wong F-L, et al. Resequencing of 31 wild and cultivated soybean genomes identifies patterns of genetic diversity and selection. Nat Genet. 2010;42:1053–9.

8. Chung WH, Jeong N, Kim J, Lee WK, Lee YG, Lee SH, et al. Population structure and domestication revealed by high-depth resequencing of Korean cultivated and wild soybean genomes. DNA Res. 2014;21(2):153–67.

9. Kim MY, Lee S, Van K, Kim T-H, Jeong S-C, Choi I-Y, et al. Whole-genome sequencing and intensive analysis of the undomesticated soybean (Glycine soja Sieb. and Zucc.) genome. Proc Natl Acad Sci U S A. 2010;107(51):22032–7.

10. Li Y-H, Zhao S-C, Ma J-X, Li D, Yan L, Li J, et al. Molecular footprints of domestication and improvement in soybean revealed by whole genome re-sequencing. BMC Genomics. 2013;14:579.

11. Doebley JF, Gaut BS, Smith BD. The molecular genetics of crop domestication. Cell. 2006;127:1309–21.

12. Qi X, Li M-W, Xie M, Liu X, Ni M, Shao G, et al. Identification of a novel salt tolerance gene in wild soybean by whole-genome sequencing. Nat Commun. 2014;5:4340.

13. Nielsen R. Molecular signatures of natural selection. Annu Rev Genet. 2005;39:197–218.

14. Innan H, Kim Y. Pattern of polymorphism after strong artificial selection in a domestication event. Proc Natl Acad Sci U S A. 2004;101:10667–72.

15. Li R, Yu C, Li Y, Lam T-W, Yiu S-M, Kristiansen K, et al. SOAP2: an improved ultrafast tool for short read alignment. Bioinformatics. 2009;25:1966–7.

16. Li R, Li Y, Fang X, Yang H, Wang J, Kristiansen K, et al. SNP detection for massively parallel whole-genome resequencing. Genome Res. 2009;19:1124–32.

17. Wang J, Wang W, Li R, Li Y, Tian G, Goodman L, et al. The diploid genome sequence of an Asian individual. Nature. 2008;456:60–5.

18. Xia QQ, Guo YY, Zhang ZZ, Li DD, Xuan ZZ, Li ZZ, et al. Complete resequencing of 40 genomes reveals domestication events and genes in silkworm (Bombyx). Science. 2009;326:433–6.

19. Xu X, Liu X, Ge S, Jensen JD, Hu F, Li X, et al. Resequencing 50 accessions of cultivated and wild rice yields markers for identifying agronomically important genes. Nat Biotech. 2012;30:105–11.

20. Hyten DL, Song Q, Zhu Y, Choi I-Y, Nelson RL, Costa JM, et al. Impacts of genetic bottlenecks on soybean genome diversity. Proc Natl Acad Sci U S A. 2006;103:16666–71.

21. Tajima F. Evolutionary relationship of DNA sequences in finite populations. Genetics. 1983;105:437–60.

22. Watterson GA. On the number of segregating sites in genetical models without recombination. Theor Popul Biol. 1975;7:256–76.

23. Tanksley SD. Seed banks and molecular maps: unlocking genetic potential from the wild. Science. 1997;277:1063–6.

24. Nielsen R, Williamson S, Kim Y, Hubisz MJ, Clark AG, Bustamante C. Genomic scans for selective sweeps using SNP data. Genome Res. 2005;15:1566–75.

25. Tian F, Stevens NM, Buckler ES. Tracking footprints of maize domestication and evidence for a massive selective sweep on chromosome 10. Proc Natl Acad Sci U S A. 2009;106:9979–86.

26. Huang X, Kurata N, Wei X, Wang Z-X, Wang A, Zhao Q, et al. A map of rice genome variation reveals the origin of cultivated rice. Nature. 2012;490:497–501.

27. Yi X, Liang Y, Huerta-Sanchez E, Jin X, Cuo ZXP, Pool JE, et al. Sequencing of 50 human exomes reveals adaptation to high altitude. Science. 2010;329:75–8.

28. Tajima F. Statistical method for testing the neutral mutation hypothesis by DNA polymorphism. Genetics. 1989;123:585–95.

29. Weir BS, Cockerham CC. Estimating F-statistics for the analysis of population structure. Evolution. 1984;1358–1370.

30. Bamshad M, Wooding SP. Signatures of natural selection in the human genome. Nat Rev Genet. 2003;4:99–111.

31. Purugganan MD, Fuller DQ. The nature of selection during plant domestication. Nature. 2009;457:843–8.

32. Yang Z, Nielsen R. Estimating synonymous and nonsynonymous substitution rates under realistic evolutionary models. Mol Biol Evol. 2000;17:32–43.

33. Rosenberg NA, Pritchard JK, Weber JL, Cann HM, Kidd KK, Zhivotovsky LA, et al. Genetic structure of human populations. Science. 2002;298:2381–5.

34. McNally KL, Childs KL, Bohnert R, Davidson RM, Zhao K, Ulat VJ, et al. Genomewide SNP variation reveals relationships among landraces and modern varieties of rice. Proc Natl Acad Sci U S A. 2009;106:12273–8.

35. Molina J, Sikora M, Garud N, Flowers JM, Rubinstein S, Reynolds A, et al. Molecular evidence for a single evolutionary origin of domesticated rice. Proc Natl Acad Sci U S A. 2011;108:8351–6.

36. Xu D, Abe J, Gai J, Shimamoto Y. Diversity of chloroplast DNA SSRs in wild and cultivated soybeans: evidence for multiple origins of cultivated soybean. Theor Appl Genet. 2002;105:645–53.

37. Guo J, Wang Y, Song C, Zhou J, Qiu L, Huang H, et al. A single origin and moderate bottleneck during domestication of soybean (Glycine max):

implications from microsatellites and nucleotide sequences. Ann Bot. 2010;106:505–14.

38. Tajima F. Relationship between DNA polymorphism and fixation time. Genetics. 1990;125:447–54.

39. Fuller DQ. Contrasting patterns in crop domestication and domestication rates: recent archaeobotanical insights from the Old World. Ann Bot. 2007;100:903–24.

40. Grant MR, Godiard L, Straube E, Ashfield T, Lewald J, Sattler A, et al. Structure of the Arabidopsis RPM1 gene enabling dual specificity disease resistance. Science. 1995;269:843–6.

41. Ashfield T, Redditt T, Russell A, Kessens R, Rodibaugh N, Galloway L, et al. Evolutionary relationship of disease resistance genes in soybean and Arabidopsis specific for the pseudomonas syringae effectors AvrB and AvrRpm1. Plant Physiol. 2014;166:235–51.

42. Diamond J. Evolution, consequences and future of plant and animal domestication. Nature. 2002;418:700–7.

43. Gepts P. A comparison between crop domestication, classical plant breeding, and genetic engineering. Crop Sci. 2002;42:1780–90.

44. Takai T. Time-related mapping of quantitative trait loci controlling grain-filling in rice (Oryza sativa L.). J Exp Bot. 2005;56:2107–18.

45. Wang E, Wang J, Zhu X, Hao W, Wang L, Li Q, et al. Control of rice grain-filling and yield by a gene with a potential signature of domestication. Nat Genet. 2008;40:1370–4.

46. Zhang D, Cheng H, Wang H, Zhang H, Liu C, Yu D. Identification of genomic regions determining flower and pod numbers development in soybean (Glycine max L). J Genet Genomics. 2010;37:545–56.

47. Grimmig B, Matern U. Structure of the parsley caffeoyl-CoA O-methyltransferase gene, harbouring a novel elicitor responsive cis-acting element. Plant Mol Biol. 1997;33:323–41.

48. Yang Z, Xin D, Liu C, Jiang H, Han X, Sun Y, et al. Identification of QTLs for seed and pod traits in soybean and analysis for additive effects and epistatic effects of QTLs among multiple environments. Mol Genet Genomics. 2013;288:651–67.

49. Li Y-H, Li W, Zhang C, Yang L, Chang R-Z, Gaut BS, et al. Genetic diversity in domesticated soybean (Glycine max) and its wild progenitor (Glycine soja) for simple sequence repeat and single-nucleotide polymorphism loci. New Phytologist. 2010;188:242–53.

50. Hwang DM, Dempsey A, Tan KT, Liew CC. A modular domain of NifU, a nitrogen fixation cluster protein, is highly conserved in evolution. J Mol Evol. 1996;43:536–40.

51. Yuvaniyama P, Agar JN, Cash VL, Johnson MK, Dean DR. NifS-directed assembly of a transient [2Fe-2S] cluster within the NifU protein. Proc Natl Acad Sci U S A. 2000;97:599–604.

52. Atta S, Maltese S, Cousin R. Protein content and dry weight of seeds from various pea genotypes. Agronomie. 2004;24:257–66.

53. Lee I, Aukerman MJ, Gore SL, Lohman KN, Michaels SD, Weaver LM, et al. Isolation of LUMINIDEPENDENS: a gene involved in the control of flowering time in Arabidopsis. Plant Cell. 1994;6:75–83.

54. Du Z, Zhou X, Ling Y, Zhang Z, Su Z. agriGO: a GO analysis toolkit for the agricultural community. Nucleic Acids Res. 2010;38(Web Server):W64–70.

55. Doebley J, Stec A, Hubbard L. The evolution of apical dominance in maize. Nature. 1997;386:485–8.

56. Li C. Rice domestication by reducing shattering. Science. 2006;311:1936–9.

57. Komatsuda T, Pourkheirandish M, He C, Azhaguvel P, Kanamori H, Perovic D, et al. Six-rowed barley originated from a mutation in a homeodomain-leucine zipper I-class homeobox gene. Proc Natl Acad Sci U S A. 2007;104:1424–9.

58. Ramsay L, Comadran J, Druka A, Marshall DF, Thomas WTB, Macaulay M, et al. INTERMEDIUM-C, a modifier of lateral spikelet fertility in barley, is an ortholog of the maize domestication gene TEOSINTE BRANCHED 1. Nat Genet. 2011;43:169–72.

59. Staskawicz BJ. Genetics of plant-pathogen interactions specifying plant disease resistance. Plant Physiol. 2000;125:73–6.

60. Dodds PN, Rathjen JP. Plant immunity: towards an integrated view of plant-pathogen interactions. Nat Rev Genet. 2010;11:539–48.

61. Bent AF, Kunkel BN, Dahlbeck D, Brown KL, Schmidt R, Giraudat J, et al. RPS2 of Arabidopsis thaliana: a leucine-rich repeat class of plant disease resistance genes. Science. 1994;265:1856–60.

62. Kunkel BN, Bent AF, Dahlbeck D, Innes RW, Staskawicz BJ. RPS2, an Arabidopsis disease resistance locus specifying recognition of Pseudomonas syringae strains expressing the avirulence gene avrRpt2. Plant Cell. 1993;5:865–75.

63. Luck JE, Lawrence GJ, Dodds PN, Shepherd KW, Ellis JG. Regions outside of the leucine-rich repeats of flax rust resistance proteins play a role in specificity determination. Plant Cell. 2000;12:1367–77.

64. Mauricio R, Stahl EA, Korves T, Tian D, Kreitman M, Bergelson J. Natural selection for polymorphism in the disease resistance gene Rps2 of *Arabidopsis thaliana*. Genetics. 2003;163:735–46.

65. Ma W. Roles of Ca2+ and cyclic nucleotide gated channel in plant innate immunity. Plant Sci. 2011;181:342–6.

66. Krogh A, Larsson B, von Heijne G, Sonnhammer EL. Predicting transmembrane protein topology with a hidden Markov model: application to complete genomes. J Mol Biol. 2001;305:567–80.

67. Chinchilla D, Bauer Z, Regenass M, Boller T, Felix G. The Arabidopsis receptor kinase FLS2 binds flg22 and determines the specificity of flagellin perception. Plant Cell. 2006;18:465–76.

68. Wittstock U, Gershenzon J. Constitutive plant toxins and their role in defense against herbivores and pathogens. Curr Opin Plant Biol. 2002;5:300–7.

69. Zhang M, Liu J, Li K, Yu D. Identification and characterization of a novel monoterpene synthase from soybean restricted to neryl diphosphate precursor. PLoS One. 2013;8:e75972.

70. Chen Y, Nelson RL. Identification and characterization of a white-flowered wild soybean plant. Crop Sci. 2004;44:339–42.

71. Buckler ES, Thornsberry JM, Kresovich S. Molecular diversity, structure and domestication of grasses. Genet Res. 2001;77(3):213–8.

72. Burger JC, Chapman MA, Burke JM. Molecular insights into the evolution of crop plants. Am J Bot. 2008;95:113–22.

73. Varshney RK, Hoisington DA, Tyagi AK. Advances in cereal genomics and applications in crop breeding. Trends Biotechnol. 2006;24:490–9.

74. Morrell PL, Buckler ES, Ross-Ibarra J. Crop genomics: advances and applications. Nat Rev Genet. 2011;13:85–96.

75. Hufford MB, Xu X, van Heerwaarden J, Pyhäjärvi T, Chia J-M, Cartwright RA, et al. Comparative population genomics of maize domestication and improvement. Nat Genet. 2012;44:808–11.

76. Morris GP, Ramu P, Deshpande SP, Hash CT, Shah T, Upadhyaya HD, et al. Population genomic and genome-wide association studies of agroclimatic traits in sorghum. Proc Natl Acad Sci U S A. 2013;110:453–8.

77. Tang H, Sezen U, Paterson AH. Domestication and plant genomes. Curr Opin Plant Biol. 2010;13:160–6.

78. Doebley J, Stec A. Inheritance of the morphological differences between maize and teosinte: comparison of results for two F2 populations. Genetics. 1993;134:559–70.

79. Doebley J. The genetics of maize evolution. Annu Rev Genet. 2004;38:37–59.

80. HILLMAN GC, DAVIES MS. Domestication rates in wild-type wheats and barley under primitive cultivation. Biol J Linn Soc. 1990;39:39–78.

81. Sabeti PC, Varilly P, Fry B, Lohmueller J, Hostetter E, Cotsapas C, et al. Genome-wide detection and characterization of positive selection in human populations. Nature. 2007;449:913–8.

82. Patterson N, Price AL, Reich D. Population structure and eigenanalysis. PLoS Genet. 2006;2:e190.

83. Felsenstein J. PHYLIP - Phylogeny inference package (version 3.2). Cladistics. 1989;5:164–6.

84. Maere S, Heymans K, Kuiper M. BiNGO: a Cytoscape plugin to assess overrepresentation of Gene Ontology categories in Biological Networks. Bioinformatics. 2005;21:3448–9.

9

High-resolution confocal imaging of wall ingrowth deposition in plant transfer cells: Semi-quantitative analysis of phloem parenchyma transfer cell development in leaf minor veins of Arabidopsis

Suong T T Nguyen and David W McCurdy[*]

Abstract

Background: Transfer cells (TCs) are *trans*-differentiated versions of existing cell types designed to facilitate enhanced membrane transport of nutrients at symplasmic/apoplasmic interfaces. This transport capacity is conferred by intricate wall ingrowths deposited secondarily on the inner face of the primary cell wall, hence promoting the potential trans-membrane flux of solutes and consequently assigning TCs as having key roles in plant growth and productivity. However, TCs are typically positioned deep within tissues and have been studied mostly by electron microscopy.

Recent advances in fluorophore labelling of plant cell walls using a modified pseudo-Schiff-propidium iodide (mPS-PI) staining procedure in combination with high-resolution confocal microscopy have allowed visualization of cellular details of individual tissue layers in whole mounts, hence enabling study of tissue and cellular architecture without the need for tissue sectioning. Here we apply a simplified version of the mPS-PI procedure for confocal imaging of cellulose-enriched wall ingrowths in vascular TCs at the whole tissue level.

Results: The simplified mPS-PI staining procedure produced high-resolution three-dimensional images of individual cell types in vascular bundles and, importantly, wall ingrowths in phloem parenchyma (PP) TCs in minor veins of Arabidopsis leaves and companion cell TCs in pea. More efficient staining of tissues was obtained by replacing complex clearing procedures with a simple post-fixation bleaching step. We used this modified procedure to survey the presence of PP TCs in other tissues of Arabidopsis including cotyledons, cauline leaves and sepals. This high-resolution imaging enabled us to classify different stages of wall ingrowth development in Arabidopsis leaves, hence enabling semi-quantitative assessment of the extent of wall ingrowth deposition in PP TCs at the whole leaf level. Finally, we conducted a defoliation experiment as an example of using this approach to statistically analyze responses of PP TC development to leaf ablation.

Conclusions: Use of a modified mPS-PI staining technique resulted in high-resolution confocal imaging of polarized wall ingrowth deposition in TCs. This technique can be used in place of conventional electron microscopy and opens new possibilities to study mechanisms determining polarized deposition of wall ingrowths and use reverse genetics to identify regulatory genes controlling TC *trans*-differentiation.

Keywords: Transfer cells, Cell wall ingrowths, Confocal imaging, Pseudo-Schiff base, Propidium iodide, Phloem parenchyma, Arabidopsis, Companion cells

* Correspondence: David.McCurdy@newcastle.edu.au
Centre for Plant Science, School of Environmental and Life Sciences, The University of Newcastle, Newcastle, NSW 2308, Australia

Background

Transfer cells (TCs) are important for plant development as they form at nutrient transport bottlenecks where an apoplasmic/symplasmic transport step is required for acquisition and/or delivery of nutrients [1]. TCs are anatomically specialized for this function as they develop extensive wall ingrowths which result in increased plasma membrane surface area which supports an increased density of nutrient transporters [1-3]. Seeds of many crop species develop TCs to facilitate seed filling [4], and TCs support both phloem loading and short and long distance transport via xylem/phloem exchange [5]. TCs develop by *trans*-differentiation of existing cell types in response to developmental or stress-induced signals [1], but despite the importance of TCs to plant development, little is known of the molecular processes responsible for their *trans*-differentiation. This situation is caused in part by TCs typically being located deep within tissues [6] and thus not readily accessible for experimental manipulation and study.

The *trans*-differentiation of TCs involves differential expression of hundreds of genes. The formation of nucellar projection and endosperm TCs in barley grains involves differential expression of at least 815 genes [7], while the development of epidermal TCs in *Vicia faba* cotyledons is predicted to involve up to 650 genes [8]. These and other observations have led to the proposition that wall ingrowth deposition in TCs involves hierarchical regulation of cascades of gene expression, presumably controlled by key transcription factors [9], a model based on the genetic regulation of secondary wall deposition in xylem tissue [10,11]. The identification of such factors putatively regulating wall ingrowth deposition in TCs is best undertaken in a genetic model such as *Arabidopsis thaliana* (Arabidopsis).

In Arabidopsis, phloem parenchyma (PP) TCs are known to form in minor veins of leaves and sepals where they are proposed to function in apoplasmic phloem loading [12-14]. Previous studies examining PP TCs in Arabidopsis have relied on transmission electron microscopy (TEM) to analyze these cells. Indeed, Amiard et al. [13] traced cell wall contours of PP TCs viewed by TEM to demonstrate a role for high light and jasmonic acid in signaling wall ingrowth development, and similar approaches were undertaken to demonstrate a relationship between photosynthetic capacity and PP TC development [15]. Analysis by electron microscopy, however, is time-consuming and clearly not compatible for high-throughput screening required to identify genetic factors controlling the *trans*-differentiation of TCs.

High-resolution imaging of cell walls by confocal microscopy has been achieved using a modified pseudo-Schiff base-propidium iodide (mPS-PI) staining procedure [16]. In this process, treatment of fixed plant tissue with

periodic acid results in the formation of aldehyde groups in the carbohydrate moieties of cells walls. These aldehyde groups can then be reacted with various fluorescent pseudo-Schiff reagents, such as propidium iodide, resulting in strong covalent fluorophore labelling of cell walls [16]. The strong covalent labelling enables the tissue to be extensively cleared and mounted in high-refractive index mounting medium, giving strong and stable fluorescence labelling of cell walls and thus enabling extensive z-stack imaging of cellular organization in complex tissues [16,17]. Wall ingrowths of TCs are rich in cellulose and other polysaccharides such as pectins [18], a feature that may provide an opportunity to use the mPS-PI procedure to image wall ingrowth deposition in PP TCs.

Here we report the successful use of mPS-PI staining of Arabidopsis leaves to visualize wall ingrowth deposition in PP TCs in minor veins of leaves, cotyledons and sepals by confocal imaging. Wall ingrowths in these cells are discernable as highly localized thickenings of wall material deposited along the face of the PP TC adjacent to neighboring cells of the sieve element/companion cell (SE/CC) complex. Depending on tissue orientation, this deposition can often be seen as a central band running along each PP TC and superimposing an underlying SE or CC. We have used this procedure to also image light-dependent wall ingrowth deposition in CC TCs of pea minor veins [19,20], and have developed a scoring method based on the extent of wall ingrowth deposition for semi-quantitative analysis of TC development. Furthermore, introduction of a simple post-fixation bleaching step as an alternative to extensive clearing procedures in the original technique has simplified the processing steps to enable more efficient staining of tissue. Collectively, this procedure now provides the opportunity to investigate the cell biology of wall ingrowth deposition of PP TCs in Arabidopsis in a semi-quantitative manner without resorting to electron microscopy, and will also enable high-throughput phenotypic screening of TC development to identify key transcriptional regulators of this process.

Results

A modified pseudo-Schiff staining technique using propidium iodide to visualize wall ingrowths in TCs

The development of high-contrast staining of cell walls in cleared plant tissue using a mPS-PI procedure has enabled improved confocal imaging throughout plant tissues generally [16] and leaf tissue in particular [17]. To develop a procedure for confocal imaging of wall ingrowths in PP TCs in Arabidopsis leaves we used the technique of Wuyts et al. [17], modified by first peeling away the abaxial epidermis of rosette leaves immediately prior to fixation. Removing the abaxial epidermal layer and most of the associated mesophyll tissue and viewing from the abaxial face of the leaf enabled clear viewing of vascular bundles (Figure 1D). Under these conditions, confocal imaging of mPS-PI-stained leaves clearly resolved bands of wall ingrowth material, seen as unevenly thickened and mottled staining, positioned along the face of PP TCs adjacent to cells of the SE/CC complex (Figure 1A). In these images PP TCs can be identified as relative thin, elongated cells sharing a common longitudinal wall with a larger bundle sheath cell and the opposite wall with neighboring cells of the SE/CC complex. The highly localized deposition of wall ingrowths in PP TCs is evidenced by their occurrence only along the wall shared with a cell of the SE/CC complex (Figure 1A,B). In Figure 1A, the PP TC labelled with a double asterisk shows localized ingrowth deposition on

the two faces of the cell neighboring different CCs, indicating that the localizing signal most likely emanates from cells of the SE/CC complex. A longitudinal y-z projection of a z-stack through the vascular bundle shown in Figure 1A resolved discrete finger-like projections of wall ingrowth material along the face of PP TCs neighboring two SEs (Figure 1B). This image is highly reminiscent of TEM views of finger-like wall ingrowth projections in these cells (see Figure six of [21]), thus supporting the conclusion that the structures being imaged are indeed wall ingrowths in PP TCs. An x-z projection of the same z-stack showing the vascular bundle in transverse section, clearly resolved localized patches of wall ingrowth material deposited in PP TCs adjacent

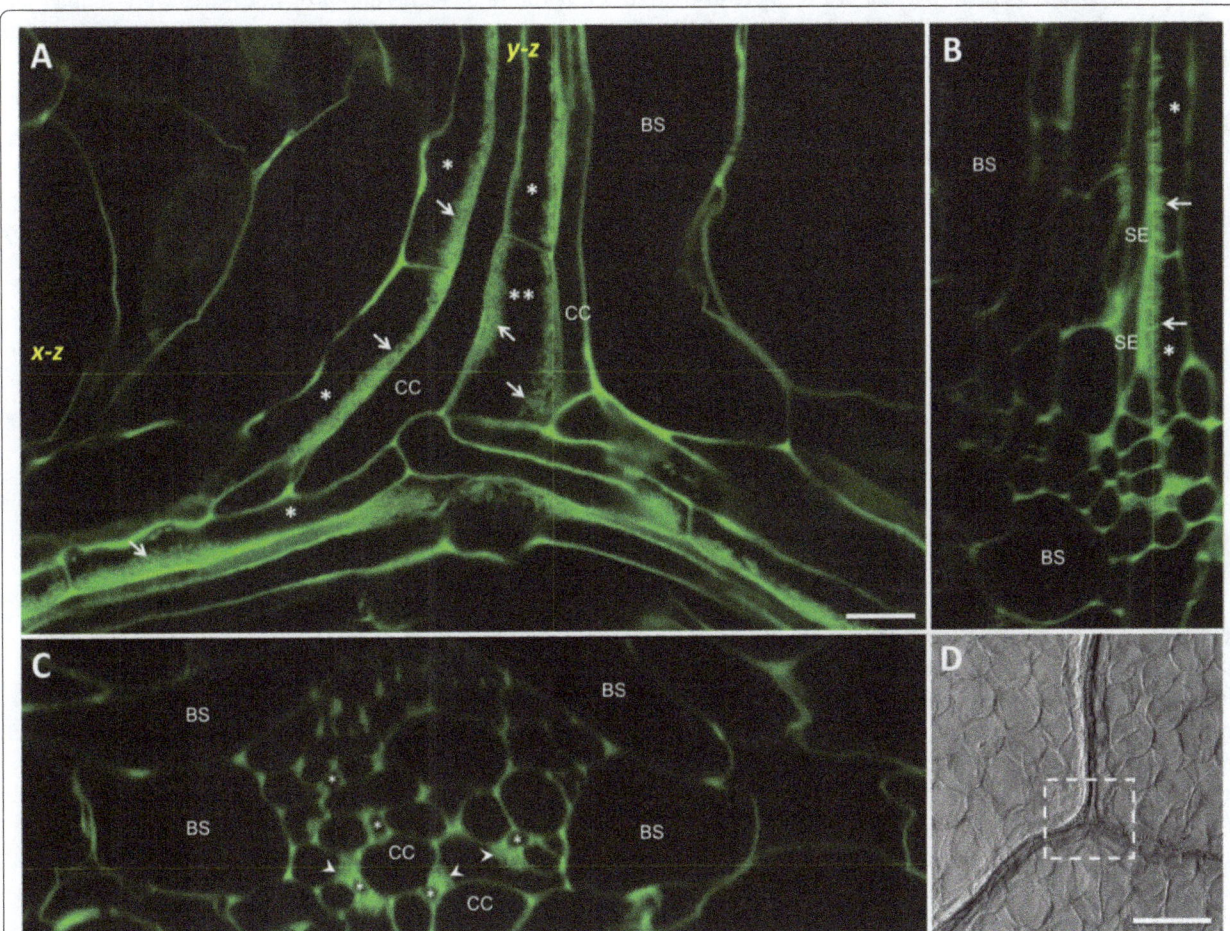

Figure 1 Confocal imaging of wall ingrowths in PP TCs of Arabidopsis leaf minor veins stained by the mPS-PI procedure. **A**. Single confocal section of a minor vein junction revealing polarized deposition of wall ingrowths (arrows) on the face of PP TCs (single asterisks) adjoining CCs. Polarized deposition of wall ingrowth material can be seen in other PP TCs, including the central PP TC (double asterisk) where ingrowth deposition is directed to opposite faces of the PP TC, each adjoining a different CC. The yellow dotted lines labelled y-z and x-z correspond to the projections shown in **B** and **C**, respectively. **B**. y-z projection of a z-stack of the image shown in **A** revealing finger-like projections (arrows) of wall ingrowth material extending from the face of two linearly-arranged PP TCs (asterisks) adjacent to neighboring SEs. **C**. x-z projection of a z-stack of the image shown in **A** revealing minor vein architecture in transverse section and the presence of highly-localized depositions of wall ingrowth material (arrowheads) adjacent to small SEs (asterisks) and larger CCs. **D**. Bright-field image of minor vein junction. The boxed area corresponds to the region shown in **A** and indicates the clarity of viewing vascular tissue when the abaxial epidermal layer is removed and the tissue is viewed from the abaxial surface of the leaf. BS, bundle sheath cell; CC, companion cell; SE, sieve element. Scale bar = 10 μm in **A**, **B** and **C**. Scale bar = 100 μm in **D**.

mostly to the smaller SEs but also to CCs (Figure 1C). A survey of PP TCs revealed that in most discernable instances, wall ingrowth deposition in a PP TC was initiated immediately opposite a SE, but consolidation of this deposition spreads to areas of the cell wall opposite CCs (data not shown). This observation supports the suggestion that the source of signals such as reactive oxygen species likely to drive wall ingrowth deposition in PP TCs is derived from SEs [5].

When rosette leaves are torn paradermally and viewed by scanning electron microscopy (SEM), wall ingrowth deposition can often be seen as a central band of reticulate ingrowth material running along the length of a given PP TC (Figure 2A; see [22]), or as discrete clumps of tangled, finger-like projections (Figure 2D). These features are also seen by confocal imaging of mPS-PI-stained leaf material, namely central bands of wall ingrowth material running along PP TCs (arrows, Figure 2B) and isolated clumps of wall ingrowths (arrows, Figure 2E,F). The central bands of ingrowth material reflect their highly localized deposition immediately adjacent to neighboring cells of the SE/CC complex (see [12,13]). This spatial relationship is clearly seen in Figure 2B where the focal plane in the middle of the image passes from a PP TC (double asterisk, Figure 2B) to an underlying SE, revealing how the band of wall ingrowth material in the PP TC superimposes the underlying SE (Figure 2B). This feature is particularly evident when viewed as a z-stack movie through these cells (Additional file 1: Movie S1). At higher magnification, confocal imaging clearly resolved the intertwined, finger-like projections of wall ingrowth material (arrow, Figure 2C), a feature that is readily evident when viewed by SEM (Figure 2A).

We surveyed different vein orders in mature leaves for the presence of PP TCs. Consistent with previous studies identifying phloem involved in assimilate loading in mature leaves [23], we observed typically substantial levels of wall ingrowth deposition in virtually all veins except the midrib and most of the secondary veins (data not shown). The exception to this observation was minor levels of wall ingrowth deposition seen in small terminating regions of secondary veins (Additional file 2: Movie S2). These observations are consistent with the conclusion that wall ingrowth deposition in PP TCs correlates with phloem loading capacity of minor veins [23], which in turn correlates inversely with vein size, as suggested by Haritatos et al. [12].

Imaging wall ingrowth deposition in CC TCs in leaf minor veins of pea

To test the general applicability of this method for imaging wall ingrowths in other species, we used the mPS-PI procedure of Wuyts et al. [17] to stain CC TCs in leaf minor veins of pea, where reticulate wall ingrowth

deposition occurs on all faces of these cells [19,20,24]. Confocal imaging of minor veins showed mottled labelling across the full face of CC TCs (arrows, Figure 3A). In contrast, wall ingrowth deposition was not detected in PP cells neighboring cells of the bundle sheath (Figure 3A). A y-z projection of a z-stack passing longitudinally through the vertically-orientated minor vein in Figure 3A revealed the presence of reticulate wall ingrowths along the longitudinal walls of a CC TC (inset A', Figure 3A), and similarly, a y-z projection passing transversely through the horizontal minor vein in Figure 3A revealed wall ingrowth deposition across all faces of the large, mostly circular CC TCs (inset A", Figure 3A), consistent with TEM images of these cells [20,24]. To verify that these structures were indeed wall ingrowths, leaves were stained from light-grown plants subjected to 4 days of dark treatment, conditions known to cause reduced wall ingrowth deposition [19,20]. Accordingly, reticulate wall ingrowth deposition in CC TCs was also greatly reduced, as shown by the y-z projection of a z-stack passing transversely through a vascular bundle (inset B', Figure 3B). This result confirms that the CC TCs shown in Figure 3A contain reticulate wall ingrowths, and that these ingrowths can be detected by confocal imaging of mPS-PI-stained pea leaves.

A simplified extraction procedure for mPS-PI staining of PP TCs using sodium hypochlorite

The collective analyses described above established that the mPS-PI staining procedure adapted from Wuyts et al. [17] can be used to image wall ingrowth deposition in TCs involved in phloem loading in both Arabidopsis and pea. The Wuyts et al. procedure is lengthy, however, involving several extractions in organic solvents and clearing in SDS/NaOH, followed by overnight treatment with amylase and pullulanase prior to mPS-PI staining (see Methods). To circumvent these lengthy procedures, we tested bleaching of fixed and ethanol-washed leaf tissue in sodium hypochlorite to clear cellular content for subsequent mPS-PI-staining of cell walls. Sodium hypochlorite was used by Sugimoto et al. [25] to extract cellular content from root tissue prior to viewing cellulose microfibrils by field emission SEM, and was used by Edwards et al. [22] to clear leaf tissue for fluorescence imaging of wall ingrowth deposition in PP TCs using Calcofluor White. In this current study, the use of bleach to clear tissue resulted in equivalent, if indeed somewhat improved, imaging of wall ingrowths in PP TCs (data not shown) compared to the procedure of Wuyts et al. The bleach method provided consistently good extraction of cellular content, with the minor exception of cotyledons (see below), and often yielded well defined cellular morphology as revealed by a z-series scan of vascular tissue (Additional file 3: Movie S3).

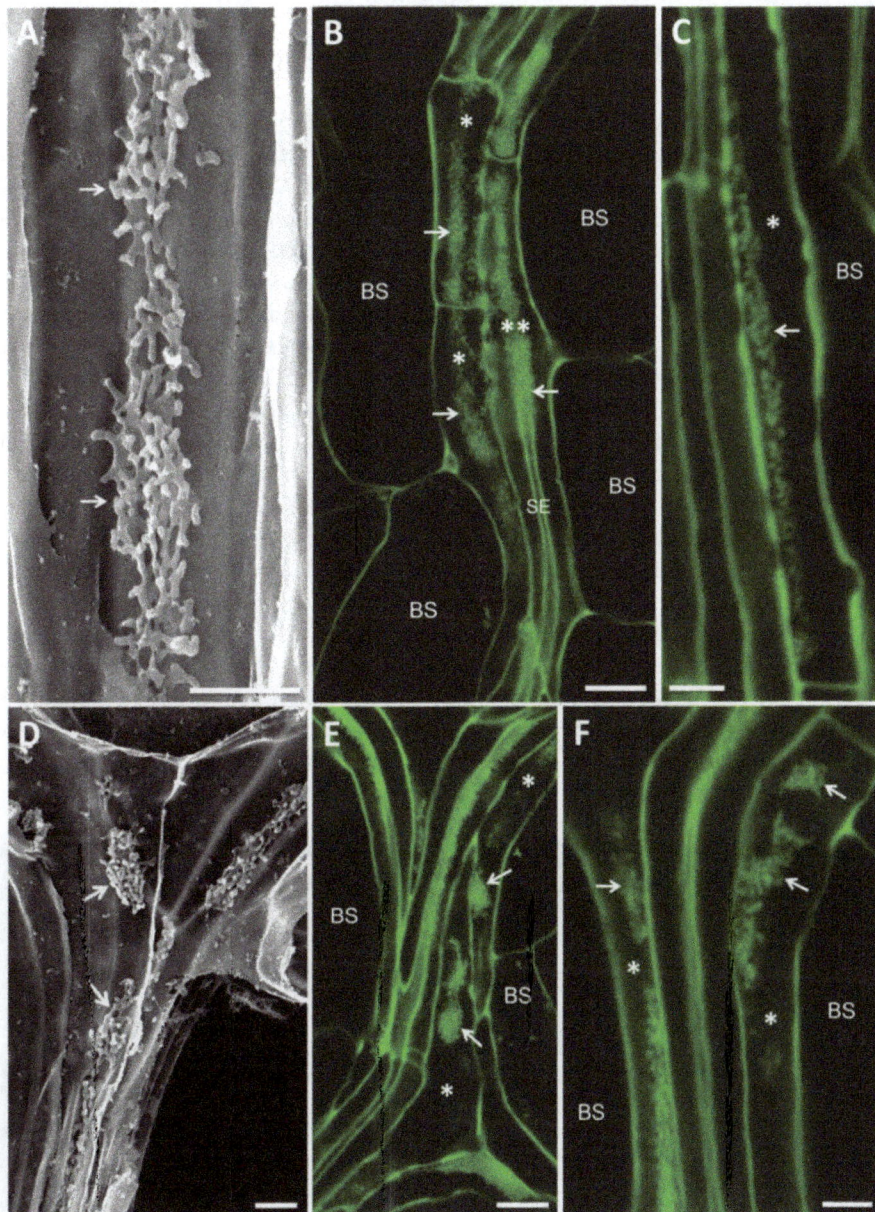

Figure 2 Comparison of wall ingrowths in PP TCs of Arabidopsis leaf minor veins by confocal imaging and SEM. **A, D**. SEM views of fresh leaf material torn paradermally then subjected to bleach extraction and viewed by SEM. **B, C, E, F**. Confocal imaging of minor veins from leaf material stained by the mPS-PI procedure. **A**. SEM image of a PP TC showing a central band of reticulate wall ingrowth material (arrows). **B**. Highly localized deposition of wall ingrowth material seen as a central band (arrows) running along the length of each PP TC (asterisks). The focal plane of the image passes from the PP TC on the right (double asterisk) into the underlying SE, indicating how the band of wall ingrowth material superimposes the underlying SE. **C**. Confocal image at higher magnification showing substructure (arrow) of the wall ingrowth material in a PP TC (asterisk). In this image the SE to the left of this PP TC is obscured. **D**. SEM view of a minor vein junction of two PP TCs showing examples of isolated patches of wall ingrowth deposition (arrows). **E**. Confocal image of minor vein junction showing discrete patches of wall ingrowth deposition (arrows). **F**. Higher magnification confocal image showing patches of wall ingrowth deposition (arrows) in two PP TCs (asterisks). BS, bundle sheath cell; CC, companion cell; SE, sieve element. Scale bars = 2 μm in **A** and **D**. Scale bars = 10 μm in **B** and **E**. Scale bars = 5 μm in **C** and **F**. The image in **A** is reproduced in part from Edwards et al. [22].

Given this outcome, we subsequently adopted the bleaching of fixed and ethanol-washed tissue as our standard method for confocal imaging of mPS-PI-stained tissue.

Wall ingrowth deposition in PP TCs from different tissues of Arabidopsis

We used our modified mPS-PI-staining procedure to survey other tissues in Arabidopsis for the presence of

PP TCs. In cotyledons, as typically seen in rosette leaves and other tissues (see below), the morphology of wall ingrowths can vary depending on age of the tissue. For example, early-stage wall ingrowth development in cotyledons of 7 day-old seedlings appears identical to early-stage wall ingrowth deposition in immature leaves of 14 day-old seedlings (data not shown). In cotyledons from 18 day-old seedlings, however, extensive deposition of wall ingrowths is seen along the face of PP TCs adjacent to cells of the SE/CC complex (Figure 4A-C). The morphology of wall ingrowth deposition in cotyledons from such plants was surprisingly varied, ranging from uniform deposition similar to that seen in rosette leaves (Figure 4A, Additional file 4: Figure S1A), to sharply pointed peaks of wall ingrowth material (Figure 4B), or very substantial deposition, albeit irregularly distributed along the length of a given PP TC and occupying a considerable volume of the cell (Figure 4B,C). This feature is similar to the manner in which dense fenestrated networks of ingrowth material protrude extensively into the outer periclinal cytoplasmic volume of abaxial epidermal TCs in *V. faba* cotyledons [26]. The images shown in Figure 4A-C are of PP TCs in vascular bundles located at the base, middle and tip regions of cotyledons, respectively, reflecting a basipetal gradient of wall ingrowth

deposition which correlates with phloem loading capacity in cotyledons [27]. Variations in wall ingrowth development are also apparent in nearby veins as seen in Additional file 4: Figure S1A. The PP TC marked with an asterisk in Figure S1A and A' developed very extensive and dense wall ingrowths, while in a nearby PP TC (double asterisk, Additional file 4: Figure S1A) wall ingrowth deposition was less developed, hence typical finger-like projections can be detected in a longitudinal view (double asterisk, Additional file 4: Figure S1A") reconstructed from the *z*-stack image shown in S1A.

Wall ingrowth deposition in PP TCs was also detected in cauline leaves (Figure 4D, Additional file 4: Figure S1B) and sepals (Figure 4E,F, Additional file 4: Figure S1C). In cauline leaves deposition of ingrowth material was typically abundant, especially in veins in the tip region of the leaf (Figure 4D). In sepal tissue, wall ingrowths were often seen as discrete clusters of wall material positioned along the length of a PP TC (Figure 4E). These discrete clusters appeared in places to merge with neighboring clusters to form localized clumps of ingrowth material along a given PP TC (Figure 4F). These features were also seen in PP TCs in other tissues (e.g., Figure 5C), but were more common in sepals. Xylem elements were often detected adjacent to PP TCs in sepals (Figure 4E,F,

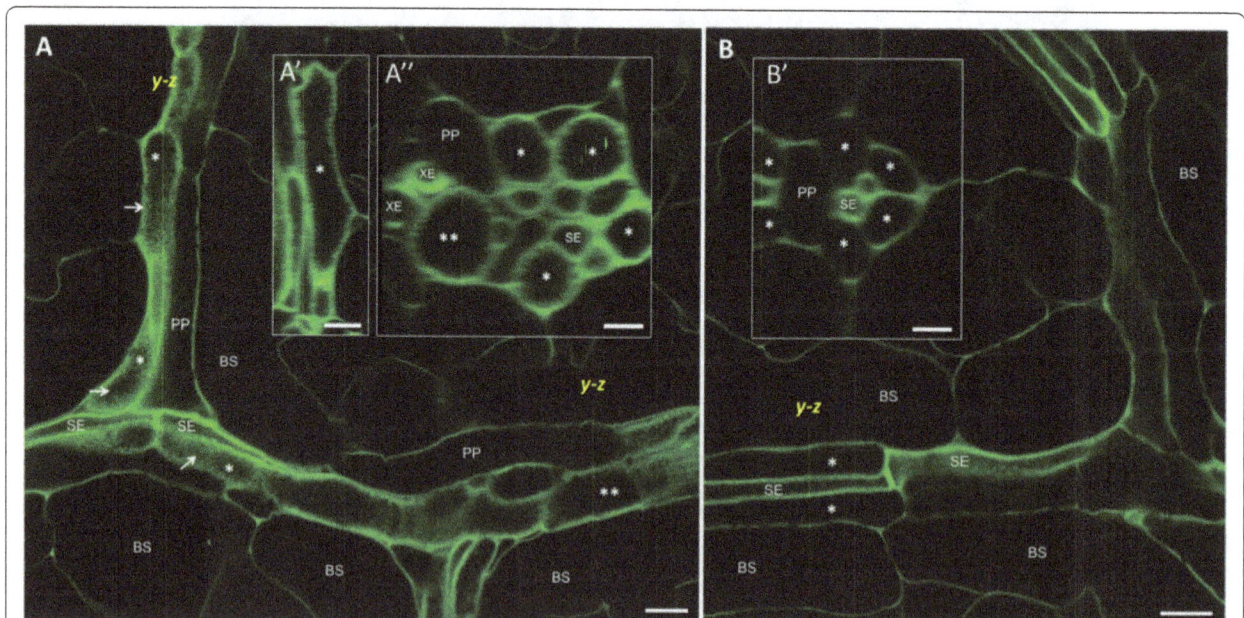

Figure 3 Confocal imaging of wall ingrowth deposition in CC TCs of leaf minor veins in pea. **A.** Minor veins from second true leaf of 13 day-old full light-grown seedlings showing mottled wall ingrowth labelling across the full face of CC TCs (asterisks). No wall ingrowth deposition is seen in PP cells. A *y-z* projection of a *z*-stack through the vertical minor vein is shown in inset A'. Wall ingrowth deposition detected as fuzzy labelling can be seen along all longitudinal walls of the CC TC (asterisk). A *y-z* projection through the horizontal minor vein at the bottom right of **A** is shown in inset A". Here, fuzzy labelling indicating wall ingrowth deposition is seen around all faces of the CC TCs seen in transverse view. The double asterisk in **A** and A" indicates a large CC TC. **B.** Minor veins from second true leaf of 9 day-old seedling subjected to 4 days of darkness. No wall ingrowth deposition is seen in CC TCs (asterisks) seen in **B** or when the minor vein is seen in transverse view as a *y-z* projection through this minor vein (inset B'). BS, bundle sheath cell; PP, phloem parenchyma; SE, sieve element; XE, xylem element. Scale bars = 10 μm in **A**, A' and **B**. Scale bars = 5 μm in A" and B'.

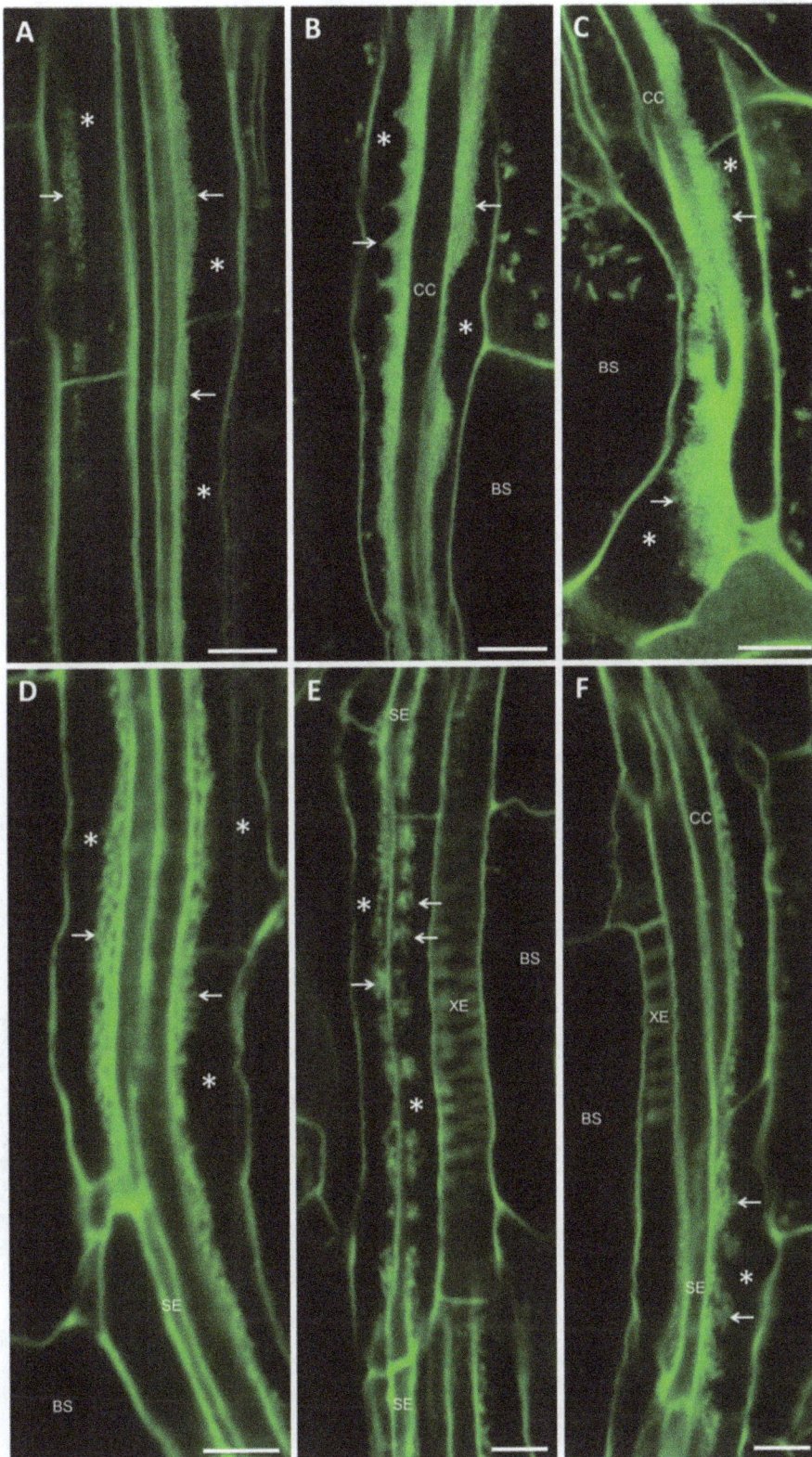

Figure 4 (See legend on next page.)

(See figure on previous page.)

Figure 4 Confocal imaging of wall ingrowth deposition in PP TCs in cotyledons, cauline leaves and sepal. **A**. Minor vein from the base of an 18 day-old cotyledon showing extensive wall ingrowth deposition (arrows) in PP TCs (asterisks). **B**. Minor vein from the mid-region of an 18 day-old cotyledon showing highly sculptured and extensive wall ingrowth deposition (arrows) in PP TCs (asterisks). **C**. Minor vein from the tip an 18 day-old cotyledon showing massive levels of wall ingrowth deposition (arrows) that occupy a considerable volume of each PP TC (asterisks). The fragments of fluorescent labelling seen in bundle sheath cells in both **B** and **C** correspond to remnant starch grains not completely extracted by the bleach treatment. **D**. Wall ingrowth deposition (arrows) in PP TCs (asterisks) in a fully expanded cauline leaf. **E**. Wall ingrowths in a minor vein of sepal, showing numerous localized patches of wall deposition (arrows) along each PP TC (asterisk). **F**. Wall ingrowths in a minor vein of sepal showing apparent consolidation or merging of localized patches of wall ingrowth material (arrows) in a PP TC (asterisk). BS, bundle sheath cell; CC, companion cell; XE, xylem element. Scale bars = 10 μm in **A**, **B** and **C**. Scale bars = 5 μm in **D**, **E** and **F**.

Additional file 4: Figure S1C) due to the simple structure of vascular bundles in this tissue (see [14]).

Semi-quantitative assessment of wall ingrowth deposition in PP TCs

The clarity of confocal imaging by the bleach-modified mPS-PI procedure enabled semi-quantitative assessment of both the extent of *trans*-differentiation of PP TCs and the abundance of wall ingrowth deposition in a given cell. To facilitate this process, we developed a scoring procedure ranging across four categories of wall ingrowth deposition as defined by our observations. Class I represents PP cells with no detectable wall ingrowths (Figure 5A, Additional file 5: Figure S2A, B). These cells, defined by their elongated, rectangular shape and connection to a neighboring bundle sheath cell, were devoid

of detectable wall ingrowths as evidenced by the thin, regular outline of their stained cell walls. In Class II, wall ingrowths were detected in early stages of development as evidenced by limited regions of patchy, mottled staining along the wall of the PP TC opposite that of a bundle sheath cell (Figure 5B, Additional file 5: Figure S2F), or visualized as discrete dots of fluorescence in face view (Additional file 5: Figure S2C, D, E). For this class, not all PP cells in a given region of vein showed evidence of wall ingrowth deposition. In Class III, wall ingrowth deposition was more obvious as wider regions of mottled fluorescence and this level of deposition was commonly detected in most but not necessarily all PP cells in a given field of view (Figure 5C, Additional file 5: Figure S2G, H, I, J). In Class IV, wall ingrowths in PP TCs were very abundant and seen as continuous thick bands

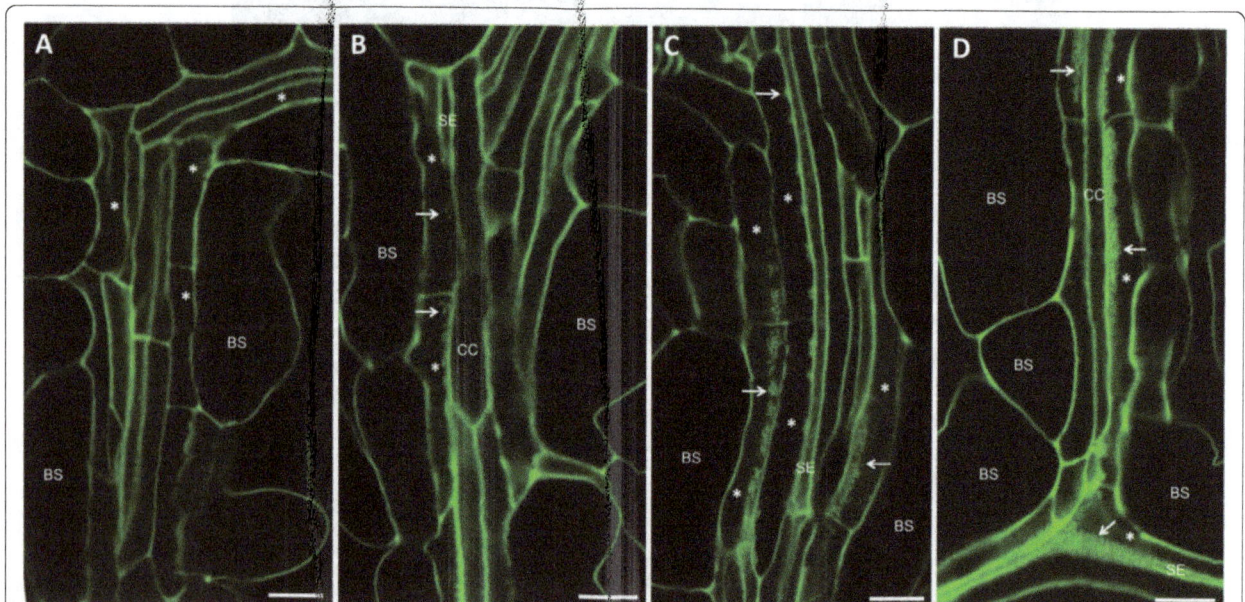

Figure 5 Classification system for the extent of wall ingrowth deposition in PP TCs in minor veins of Arabidopsis leaves. **A**. Class I - no wall ingrowths visible in PP cells (asterisks), which are identified by their sharing a common wall with a large bundle sheath cell. **B**. Class II - evidence of discrete, punctate-like wall ingrowth deposition seen as discrete fluorescent dots (arrows) distributed along the face of PP TCs adjacent to either a CC or SE. Not all PP cells in a given field of view contain wall ingrowths. **C**. Class III - substantial levels of reticulate wall ingrowth deposition is seen as clusters or continuous stretches of fluorescent labelling (arrows) on the face of PP TCs (asterisks) neighboring cells of the SE/CC complex. This level of labelling is seen in most PP cells in a given field of view, but can be somewhat variable. **D**. Class IV - extensive wall ingrowth deposition seen as thick bands of fluorescence labelling (arrows) seen in essentially all PP TCs in a given field of view. BS, bundle sheath cell; CC, companion cell; SE, sieve element. Scale bars = 10 μm in **A**, **B**, **C** and **D**.

of mottled fluorescence present in essentially all PP TCs in the field of view (Figure 5D, Additional file 5: Figure S2K, L).

Using this classification system, we qualitatively surveyed the abundance of wall ingrowth deposition in PP TCs in terminating minor veins across leaf development, selecting leaf 11, 8 and 5 as representative of immature, intermediate and mature leaves, respectively (Figure 6A). A representation of this survey is shown in Figure 6B. Immature sink leaves contain predominantly Class I PP cells with no wall ingrowths, while intermediate leaves are characterized by a basipetal gradient with Class III PP TCs in minor veins at the tip of the leaf and predominantly Class I PP cells at the leaf base. In contrast, mature source leaves are dominated by Class IV PP TCs with highly abundant wall ingrowths in minor veins across virtually the entire leaf (Figure 6B). The distribution of PP TCs seen in this analysis is consistent with the known sink-source transition that occurs in maturing leaves [28], as well as the development of apoplasmic loading that occurs in a basipetal gradient within a dicot leaf [23,27]. The value of mPS-PI staining in this case, however, is that it provides a rapid means to assess the development of PP TCs across an entire leaf in response to different biotic or abiotic signals and within different genetic backgrounds, with high spatial resolution without relying on time-consuming procedures such as TEM.

Manipulating wall ingrowth deposition in PP TCs by defoliation

We next examined the response of PP TC development to altered sink-source status in Arabidopsis leaves as a test of the mPS-PI staining procedure to provide semi-quantitative assessment of wall ingrowth development. For this analysis we conducted a defoliation experiment on 3-week-old plants by removing all leaves except leaf 9, 10 and 11 (Figure 7A). After five days further growth, leaves from control and defoliated plants (Figure 7B) were harvested and stained by the bleach-modified mPS-PI procedure. In leaf 10 from control plants, typically no wall ingrowths were visible in PP cells (Class I; Figure 7C), but in a few cases Class II wall ingrowth deposition was occasionally seen (Figure 7D). In this case, the ingrowths were small discrete clusters of wall material. In contrast to the typical lack of ingrowth deposition in control leaves, the extent of wall ingrowth deposition in leaf 10 from defoliated plants was substantially increased (Figure 7E-G). In these cases, Class III deposition was commonly seen, either in early stages of deposition where discrete clusters of tangled wall ingrowths were detected (Figure 7E), or as more dense clusters or bulges of ingrowth wall deposition (Figure 7F). In the tip region of leaf 10 from a defoliated plant, numerous cases of Class IV deposition were detected in these leaves 5 days-post defoliation (Figure 7G).

To provide a semi-quantitative analysis of the response of wall ingrowth development to defoliation, we used a scoring system whereby Class I, II, III and IV were assigned 0, 2, 4 and 6 points, respectively. Using these values, we then scored PP TC development in at least five terminating veins within the base, middle and tip regions of each leaf on either side of the mid-vein (see Figure 6B). The results of this analysis showed a rapid and significant increase in wall ingrowth

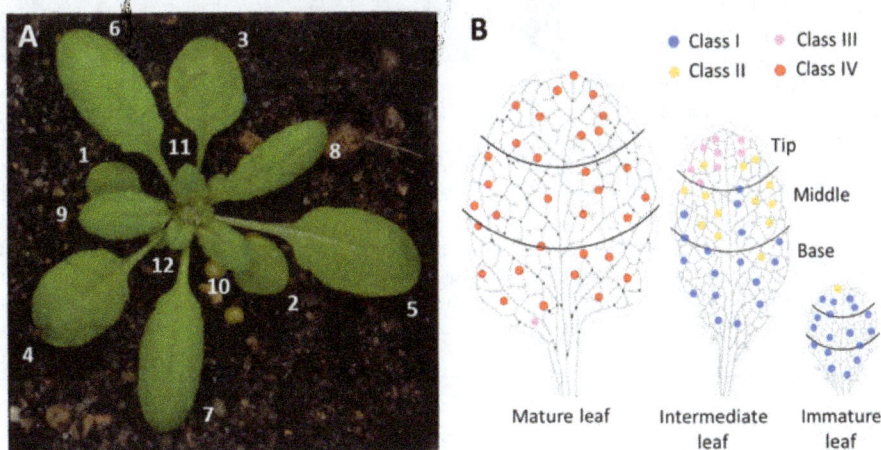

Figure 6 Survey of wall ingrowth deposition in PP TCs in minor veins across leaf development in Arabidopsis. **A**. Leaf numbering of 3.5 week-old rosette leaf. **B**. Representation of wall ingrowth deposition in immature, intermediate and mature leaves as represented by color-coding of the four classes of wall ingrowth deposition as described in Results. The result shown here is representative of three replicate leaves for each stage of development. Immature leaves contain little if any wall ingrowth deposition in PP cells. Intermediate leaves show a basipetal gradient from typically Class III at the tip to Class I at the leaf base. Mature leaves contain Class IV wall ingrowth deposition in PP TCs of minor veins throughout the entire leaf, with the exception of a few examples of Class III at the very base of the leaf. Skeletonized images of leaves shown in Figure 6B were adapted from Alonso-Peral et al. [34].

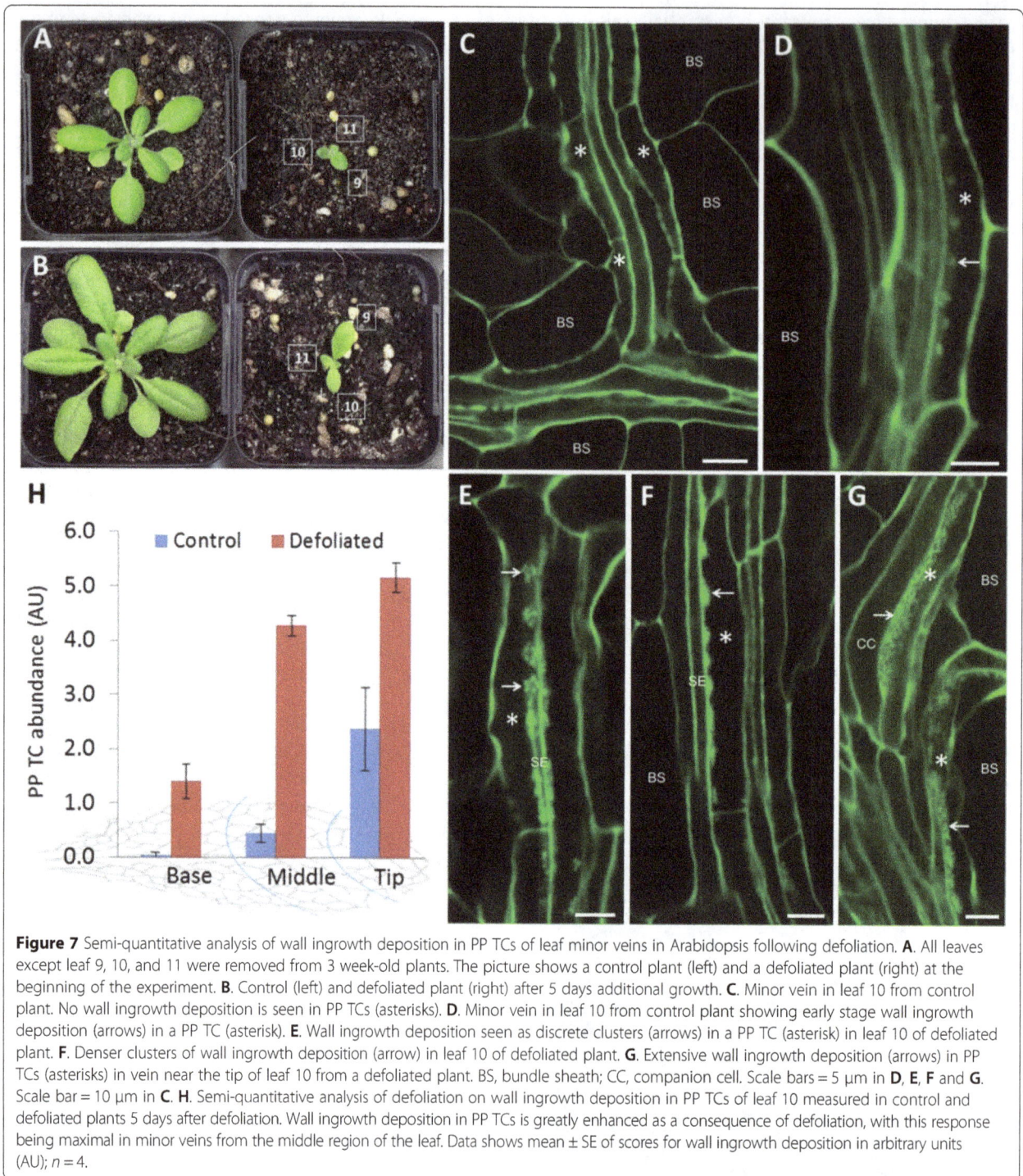

Figure 7 Semi-quantitative analysis of wall ingrowth deposition in PP TCs of leaf minor veins in Arabidopsis following defoliation. **A**. All leaves except leaf 9, 10, and 11 were removed from 3 week-old plants. The picture shows a control plant (left) and a defoliated plant (right) at the beginning of the experiment. **B**. Control (left) and defoliated plant (right) after 5 days additional growth. **C**. Minor vein in leaf 10 from control plant. No wall ingrowth deposition is seen in PP TCs (asterisks). **D**. Minor vein in leaf 10 from control plant showing early stage wall ingrowth deposition (arrows) in a PP TC (asterisk). **E**. Wall ingrowth deposition seen as discrete clusters (arrows) in a PP TC (asterisk) in leaf 10 of defoliated plant. **F**. Denser clusters of wall ingrowth deposition (arrow) in leaf 10 of defoliated plant. **G**. Extensive wall ingrowth deposition (arrows) in PP TCs (asterisks) in vein near the tip of leaf 10 from a defoliated plant. BS, bundle sheath; CC, companion cell. Scale bars = 5 μm in **D**, **E**, **F** and **G**. Scale bar = 10 μm in **C**. **H**. Semi-quantitative analysis of defoliation on wall ingrowth deposition in PP TCs of leaf 10 measured in control and defoliated plants 5 days after defoliation. Wall ingrowth deposition in PP TCs is greatly enhanced as a consequence of defoliation, with this response being maximal in minor veins from the middle region of the leaf. Data shows mean ± SE of scores for wall ingrowth deposition in arbitrary units (AU); $n = 4$.

deposition in PP TCs in leaf 10 of defoliated plants relative to leaf 10 in control (non-defoliated) plants, with this response being more pronounced in PP TCs in minor veins in the middle sector of leaves (ca. 8.6-fold change) compared to the tip (ca. 2.2-fold change; Figure 7H).

Discussion

We have developed a simplified mPS-PI staining procedure for confocal imaging of wall ingrowths in vascular TCs involved in phloem loading. This procedure, involving clearing of fixed tissue using sodium hypochlorite followed by mPS-PI staining and mounting in high refractive index mounting medium provides a rapid means to image wall ingrowth deposition in TCs without the need to use more time-consuming electron microscopy techniques. The clarity of the mPS-PI staining enables high-resolution imaging of wall ingrowths in vascular tissue buried deep within leaves, and the use of sodium hypochlorite as a clearing step simplifies the original methods of Truernit et al. [16] and Wuyts et al. [17] to enable high-throughput processing of samples to suite semi-quantitative assessment of wall ingrowth deposition in TCs. Furthermore, the ability to optically section throughout an entire vascular bundle and reconstruct from a series of z-axis images enables the three-dimensional reconstruction of wall ingrowth deposition to analyse the highly polarised nature of this process in TCs.

We used this procedure to analyse the distribution of PP TCs both within individual leaves and in leaves of different developmental stages. Haritatos et al. [12] noted that veins of different size from mature leaves have overall similar cellular structure and organization, and since most veins are in close proximity to mesophyll, they can be considered to participate in phloem loading and thus physiologically defined as "minor" veins. Our observations support this conclusion since virtually all veins examined from mature leaves, except the midrib and larger regions of the secondary vein network, contained PP TCs with typically substantial levels of wall ingrowth deposition. Furthermore, within individual developing leaves, a basipetal gradient of PP TCs was detected (Figure 6). Both observations are consistent with development of phloem loading capacity in leaves [27,28] and demonstrate a presumed correlation between wall ingrowth deposition in PP TCs and their role in phloem loading.

The general applicability of this method for confocal imaging of wall ingrowth formation in TCs was demonstrated by visualizing CC TCs in minor veins of pea leaves. In this case, light-dependent deposition of wall ingrowths to all faces of CC TCs was clearly detected (Figure 3), consistent with earlier studies using TEM [19,20]. Thus, it is likely that this approach can be used to investigate TC development in other species and tissue locations. In Arabidopsis, we observed PP TCs with extensive wall ingrowths in sepals, cotyledons and cauline leaves (Figure 4) as well as the first true leaves of young seedlings (data not shown). Interestingly, in both cotyledons and cauline leaves, the presence of PP TCs with abundant wall ingrowths correlates with the high relative expression of both *AtSWEET11* and *AtSWEET12* (eFP Browser; bar.toronto.ca). These sucrose effluxers have recently been demonstrated to be involved in a two-step phloem loading strategy used in Arabidopsis leaves, namely unloading of sucrose into the apoplasm

by PP TCs driven by AtSWEET transporters, followed by active uptake into cells of the SE/CC complex by AtSUC2 [14]. The extensive wall ingrowth deposition observed in PP TCs of cotyledons is also consistent with cotyledons acting as a strong source of photosynthesis-derived sucrose required to sustain root growth in response to light [29].

An intriguing observation seen most clearly in sepal tissue was the initial deposition of wall ingrowths as numerous discrete clusters along the length of a PP TC (Figure 4E). A similar pattern of deposition was seen in young leaves responding to defoliation (Figure 7E). These structures are presumably equivalent to the isolated patches of wall deposition observed by SEM (Figure 2D). These observations suggest that early stages of reticulate ingrowth deposition can be highly localized to discrete regions within an individual PP TC, and then continued deposition causes consolidation of these patches into more continuous regions of ingrowth deposition. The signals directing such localized patches of ingrowth deposition are unknown, but in non-vascular TC types the reactive oxygen species hydrogen peroxide has been implicated as a polarizing signal directing wall ingrowth deposition [5,30,31]. Recently, localized plumes of Ca^{2+} have been implicated in directing the highly localized deposition of individual papillae wall ingrowths in epidermal TCs of *V. faba* cotyledons [32]. A similar mechanism may operate in PP TCs, however the larger clusters of wall deposition seen by SEM and confocal imaging (Figures 2, 4 and 7) imply a higher level of organization may be operating, possibly aggregation of Ca^{2+} channels, to direct the deposition of wall ingrowths into such clusters. The ability to clearly image wall ingrowths by confocal microscopy in Arabidopsis PP TCs will enable a genetic approach to investigate signaling mechanisms driving this process.

The defoliation experiment (Figure 7) illustrates the value of the bleach-modified mPS-PI method in combination with semi-quantitative scoring to provide high-throughput and high-resolution assessment of PP TC development in leaves. The significant increase in wall ingrowth deposition in PP TCs in leaf 10 remaining after defoliation suggests a rapid switch from sink to source status within this leaf [33], and a requirement for wall ingrowth deposition for this to occur. The predicted concomitant changes in gene expression required for wall ingrowth deposition amid other processes associated with this transition provides an opportunity to identify these genes by transcriptional profiling.

Conclusion
We have developed a simple method for confocal imaging of wall ingrowth deposition in TCs using mPS-PI staining. This method was used to image wall ingrowth

deposition in PP TCs in rosette leaves, cauline leaves, cotyledons and sepals of Arabidopsis as well as CC TCs in leaf minor veins of pea. The clarity of the staining provides cellular detail of wall ingrowth deposition in these diverse tissues, thus enabling future studies investigating the cellular mechanisms directing the highly polarized deposition of wall ingrowths in TCs without the need to use electron microscopy. The high-throughput potential of this procedure also offers the opportunity to apply reverse genetics to identify genes involved in wall ingrowth deposition in TCs.

Methods

Plant growth conditions

Arabidopsis thaliana (Col-0) seeds were sown directly onto pasteurised soil mix and stratified for three days in darkness at 4°C. Plants were then transferred to a growth cabinet (100–120 μmol m^{-2} sec^{-1}, 22°C day/18°C night, 16 h photoperiod) for 2–3 weeks or until stated, and cotyledons, rosette leaves, cauline leaves and sepals were then collected for analysis. Peas (*Pisum sativum*) were raised in potting mix in a glasshouse maintained at 20-24°C and approximately 800–900 μmol m^{-2} sec^{-1} during daytime. Nine days after sowing, some seedlings were covered with aluminium foil to provide dark treatment and then the second pair of true leaves from both control (full light) or dark-treated plants were harvested after 4 days further growth. For defoliation experiments, all rosette leaves except leaf 9, 10 and 11 were removed from 3-week-old Arabidopsis plants. After 5 days of additional growth, leaf 10 from both control and defoliated plants were collected and processed for mPS-PI staining.

Pseudo-Schiff-propidium iodide staining of tissues

Rosette and cauline leaves from Arabidopsis and the second pair of true leaves from pea seedlings were pressed firmly onto clear sticky tape and the abaxial epidermal layer and associated mesophyll tissue from each leaf was peeled away using Scotch 3 M™ magic tape. Sepal and cotyledon tissue was processed without epidermal peeling. Tissue was then fixed overnight at 4°C in ethanol:acetic anhydride (3:1), then washed in 70% (v/v) ethanol and processed at room temperature as described below or stored in 70% ethanol at 4°C for several months. Tissue was washed in chloroform for 10 min, then progressively rehydrated and cleared in 1% (w/v) SDS in 0.2 N NaOH for 10 min. The extracted tissue was washed extensively in water and then incubated overnight at 37°C in 0.5% (v/v) amylase and 0.5% (v/v) pullulanase (Sigma, Australia) to remove starch. Tissue was then washed in water and incubated in 1% (v/v) periodic acid for 15 min, then washed in water again and stained in pseudo-Schiff propidium iodide for 1 h (100 mM Na$_2$S$_2$O$_5$, 0.15 N HCl with propidium iodide added to a final concentration of 100 μg/mL at the time of staining). Stained leaves were washed briefly in water and then mounted in chloral hydrate (4 g chloral hydrate, 1 mL glycerol, and 2 mL water) with the abaxial surface of the leaf facing up. The mounted tissue was covered with a coverslip and left overnight in darkness at room temperature before viewing.

Simplified extraction of tissue using sodium hypochlorite

Tissue was processed and fixed as described above and washed in 70% (v/v) ethanol. Cellular content of tissue was cleared by extracting tissue in 0.25% (v/v) White King™ bleach (4% (v/v) effective hypochlorite concentration) with gentle shaking at room temperature for at least 2 h depending on the tissue type. Cleared tissue was washed extensively in water and subjected to mPS-PI staining and mounting as described above.

Confocal microscopy and image acquisition

Confocal imaging of stained tissues was performed using an Olympus FluoView FV1000 confocal microscope. Imaging used 488 nm Argon-ion laser excitation and a 60 × 1.35NA Olympus oil-immersion objective. Emission wavelengths were collected at 522–622 nm. Image pixel resolution was set at 1600 × 1600, and used pixel dwell time of 4 μs and one-way scanning. Kalman average filtering of 4 was used during image acquisitions to improve signal-to-noise ratio of the acquired images.

Scanning electron microscopy

SEM analysis of PP TCs in mature rosette leaves was performed as described in Edwards et al. [22].

Additional files

Additional file 1: Movie S1. *z*-stack scan of Arabidopsis leaf minor vein showing wall ingrowth deposition in PP TCs. The scan moves through the tissue in an abaxial to adaxial direction. Wall ingrowths (red arrows) showing polarized deposition to the face of two PP TCs neighboring a common companion cell (CC) are seen. In some regions of the minor vein, wall ingrowth deposition can be seen as a central band (yellow arrows) which co-aligns with a neighboring sieve element (SE).

Additional file 2: Movie S2. *z*-stack scan of Arabidopsis second order vein showing a rare example of wall ingrowth deposition (purple arrows) in a PP TC in the vein. The scan moves through the tissue in an abaxial to adaxial direction. Most second order veins do not contain PP TCs. Wall ingrowth deposition (yellow arrows) is seen in a PP TC in the third order vein at the top of the image. M, myrosin cell; SE, sieve element.

Additional file 3: Movie S3. *z*-stack scan of Arabidopsis leaf minor vein from fixed tissue extracted with bleach. The scan moves through the tissue in an abaxial to adaxial direction. Bleach extraction provides good preservation of cellular architecture and imaging of PP TCs with imaging of wall ingrowths in PP TCs being comparable if not superior to the more complex extraction procedures employed by Truernit et al. [16] and Wuyts et al. [17].

Additional file 4: Figure S1. Additional examples of wall ingrowth deposition in PP TCs in cotyledons, cauline leaves and sepal. A, A' and A". Cotyledons. A. Single confocal section of a minor vein junction revealing polarized deposition of wall ingrowths (arrows). The dotted lines labelled *x-z* and *x'-z* correspond to the projection shown in A' and A'', respectively. A'. *x-z* projection of a *z*-stack of the image shown in A revealing minor vein architecture in transverse section and the presence of highly-localized and very substantial deposition of wall ingrowth material occupying nearly half the cell volume (asterisk). A''. *x'-z* projection of a *z*-stack of the image shown in A revealing the longitudinal section of a PP TC (double asterisks) with less extensive wall ingrowth deposition but with finger-like projections (arrow). B, B' and B''. Cauline leaves. B. Single confocal section of a minor vein junction showing polarized deposition of wall ingrowths (arrows). The dotted lines labelled *y-z* and *x-z* correspond to the projections shown in B' and B'', respectively. B'. *x-z* projection of a *z*-stack of the image shown in B revealing minor vein architecture in transverse section and the presence of highly-localized wall ingrowth deposition (arrow). B''. *y-z* projection of a *z*-stack of the image shown in B revealing finger-like projections (arrows) of wall ingrowths in a PP TC (double asterisks). C and C'. Sepals. C. Single confocal section of a minor vein revealing polarized deposition of wall ingrowths (arrows). Note that xylem elements (XE) were also detected in this small minor vein. The dotted line labelled *y-z* corresponds to the projections shown in C'. C'. *y-z* projection of a *z*-stack of the image shown in C revealing finger-like wall ingrowth projections (arrows) in a PP TC (asterisk). Scale bars = 10 μm.

Additional file 5: Figure S2. Additional examples of the four classes of wall deposition in PP TCs in Arabidopsis leaf veins. Asterisks in all figures represent PP or PP TCs. The double asterisks in I and J represent PP TCs with Class II deposition in a region of minor vein otherwise defined as Class III. Arrows point to wall ingrowth deposition in PP TCs. See Figure 5 for description of each class. Scale bars = 10 μm.

Abbreviations
CC: Companion cell; mPS-PI: Modified pseudo-Schiff-propidium iodide; PP: Phloem parenchyma; SE: Sieve element; SEM: Scanning electron microscopy; TCs: Transfer cells; TEM: Transmission electron microscopy.

Competing interests
The authors declare they have no competing interests.

Authors' contributions
DMcC and STTN conceived and designed the study. STTN performed the experiments and conducted the confocal imaging, and both DMcC and STTN discussed and interpreted the data. DMcC wrote the manuscript with assistance from STTN. Both authors read and approved the final manuscript.

Acknowledgements
We thank Felicity Andriunas for providing the SEM image shown in Figure 2D and Joe Enright for assistance with plant growth. This work was supported by grants to DMcC from the Australian Research Council (DP110100770) and the University of Newcastle Faculty of Science and Information Technology. STTN is supported by a VIED scholarship from the Vietnam Ministry of Agriculture and Rural Development.

References
1. Offler CE, McCurdy DW, Patrick JW, Talbot MJ. Transfer cells. Specialized cells for a special purpose. Ann Rev Plant Biol. 2003;54:431–54.
2. McCurdy DW, Patrick JP, Offler CE. Wall ingrowth formation in transfer cells: novel examples of localized wall deposition in plant cells. Curr Opin Plant Biol. 2008;11:653–61.
3. Harrington GN, Nussbaumer Y, Wang X-D, Tegeder M, Franceschi VR, Frommer WB, et al. Spatial and temporal expression of sucrose transport-related genes in developing cotyledons of Vicia faba L. Protoplasma. 1997;200:35–50.
4. Thompson RD, Hueros G, Becker H-A, Maitz M. Development and functions of seed transfer cells. Plant Sci. 2001;160:775–83.
5. Andriunas FA, Zhang HM, Xia X, Patrick JW, Offler CE. Intersection of transfer cells with phloem biology-broad evolutionary trends, function, and induction. Front Plant Sci. 2013;4:221.
6. Gunning BES, Pate JS. Transfer cells. In: Robards AW, editor. Dynamic Aspects of Plant Ultrastructure. London: McGraw-Hill; 1974. p. 441–79.
7. Thiel J, Weier D, Sreenivasulu N, Strickert M, Weichert N, Melzer M, et al. Different hormonal regulation of cellular differentiation and function in nucellar projection and endosperm transfer cells: a microdissection-based transcriptome study of young barley grains. Plant Physiol. 2008;148:1436–52.
8. Dibley SJ, Zhou Y, Andriunas FA, Talbot MJ, Offler CE, Patrick JW, et al. Early gene expression programs accompanying trans-differentiation of epidermal cells of Vicia faba cotyledons into transfer cells. New Phytol. 2009;182:863–77.
9. Chinnappa KSA, Nguyen STT, Hou J, Wu Y, McCurdy DW. Phloem parenchyma transfer cells in Arabidopsis - an experimental system to identify transcriptional regulators of wall ingrowth formation. Front Plant Sci. 2013;4:102.
10. Zhong R, McCarthy RL, Lee C, Ye Z-H. Dissection of the transcriptional program regulating secondary wall biosynthesis during wood formation in poplar. Plant Physiol. 2011;157:1452–68.
11. Zhong R, Ye Z-H. Complexity of the transcriptional network controlling secondary wall biosynthesis. Plant Sci. 2014;229:193–207.
12. Haritatos E, Medville R, Turgeon R. Minor vein structure and sugar transport in Arabidopsis thaliana. Planta. 2000;211:105–11.
13. Amiard V, Demmig-Adams B, Mueh KE, Turgeon R, Combs AF, Adams III WW. Role of light and jasmonic acid signaling in regulating foliar phloem cell wall ingrowth development. New Phytol. 2007;173:722–31.
14. Chen L-Q, Qu X-Q, Hou B-H, Sosso D, Osorio S, Fernie AR, et al. Sucrose efflux mediated by SWEET proteins as a key step for phloem transport. Science. 2012;335:207–10.
15. Adams III WW, Cohu CM, Amiard V, Demmig-Adams B. Associations between phloem-cell wall ingrowths in minor veins and maximal photosynthesis rate. Front Plant Sci. 2014;5:24.
16. Truernit E, Bauby H, Dubreucq B, Grandjean O, Runions J, Barthelemy J, et al. High-resolution whole-mount imaging of three-dimensional tissue organization and gene expression enables the study of phloem development and structure in Arabidopsis. Plant Cell. 2008;20:1494–503.
17. Wuyts N, Palauqui JC, Conejero G, Verdeil JL, Granier C, Massonnet C. High-contrast three-dimensional imaging of the Arabidopsis leaf enables the analysis of cell dimensions in the epidermis and mesophyll. Plant Methods. 2010;6:17–30.
18. Vaughn KC, Talbot MJ, Offler CE, McCurdy DW. Wall ingrowths in epidermal transfer cells of Vicia faba cotyledons are modified primary walls marked by localized accumulations of arabinogalactan proteins. Plant Cell Physiol. 2007;48:159–68.
19. Henry Y, Steer MW. A re-examination of the induction of phloem transfer cell development in pea leaves (Pisum sativum). Plant Cell Environ. 1980;3:377–80.
20. Wimmers LE, Turgeon R. Transfer cells and solute uptake in minor veins of Pisum sativum leaves. Planta. 1991;186:2–12.
21. Maeda H, Song W, Sage T, DellaPenna D. Role of callose synthases in transfer cell wall development in tocopherol deficient Arabidopsis mutants. Front Plant Sci. 2014;5:46.
22. Edwards J, Martin AP, Andriunas F, Offler CE, Patrick JW, McCurdy DW. GIGANTEA is a component of a regulatory pathway determining wall ingrowth deposition in phloem parenchyma transfer cells of Arabidopsis thaliana. Plant J. 2010;63:651–61.
23. Haritatos E, Ayre BG, Turgeon R. Identification of phloem involved in assimilate loading in leaves by the activity of the galactinol synthase promoter. Plant Physiol. 2000;123:929–37.
24. Gunning B, Pate J, Briarty L. Specialized "transfer cells" in minor veins of leaves and their possible significance in phloem translocation. J Cell Biol. 1968;37:C7–12.
25. Sugimoto K, Williamson RE, Wasteneys GO. New techniques enable comparative analysis of microtubule orientation, wall texture, and growth rate in intact roots of Arabidopsis. Plant Physiol. 2000;124:1493–506.
26. Talbot MJ, Franceschi VR, McCurdy DW, Offler CE. Wall ingrowth architecture in epidermal transfer cells of Vicia faba cotyledons. Protoplasma. 2001;215:191–203.
27. Wright KM, Roberts AG, Martins HJ, Sauer N, Oparka KJ. Structural and functional vein maturation in developing tobacco leaves in relation to AtSUC2 promoter activity. Plant Physiol. 2003;131:1555–65.
28. Turgeon R. The sink-source transition in leaves. Annu Rev Plant Physiol Plant Mol Biol. 1989;40:119–38.
29. Kircher S, Schopfer P. Photosynthetic sucrose acts as cotyledon-derived long-distance signal to control root growth during early seedling development in Arabidopsis. Proc Natl Acad Sci U S A. 2012;109:11217–21.
30. Andriunas FA, Zhang HM, Xia X, Offler CE, McCurdy DW, Patrick JW. Reactive oxygen species form part of a regulatory pathway initiating

trans-differentiation of epidermal transfer cells in *Vicia faba* cotyledons. J Exp Bot. 2012;63:3617–29.

31. Xia X, Zhang HM, Andriunas FA, Offler CE, Patrick JW. Extracellular hydrogen peroxide, produced through a respiratory burst oxidase/superoxide dismutase pathway, directs ingrowth wall formation in epidermal transfer cells of Vicia faba cotyledons. Plant Signal Behav. 2012;7:1125–8.

32. Zhang H, Imtiaz MS, Laver DR, McCurdy DW, Offler CE, van Helden DF, et al. Polarized and persistent Ca^{2+} plumes define loci for formation of wall ingrowth papillae in transfer cells. J Exp Bot. 2014 doi:10.1093/jxb/eru460.

33. Berthier A, Desclos M, Amiard V, Morvan-Bertrand A, Demmig- Adams B, Adams WWL, et al. Activation of sucrose transport in defoliated *Lolium perenne* L.: an example of apoplastic phloem loading plasticity. Plant Cell Physiol. 2009;50:1329–44.

34. Alonso-Peral MM, Candela H, del Pozo JC, Martinez-Laborda A, Ponce MR, Micol JL. The HVE/CAND1 gene is required for the early patterning of leaf venation in *Arabidopsis*. Development. 2006;133:3755–66.

Establishment of *Anthoceros agrestis* as a model species for studying the biology of hornworts

Péter Szövényi[1,2,3,4†], Eftychios Frangedakis[5,8†], Mariana Ricca[1,3], Dietmar Quandt[6], Susann Wicke[6,7] and Jane A Langdale[5*]

Abstract

Background: Plants colonized terrestrial environments approximately 480 million years ago and have contributed significantly to the diversification of life on Earth. Phylogenetic analyses position a subset of charophyte algae as the sister group to land plants, and distinguish two land plant groups that diverged around 450 million years ago – the bryophytes and the vascular plants. Relationships between liverworts, mosses hornworts and vascular plants have proven difficult to resolve, and as such it is not clear which bryophyte lineage is the sister group to all other land plants and which is the sister to vascular plants. The lack of comparative molecular studies in representatives of all three lineages exacerbates this uncertainty. Such comparisons can be made between mosses and liverworts because representative model organisms are well established in these two bryophyte lineages. To date, however, a model hornwort species has not been available.

Results: Here we report the establishment of *Anthoceros agrestis* as a model hornwort species for laboratory experiments. Axenic culture conditions for maintenance and vegetative propagation have been determined, and treatments for the induction of sexual reproduction and sporophyte development have been established. In addition, protocols have been developed for the extraction of DNA and RNA that is of a quality suitable for molecular analyses. Analysis of haploid-derived genome sequence data of two *A. agrestis* isolates revealed single nucleotide polymorphisms at multiple loci, and thus these two strains are suitable starting material for classical genetic and mapping experiments.

Conclusions: Methods and resources have been developed to enable *A. agrestis* to be used as a model species for developmental, molecular, genomic, and genetic studies. This advance provides an unprecedented opportunity to investigate the biology of hornworts.

Keywords: Bryophytes, Non-seed plants, Model species, Development, Evolution, Sporophyte, Genetically divergent strains

Background

Plants colonized terrestrial environments approximately 480 million years ago [1,2]. Phylogenetic analyses position one or more groups of charophyte algae as the sister group to land plants and reveal two distinct groups of land plants: the bryophytes and the monophyletic group of vascular plants [3]. The bryophytes comprise three monophyletic lineages, the liverworts, the mosses and the hornworts. Although subject to much scrutiny, the phylogenetic

relationship between these three lineages remains fiercely debated [3-9]. The widely accepted view, supported by phylogenomic analyses [3], is that liverworts, mosses and hornworts branch as successive sister groups such that hornworts are the sister to vascular plants. However, more recent analyses based on protein sequences suggested that the position of hornworts as vascular plant sister group is an artefact of convergent codon usage in the two lineages [8]. Moreover, the data supported monophyly of liverworts and mosses, a relationship that is further validated by phylotranscriptomic analyses of a much larger taxon group [9]. Depending on the phylogenetic method used, this latest study identified hornworts as either sister to all land

* Correspondence: jane.langdale@plants.ox.ac.uk
†Equal contributors
5Department of Plant Sciences, University of Oxford, South Parks Rd, Oxford, UK
Full list of author information is available at the end of the article

plants, in a clade with mosses and liverworts, or sister to vascular plants [9].

The uncertainty over the phylogenetic position of hornworts is compounded by our relatively limited understanding of hornwort biology. As land plants evolved, the modification of various character traits led to a general increase in size and complexity such that the bryophytes are relatively simple, both in terms of morphology and physiology, as compared to flowering plants. An understanding of how this complexity evolved can be obtained through comparative analyses of developmental processes in extant land plant species. To date, the liverwort *Marchantia polymorpha* and the moss *Physcomitrella patens* have been used to reveal evolutionary trajectories of developmental mechanisms that regulate morphological traits such as root hairs [10], and both endogenous (e.g. hormone signaling [11]) and environmentally-induced (e.g. chloroplast function [12]) physiological traits. However, such analyses have not been possible in hornworts because no species has thus far proved amenable to experimental manipulation in the laboratory.

Regardless of whether hornworts are sister to all other land plants, sister to vascular plants, or part of a bryophyte clade, their phylogenetic position is key to understanding the evolution of land plant body plans [13-15]. Notably, hornworts exhibit a number of morphological features that are distinct from those in liverworts and mosses, and thus they represent the only bryophyte lineage that can be effectively utilized for comparative analyses [16]. For example, the first zygotic division in hornworts is longitudinal whereas it is transverse in liverworts and mosses [17,18]; hornworts are the only land plants to develop chloroplasts with algal-like pyrenoids [19,20]; and hornworts characteristically have a symbiotic relationship with *Nostoc* cyanobacteria [16]. An understanding of how these biological processes are regulated and have evolved can only be achieved using a hornwort model system that can be easily grown throughout the entire life-cycle in laboratory conditions.

Here we introduce *Anthoceros agrestis* as a tractable hornwort experimental system. *Anthoceros* was the first hornwort genus described [21], it has worldwide distribution [22], most species have small genomes [23] with *A. agrestis* having the smallest genome of all bryophytes investigated so far (1C = 0.085 pg ca. 83 Mbp [Megabase pairs]; [24]). Similar to all bryophytes, the haploid gametophyte generation of *A. agrestis* is the dominant phase of the life cycle (Figure 1A). Spores germinate to produce a flattened thallus that generally lacks specialized internal tissue differentiation with the exception of cavities that contain mucilage (Figure 1B-D, [16]). Each cell of the thallus (including the epidermal cells) contains one to four chloroplasts [16]. Gametophytes are monoecious with both male (antheridia) and female (archegonia) reproductive

organs developing on the same thallus. Antheridia develop in chambers (up to 45 per chamber) (Figure 1E, [16]) and produce motile sperm, whereas archegonia contain a single egg that is retained in the thallus. After fertilization, the diploid embryo develops within the archegonium to produce the sporophyte, in which spores are produced via meiosis. At maturity the *A. agrestis* sporophyte is an elongated cylindrical structure (Figure 1F) that is composed of the columella, a spore layer, a multicellular jacket and elaters for spore dispersal [16]. The meristem at the base of the sporophyte (basal meristem) remains active throughout the life of the sporophyte, a feature that is unique to hornworts [16]. The propagation of *A. agrestis* callus and suspension cultures has previously been reported for biochemical analyses [25]. Here, we report the development of methods and resources to grow and propagate *A. agrestis* axenically, to facilitate molecular analysis, and to generate populations for genetic analysis

Results and discussion
A. agrestis strains

Two different *A. agrestis* strains have been propagated. The first was established from plant material collected near Fogo in Berwickshire, UK (hereafter referred to as the "Oxford strain") and the second from plant material collected near Hirschbach, Germany (hereafter referred to as the "Bonn strain"). All existing material of both the Oxford and Bonn strains originate from a single spore. Attempts to establish *Anthoceros punctatus* strains were carried out in parallel, and although vegetative propagation was successful, conditions for reproductive propagation proved elusive. As such, *A. punctatus* was rejected as a potential model organism.

Establishment of axenic cultures

To initiate axenic cultures, several sterilization protocols were tested. Bacterial and fungal contamination of spores was successfully eliminated using bleach, and thus a simple three-minute treatment followed by washing was adopted (see Methods). Following sterilization, spores were germinated on Lorbeer's medium, a substrate that has previously been used for hornwort cultivation [26]. Germination occurred after approximately 7 days when plates were incubated at 23°C, with a diurnal cycle of 16 h light (300 μEm^2sec^{-1})/8 h dark. Young gametophytes were large enough to be sub-cultured 1–2 months after spore germination.

Gametophyte cultures and vegetative propagation

Three different media were tested for their ability to support vegetative growth of gametophytes. In addition to Lorbeer's medium, gametophytes were transferred to 1/10 KNOP medium [27] and to BCD [28] medium, both of which have been previously used to culture the

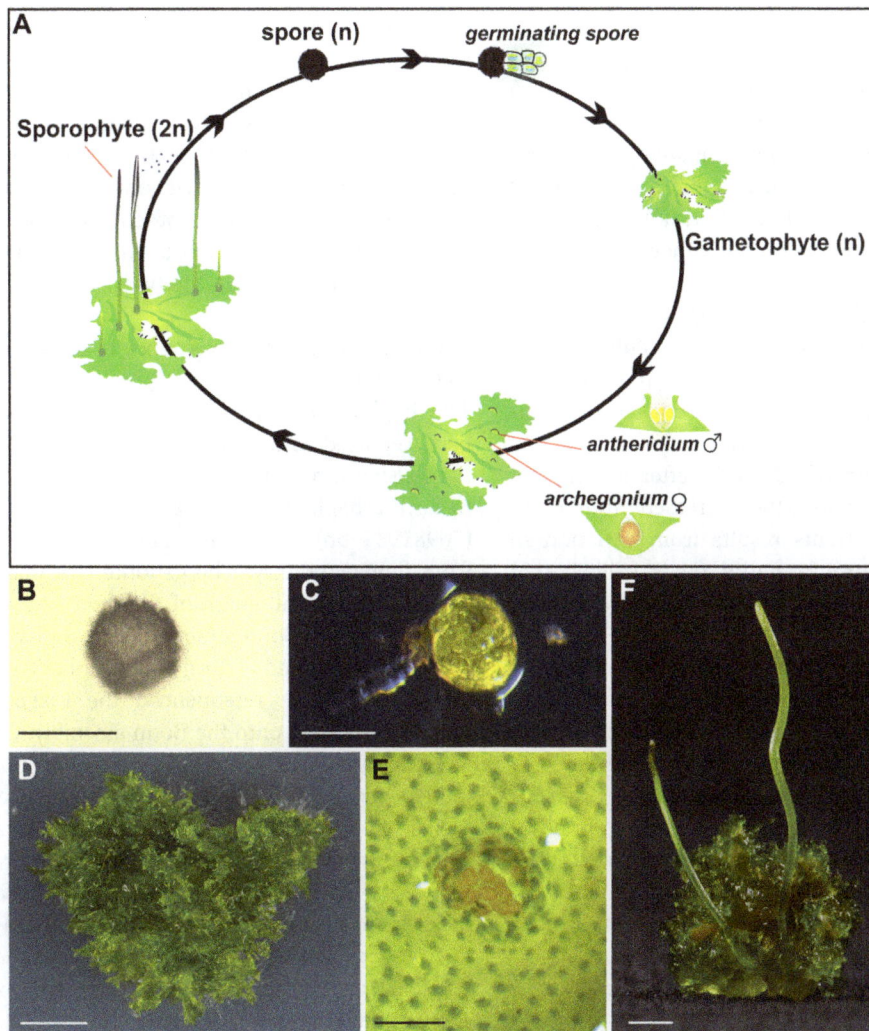

Figure 1 Life cycle of the hornwort *Anthoceros agrestis*. The life cycle of *A. agrestis* **(A)** starts with the spore **(B)** that germinates **(C)** and gives rise to the gametophyte **(D)**. Gametophytes are monoecious and thus individual plants bear both male antheridia **(E)** and female archegonia. After fertilization of the egg by sperm from the antheridia, the zygote is retained within the archegonium. The resultant embryo develops into the sporophyte **(F)** in which spores are produced via meiosis. Scale bars = B: 40 µm; C: 100 µm; D: 2 mm; E: 200 µm; F: 2 mm.

moss *P. patens*. Plates were incubated at 23°C, either under a diurnal cycle of 16 h light (300 µEm^{2}sec^{-1})/8 h dark or under continuous light (300 µEm^{2}sec^{-1}). In all cases, cultures were propagated and maintained by monthly sub-culturing, in which a small fragment of thallus tissue (~5-7 mm in diameter) was cut and placed on fresh medium. In general, the Oxford strain grew better (faster, greener and healthier) on Lorbeer's or 1/10 KNOP media whereas the Bonn strain grew better on BCD medium. In all cases, plants grew faster under continuous light than with long day photoperiods, as long as the light intensity was kept at, or below, 300 µEm^{2}sec^{-1}.

Sporophyte induction and sexual reproduction

In natural ecosystems, sexual reproduction in hornworts is initiated by the formation of antheridia on the thallus, and then after approximately one month archegonia develop [29]. To determine the conditions under which this developmental transition towards gametangia formation can be induced in the laboratory, growth parameters were varied. Cultures were initiated by either sub-culturing thallus fragments (as above) or by germinating spores, with thallus fragments being preferable starting material because the time from spore germination to the development of thallus that was mature enough for reproductive induction was around 2–3 months. The most significant factor that influenced whether gametophytes grew vegetatively or formed gametangia was growth temperature. Effective induction of gametangia was achieved by dropping the growth temperature of gametophyte cultures from 23°C to 16°C.

To optimize induction conditions, growth at 16°C was next compared on different media and under different light regimes. Gametangia were successfully induced on both 1/10 KNOP and BCD media but not on Lorbeer's medium, and in both continuous light (150 μEm^2sec^{-1}) or long day photoperiod 16 h light (150 $\mu Em^2sec^{-}1$)/8 h dark. In all cases, antheridia appeared as reddish dots on the surface of the thallus after approximately one month. Given that archegonia are colourless and are embedded within the thallus, their formation could not be easily visualized, and thus the appearance of antheridia was used as a prompt to induce fertilization.

Fertilization was facilitated by adding 5–10 mL of either water or liquid culture media to each culture. Sporophytes were visible after another month of growth. However, the number of sporophytes produced per thallus was increased if the liquid addition step was repeated 3–5 times over a period of ~2 weeks after addition of the first aliquot. Presumably the increased number of successful fertilization events results from variation in the timing of archegonium formation (i.e. it is likely that when the first aliquot was added very few archegonia were present). This variability is also reflected in the fact that even with the extra liquid addition steps, the number of sporophytes produced by each thallus ranged from 5 to over 100. There is no apparent way in which this variation can be more carefully controlled given that the development of archegonia is difficult to monitor. Emerging sporophytes went through the normal cycle of sporophyte maturation and contained hundreds of spores. Spores were viable and were regularly used to initiate new cultures.

Nucleic acid extraction

Although the extraction of nucleic acids from any organism is generally considered to be straightforward, hornwort gametophyte tissue is rich in polysaccharides (mucilage) [16], and was also found to be rich in polyphenolics. Both compounds pose a problem for DNA and RNA extraction. A range of DNA and RNA extraction protocols were therefore tested to optimize the procedure and to reduce contamination levels as much as possible. A modified CTAB protocol, adapted from Porebski et al. [30], was found to be optimal for genomic DNA extraction in that yields were approximately ten times higher than standard CTAB protocols. This protocol uses polyvinylpyrrolidone to remove polyphenolics and contains an extra ethanol precipitation step with a relatively high NaCl concentration compared to standard DNA extraction protocols. At NaCl concentrations higher than 0.5 M, polysaccharides remain in solution and do not co-precipitate with DNA. The overall yield of DNA extracted was also highly dependent on the conditions under which the thalli were grown. Thalli grown on petri dishes in which extra liquid medium (~5-10 mL per 9 cm diameter petri dish) was added every 2–3 weeks to maintain a liquid film (1–2 mm thick) connecting the thalli on the surface of the agar

yielded the greatest amounts of DNA. In addition, less, rather than more plant material led to the highest yields. Optimal yields were obtained in extractions that used 1–2 thalli, each of ~0.5 cm in diameter, that had been grown under wet conditions. DNA extracted with this protocol was successfully used in next-generation sequencing library preparation, for restriction enzyme digests, and in PCR reactions. The same protocol could be used for RNA extraction with the addition of an overnight RNA precipitation step with LiCl (see Methods).

Genome-wide genetic divergence of the Oxford and Bonn strains of A. agrestis

The haploid genome size of A. agrestis has previously been reported as 83Mbp on the basis of flow cytometry [24]. Using k-mer analysis we estimated the Bonn strain to have a haploid genome size of approximately 71Mbp (70981934 bp), a number consistent with that derived from flow cytometry. The genome size was further confirmed by the total length of the draft assembly (approximately 90 Mbp, Bonn strain). To determine the extent to which the Bonn and Oxford strains are different at the nucleotide level, we resequenced the Oxford strain and mapped the reads onto the Bonn assembly. On average we found approximately 2 single nucleotide polymorphisms (SNPs) per 1 Kbp (Kilobase pairs) sequence data (1.996 SNPs). This is less than that reported for accessions of Arabidopsis thaliana (5 SNPs/1 Kbp) [31] or Populus tremula (2–6 SNPs/1 Kbp) [32], but is of the same order of magnitude. This level of variation is likely to be sufficient to conduct classical genetic work and gene mapping by sequencing, as reported for the moss P. patens where strains show a similar level of genetic divergence [33].

Conclusions

Methods and resources have been developed to enable A. agrestis to be used as a model species for developmental, molecular, genomic and genetic studies. Axenic cultures have been established, conditions for sexual propagation and nucleic acid extraction have been optimised, and two strains with sufficient genetic divergence have been identified for genetic analyses. This advance provides an unprecedented opportunity to investigate the biology of hornworts.

Availability of supporting data

Raw sequence data for the Bonn and Oxford strains have been deposited in the European Nucleotide Archive and are available under study accession number PRJEB8683 (http://www.ebi.ac.uk/ena/data/view/PRJEB8683).

Methods
Plant material

The Anthoceros agrestis Oxford strain was obtained from Berwickshire (Berwickshire, near Fogo, Grid: NT 7700

4894, v.-c. 81, Alt. c. 115 m) on 30th October 2012 by Dr David Long (Royal Botanic Garden Edinburgh). Voucher specimens have been deposited in the Fielding Druce Herbarium, University of Oxford (OXF). The *A. agrestis* Bonn strain was obtained between Hirschbach and Reinhardtsgrimma, on a crop field approximately 500 m from the street (K9022), near a small copse on 15th November 2006 by Dr Susann Wicke and Dr Dietmar Quandt. Voucher specimens have been deposited in the Herbarium of the University of Bonn (H015-H018).

Growth media

Three different media were used: Lorbeer's medium [26] (0.1 g/L $MgSO_4.7H_2O$, 0.1 g/L KH_2PO_4, 0.2 g/L NH_4NO_3, 0.1 g/L $CaCl_2$) supplemented with 1 mL of Hutner's trace elements [34] (50 g/L EDTA disodium salt, 22 g/L $ZnSO_4.7H_2O$, 11.4 g/L H_3BO_3, 5.06 g/L $MnCl_2.4H_2O$, 1.61 g/L $CoCl_2.6H_2O$, 1.57 g/L $CuSO_4.5H_2O$, 1.1 g/L $(NH_4)_6Mo_7O_{24}.4H_2O$, 4.99 g/L $FeSO_4.7H_2O$) adjusted to pH6.5 and solidified with 6.5 g/L agar; 1/10 KNOP medium [27] (0.025 g/L K_2HPO_4, 0.025 g/L KH_2PO_4, 0.025 g/L KCl, 0.025 g/L $MgSO_4.7H_2O$, 0.1 g/L $Ca(NO_3)_2.4H_2O$, 37 mg/L $FeSO_4.7H_2O$) adjusted to pH6.5 and solidified with 6.5 g/L agar; and BCD medium [28] (0.25 g/L $MgSO_4.7H_2O$, 0.25 g/L KH_2PO_4 (pH6.5), 1.01 g/L KNO_3, 0.0125 g/L $FeSO_4.7H_2O$ and 0.001% Trace Element Solution (0.614 mg/L H_3BO_3, 0.055 mg/L $AlK(SO_4)2.12H_2O$, 0.055 mg/L $CuSO_4.5H_2O$, 0.028 mg/L KBr, 0.028 mg/L LiCl, 0.389 mg/L $MnCl_2.4H_2O$, 0.055 mg/L $CoCl_2.6H_2O$, 0.055 mg/L $ZnSO_4.7H_2O$, 0.028 mg/L KI and 0.028 mg/L $SnCl_2.2H_2O$) supplemented with 1mM $CaCl_2$ and solidified with 8 g/L agar.

Tissue sterilization

Isolated sporophytes were left to dry before removing the spore contents. Spores were sterilized by gentle agitation in 5% (v/v) bleach (sodium hypochlorite solution, ~10%) in microcentrifuge tubes, followed by three washes in sterile water with brief centrifugation steps between each wash.

Genomic DNA extraction

DNA was extracted from 1 gram of ground frozen tissue using 10 mL prewarmed (60°C) extraction buffer (100 mM Tris–HCl pH8, 1.4M NaCl, 20 mM EDTA pH8, 2% (w/v) CTAB, 0.3% (v/v) β-mercaptoethanol, 100 mg of polyvinylpyrrolidone-40 (PVP) per 1 g tissue) plus 5 µl of 100 mg/mL RNAase A. After incubation at 60°C for 30 min, samples were cooled to room temperature and then extracted with chloroform:isoamylalcohol (24:1). A second chloroform:isoamylalcohol (24:1) step was carried out to remove any remaining PVP. DNA was precipitated from the aqueous phase with 0.5 volumes 5M NaCl and 2 volumes of cold (−20°C) 95% ethanol. After resuspension in 2 mL 10 mM Tris pH8, 1mM EDTA

(TE), a second ethanol precipitation was carried out and then the DNA was dissolved in TE for storage and subsequent analyses.

RNA extraction

RNA was extracted in two different ways. For large scale RNA extractions, samples were treated as for DNA extractions with the exception that all solutions were prepared with water that had been autoclaved after treatment with 0.1% diethylpyrocarbonate (DEPC) and RNAase was omitted from the extraction buffer. In addition, after the second ethanol precipitation, the pellet was resuspended in DEPC-treated dH_2O instead of TE. RNA was then precipitated overnight at 4°C after the addition of 0.25 volumes of 8 M LiCl. After resuspension and a third ethanol precipitation, RNA was resuspended in DEPC-treated water for storage at −80°C and subsequent analyses.

For extractions where the recovery of small RNAs was required, the Spectrum™ Plant Total RNA Kit (Sigma) was used. Before each extraction residual water was removed from ~2-3 thalli (each ~0.5 cm diameter) using paper towel. Tissue was flash-frozen in liquid N_2, ground into a fine powder and resuspended in 750 µL binding buffer. RNA was eluted in 30 + 30 µL of nuclease free water and stored at −80°C.

Sequence analysis

To generate a low-coverage reference sequence for the *A. agrestis* Bonn strain, DNA was extracted from one month old thalli using the protocol detailed above. The draft genome sequence data are derived from the haploid phase, a significant advantage over vascular-plant genomes, which are all based on diploid individuals. Paired-end libraries were prepared for next generation sequencing using the Nextera XT kit (Illumina inc.) with 1 to 10 ng DNA. Nextera DNA libraries were sequenced on 1/3rd of a Miseq flow cell with 250 cycles. After sequencing and de-multiplexing, approximately 4.99 million paired-end reads were obtained. Reads were trimmed using Trimmomatic [35] and all reads that were 36 bp or longer after quality trimming and filtering (−phred33 ILLUMINACLIP: NexteraPE-PE.fa:2:30:10:8:true LEADING:9 TRAILING:3 SLIDINGWINDOW:4:15 MINLEN:36) were retained. The resultant 4.94 million paired-end reads were assembled using the udba500 code (part of the A5 pipeline; [36]) with k-mer values ranging from 20 to 230 and a step size of 20. To verify the validity of previous estimates of genome size [24] k-mer analysis was used as implemented in the code kmergenie (version 1.6950; [37]). To identify SNPs between the Bonn and the Oxford strains, the Oxford strain was resequenced as above. We obtained approximately 2.33 million raw paired-end reads of which 2.29 million reads survived quality filtering and trimming as described above. This sequencing depth corresponds to a theoretical average coverage of 8x. Raw sequence data of the Bonn and Oxford strains will be deposited in the SRA archive upon acceptance of the manuscript for publication.

SNP discovery

GATK (Genome analysis toolkit) best practice was followed to identify SNPs with high-confidence [38]. Briefly, we mapped cleaned and trimmed reads to the Bonn strain`s preliminary assembly using bowtie2 (bowtie2_2.1.0, using the −sensitive option; [39]). Duplicates were then marked and removed using the picard tool MarkDuplicates module (http://broadinstitute.github.io/picard/) and reads re-aligned using the GATK IndelRealigner [40]. Finally, we used SNVer [41] to extract SNPs between the Bonn and the Oxford strains (−n 1 -mq 20 -bq 17 -b 0.75 -het 0.0001 -a 1 -s 0.0001). Because the Oxford strain was resequenced with low coverage, SNPs were called at all positions with a coverage value greater than five. Finally, we used vcftools [42] to calculate the density of SNPs in 1 Kbp windows. For this analysis we excluded all contigs from the Bonn strain assembly that were shorter than 1 Kbp.

Abbreviations

CTAB: Cetyltrimethylammonium bromide; GATK: Genome analysis toolkit; Mbp: Megabase pairs; Kbp: Kilobase pairs; SNPs: Single nucleotide polymorphisms; PVP: Polyvinylpyrrolidone; TE: Tris EDTA; EDTA: Ethylenediaminetetraacetic acid; DEPC: Diethylpyrocarbonate.

Competing interests

The authors declare that they have no competing interests.

Authors' contributions

EF, JAL and PS designed and conceived the experiments; EF, DQ, SW, and PS established culture conditions for gametophytes and sporophytes; EF and PS developed nucleic acid extraction protocols; and PS and MR carried out genome sequence analysis. JAL, EF and PS wrote the manuscript. All authors read and approved the final manuscript.

Acknowledgements

We are grateful to Juan Carlos Villarreal for fuelling our interest in hornworts; to David Long for providing spores of the Oxford strain; to John Baker, Julie Bull, Ester Rabbinowitsch, Mary Saxton, Zoe Bont, Martina Schenkel, Karola Maul and Monika Ballmann for technical assistance; to Lucy Poveda Mozolowski (Functional Genomic Center Zurich) for next-generation sequencing. This work was funded by an ERC Advanced Investigator Grant (EDIP) to JAL, by an SNSF Ambizione grant (#131726) to PS, by a FCT post-doctoral fellowship (SFRH/BPD/78814/2011), Plant Fellows Fellowship (#267423) and Forschungskredit der Universität Zurich to MR and by TU Dresden (Special grant for innovation in research) to DQ. Comments of two anonymous reviewers to an earlier version of the manuscript are also acknowledged.

Author details

[1]Institute of Evolutionary Biology and Environmental Studies, University of Zurich, Zurich, Switzerland. [2]Institute of Systematic Botany, University of Zurich, Zurich, Switzerland. [3]Swiss Institute of Bioinformatics, Quartier Sorge-Batiment Genopode, Lausanne, Switzerland. [4]MTA-ELTE-MTM Ecology Research Group, ELTE, Biological Institute, Budapest, Hungary. [5]Department of Plant Sciences, University of Oxford, South Parks Rd, Oxford, UK. [6]Nees-Institut für Biodiversität der Pflanzen, University of Bonn, Meckenheimer Allee 170, D – 53115 Bonn, Germany. [7]Institute for Evolution and Biodiversity, University of Muenster, Huefferstr. 1, 48149 Muenster, Germany. [8]Current Address: Graduate School of Science, University of Tokyo, 7-3-1 Hongo, Bunkyo-ku, Tokyo 113 0033, Japan.

References

1. Gensel PG. The earliest land plants. Ann Rev Ecol Evol. 2008;39:459–77.
2. Kenrick P, Crane PR. The origin and early evolution of plants on land. Nature. 1997;389:33–9.
3. Qiu YL, Li L, Wang B, Chen Z, Knoop V, Groth-Malonek M, et al. The deepest divergences in land plants inferred from phylogenomic evidence. Proc Natl Acad Sci. 2006;103:15511–6.
4. Chang Y, Graham SW. Inferring the higher-order phylogeny of mosses (Bryophyta) and relatives using a large, multigene plastid data set. Am J Bot. 2011;98:839–49.
5. Nickrent DL, Parkinson CL, Palmer JD, Duff RJ. Multigene phylogeny of land plants with special reference to bryophytes and the earliest land plants. Mol Biol Evol. 2000;17:1885–95.
6. Nishiyama T, Wolf PG, Kugita M, Sinclair RB, Sugita M, Sugiura C, et al. Chloroplast phylogeny indicates that bryophytes are monophyletic. Mol Biol Evol. 2004;21:1813–9.
7. Qiu YL, Cho Y, Cox JC, Palmer JD. The gain of three mitochondrial introns identifies liverworts as the earliest land plants. Nature. 1998;394:671–4.
8. Cox CJ, Li B, Foster PG, Embley TM, Civan P. Conflicting phylogenies for early land plants are caused by composition biases among synonymous substitutions. Syst Biol. 2014;63:272–9.
9. Wickett NJ, Mirarab S, Nguyen N, Warnow T, Carpenter E, Matasci N, et al. Phylotranscriptomic analysis of the origin and early diversification of land plants. Proc Natl Acad Sci. 2014;111:E4859–68.
10. Menand B, Yi K, Jouannic S, Hoffmann L, Ryan E, Linstead P, et al. An ancient mechanism controls the development of cells with a rooting function in land plants. Science. 2007;316:1477–80.
11. Yasumura Y, Crumpton-Taylor M, Fuentes S, Harberd NP. Step-by-step acquisition of the gibberellin-DELLA growth-regulatory mechanism during land-plant evolution. Curr Biol. 2007;17:1225–30.
12. Yasumura Y, Moylan E, Langdale J. A conserved transcription factor mediates nuclear control of organelle biogenesis in anciently diverged land plants. Plant Cell. 2005;17:1894–907.
13. Ligrone R, Duckett JG, Renzaglia KS. Major transitions in the evolution of early land plants: a bryological perspective. Ann Bot. 2012;109:851–71.
14. Tomescu AM, Wyatt SE, Hasebe M, Rothwell GW. Early evolution of the vascular plant body plan - the missing mechanisms. Curr Opin Plant Biol. 2014;17:126–36.
15. Rothwell GW, Wyatt SE, Tomescu AM. Plant evolution at the interface of paleontology and developmental biology: An organism-centered paradigm. Am J Bot. 2014;101:899–913.
16. Renzaglia KS, Villarreal JC, Duff RJ. New Insights into Morphology, Anatomy and Systematics of Hornworts. In: Bryophyte Biology II. Cambridge: Cambridge University Press; 2009. p. 139–71.
17. Renzaglia KS. A comparative morphology and developmental anatomy of the anthocerotophyta. J Hattori Bot Lab. 1978;44:31–90.
18. Ligrone R, Duckett JG, Renzaglia KS. The origin of the sporophyte shoot in land plants: a bryological perspective. Ann Bot. 2012;110:935–41.
19. Villarreal JC, Renner SS. Hornwort pyrenoids, carbon-concentrating structures, evolved and were lost at least five times during the last 100 million years. Proc Natl Acad Sci. 2012;109:18873–8.
20. Duckett JG, Renzaglia KS. Ultrstructure and development of plastids in bryophytes. Adv Bryology. 1988;3:33–93.
21. Merrett C. Pinax rerum naturalium Britannicarum: continens vegetabilia, animalia, et fossilia, in hac insula reperta inchoatus, London; 1667.
22. Villarreal JC, Cargill DC, Hagborg A, Soderstrom L, Renzaglia KS. A synthesis of hornwort diversity: patterns, causes and future work. Phytotaxa. 2010;9:150–66.
23. Bainard JD, Villarreal JC. Genome size increases in recently diverged hornwort clades. Genome. 2013;56:431–5.
24. Leitch IJ, Bennett MD. Genome Size and Its Uses: The Impact of Flow Cytometry. In: Flow Cytometry with Plant Cells: Analysis of Genes Chromosomes and Genomes. Weinheim: John Wiley & Sons; 2007. p. 153–76.
25. Vogelsang K, Schneider B, Petersen M. Production of rosmarinic acid and a new rosmarinic acid 3'-O-ß-D-glucoside in suspension cultures of the hornwort Anthoceros agrestis Paton. Planta. 2006;223:369–73.
26. Proskauer JM. Studies on Anthocerotales. VIII Phytomorphology. 1969;19:52–66.

27. Reski R, Abel WO. Induction of budding on chloronemata and caulonemata of the moss, *Physcomitrella patens*, using isopentenyladenine. Planta. 1985;165:354–8.

28. Cove DJ, Perroud PF, Charron AJ, McDaniel SF, Khandelwal A, Quatrano RS. Culturing the moss Physcomitrella patens. Cold Spring Harb Protoc 2009, 2009:pdb prot5136.

29. Proskauer JM. Studies on Anthocerotales VII. Phytomorphology. 1967;17:61–70.

30. Porebski S, Bailey LG, Baum B. Modification of a CTAB DNA extraction protocol for plants containing high polysaccharide and polyphenol components. Plant Mol Biol Rep. 1997;15:8–15.

31. Gan X, Stegle O, Behr J, Steffen JG, Drewe P, Hildebrand KL, et al. Multiple reference genomes and transcriptomes for *Arabidopsis thaliana*. Nature. 2011;477:419–23.

32. Zhou L, Bawa R, Holliday JA. Exome resequencing reveals signatures of demographic and adaptive processes across the genome and range of black cottonwood (*Populus trichocarpa*). Mol Ecol. 2014;23:2486–99.

33. Kamisugi Y, von Stackelberg M, Lang D, Care M, Reski R, Rensing SA, et al. A sequence-anchored genetic linkage map for the moss, *Physcomitrella patens*. Plant J. 2008;56:855–66.

34. Hutner SH, Provasoli L, Schatz A, Haskins CP. Some approaches to the study of the role of metals in the metabolism of microrganisms. Proc Am Phil Soc. 1950;94:152–70.

35. Bolger AM, Lohse M, Usadel B. Trimmomatic: a flexible trimmer for Illumina sequence data. Bioinformatics. 2014;30:2114–20.

36. Coil D, Jospin G, Darling AE. A5-miseq: an updated pipeline to assemble microbial genomes from Illumina MiSeq data. Bioinformatics 2014, doi:10.1093/bioinformatics/btu661.

37. Chikhi R, Medvedev P. Informed and automated *k*-mer size selection for genome assembly. Bioinformatics. 2014;30:31–7.

38. McKenna A, Hanna M, Banks E, Sivachenko A, Cibulskis K, Kernytsky A, et al. The Genome Analysis Toolkit: a MapReduce framework for analyzing next-generation DNA sequencing data. Genome Res. 2010;20:1297–303.

39. Langmead B, Salzberg S. Fast gapped-read alignment with Bowtie 2. Nat Methods. 2012;9:357–9.

40. DePristo M, Banks E, Poplin R, Garimella K, Maguire J, Hartl C, et al. A framework for variation discovery and genotyping using next-generation DNA sequencing data. Nat Genet. 2011;43:491–8.

41. Wei Z, Wang W, Hu P, Lyon GJ, Hakonarson H. SNVer: a statistical tool for variant calling in analysis of pooled or individual next-generation sequencing data. Nucl Acids Res. 2011;39:e32.

42. Danecek P, Auton A, Abecasis G, Albers CA, Banks A, DePristo MA, et al. The variant call format and VCFtools. Bioinformatics. 2011;27:2156–8.

Association mapping in sunflower (Helianthus annuus L.) reveals independent control of apical vs. basal branching

Savithri U Nambeesan[1,2], Jennifer R Mandel[1,3], John E Bowers[1], Laura F Marek[4], Daniel Ebert[5], Jonathan Corbi[1,6], Loren H Rieseberg[5], Steven J Knapp[7] and John M Burke[1*]

Abstract

Background: Shoot branching is an important determinant of plant architecture and influences various aspects of growth and development. Selection on branching has also played an important role in the domestication of crop plants, including sunflower (*Helianthus annuus* L.). Here, we describe an investigation of the genetic basis of variation in branching in sunflower via association mapping in a diverse collection of cultivated sunflower lines.

Results: Detailed phenotypic analyses revealed extensive variation in the extent and type of branching within the focal population. After correcting for population structure and kinship, association analyses were performed using a genome-wide collection of SNPs to identify genomic regions that influence a variety of branching-related traits. This work resulted in the identification of multiple previously unidentified genomic regions that contribute to variation in branching. Genomic regions that were associated with apical and mid-apical branching were generally distinct from those associated with basal and mid-basal branching. Homologs of known branching genes from other study systems (i.e., *Arabidopsis*, rice, pea, and petunia) were also identified from the draft assembly of the sunflower genome and their map positions were compared to those of associations identified herein. Numerous candidate branching genes were found to map in close proximity to significant branching associations.

Conclusions: In sunflower, variation in branching is genetically complex and overall branching patterns (i.e., apical vs. basal) were found to be influenced by distinct genomic regions. Moreover, numerous candidate branching genes mapped in close proximity to significant branching associations. Although the sunflower genome exhibits localized islands of elevated linkage disequilibrium (LD), these non-random associations are known to decay rapidly elsewhere. The subset of candidate genes that co-localized with significant associations in regions of low LD represents the most promising target for future functional analyses.

Keywords: Apical dominance, Association mapping, Branching, *Helianthus annuus*, Linkage disequilibrium, Plant architecture, Sunflower

Background

Shoot branching is a major determinant of plant architecture and plays an important role in the adaptation of plants to their environment. Variation in branching helps plants compete with their neighbors and also offers protection against herbivory [1-4]. Shoot branching can also affect developmental phenotypes such as flowering time and reproductive success [5]. Moreover, this trait is an important component of the so-called "domestication syndrome" [6], with many crops exhibiting reduced branching (i.e., increased apical dominance) relative to their wild progenitors.

In cultivated sunflower (Helianthus annuus L.), selection during domestication resulted in the production of an apically dominant, unbranched growth form that differs markedly from its highly branched wild progenitor (common sunflower; also *H. annuus*) [7-10]. During the transition to hybrid breeding in the mid-20th century,

* Correspondence: jmburke@uga.edu
[1]Department of Plant Biology, Miller Plant Sciences, University of Georgia, Athens, GA 30602, USA
Full list of author information is available at the end of the article

however, recessive branching was reintroduced to the sunflower gene pool to produce male-fertile restorer (R) lines that can be crossed with unbranched, cytoplasmic male-sterile (i.e., female; A) lines to produce unbranched, fully fertile hybrids. The apical branching of the R-lines is desirable because it provides fertile pollen for a longer time period, resulting in a longer window for pollination and hybrid production [11]. The modern-day cultivated sunflower gene pool thus exhibits substantial variation in plant architecture, making it an ideal system to study the genetics of branching.

In general terms, branching is initiated from axillary meristems in leaf axils on the primary shoot. These meristems give rise to axillary buds which remain dormant or grow out into a branch that can be influenced by environmental conditions or developmental signals such as hormones [5]. Three phytohormones (auxin, cytokinin [CK], and strigolactone [SL]) and genes associated with their homeostasis and signaling are thought to be largely responsible for the regulation of branching [12-15]. Bud outgrowth is inhibited by basipetal transport of auxin produced at the shoot apical meristem. SL is a carotenoid-derived phytohormone that also inhibits bud outgrowth. It is produced in the roots and transported acropetally in the stem [12,16]. In contrast to auxin and SL, CKs are locally synthesized in the bud and promote the outgrowth of the axillary bud. Ultimately, cross talk by these phytohormone related pathways regulates branching [12]. Additionally, genes related to gibberellic acid (GA) and polyamine metabolism, and genes encoding transcription factors, at least one MAP kinase, and cytochrome P450 all play important roles in axillary bud initiation and branch growth [15-18].

In crosses between cultivated and wild sunflower, branching has been found to be a genetically complex trait influenced by numerous small effect loci distributed throughout the genome [19,20]. Classical genetic analyses in cultivated sunflower have, however, revealed the existence of loci with major effects on both apical and basal branching [11,21-23]. More recently, quantitative trait locus (QTL) mapping has been used to localize the recessive apical branching of restorer lines to a region (known as the B locus) on the upper half of linkage group (LG) 10 [24,25]. The unbranched phenotype characteristic of female lines and hybrids is thought to be controlled by the dominant B allele, while the branched R-lines are homozygous for the recessive b allele. While traditional QTL analyses have proven to be useful in identifying genomic regions that influence plant architecture in sunflower, this general approach suffers some limitations. Most notably, the use of biparental populations only enables the analysis of two alleles per gene, and also limits the number of recombination events, thereby providing relatively limited genetic resolution [26].

Association mapping (also known as LD mapping) has emerged as an alternative to QTL mapping for investigating the genetic basis of quantitative traits [27]. Because it involves the analysis of a diverse collection of more or less unrelated individuals, association mapping allows for the simultaneous evaluation of the effects of multiple haplotypes across diverse genetic backgrounds. Moreover, because association populations typically capture numerous generations of historical recombination, this approach provides much higher resolution than is possible with a family-based mapping population. Herein, we report the results of a detailed analysis of variation in branching in an association mapping population that captures nearly 90% of the allelic diversity present within the cultivated sunflower gene pool [28,29]. We evaluated this population for various branching-related traits at three different locations and tested for genetic associations across the genome using genotypic data derived from a high-density SNP array [30,31]. We also identified candidate genes involved in hormonal or transcriptional regulation that mapped in close proximity to significant associations.

Methods
Development of the association mapping population
The development and initial characterization of the sunflower association mapping population utilized herein are described by Mandel et al. [28,29]. Briefly, a diverse collection of cultivated sunflower lines was obtained from the USDA North Central Regional Plant Introduction Station (NCRPIS; Ames, IA, USA) and from the French National Institute for Agricultural Research (INRA; France). These lines were genotyped with simple-sequence repeat (SSR) markers distributed across all 17 LGs and the resulting data were used to identify hierarchical subsets of lines that captured maximum diversity [28]. The present study was based on the same subset of 288 lines employed by Mandel et al. [29], which differed slightly from the core 288 described by Mandel et al. [28] due to limited seed availability for some lines. This population, which is available for distribution from the Germplasm Resources Information Network (GRIN) of the National Plant Germplasm System, is known as UGA-SAM1. Of the full set of 288 lines in this population, only 271 were included in our final analyses due to germination difficulties and plant loss during the growing season. These lines capture nearly 90% of the allelic diversity and include lines that are oil and confectionery types from the two major heterotic groups in cultivated sunflower as well as select open-pollinated varieties (OPVs) and land races. Many of the lines were advanced via single-seed descent for one or two cycles to reduce residual heterozygosity prior to the start of this experiment.

Field design and phenotypic analysis of branching traits

All 288 lines were planted in two replicates using an alpha lattice at three different locations during the spring of 2010. The three locations were: the UGA Plant Sciences Farm, (Oconee County, GA, USA), the NCRPIS at Iowa State University (Ames, IA, USA), and the University of British Columbia's Botany Gardens (Vancouver, BC, Canada) [29]. Individuals of each line (3–4 individuals per replicate per location) were evaluated at the R9 reproductive stage, which represents physiological maturity [32].

At maturity, each plant was divided into four quarters (1st, 2nd, 3rd, and 4th, from top to bottom) and the number of primary branches was counted in each quarter. Primary branches were recorded as present if they were longer than 2 cm and had developed a terminal inflorescence. The numbers of primary branches in each quarter were then used to estimate the extent and type of branching on each plant. Apical branching was estimated as the number of primary branches in the 1st quarter, mid-apical branching was estimated as total number of branches in the 1st and 2nd quarters, and so forth (Figure 1). If a particular quarter did not branch, the leaf axils were examined to determine if axillary bud initiation had occurred. If more than 50% of the nodes in a particular quarter displayed initiation, it was recorded as branch initiation. In addition, the presence or absence of secondary branches was recorded.

For each line, branch numbers and lengths were averaged within replicates prior to analysis. For secondary branching, scores of 0 for absence or 1 for presence were

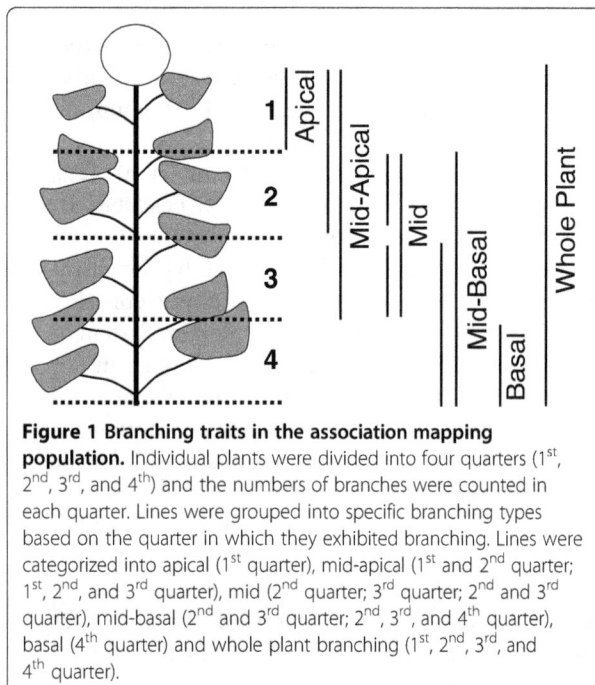

Figure 1 Branching traits in the association mapping population. Individual plants were divided into four quarters (1st, 2nd, 3rd, and 4th) and the numbers of branches were counted in each quarter. Lines were grouped into specific branching types based on the quarter in which they exhibited branching. Lines were categorized into apical (1st quarter), mid-apical (1st and 2nd quarter; 1st, 2nd, and 3rd quarter), mid (2nd quarter; 3rd quarter; 2nd and 3rd quarter), mid-basal (2nd and 3rd quarter; 2nd, 3rd, and 4th quarter), basal (4th quarter) and whole plant branching (1st, 2nd, 3rd, and 4th quarter).

assigned and these values were subsequently averaged, as above. Using PROC GLM in SAS (ver. 9.3; SAS Institute, Cary NC), we found a significant genotype x environment (G x E) interaction ($P < 0.001$). Therefore, all subsequent analyses were performed separately by location, which also made possible the identification of loci that appear to be susceptible to environmental effects (i.e., those that were significant in at least one, but not all locations). In each location, genotypes were treated as fixed effects and block and replicates as random effects. For association mapping of the various branching traits, we used least-squares means (LS means) since block and rep effects were found to be statistically significant ($P < 0.05$).

Correlations between various branching traits and among locations were determined using Spearman's rank correlation coefficients (ρ) in JMP (ver. 9; SAS Institute), and corrected for multiple tests using the sequential Bonferroni correction [33]. Principal component analysis (PCA) was performed to visualize the various branching types (e.g., apical vs. basal branching) within the SAM association mapping population using the *pca* function implemented in the FactoMineR package ver. 1.16 [34] available in the R statistical computing language (ver. 3.1.0) [35].

Genotyping

As described by Mandel et al. [29], total DNA was extracted from pooled leaf tissue from four individuals of each line using a CTAB extraction protocol [36]. These samples were then genotyped using an Infinium SNP array (Illumina, San Diego, CA) at the Emory University Biomarker Service Center. This array was designed to target polymorphic SNPs from across the sunflower genome. Details about the development of this array have been previously provided by Bachlava et al. [30]. Genome studio ver. 2011.1 (Illumina) was used to make SNP calls and map positions were assigned based on the consensus genetic map of Bowers et al. [31]. Of a total of 9,480 SNPs on the array, 5,788 genetically mapped polymorphic SNPs could be reliably scored as apparently single copy loci in our population [29]. Further, only SNPs with a minor allele frequency of ≥10% (5,359) was used for association mapping analyses [30]. The previously identified *B* locus was not directly used as a marker in this study since its presence or absence was not known for all the lines used in this study. However, markers spanning this region (as determined based on marker position on the sunflower consensus map; Bowers et al. [31]) were included in our analyses.

Association mapping

Association analyses were performed using TASSEL ver. 3.0 [37]. Because such analyses are prone to false positives due to unrecognized kinship and/or population structure [38], three different mixed models were employed. The first model corrected for kinship (K, estimated using SPAGeDi) [39] only. The second and third models corrected for both kinship and population structure. In the Q + K model, population structure (Q) was estimated using STRUCTURE ver. 2.2 [40]. In the P + K model, population structure (P) was estimated via a principal coordinate analysis (PCoA) using GenAlEx ver. 6.41 [41]. Details of the underlying SPAGeDi, STRUCTURE, and PCoA analyses can be found in Mandel et al. [29].

Following the association analyses, Q-Q (quantile-quantile) plots were constructed for each of the three models and compared to the results of a naïve model to select the most appropriate model for analysis. This was performed for apical (1st quarter), mid (2nd and 3rd quarter) and basal (4th quarter) branching. For association mapping of apical branching, the total number of branches in the first quarter from lines that displayed apical, mid-apical, and whole plant branching were included in the analysis. Similarly for all other branching types, the number of branches from all lines that displayed branching in the respective quarter were included in our analyses. Because non-independence of linked markers can result in highly conservative significance thresholds [42], we set a threshold -log(P) value of 3.6 (alpha=0.05, P=0.00025, log 1/P=3.60) to identify significant associations using a multiple testing correction method that accounts for correlation among markers while also controlling the type I error rate [43].

Identification and mapping of candidate branching genes

In order to identify putative sunflower orthologs of branching genes identified in other plant species (i.e., Arabidopsis, rice, pea, and petunia), we searched the literature for genes involved in axillary meristem initiation and outgrowth. These genes included transcription factors such as REVOLUTA (REV), LATERAL SUPPRESSOR (LAS) and REGULATORS OF AXILLARY MERSITEMS (RAX 1, 2, and 3) from Arabidopsis that have been shown to play important roles during initiation of axillary meristem and bud formation [44-46]. In addition, genes associated with homeostasis and signaling of phytohormones and growth regulators such as auxin, CK, SL, GA, and polyamines were included [12-15,17,18]. Many other genes involved in branch outgrowth that encode transcription factors, cytochrome P450, MAP KINASE KINASE 7 (MAPKK7), arabinogalactan proteins, and other DNA binding proteins were also included. For genes that were a part of a multigene family, only the genes that have been implicated in branching have been included. This resulted in the identification of 48 candidate genes (Additional file 1).

Once identified, the sequence of each candidate gene was searched against v0.1 of the draft sunflower genome assembly (http://www.sunflowergenome.org/) using tblastx

with an E-value threshold of 1e-6. Since sunflower has undergone at least three whole genome duplication events [47] and is also likely to have experienced numerous segmental duplications, up to eight sunflower blast hits were considered for each of the candidate genes to allow for multiple possible homologs. Map positions of as many of these genes as possible were then determined by comparison to genetically mapped contigs from the whole genome assembly using the whole genome shotgun (WGS) sequence-based sunflower genetic map [48]. Genetic positions from the WGS map were translated into map positions on the Bowers et al. [31] consensus genetic map using common markers. This allowed us to determine if any of the candidate genes mapped in close proximity to SNPs associated with the branching traits. We used a window size of 2.5 cM to determine co-localization since the SNPs in question had been previously ordered using multiple mapping populations differing from the population used to order the WGS map. As a result, an uncertainty of several cM in the exact genetic position of markers remained after map integration. The sunflower contigs containing the candidate genes (based on the blast results) and their position on LGs are listed in Additional file 2. To determine if candidate genes and significant associations occurred in regions of high or low LD, r^2 values were computed using the diversity panel between all SNP pairs within 2.5 cMs of each other. An overall r^2 value was then computed for each SNP by averaging the individual pairwise r^2 values.

Results
Association mapping of branching patterns
Phenotypic analyses revealed extensive variation in the type and extent of branching within the association mapping population (Table 1, Figure 2, Additional file 3). Depending on the location, 89–102 lines exhibited whole-plant branching, while 70–110 lines exhibited no

Table 1 Branching patterns across locations (GA, IA, and BC)

BRANCHING	GA	IA	BC
None	79	110	70
Basal	53	24	60
Mid-basal	15	15	18
Mid	6	12	9
Mid-apical	15	16	6
Apical	1	1	1
Whole plant	90	89	102
Other types	8	1	5
Missing data	4	3	0

Total number of lines exhibiting a specific branching type was calculated for each location. "Other types" includes lines that did not fall into standard branching categories such as branching in 1st and 4th quarter or 2nd and 4th quarter or 1st, 2nd and 3rd quarter or 1st, 2nd and 4th quarter.

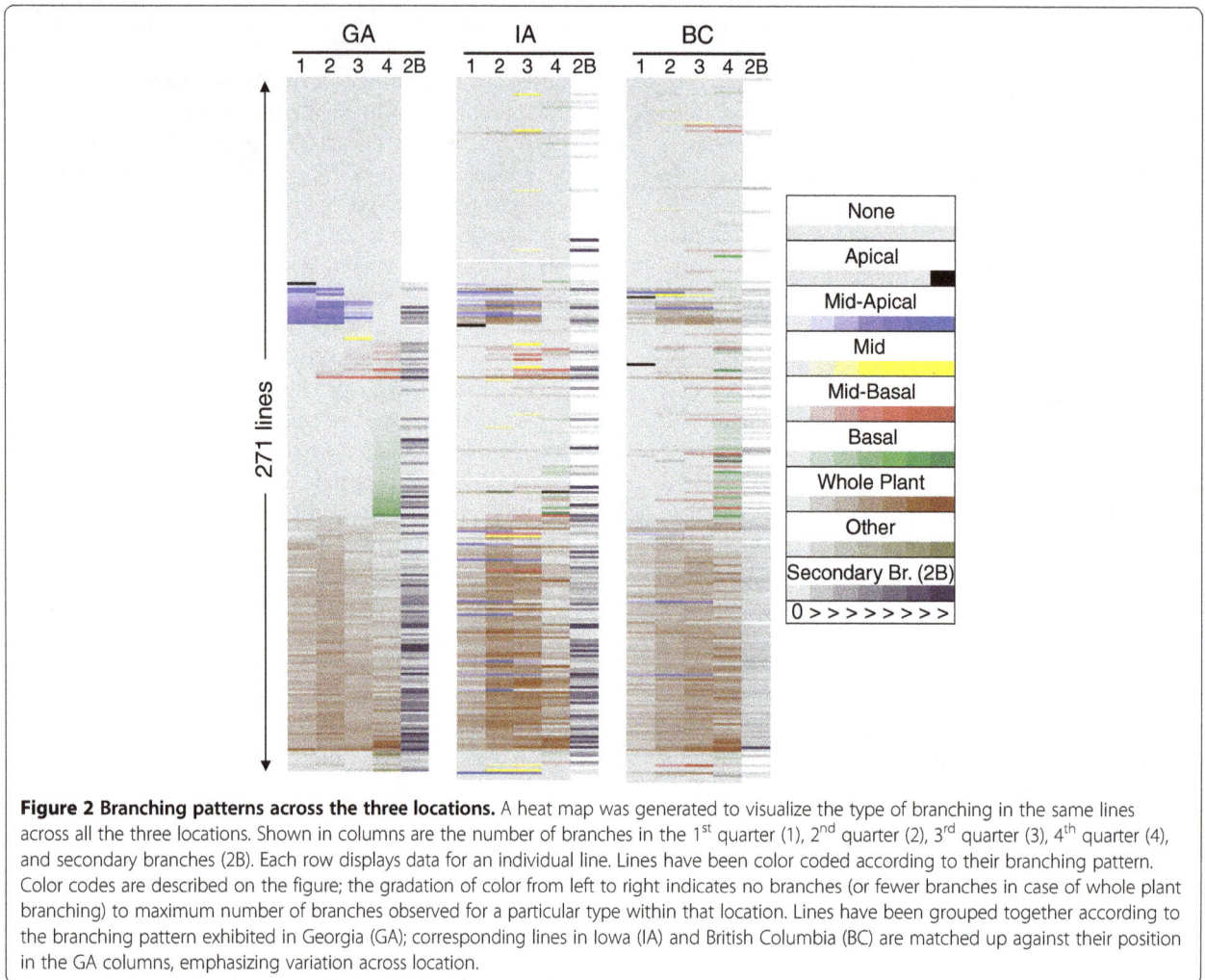

Figure 2 Branching patterns across the three locations. A heat map was generated to visualize the type of branching in the same lines across all the three locations. Shown in columns are the number of branches in the 1st quarter (1), 2nd quarter (2), 3rd quarter (3), 4th quarter (4), and secondary branches (2B). Each row displays data for an individual line. Lines have been color coded according to their branching pattern. Color codes are described on the figure; the gradation of color from left to right indicates no branches (or fewer branches in case of whole plant branching) to maximum number of branches observed for a particular type within that location. Lines have been grouped together according to the branching pattern exhibited in Georgia (GA); corresponding lines in Iowa (IA) and British Columbia (BC) are matched up against their position in the GA columns, emphasizing variation across location.

branching. The numbers of lines that exhibited other types of branching are listed in Table 1. There was high overlap of 77 lines exhibiting whole plant branching at all three locations (Table 2). However, all other branching patterns exhibited extensive variation across locations also displayed as a heat map (Figure 2). There was a greater overlap of lines exhibiting basal branching in GA and BC

compared to IA, where many did not branch. Collectively these data illustrate that the environment can exert a strong influence on branching patterns. Interestingly, unbranched lines did not display any axillary bud initiation.

As expected, the correlation analyses revealed positive correlations amongst branching types and across locations (Table 3). However, basal branching tended to

Table 2 Similarity in branching patterns across locations

Location	Branching traits							
	None	Apical	Mid-apical	Mid	Mid-basal	Basal	Whole plant	Other
GA, IA, BC	46	-	2	-	1	9	77	-
GA, BC	5	-	1	1	2	24	8	-
GA, IA	19	-	4	-	5	4	-	-
IA, BC	13	-	2	-	2	6	8	-
GA	9	1	8	5	7	16	5	8
BC	6	1	1	8	13	21	9	5
IA	32	1	8	12	7	5	4	1

Summary of the numbers of lines exhibiting similar branching patterns across the three locations (GA, Georgia; IA, Iowa; BC, British Columbia). Several lines had similar branching patterns only across two locations or had a different pattern of branching at every location.

Table 3 Correlation of apical, mid, and basal branching across GA, IA, and BC

Location	GA				IA				BC			
Branching	Apical (1Q)	Mid (2Q)	Mid (3Q)	Basal (4Q)	Apical (1Q)	Mid (2Q)	Mid (3Q)	Basal (4Q)	Apical (1Q)	Mid (2Q)	Mid (3Q)	Basal (4Q)
GA Apical (1Q)	-	0.91	0.81	0.51	0.90	0.89	0.82	0.60	0.90	0.88	0.85	0.63
Mid (2Q)		-	0.88	0.57	0.88	0.89	0.85	0.64	0.88	0.89	0.87	0.65
Mid (3Q)			-	0.65	0.81	0.86	0.85	0.67	0.84	0.84	0.82	0.65
Basal (4Q)				-	0.52	0.59	0.60	0.69	0.56	0.56	0.58	0.74
IA Apical (1Q)					-	0.93	0.86	0.64	0.90	0.89	0.86	0.64
Mid (2Q)						-	0.93	0.72	0.88	0.89	0.88	0.70
Mid (3Q)							-	0.78	0.83	0.86	0.87	0.71
Basal (4Q)								-	0.62	0.64	0.68	0.76
BC Apical (1Q)									-	0.92	0.88	0.66
Mid (2Q)										-	0.92	0.66
Mid (3Q)											-	0.74
Basal (4Q)												-

The correlation between each branching type was calculated within and across locations (GA, Georgia; IA, Iowa; BC, British Columbia). All values were positive and significant (P < 0.0001) after correcting for multiple comparisons. Q: plant quarter.

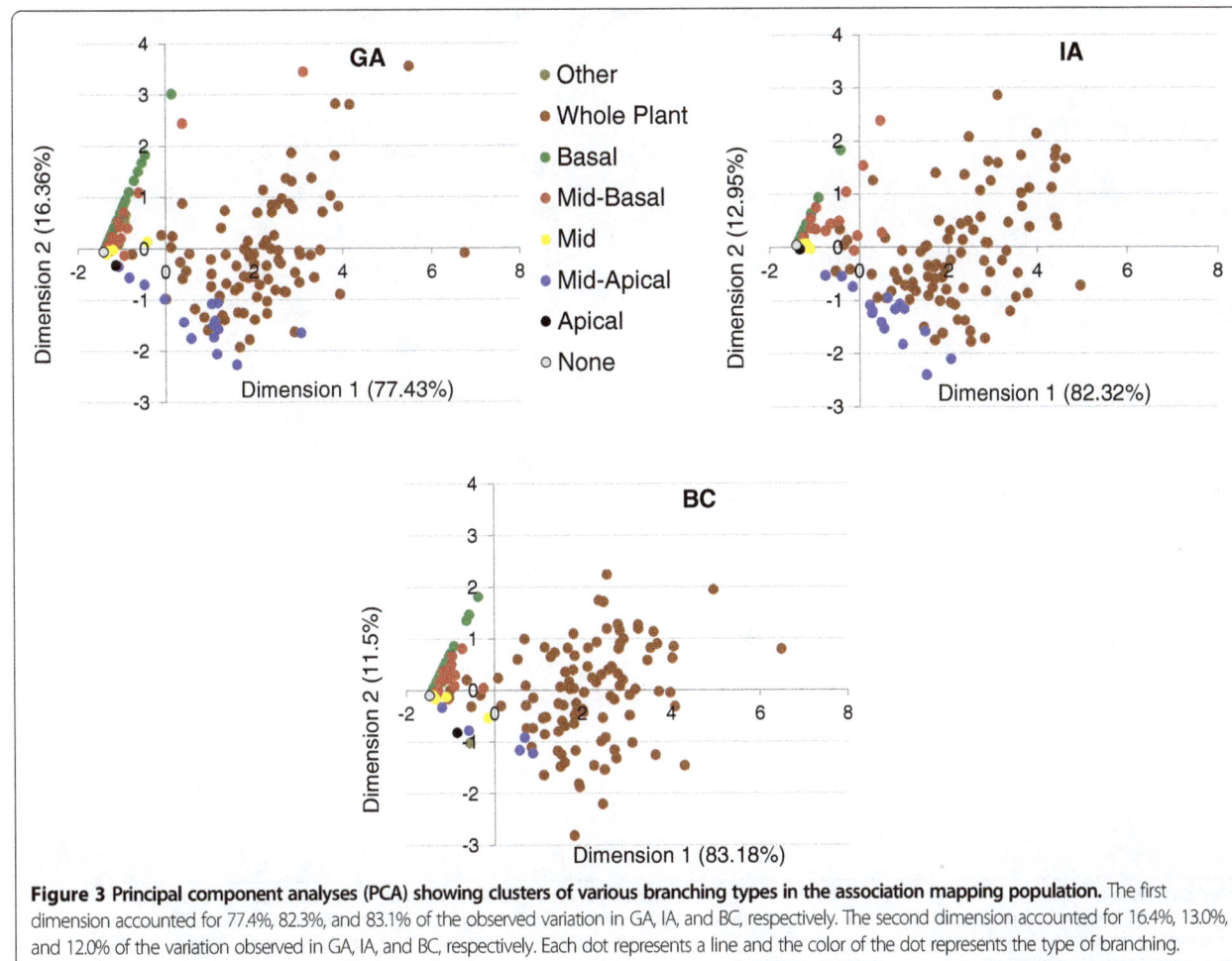

Figure 3 Principal component analyses (PCA) showing clusters of various branching types in the association mapping population. The first dimension accounted for 77.4%, 82.3%, and 83.1% of the observed variation in GA, IA, and BC, respectively. The second dimension accounted for 16.4%, 13.0%, and 12.0% of the variation observed in GA, IA, and BC, respectively. Each dot represents a line and the color of the dot represents the type of branching.

exhibit lower correlations overall, both within and across locations. This suggests that basal branching may be under different genetic control, and might also be more subject to environmental influences than apical and mid branching. In the PCA (Figure 3), the first dimension accounted for 77-83% (depending on location) of the underlying variation, and largely reflected variation in the extent of branching (i.e., the number of branches produced). The second dimension captured 12-16% of the phenotypic variation and primarily reflected differences in apical vs. basal branching.

Association analyses

The P + K model, which corrected for both kinship and population structure (using PCoA results) appeared to be the most conservative model across traits and showed the lowest tendency toward false positives (Additional file 4). This model was thus selected for all subsequent association analyses. Our analyses detected significant associations on multiple LGs (Table 4). In total, SNPs on

14 of the 17 sunflower LGs groups (all but LGs 7, 11, and 14) showed significant associations with various branching types. The Manhattan plots (Figure 4) illustrate our results for the various branching types at all three locations. Overall, the largest number of significant SNPs could be found on LG 10 in a region that was mostly associated with branching in the apical and mid regions of the plant. The broad peak of associations that is visible on the upper portion of LG 10 (referred to herein as 10a; Table 4; Figure 4) corresponds to the so-called *B* locus. In general, we were able to identify SNPs associated with basal and mid-basal branching which were distinct from apical, mid-apical, and mid branching (Table 4; Figure 4). For example, certain LGs (such as LG 4 in GA and BC and LG 8 in GA and IA) were associated with basal and mid-basal branching whereas SNPs near the bottom of LG 6 (in GA, IA and BC) and 13 (IA and BC) were associated with branching in the apical and mid regions. Only one secondary branching association was observed on LG 10 (in BC). The results for

Table 4 Linkage groups associated with distinct branching patterns

Quarter	1	12	123	2	3	23	34	234	4	2B[a]
Branching	Apical	Mid-apical	Mid-apical	Mid	Mid	Mid	Mid-basal	Mid-basal	Basal	
1									XX	
2a							X	X	X	
2b	X									
3					X		XX	X	X	
4							XX	X	XX	
5							X	X		
6a			X				X	X		
6b	X	XX			X					
8a									XX	
8b							X			
9							X			
10a	XX	XX	XX	XX	XX	XX	XX	XX	XX	X
10b		X	X	X		XX		XX		
12	X									
13a			X			X				
13b	X	X								
13c	X	XX	XX	XX						
15a	X									
15b									X	
16a					X			X	X	
16b	X									
16c		X	X	X						
17		X								

Plants were classified into various branching types depending on the presence of branches in a given quarter (see Figure 1 for more details). [a]2B indicates secondary branching. The orange (Georgia), black (Iowa), and blue (British Columbia) Xs indicate significant SNPs at the three different locations. When more than one association was detected on a single linkage group (LG), the associations were labeled alphabetically based on their position in the LG. The position of the *B* locus corresponds to the associations on LG 10a (see methods for details).

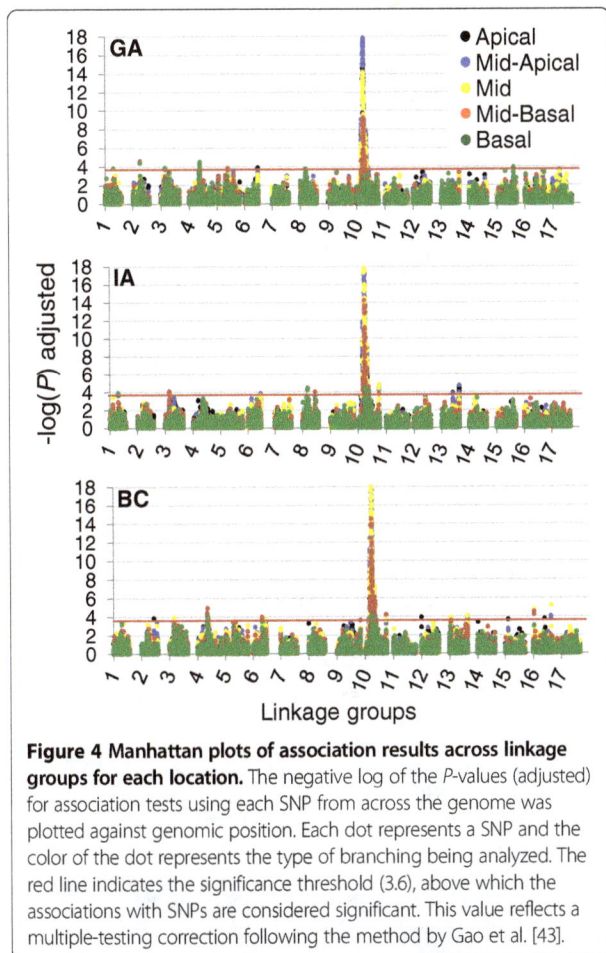

Figure 4 Manhattan plots of association results across linkage groups for each location. The negative log of the *P*-values (adjusted) for association tests using each SNP from across the genome was plotted against genomic position. Each dot represents a SNP and the color of the dot represents the type of branching being analyzed. The red line indicates the significance threshold (3.6), above which the associations with SNPs are considered significant. This value reflects a multiple-testing correction following the method by Gao et al. [43].

included an auxin biosynthetic gene (*YUCCA*; *YUC*), an auxin induced gene (*AUXIN/INDOLE-3-ACETIC ACID*; *Aux/IAA*), and genes involved in CK biosynthesis (*ISOPENTYLTRANSFERASE*; *IPT*) and degradation (*CK OXIDASE/DEHYDROGENASE*; *CKX*), GA catabolism (*GIBBERELLIN 2-OXIDASES*; *GA2ox*), and SL biosynthesis (*DWARF27*; *D27*). Several transcription factors also co-localized with significant branching associations, including *CUP-SHAPED COTYLEDON* (*CUC*), *LATERAL SHOOT INDUCING FACTOR* (*LIF*), *BRANCHED* (*BRC*), as well as a histone methyltransferase, *SET DOMAIN GROUP8* (*SDG8*). Of these, eight candidate genes were in regions of elevated LD (Figures 5 and 6; Additional file 6), including all genes on LG 10, one on the lower half of LG 13, and one on LG 16. The remaining genes on LGs 4, 5, 6, and the upper half of 13 were present in regions of lower LD (i.e., $r^2 < 0.20$).

Discussion

The identification of loci influencing specific branching patterns has the potential to facilitate the manipulation of plant architecture, which can influence yield and seed/fruit quality in crops [15,49-52]. Here, we have identified numerous distinct loci that influence apical vs. basal branching in sunflower. It is important to note, however, that the occurrence of variation in branching patterns within lines across the three locations suggests that environmental variation also plays an important role in determining sunflower branching architecture. This conclusion is consistent with the results of previous studies on the effect of external factors on branch formation. For example, increased planting density can result in a suppression of branching [53,54], and photoperiod is also known to affect branching [55-58]. Other environmental factors influencing branching include light levels and quality [59-61], plant nutrition status, and availability of nutrients [62,63]. It seems likely that some combination of these factors influenced our results across locations.

The formation of a branch involves two developmental processes: the initiation of an axillary bud and its elongation into a branch [5,64]. As noted above, genes such as *REV*, *LAS*, and *RAX* in *Arabidopsis* and their orthologs in rice and tomato play a role during axillary meristem initiation and bud formation [44-46,65-67]. Because the non-branching lines in our study also did not exhibit axillary bud initiation, we were unable to separate initiation from outgrowth and were thus unable to identify loci contributing specifically to one process or the other. The association mapping population did, however, exhibit substantial diversity in branching patterns which facilitated the identification of multiple branching-related loci that had not been previously identified in sunflower. These included associations on LGs 1, 2b, 5, 12, 14, 15, and 16 not previously identified via

whole plant branching are not presented here, as that information was analyzed by Mandel et al. [29].

Identification of candidate genes associated with various branching phenotypes

A total of 48 candidate branching genes were identified from other species (Additional file 1), and homologs to 39 of these genes could be identified in sunflower. This attrition is likely due to the use of an incomplete, draft assembly of the genome. We allowed up to eight of the best blast hits for each gene which resulted in 278 homologs. Of these, 153 (corresponding to 38 genes) could be placed on the sunflower genetic map (Additional files 2 and 5). Among these 153 blast hits, 92 unique contigs were identified. In many cases, multiple genes (belonging to gene families) were found within the same sunflower contig.

In comparing the map positions of these genes to the locations of SNPs exhibiting significant associations, we identified 13 potential candidate branching genes (from the larger set of 92 unique genes) that mapped in close proximity to various branching traits (Figure 5). Genes of interest involved in hormone related pathways

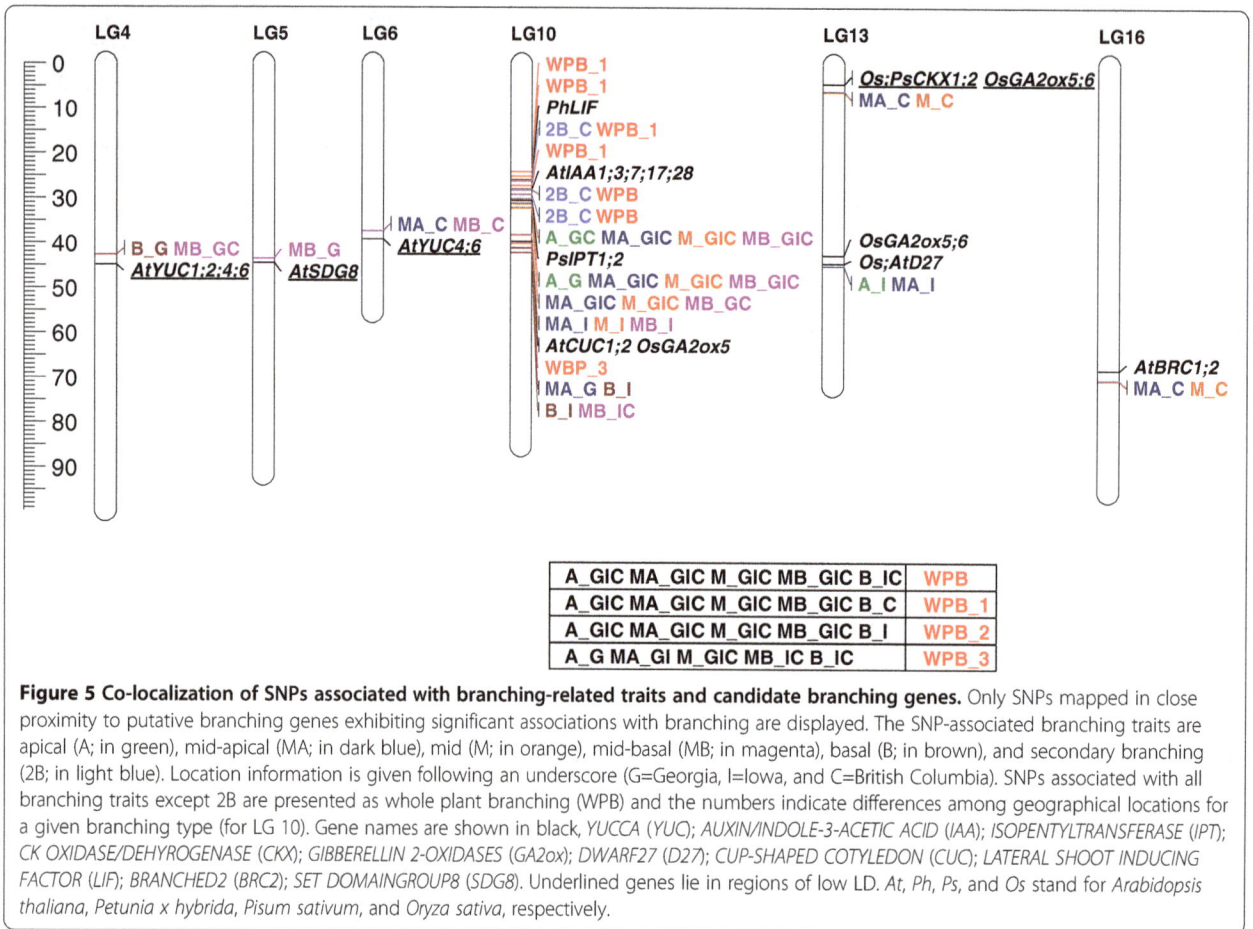

Figure 5 Co-localization of SNPs associated with branching-related traits and candidate branching genes. Only SNPs mapped in close proximity to putative branching genes exhibiting significant associations with branching are displayed. The SNP-associated branching traits are apical (A; in green), mid-apical (MA; in dark blue), mid (M; in orange), mid-basal (MB; in magenta), basal (B; in brown), and secondary branching (2B; in light blue). Location information is given following an underscore (G=Georgia, I=Iowa, and C=British Columbia). SNPs associated with all branching traits except 2B are presented as whole plant branching (WPB) and the numbers indicate differences among geographical locations for a given branching type (for LG 10). Gene names are shown in black, YUCCA (YUC); AUXIN/INDOLE-3-ACETIC ACID (IAA); ISOPENTYLTRANSFERASE (IPT); CK OXIDASE/DEHYROGENASE (CKX); GIBBERELLIN 2-OXIDASES (GA2ox); DWARF27 (D27); CUP-SHAPED COTYLEDON (CUC); LATERAL SHOOT INDUCING FACTOR (LIF); BRANCHED2 (BRC2); SET DOMAINGROUP8 (SDG8). Underlined genes lie in regions of low LD. At, Ph, Ps, and Os stand for Arabidopsis thaliana, Petunia x hybrida, Pisum sativum, and Oryza sativa, respectively.

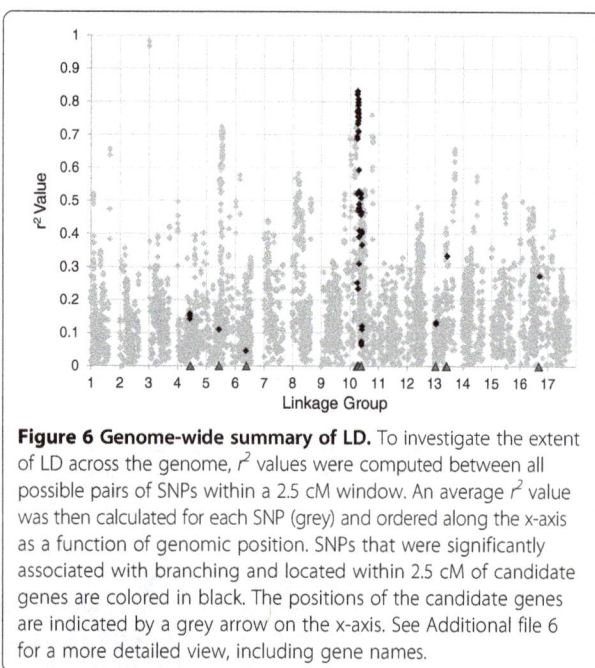

Figure 6 Genome-wide summary of LD. To investigate the extent of LD across the genome, r^2 values were computed between all possible pairs of SNPs within a 2.5 cM window. An average r^2 value was then calculated for each SNP (grey) and ordered along the x-axis as a function of genomic position. SNPs that were significantly associated with branching and located within 2.5 cM of candidate genes are colored in black. The positions of the candidate genes are indicated by a grey arrow on the x-axis. See Additional file 6 for a more detailed view, including gene names.

traditional QTL mapping [19,20,68] or via association mapping [29].

Importantly, our results also revealed that apical and basal branching are under largely independent genetic control (Figure 4; Table 4). Previous studies have identified a small number of loci that influence either apical or basal branching [22,23] and suggested that these traits are controlled by different loci, but our results indicate that the effects of branching-related loci are largely compartmentalized in sunflower. In fact, with the exception of loci on LGs 6a, 10a, and 10b, all three of which are associated with whole plant branching, the remaining 20 significant associations had primarily apical or basal effects. As mentioned above, the broad peak of associations on LG 10a (Figure 4) corresponds to the so-called B locus [11]. This locus is responsible for the branching that was reintroduced into the sunflower gene pool to extend the flowering time in R lines, and is present as a large haplotypic block within many of these lines [29], resulting in a sizable island of elevated LD (Figure 6). Similar instances of elevated LD, mostly associated with past episodes of selection during the evolution of cultivated sunflower, are also visible elsewhere in the genome [29]. While multiple novel branching loci were identified in this study, it is possible that segregation of the *B* locus masked the effect of other loci contributing to the variation observed within this population. The analysis of a population that varies for branching while being fixed for the recessive, unbranched *b/b* genotype at this locus has the potential to shed light on this possibility.

We further identified candidate branching genes that are co-localized with significant branching associations. Similar approaches using association mapping have been used to identify positional candidates for important agronomic traits in crops such as rice (e.g., [69,70]) and maize (e.g., [71]), to identify candidate genes associated with flowering time, pathogen resistance, and tolerance to salinity in *Arabidopsis* (e.g., [72,73]). Of course, the resolution afforded by association mapping varies across the genome due to localized variation in the extent of LD. We thus focused on the identification of candidate genes that co-localized with branching associations in regions of low LD (i.e., $r^2 < 0.20$). These genes included homologs of *AtYUC* that were closely linked to mid-basal and basal branching on LG 4 and mid-apical and mid-basal branching on LG 6; a homolog of *CKX*, which co-localized with apical and mid-branching on LG 13; and homologs of *GA2ox5* and *GA2ox6* in that same region of LG 13. In addition to these hormone-related genes, a homolog of *SDG8* that was associated with mid-basal branching co-localized on LG 5. *SDG8* is thought to epigenetically regulate other branching genes, and loss of this gene in *Arabidopsis* exhibit increased shoot branching [74]. The above genes are thus excellent candidates for further functional characterization. In maize, the primary determinant of branched vs. unbranched is *teosinte branched1* (*tb1*), which is a transcription factor within the *CYCLOIDEA* (*CYC*)/*TB1* subfamily of the TEOSINTE-BRANCHED1/CYCLOIDEA/PCF (TCP) transcription factor family [75-77]. While homologs of this gene are known to influence branching in other species such as rice [78] and *Arabidopsis* [54], we found no evidence for a similar role of *CYC*-like genes in sunflower (see also Chapman et al. [79]).

Conclusions

In this study, we identified numerous loci with significant effects on branching in sunflower, many of which had variable effects across environments. This includes multiple genomic regions that had not previously been implicated in branching. Interestingly, the majority of these loci primarily affected either apical or basal branching, as opposed to influencing branching at the whole-plant level. We also identified a collection of branching-related candidate genes that co-localized with significant association in regions of low LD, providing us with a pool of promising candidates for future functional validation.

Additional files

Additional file 1: Candidate branching genes. Genes involved in axillary meristem initiation and branch outgrowth in *Arabidopsis thaliana* (At), *Petunia* x *hybrida* (Ph), *Pisum sativum* (Ps), and *Oryza sativa* (Os).

Additional file 2: List of candidate genes with significant hits to contigs within the draft sunflower genome assembly. For each candidate gene, up to eight sunflower blast hits were considered, with the actual number being indicated as the hit number. The positions of the mapped genes within the contigs are also listed. See text for details. Additional file 1 about the candidate genes.

Additional file 3: Phenotypic data for branching in the association mapping population. Also included are the classification (i.e., oil vs. nonoil, breeding group, etc.) and USDA designation of each line. For more specific descriptions of these lines, refer to Mandel et al. [28]. Locations are G=Georgia, I=Iowa, and C=British Columbia. Q indicates quarter. The various branching types are illustrated in Figure 1.

Additional file 4: Quantile-quantile (Q-Q) plots of branching associations based on models accounting for the effects of kinship and/or population structure vs. a naive model. Results for branching in each of the four quarters at all three locations were computed using three models, K (red), P + K (blue), and Q + K (grey) and compared against the results of a naive model (brown). The expected values are shown in green. Locations are indicated as G=Georgia, I=Iowa, and C=British Columbia. Q indicates quarter. The branching types are illustrated in Figure 1.

Additional file 5: Sunflower genetic map depicting the positions of significant SNPs associated with branching traits and candidate branching genes. Shown are all 17 LGs and positions of all significant SNPs found in the study. Candidate branching genes are in black. Branching traits associated with the SNPs are color coded as in Figure 5 with apical (A; in green), mid-apical (MA; in dark blue), mid (M; in orange), mid-basal (MB; in magenta), basal (B; in brown), and secondary branching (2B; in light blue). Locations are indicated as follows: G=Georgia, I=Iowa, and C=British Columbia. For LG 10 only, SNPs associated with all branching traits other than 2B are indicated as whole plant branching (WPB) and the numbers denote differences among geographical locations for a given branching type.

Additional file 6: Average r^2 values for SNPs along LGs that contained candidate branching genes that co-localized with significant branching associations. Average r^2 values were calculated as described in Figure 6. SNPs exhibiting significant branching associations and located in close proximity to a candidate branching gene are depicted in red. The positions of the putative candidate branching genes are indicated by blue arrows on the x-axis; letters refer to specific genes.

Abbreviations

A lines: Cytoplasmic male-sterile female lines; *Aux/IAA: AUXIN/INDOLE-3-ACETIC ACID*; *BRC: BRANCHED*; CK: Cytokinin; *CKX: CYTOKININ OXIDASE/ DEHYDROGENASE*; *CUC: CUP-SHAPED COTYLEDON*; *CYC: CYCLOIDEA*; *D27: DWARF27*; GA: Gibberellic acid; *GA2ox: GIBBERELLIN 2-OXIDASES*; GRIN: Germplasm Resources Information Network; INRA: French National Institute for Agricultural Research; K: Kinship; *LAS: LATERAL SUPPRESSOR*; LD: Linkage disequilibrium; LG: Linkage group; *LIF: LATERAL SHOOT INDUCING FACTOR*; LS Means: Least-squares means; *MAPKK7: MAP KINASE KINASE 7*; NCRPIS: North Central Regional Plant Introduction Station; OPVs: Open-pollinated varieties; P: Population structure estimated using principal coordinate analysis; PCA: Principal component analysis; PCoA: Principal coordinate analysis; Q: Population structure estimated using STRUCTURE; Q-Q plots: quantile-quantile plots; QTL: Quantitative trait locus; R lines: Male-fertile restorer lines; *RAX: REGULATORS OF AXILLARY MERSITEMS*; *REV: REVOLUTA*; *SDG8: SET DOMAIN GROUP8*; SL: Strigolactone; SSR: Simple-sequence repeat; *tb1: teosinte branched 1*; TCP: TEOSINTE-BRANCHED1/CYCLOIDEA/PCF; WGS: Whole genome shotgun; YUC: YUCCA.

Competing interests

The authors declare that they have no competing interests.

Authors' contributions
SUN, JRM, LHR, SJK, and JMB conceived the study and designed the experiments. JRM and JMB were involved in the development of the association mapping population. SUN, JRM, LFM, and DE performed phenotypic analyses in the field. Data analyses were performed by SUN, JRM, JEB, LFM, DE, and JC. The manuscript was prepared by SUN, JRM, and JMB. All the authors read and approved the final manuscript.

Acknowledgements
We thank Eric Elsner, Jenny Leverett, Irvine Larsen, Tanya MacInnes, Anh Nguyen, and many undergraduate workers for their assistance in the field. This research was supported by USDA National Institute of Food and Agriculture (2008-35300-19263), the NSF Plant Genome Research Program (DBI-0820451), the USDA/DOE Plant Feedstocks Genomics Joint Program (ER 64664), Genome BC, and Genome Canada (http://www.genomecanada.ca).

Author details
[1]Department of Plant Biology, Miller Plant Sciences, University of Georgia, Athens, GA 30602, USA. [2]Present address: Department of Horticulture, University of Georgia, Athens, GA 30602, USA. [3]Present address: Department of Biological Sciences, University of Memphis, Memphis, TN 38152, USA. [4]North Central Regional Plant Introduction Station, Iowa State University/USDA-ARS, Ames, IA 50014, USA. [5]Department of Botany, University of British Columbia, Vancouver, BC V6T 1Z4, Canada. [6]Present address: Department of Crop and Soil Sciences, University of Georgia, Athens, GA 30602, USA. [7]Department of Plant Sciences, University of California, Davis, CA 95616, USA.

References
1. Irwin DL, Aarssen LW. Effects of nutrient level on cost and benefit of apical dominance in *Epilobium ciliatum*. Am Midl Nat. 1996;136:14–28.
2. Jaremo J, Nilsson P, Tuomi J. Plant compensatory growth: herbivory or competition? Oikos. 1996;77:238–47.
3. Fay PA, Throop HL. Branching responses in *Silphium Integrifolium* (Asteraceae) following mechanical or gall damage to apical meristems and neighbor removal. Am J Bot. 2005;92:954–9.
4. Evers JB, Van der Krol AR, Vos J, Struik PC. Understanding shoot branching by modelling form and function. Trends Plant Sci. 2011;16:464–7.
5. Ward SP, Leyser O. Shoot branching. Curr Opin Plant Biol. 2004;7:73–8.
6. Hammer K. Das Domestikationssyndrom. Kulturpflanze. 1984;32:11–34.
7. Heiser CB. The sunflower among the North American Indians. Proc Am Philos Soc. 1951;95:432–48.
8. Heiser CB, Smith DM, Clevenger SB, Martin WC. The North American sunflowers (*Helianthus*). Mem Torrey Bot Club. 1969;22:1–218.
9. Rogers CE, Thompson TE, Seiler GJ. Sunflower species of the United States. In *National Sunflower Association, Bismarck, North Dakota*; 1982:75.
10. Rieseberg LH, Seiler GJ. Molecular evidence and the origin and development of the domesticated sunflower (*Helianthus annuus*, Asteraceae). Econ Bot. 1990;44:79–91.
11. Fick GN, Miller JF. Sunflower breeding. In: AA S, editor. Sunflower technology and production. Madison: American Society of Agronomy; 1997. p. 395–440.
12. Ongaro V, Leyser O. Hormonal control of shoot branching. J Exp Bot. 2008;59:67–74.
13. Ferguson BJ, Beveridge CA. Roles for auxin, cytokinin, and strigolactone in regulating shoot branching. Plant Physiol. 2009;149:1929–44.
14. Shimizu-Sato S, Tanaka M, Mori H. Auxin-cytokinin interactions in the control of shoot branching. Plant Mol Biol. 2009;69:429–35.
15. Yaish MWF, Guevara DR, El-Kereamy A, Rothstein SJ. Axillary shoot branching in plants. In: Pua EC, Davey MR, editors. Plant developmental biology–biotechnological perspectives:. Volume 1. Berlin, Heidelberg: Springer Berlin Heidelberg; 2010. p. 37–52.
16. McSteen P. Hormonal regulation of branching in grasses. Plant Physiol. 2009;149:46–55.
17. Lo SF, Yang SY, Chen KT, Hsing YI, Zeevaart JA, Chen LJ, et al. A novel class of gibberellin 2-oxidases control semidwarfism, tillering, and root development in rice. Plant Cell. 2008;20:2603–18.
18. Ge C, Cui X, Wang Y, Hu Y, Fu Z, Zhang D, et al. *BUD2*, encoding an S-adenosylmethionine decarboxylase, is required for *Arabidopsis* growth and development. Cell Res. 2006;16:446–56.
19. Burke JM, Tang S, Knapp SJ, Rieseberg LH. Genetic analysis of sunflower domestication. Genetics. 2002;161:1257–67.
20. Wills DM, Burke JM. Quantitative trait locus analysis of the early domestication of sunflower. Genetics. 2007;176:2589–99.
21. Putt ED. Recessive branching in sunflowers. Crop Sci. 1964;4:444–5.
22. Gentzbittel L, Mestries E, Mouzeyar S, Mazeyrat F, Badaoui S, Vear F, et al. A composite map of expressed sequences and phenotypic traits of the sunflower (*Helianthus annuus* L.) genome. Theor Appl Genet. 1999;99:218–34.
23. Hockett EA, Knowles PF. Inheritance of branching in sunflowers, *Helianthus annuus* L. Crop Sci. 1970;10:432–6.
24. Tang S, Leon A, Bridges WC, Knapp SJ. Quantitative trait loci for genetically correlated seed traits are tightly linked to branching and pericarp pigment loci in sunflower. Crop Sci. 2006;46:721–34.
25. Tang S, Yu JK, Slabaugh B, Shintani D, Knapp J. Simple sequence repeat map of the sunflower genome. Theor Appl Genet. 2002;105:1124–36.
26. Zhu C, Gore M, Buckler ES, Yu J. Status and prospects of association mapping in plants. Plant Genome. 2008;1:5–20.
27. Yu J, Buckler ES. Genetic association mapping and genome organization of maize. Curr Opin Biotechnol. 2006;17:155–60.
28. Mandel JR, Dechaine JM, Marek LF, Burke JM. Genetic diversity and population structure in cultivated sunflower and a comparison to its wild progenitor, *Helianthus annuus* L. Theor Appl Genet. 2011;123:693–704.
29. Mandel JR, Nambeesan S, Bowers JE, Marek LF, Ebert D, Rieseberg LH, et al. Association mapping and the genomic consequences of selection in sunflower. PLoS Genet. 2013;9:e1003378. 10.1371/journal.pgen.1003378.
30. Bachlava E, Taylor CA, Tang S, Bowers JE, Mandel JR, Burke JM, et al. SNP discovery and development of a high-density genotyping array for sunflower. PLoS One. 2012;7:e29814. 10.1371/journal.pone.0029814.
31. Bowers JE, Bachlava E, Brunick RL, Rieseberg LH, Knapp SJ, Burke JM. Development of a 10,000 locus genetic map of the sunflower genome based on multiple crosses. Genes Genomes Genet. 2012;2:721–9.
32. Schneiter AA, Miller JF. Description of sunflower growth stages. Crop Sci. 1981;21:901–3.
33. Holm S. A simple sequentially rejective multiple test procedure. Scand J Stat. 1979;6:65–70.
34. Le S, Josse J, Husson F. FactoMineR: An R package for multivariate analysis. J Stat Softw. 2008;25:1–18.
35. R Core Team: R: A language and environment for statistical computing. 2014.
36. Doyle JJ, Doyle JL. A rapid DNA isolation procedure for small quantities of fresh leaf tissue. Phytochem Bull. 1987;19:11–5.
37. Bradbury PJ, Zhang Z, Kroon DE, Casstevens TM, Ramdoss Y, Buckler ES. TASSEL: software for association mapping of complex traits in diverse samples. Bioinformatics. 2007;23:2633–5.
38. Pritchard JK, Stephens M, Rosenberg N, Donnelly P. Association mapping in structured populations. Am J Hum Genet. 2000;67:170–81.
39. Hardy OJ, Vekemans X. SPAGeDi: a versatile computer program to analyse spatial genetic structure at the individual or population levels. Mol Ecol Notes. 2002;2:618–20.
40. Pritchard JK, Stephens M, Donnelly P. Inference of population structure using multilocus genotype data. Genetics. 2000;155:945–59.
41. Peakall R, Smouse PE. Genalex 6: genetic analysis in Excel. Population genetic software for teaching and research. Mol Ecol Notes. 2006;6:288–95.
42. De Silva H, Ball R. Linkage disequilibrium mapping concepts. In: Oraguzie NC, Rikkerink EHA, Gardiner SE SH, editors. Association mapping in plants. New York: Springer; 2007. p. 103–32.
43. Gao X, Starmer J, Martin ER. A multiple testing correction method for genetic association studies using correlated single nucleotide polymorphisms. Genet Epidemiol. 2008;32:361–9.
44. Greb T, Clarenz O, Schafer E, Muller D, Herrero R, Schmitz G, et al. Molecular analysis of the *LATERAL SUPPRESSOR* gene in *Arabidopsis* reveals a conserved control mechanism for axillary meristem formation. Genes Dev. 2003;17:1175–87.
45. Otsuga D, DeGuzman B, Prigge MJ, Drews GN, Clark SE. REVOLUTA regulates meristem initiation at lateral positions. Plant J. 2001;25:223–36.
46. Muller D, Schmitz G, Theres K. *Blind* homologous *R2R3 Myb* genes control the pattern of lateral meristem initiation in *Arabidopsis*. Plant Cell. 2006;18:586–97.

47. Barker MS, Kane NC, Matvienko M, Kozik A, Michelmore RW, Knapp SJ, et al. Multiple paleopolyploidizations during the evolution of the Compositae reveal parallel patterns of duplicate gene retention after millions of years. Mol Biol Evol. 2008;25:2445–55.

48. Renaut S, Grassa CJ, Yeaman S, Moyers BT, Lai Z, Kane NC, et al. Genomic islands of divergence are not affected by geography of speciation in sunflowers. Nat Commun. 2013;4:1827.

49. Ledger SE, Janssen BJ, Karunairetnam S, Wang T, Snowden KC. Modified CAROTENOID CLEAVAGE DIOXYGENASE8 expression correlates with altered branching in kiwifruit (Actinidia chinensis). New Phytol. 2010;188:803–13.

50. Cieslak M, Seleznyova AN, Hanan J. A functional-structural kiwifruit vine model integrating architecture, carbon dynamics and effects of the environment. Ann Bot. 2011;107:747–64.

51. Lauri PA, Costes E, Regnard JL, Brun L, Simon S, Monney P, et al. Does knowledge on fruit tree architecture and its implications for orchard management improve horticultural sustainability? An overview of recent advances in the apple. Acta Hortic. 2009;817:243–50.

52. Upadyayula N, Da Silva HS, Bohn MO, Rocheford TR. Genetic and QTL analysis of maize tassel and ear inflorescence architecture. Theor Appl Genet. 2006;112:592–606.

53. Casal JJ, Sanchez RA, Deregibus VA. The effect of plant density on tillering: The involvement of R/FR ratio and the proportion of radiation intercepted per plant. Environ Exp Bot. 1986;26:365–71.

54. Aguilar-Martínez JA, Poza-Carrión C, Cubas P. Arabidopsis BRANCHED1 acts as an integrator of branching signals within axillary buds. Plant Cell. 2007;19:458–72.

55. Arumingtyas EL, Floyd R, Gregory M, Murfet I. Branching in Pisum: inheritance and allelism tests with 17 ramosus mutants. Pisum Genet. 1992;24:17–31.

56. Napoli CA, Beveridge CA, Snowden KC. Reevaluating concepts of apical dominance and the control of axillary bud outgrowth. Curr Top Dev Biol. 1999;44:127–69 [Current Topics in Developmental Biology].

57. Grbić V, Bleecker AB. Axillary meristem development in Arabidopsis thaliana. Plant J. 2000;21:215–23.

58. Stirnberg P, Van De Sande K, Leyser HMO. MAX1 and MAX2 control shoot lateral branching in Arabidopsis. Development. 2002;129:1131–41.

59. Kebrom TH, Burson BL, Finlayson SA. Phytochrome B represses teosinte branched1 expression and induces sorghum axillary bud outgrowth in response to light signals. Plant Physiol. 2006;140:1109–17.

60. Snowden KC, Napoli CA. A quantitative study of lateral branching in petunia. Funct Plant Biol. 2003;30:987–94.

61. Cline MG. Exogenous auxin effects on lateral bud outgrowth in decapitated shoots. Ann Bot. 1996;78:255–66.

62. Napoli CA, Ruehle J. New mutations affecting meristem growth and potential in Petunia hybrida Vilm. J Hered. 1996;87:371–7.

63. Cline MG. Apical dominance. Bot Rev. 1991;57:318–58.

64. Shimizu-Sato S, Mori H. Control of outgrowth and dormancy in axillary buds. Plant Physiol. 2001;127:1405–13.

65. Schumacher K, Schmitt T, Rossberg M, Schmitz G, Theres K. The Lateral suppressor (Ls) gene of tomato encodes a new member of the VHIID protein family. Proc Natl Acad Sci U S A. 1999;96:290–5.

66. Li X, Qian Q, Fu Z, Wang Y, Xiong G, Zeng D, et al. Control of tillering in rice. Nature. 2003;422:618–21.

67. Schmitz G, Tillmann E, Carriero F, Fiore C, Cellini F, Theres K. The tomato Blind gene encodes a MYB transcription factor that controls the formation of lateral meristems. Proc Natl Acad Sci U S A. 2002;99:1064–9.

68. Dechaine JM, Burger JC, Chapman MA, Seiler GJ, Brunick R, Knapp SJ, et al. Fitness effects and genetic architecture of plant-herbivore interactions in sunflower crop-wild hybrids. New Phytol. 2009;184:828–41.

69. Huang X, Wei X, Sang T, Zhao Q, Feng Q, Zhao Y, et al. Genome-wide association studies of 14 agronomic traits in rice landraces. Nat Genet. 2010;42:961–7.

70. Zhao K, Tung CW, Eizenga GC, Wright MH, Ali ML, Price AH, et al. Genome-wide association mapping reveals a rich genetic architecture of complex traits in Oryza sativa. Nat Commun. 2011;2:467.

71. Weng J, Xie C, Hao Z, Wang J, Liu C, Li M, et al. Genome-wide association study identifies candidate genes that affect plant height in Chinese elite maize (Zea mays L.) inbred lines. PLoS One. 2011;6:e29229. doi:10.1371/journal.pone.0029229.

72. Aranzana MJ, Kim S, Zhao K, Bakker E, Horton M, Jakob K, et al. Genome-wide association mapping in Arabidopsis identifies previously known flowering time and pathogen resistance genes. PLoS Genet. 2005;1:e60. 10.1371/journal.pgen.0010060.

73. DeRose-Wilson L, Gaut BS. Mapping salinity tolerance during Arabidopsis thaliana germination and seedling growth. PLoS One. 2011;6:e22832. 10.1371/journal.pone.0022832.

74. Dong G, Ma D, Li J. The histone methyltransferase SDG8 regulates shoot branching in Arabidopsis. Biochem Biophys Res Commun. 2008;373:659–64.

75. Doebley J, Stec A, Gustus C. teosinte branched1 and the origin of maize: Evidence for epistasis and the evolution of dominance. Genetics. 1995;141:333–46.

76. Doebley J, Stec A, Hubbard L. The evolution of apical dominance in maize. Nature. 1997;386:485–8.

77. Cubas P, Lauter N, Doebley J, Coen E. The TCP domain: a motif found in proteins regulating plant growth and development. Plant J. 1999;18:215–22.

78. Takeda T, Suwa Y, Suzuki M, Kitano H, Ueguchi-Tanaka M, Ashikari M, et al. The OsTB1 gene negatively regulates lateral branching in rice. Plant J. 2003;33:513–20.

79. Chapman MA, Leebens-Mack JH, Burke JM. Positive selection and expression divergence following gene duplication in the sunflower CYCLOIDEA gene family. Mol Biol Evol. 2008;25:1260–73.

Baseline study of morphometric traits of wild *Capsicum annuum* growing near two biosphere reserves in the Peninsula of Baja California for future conservation management

Bernardo Murillo-Amador[1], Edgar Omar Rueda-Puente[2], Enrique Troyo-Diéguez[1], Miguel Víctor Córdoba-Matson[1], Luis Guillermo Hernández-Montiel[1] and Alejandra Nieto-Garibay[1*]

Abstract

Background: Despite the ecological and socioeconomic importance of wild *Capsicum annuum* L., few investigations have been carried out to study basic characteristics. The peninsula of Baja California has a unique characteristic that it provides a high degree of isolation for the development of unique highly diverse endemic populations. The objective of this study was to evaluate for the first time the growth type, associated vegetation, morphometric traits in plants, in fruits and mineral content of roots, stems and leaves of three wild populations of *Capsicum* in Baja California, Mexico, near biosphere reserves.

Results: The results showed that the majority of plants of wild *Capsicum annuum* have a shrub growth type and were associated with communities consisting of 43 species of 20 families the most representative being Fabaceae, Cactaceae and Euphorbiaceae. Significant differences between populations were found in plant height, main stem diameter, beginning of canopy, leaf area, leaf average and maximum width, stems and roots dry weights. Coverage, leaf length and dry weight did not show differences. Potassium, sodium and zinc showed significant differences between populations in their roots, stems and leaves, while magnesium and manganese showed significant differences only in roots and stems, iron in stems and leaves, calcium in roots and leaves and phosphorus did not show differences. Average fruit weight, length, 100 fruits dry weight, 100 fruits pulp dry weight and pulp/seeds ratio showed significant differences between populations, while fruit number, average fruit fresh weight, peduncle length, fruit width, seeds per fruit and seed dry weight, did not show differences.

Conclusions: We concluded that this study of traits of wild *Capsicum*, provides useful information of morphometric variation between wild populations that will be of value for future decision processes involved in the management and preservation of germplasm and genetic resources.

Keywords: Solanaceae, Mineral content, Growth type, Vegetation associated

* Correspondence: anieto04@cibnor.mx
[1]Centro de Investigaciones Biológicas del Noroeste, S.C. La Paz, La Paz, Baja California Sur, México
Full list of author information is available at the end of the article

Background

The genus Capsicum (Solanaceae) contains a large number of cultivated species as well as wild species that are grown for their fruits, and are an important vegetable consumed throughout the world. There are about 30 species of *Capsicum*, but only *C. annuum*, *C. frutescens*, *C. chinense* Jacq., *C. baccatum*, and *C. pubescens* Ruiz et Pav are presently domesticated.

Capsicum annuum has the highest morphometric diversity and is widely cultivated in America, Asia, Africa, and Mediterranean countries for their fruits that have numerous uses in culinary preparations. It is a good source of starch, dietary fiber, protein, lipids, and minerals. In addition to their nutritive value, they contain phytochemicals with antioxidant properties that are beneficial to human health [1].

In general, wild *Capsicum* species are found at low altitudes, rarely exceeding 1000 m.a.s.l. [2,3]. Botanically, *C. annuum* species are tender perennials when grown in their native tropical habitats but are also commonly grown as annual crops in parts of the world where frost and freezing temperatures preclude year-round field production [4], they range extends from USA to Peru.

In México, *Capsicum* peppers are cultivated and can be found in the wild. Wild populations of *C. annuum* are widely distributed in Mexico, growing in dry tropical forests, in desert scrubs, near roads, home gardens, pasturelands, and around crop fields [5]. They produce small round berries held erect on long pedicels, that are deciduous, brilliant red when ripe. They are extremely hot to the taste, and they stand out of the foliage allowing for easy harvesting during ripeness [2,6,7]. They are very attractive for birds and are consumed by frugivorous birds species, which are the main seed dispersers [7-9]. Therefore it is necessary to harvest berries before they mature. Moreover the berries tend to fall from the plant when they mature [10].

In northeastern Mexico wild *Capsicum* species are important resources for people living in rural communities because there is little farm work and employment is scarce [11]. Fruits of this species are consumed fresh,- dried or processed in vinegar or sauce representing a promising potential market both in Mexico and USA [12,13]. In Baja California Sur, México, wild *C. annuum* is called "chilpitín" or "chiltepín". and represent a wild chili that come from small shrubs with highly branched stems, with alternate petiole leaves. Flowering occurs almost year round, with white flowers and five lobes. The fruit grows in streams and is distributed in tropical areas of the Cape Region of the Baja California Peninsula [14] and is well accepted for different culinary [14] and medicinal [15] purposes. According to the Missouri Botanical Garden, the wild *Capsicum* species found in Baja California Sur, México is *C. annuum* var. *aviculare* (Dierb.)

D'Arcy & Eshbaugh, native from Mesoamerica with a distribution range extending from the south of the United States to the north of South America [7,16]. However, Kraft *et al.* [17] reported that some accessions were a different phenotype although collected in Baja California Sur. Generally speaking, these accessions collected were morphometrically similar (with similar cultural use, but not commercialized in any significant manner) to those found in Sonora and Arizona (*C. annuum* var. *glabriusculum*). However, according to the Missouri Botanical Garden, *aviculare* and *glabrisuculum* are accepted synonyms.

Chiltepín production in Mexico has been estimated to be 50 t yr^{-1}, it is an important crop product for subsistence farmers of the central and northern regions of the country [7,18-20]. The agronomic interest of chiltepín exceeds its value as a local commodity, as it is genetically compatible with the domesticated varieties of *C. annuum*. Wild *Capsicum* species are important reservoirs of genes and sources of genetic diversity for breeding programs of cultivated pepper, as sources of resistance against pests, pathogens [21,22], adverse environmental factors, and for increasing quality and quantity of production [23,24]. Maiti *et al.* [25] stated that piquín pepper might be considered as a new crop because it has been exploited for many years in its wild form. Extensive commercial farming of piquín pepper does not exist. Almost all piquín production comes from harvesting of wild plants, usually with overexploitation conditions, causing loss of biodiversity [11].

The current main limitation for planting piquín as a commercial crop is its low seed germination (dormancy). In addition, research on developing production technology for piquín is limited. Although a perennial plant it can die in times of drought or even in the winter. It sprouts with the first rains and full production occurs at the end of the rainy season from August to December, depending of locality. When it is fresh it is of green color and when dry color changes to red. The piquín pepper is found in the local markets at the end of the season of rains [26]. Domestication causes dispersal from center of origin [27,28] causing artificial selection that has led to changes in their mating systems, dispersal mechanisms, physiology, and their genetic structure [23,29].

For this reason, it is important to know the extent and distribution of genetic variation among populations since it is crucial for understanding the origin and evolution of plant populations in natural conditions. The information about where it grows, its commercial variants and their wild relatives is important for potential breeders, population geneticists, and conservation biologists concerned with the use, management and conservation of plant genetic resources [30].

Based on the aforementioned lack of biological information such as knowledge of morphometric traits, and the relatively little research available, the objective of this study was to analyze three populations of wild *C. annuum* growing near two biosphere reserves in Baja California Sur, Mexico. The purpose of which is to generate fundamental baseline data of the chili chiltepín useful for providing a framework for germplasm use for crop management domestication and species conservation. Four specific questions were addressed: (1) what is the growth type of wild *Capsicum* plants in each population? (2) which wild species and families are more associated with wild *Capsicum* plants? (3) are there differences between mineral content and morphometric traits in plants and fruits between populations? and (4) how some environment conditions affect the growth of wild *Capsicum* plants? Undoubtedly, the results of the present study will be valuable in providing a better understanding of some of the wild *C. annuum* populations growing near two biosphere reserves in Baja California Sur, Mexico.

Results

The MANOVA analysis for variables measured in plants (*in-situ*) showed significant differences between sample populations (Wilks = 0.155, F = 3.45, p = 0.01). This analysis included the variables plant height, plant coverage, main stem diameter and height of the beginning of canopy. The MANOVA analysis for morphometric traits from plants measured in laboratory such as leaf area, leaf length, average and maximum width of leaf, leaves, roots and stems dry weights showed significant differences between sample populations (Wilks = 0.036, F = 3.64, p = 0.01). The MANOVA analysis for those variables of fruits measured from collected plants (number of fruits per plant, average fresh and dry of fruits and peduncle length) showed significant differences between sample populations (Wilks = 0.062, F = 4.52, p = 0.009). The MANOVA analysis for the variables, fruit length and width, seeds per fruit, dry weight of 100 fruits, dry weight of seeds and pulp of 100 fruits, dry weight of 1000 seeds and index of pulp/seeds, measured in 400 fruits collected per population, showed significant differences between sample populations (Wilks = 0.00019, F = 30.00, p = 0.0002). The MANOVA analysis for mineral content in roots, stems and leaves (Ca, Mg, K, Na, Fe, Mn, Cu, Zn and P) showed significant differences between populations for roots (Wilks = 0.013, F = 3.37, p = 0.04), stems (Wilks = 0.022, F = 2.54, p = 0.05) and leaves (Wilks = 0.00078, F = 15.43, p = 0.00024). According with MANOVA analysis, it can be seen that the relationship of Wilks possibilities is significant at the level of $p \leq 0.01$ or $p \leq 0.05$.

Vegetation associated to wild Capsicum

The results from the first study estimation indicate that *Capsicum* in the sample populations is associated with twenty wild vegetal families where Fabaceae (21.4%), Cactaceae (16.1%) and Euphorbiaceae (12.5%) are the most representative (Table 1). The results showed 43 species associated to *Capsicum* ecotypes in the populations, these being *Jatropha cinerea* (5%) the most abundant, followed by *Prosopis glandulosa* var. *torreyana*, *Erythrina flabelliformis*, *Mimosa dystachia*, *Stenocereus thurberii*, *Tecoma stands*, *Pachycereus pecten-aboriginum*, *Ambrosia ambrosioides*, *Opuntia tapona*, *Celtis reticulata*, *Bignonia unguis-cati* and *Schaeferia shrevei* (all species with 4%) the most representatives. The rest of species showed 2% of presence (Table 1). The analysis of vegetation among collection sites showed some differences in the predominant vegetation on each site, i.e. in Los Gatos, the three species most abundant from most to least were *Jatropha cinerea* > *Prosopis glandulosa* var. *torreyana* > *Pachycereus pringleii*. In San Bartolo, the predominant species were *Prosopis glandulosa* var. *torreyana* > *Pachycereus pecten-aboriginum* > *Jatropha cinerea*, while in Santiago, the three most abundant species were in the following order *Celtis reticulata* > *Tecoma stands* > *Pachycereus pecten-aboriginum*.

Morphometric traits measured in plants (*in-situ*)
Plant height, coverage, stems diameter and height of the beginning of canopy

Significant differences between populations were observed in plant height (Table 2). The plants of San Bartolo showed higher height, while lower were showed by plants of Santiago (Table 3). The ANOVA showed no significant differences (Table 2) of plant coverage between populations. Significant differences between populations were observed for main stem diameter (Table 2). Higher values of main stem diameter were found in plants collected in Santiago, followed by San Bartolo plants and the lower values where in plants from Los Gatos (Table 3). The ANOVA showed significant differences between populations for height of the beginning of canopy (Table 2). The plants from San Bartolo showed higher values of this variable respect the plants from Los Gatos and Santiago (Table 3).

Growth type

In Los Gatos, 100% of the total plants identified in the population had erect growth (shrub type). In San Bartolo, 73% of the total plants identified had erect growth, while the rest (27%) had climbing growth. In Santiago, 90% of the total plants identified had climbing growth (vine type) while 10% had erect growth.

Table 1 Main species of vegetation associated to wild *Capsicum* chili ecotypes collected in three populations near two biosphere reserves in Mexico

Population	Common name	Scientific name	Life form	Family
Los Gatos	Lomboy blanco	*Jatropha cinerea*	Bush	Euphorbiaceae
Los Gatos	Mezquite	*Prosopis glandulosa* var. *torreyana*	Tree	Fabaceae
Los Gatos	Colorín or chilicote	*Erythrina flabelliformis*	Tree	Fabaceae
Los Gatos	Huerivo	*Populus brandegeei*	Tree	Salicaceae
Los Gatos	Cardón	*Pachycereus pringleii*	Cactus tree	Cactaceae
Los Gatos	Uña de gato	*Mimosa dystachia*	Tree	Fabaceae
Los Gatos	Pitahaya dulce	*Stenocereus thurberii*	Cactus tree	Cactaceae
Los Gatos	Palo adan	*Fouquieria diguetti*	Bush	Fouquieriaceae
San Bartolo	Mezquite	*Prosopis glandulosa* var. *torreyana*	Tree	Fabaceae
San Bartolo	Palo de arco	*Tecoma stands*	Tree	Bignoniaceae
San Bartolo	Cardón barbón	*Pachycereus pecten-aboriginum*	Cactus tree	Cactaceae
San Bartolo	Chicura	*Ambrosia ambrosioides*	Bush	Compositae
San Bartolo	Lomboy blanco	*Jatropha cinerea*	Bush	Euphorbiaceae
San Bartolo	Pitahaya dulce	*Stenocereus thurberii*	Cactus tree	Cactaceae
San Bartolo	Uña de gato	*Mimosa dystachia*	Tree	Fabaceae
San Bartolo	Nopal	*Opuntia tapona*	Cactus bush	Cactaceae
San Bartolo	Vainoro	*Celtis reticulata*	Tree	Ulmaceae
San Bartolo	Huirote de corral	*Bignonia unguis-cati*	Vine, annual herb	Bignoniaceae
San Bartolo	Hierba del cuervo	*Schaeferia shrevei*	Tree	Celastraceae
San Bartolo	Lentejilla	*Senna villosa*	Bush	Fabaceae
San Bartolo	Palo zorrillo	*Cassia emarginata*	Tree	Fabaceae
San Bartolo	Bernardia	*Bernardia mexicana*	Bush	Euphorbiaceae
San Bartolo	Alcager	*Pereskiopsis porterii*	Bush, cactus scandent	Cactaceae
San Bartolo	Ventamanta	*Coursetia caribaea*	Perennial herb	Fabaceae
San Bartolo	Cardoncillo	*Elytraria imbricata*	Annual herb	Acanthaceae
San Bartolo	Abutilón	*Abutilon palmeri*	Annual herb	Malvaceae
San Bartolo	Brikelia	*Brickelia coulteri*	Bush	Asteraceae
San Bartolo	Naranjillo	*Zantoxylon sonorensis*	Tree	Rutaceae
San Bartolo	Not available	*Carlowrightia arizonica*	Perennial herb	Acanthaceae
San Bartolo	Huirote de corral	*Bignonia unguis-cati*	Vine, annual herb	Bignoniaceae
Santiago	Lomboy blanco	*Jatropha cinerea*	Bush	Euphorbiaceae
Santiago	Palo de arco	*Tecoma stands*	Tree	Bignoniaceae
Santiago	Chicura	*Ambrosia ambrosioides*	Bush	Compositae
Santiago	Vainoro	*Celtis reticulata*	Tree	Ulmaceae
Santiago	Cardón barbón	*Pachycereus pecten-aboriginum*	Cactus tree	Cactaceae
Santiago	Palo chino	*Acacia peninsularis*	Tree	Fabaceae
Santiago	Bledo	*Celosia floribunda*	Tree	Amaranthaceae
Santiago	Guayparin	*Diospyros californica*	Tree	Ebenaceae
Santiago	Hierba del cuervo	*Schaeferia shrevei*	Tree	Celastraceae
Santiago	Mauto	*Lysiloma divaricata*	Tree	Fabaceae
Santiago	Aretito, hierba del alacrán	*Plumbago scandens*	Perennial herb	Plumbaginaceae
Santiago	Cacachila	*Karswinskia humboldtiana*	Tree	Rhamnaceae
Santiago	Crotón	*Croton boregensis*	Shrub	Euphorbiaceae

Table 1 Main species of vegetation associated to wild *Capsicum* chili ecotypes collected in three populations near two biosphere reserves in Mexico *(Continued)*

Santiago	Lomboy rojo	*Jatropha vernicosa*	Bush	Euphorbiaceae
Santiago	Sida	*Sida glutinosa*	Annual herb	Malvaceae
Santiago	Ayenia	*Ayenia glabra*	Annual herb	Malvaceae
Santiago	Caribe	*Cnidosculus angustidens*	Annual herb	Euphorbiaceae
Santiago	Malva colorada, malva rosa	*Melochia tomentosa*	Bush	Sterculiaceae
Santiago	Not available	*Aphanosperma sinaloensis*	Annual herb	Acanthaceae
Santiago	Rama parda	*Ruelia peninsularis*	Shrub	Acanthaceae
Santiago	Not available	*Cissus trifoliata*	Herbaceous vine	Vitaceae
Santiago	Papache	*Randia armata*	Shrub	Rubiaceae
Santiago	Nopal	*Opuntia tapona*	Cactus bush	Cactaceae
Santiago	Choya	*Opuntia cholla*	Cactus bush	Cactaceae
Santiago	Celosa	*Mimosa xantii*	Shrub	Fabaceae
Santiago	Colorin or chilicote	*Erythrina flabelliformis*	Tree	Fabaceae

Morphometric traits measured in collected plants and fruits (laboratory)

Leaf area, leaf length, average and maximum width of leaf

Significant differences between populations were observed for leaf area (Table 2). The higher values of leaf area were in plants from Los Gatos > Santiago > San Bartolo (Table 3). In leaf length, not significant differences between populations were observed. Significant differences between populations were observed in leaf average width (Table 2). High leaf average width was showed in plants collected in Los Gatos followed by plants from Santiago (Table 3). Significant differences between populations were observed for leaf maximum width (Table 2). The higher values of this variable were showed in leaves collected in plants from Los Gatos and Santiago (Table 3).

Leaves, roots and stems dry weight

From these variables, leaves and roots dry weights not showed significant differences between populations and only stems dry weight showed significant differences (Table 2) with higher values the plants collected in San Bartolo followed by Santiago (Table 3).

Number of fruits per plant, peduncle length and fruit average fresh and dry weights

From these variables, number of fruits per plant, peduncle length and fruit average fresh weight not showed significant differences between populations but only fruit average dry weight showed significant differences (Table 2) with higher values the fruits collected in Los Gatos plants, followed by Santiago (Table 3).

Number of seeds per fruit, fruit length and width

Only fruit length showed significant differences between populations (Table 2) with higher length the fruits collected in Santiago, followed by San Bartolo fruits (Table 3).

100 fruits dry weight, seeds and pulp dry weight of 100 fruits, 1000 seeds dry weight and pulp/seeds ratio

One hundred fruits in terms of dry weight showed significant differences between populations (Table 2) with higher values the fruits collected in San Bartolo (Table 3). One hundred seeds dry weight not showed significant differences between populations (Table 2). The variables 100 fruits pulp dry weight, 1000 seeds dry weight and pulp/seeds ratio showed significant differences between populations (Table 2). The fruits collected in San Bartolo showed higher values of 100 fruits pulp dry weight and 1000 seeds dry weight, while the fruits collected in Santiago showed the higher pulp/seeds ratio (Table 3).

Mineral content of roots, stems and leaves

The ANOVA of mineral content in roots showed significant differences between populations for Ca, Mg, K, Na, Mn and Zn but not for Fe, Cu and P (Table 2). Calcium, K, Na and Zn was higher in roots of plants collected in Santiago, while the roots of plants from Los Gatos showed higher values of Mg and Mn (Table 3). Significant differences between populations had differences for Mg, K, Na, Fe, Mn, Cu and Zn content in stems and only Ca and P did not show differences (Table 2). The stems of plants collected in Santiago had higher values of K, Na, Fe, Mn, Cu and Zn and only the stems of plants collected in Los Gatos showed higher values of

Table 2 ANOVA (mean squares) of plant, fruits characteristics and mineral content in tissues (roots, stems and leaves) of wild *Capsicum* ecotypes collected in three populations near two biosphere reserves in Mexico

Plant

Source	d.f.	Height	Coverage	Main stem diameter	Beginning of canopy
Populations	2	0.782*	3.81 ns	356.2**	728.46**
Error	12	0.151	1.11	18.9	118.00
CV (%)		32.40	99.93	34.07	62.91

Leaf

Source	Area	Maximum width	Average width	Length	Width
Populations	194418.39**	1.09**	0.41**	0.49 ns	0.19 ns
Error	31321.11	0.10	0.04	0.24	0.08
CV (%)	24.88	12.32	15.32	10.43	4.05

Dry weight

Source	Roots	Stems	Leaves
Populations	54.22 ns	129237.91**	468.83 ns
Error	357.51	29358.98	349.89
CV (%)	62.47	122.37	78.97

Fruits from collected plants

Source	d.f.	Number	Average FW	Average DW	Peduncle length
Populations	2	14.08 ns	0.0009 ns	0.0009**	0.05 ns
Error	9	10.38	0.002	0.0001	0.06
CV (%)		39.46	23.20	19.04	10.28

Four hundred fruits from not collected plants

Source	Pulp/seeds ratio	1000 seeds DW	Pulp DW 100 fruits	Seeds DW 100 fruits	DW 100 fruits	Seeds per fruit	Length	Width
Populations	0.13**	0.33**	0.81**	0.02 ns	0.69**	2.04 ns	3.59**	
Error	0.004	0.05	0.03	0.01	0.06	3.90	0.23	
CV (%)	5.73	5.54	3.93	5.08	3.77	12.33	6.27	

Roots

Source	d.f.	Ca	Mg	K	Na	Fe	Mn	Cu	Zn	P
Populations	2	21.60**	1.15**	49.24*	0.10**	1.37 ns	0.001**	0.00002 ns	0.0001**	0.05 ns
Error	12	2.53	0.19	9.53	0.005	0.66	0.0001	0.00001	0.000006	0.03
CV (%)		15.66	23.43	18.24	22.32	77.66	25.30	9.13	5.14	24.47

Stems

Source	d.f.	Ca	Mg	K	Na	Fe	Mn	Cu	Zn	P
Populations	2	6.92 ns	2.89**	239.19**	0.044**	0.001*	0.0002**	0.00002**	0.00006**	0.25 ns
Error	12	3.92	0.36	18.43	0.009	0.0003	0.00003	0.000004	0.000005	0.11
CV (%)		20.61	17.95	12.30	36.97	45.62	20.15	6.23	4.01	17.44

Leaves

Source	d.f.	Ca	Mg	K	Na	Fe	Mn	Cu	Zn	P
Populations	2	67.84**	0.29 ns	224.40**	0.054**	0.00006 ns	0.0002**	0.00005**	0.69 ns	
Error	12	6.28	0.99	24.66	0.71	0.004	0.0001	0.00001	0.000002	0.42
CV (%)		16.59	11.00	7.29	51.96	58.57	24.55	10.85	2.06	25.14

FW = fresh weight. DW = dry weight. d.f. = degree freedom. *Significant probability level $p \leq 0.05$. **Significant probability level $p \leq 0.01$. ns = not significant. CV = coefficient of variation.

Table 3 Means of plant, fruits characteristics and mineral content (g kg⁻¹ dry-weight) in tissues (roots, stems and leaves) of wild *Capsicum* ecotypes collected in three populations near two biosphere reserves in Mexico

Populations	Plant					Leaf				Dry weight (g)		
	Height (m)	Coverage (m²)	Beginning of canopy (cm)	Stems diameter (mm)	Peduncle length (cm)	Length (cm)	Average width (cm)	Maximum width (cm)	Area (cm²)	Leaves	Stems	Roots
San Bartolo	1.57 a	2.03 a	31.20 a	12.23 ab	2.47 a	4.36 a	1.08 b	2.09 b	489.25 b	34.29 a	324.75 a	29.32 a
Los Gatos	1.23 ab	0.77 a	10.60 b	8.61 b	2.52 a	4.91 a	1.63 a	3.03 a	866.45 a	21.45 a	31.80 b	27.54 a
Santiago	0.78 b	0.35 a	10.00 b	17.58 a	2.69 a	4.89 a	1.49 a	2.64 a	777.55 ab	15.31 a	63.47 ab	33.92 a

	Fruits from collected plants*			Four hundred fruits from not collected plants**							
	Number	Average FW (g)	Average DW (g)	Length (mm)	Width (mm)	Seeds per fruit	DW 100 fruits (g)	Seeds DW 100 fruits (g)	Pulp DW 100 fruits (g)	1000 seeds DW (g)	Pulp/seeds ratio
San Bartolo	8.25 a	0.21 a	0.047 b	7.69 a	7.47 a	16.60 a	7.35 a	3.30 a	4.05 a	4.39 a	1.22 a
Los Gatos	10.0 a	0.22 a	0.078 a	6.67 b	7.04 a	16.23 a	6.53 b	3.38 a	3.15 c	4.07 ab	0.93 b
Santiago	6.25 a	0.19 a	0.061 ab	8.56 a	7.32 a	15.22 a	6.82 b	3.22 a	3.60 b	3.82 b	1.25 a

Roots

	Ca	Mg	K	Na	Fe	Mn	Cu	Zn	P
San Bartolo	10.81 a	1.53 b	14.58 b	0.25 b	0.67 b	0.45 ab	0.035 a	0.048 b	0.69 a
Los Gatos	7.83 b	2.41 a	15.68 ab	0.24 b	1.65 a	0.06 a	0.033 a	0.045 b	0.67 a
Santiago	11.83 a	1.62 b	20.48 a	0.49 a	0.83 a	0.37 b	0.037 a	0.054 a	0.87 a

Stems

	Ca	Mg	K	Na	Fe	Mn	Cu	Zn	P
San Bartolo	9.24 a	2.90 b	30.98 b	0.201 b	0.022 b	0.028 ab	0.029 ab	0.058 a	1.65 a
Los Gatos	8.73 a	4.24 a	30.76 b	0.19 b	0.041 ab	0.025 b	0.028 b	0.054 b	1.95 a
Santiago	10.98 a	2.93 b	42.85 a	0.36 a	0.058 a	0.038 a	0.032 a	0.061 a	2.10 a

Leaves

	Ca	Mg	K	Na	Fe	Mn	Cu	Zn	P
San Bartolo	14.80 ab	8.77 a	71.14 a	0.771	0.06 b	0.055 a	0.036 b	0.064 b	2.97 a
Los Gatos	11.57 b	9.09 a	60.37 b	0.31 b	0.05 b	0.053 a	0.032 b	0.061 c	2.22 a
Santiago	18.92 a	9.25 a	72.66 a	0.45 ab	0.23 a	0.048 a	0.045 a	0.067 a	2.60 a

FW = fresh weight. DW = dry weight. *Each value represents the average of 3 or 10 data set. **Each value represents the average of 100 data set. Means followed by the same letter in each column are not significantly different (Tukey HSD; p = 0.05). For mineral content, each value represents the average of five data.

Mg (Table 3). The ANOVA of mineral content in leaves showed significant differences between populations for Ca, K, Na, Fe, Cu and Zn, while Mg, Mn and P did not show significant differences (Table 2). The leaves from plants collected in Santiago had higher values of Ca, K, Fe, Cu and Zn, while the leaves of plants from San Bartolo had higher values of Na (Table 3).

Relationship of environmental conditions and morphometric traits

Solar radiation of Santiago showed significant correlation ($r = -0.89$, $p = 0.04$) with root dry weight, decreasing as radiation increased. Evapotranspiration was correlated significantly with main stem dry weight in plants collected in Santiago ($r = -0.87$, $p = 0.05$), showing a decresing trend as evapotranspiration increased. The minimum temperature was correlated significantly with leaf average width in Los Gatos ($r = 0.88$, $p = 0.04$) showing an increasing trend as minimum temperature increased. In Santiago, the beginning of canopy decreased as precipitation increased; however, the correlation coefficient was not significant. Also leaf length showed increased as relative humidity increased though the correlation was not significant. In Los Gatos, the maximum leaf width decreased as evapotranspiration increased; however, the correlation was non-significant. Similarly, leaf area showed a trend to increase as minimum temperature increased; however, this correlation was not significant .

Discussion

The results of MANOVA confirms that there are morphological differences between the three sample populations of wild *Capsicum* plants at the sites studied of Los Gatos, San Bartolo and Santiago in the southern part of Peninsula of Baja California in some of the measured variables. This result strengthens the likelihood that the differences observed in the univariate analysis (ANOVA) performed on the variables, are real differences and not false positives or differences that occur simply by randomized chance [31].

The wild *Capsicum* plants collected in the three populations, showed two types of growth (erect or climbing) in agreement with Vázquez-Dávila [9], and Medina-Martínez *et al.* [32]. Villalón-Mendoza *et al.* [34] reported that some of the species which are associated with wild *Capsicum* plants are nurse plants such as *Helietta parvifolia*, *Diospyros palmeri*, *Acacia rigidula*, *Cordia boissieri*, *Leucophyllum texanum*, *Pithecellobium pallens*. They described that the main vegetation types associated with the *C. annuum* ecotypes in northeastern Mexico were thorny shrubs, followed by not thorny shrubs, forests of *Prosopis*, forest of oak-pine and medium size plants that are not thorny shrubs. Lack of abundant rains does not allow for growth of many vegetation types. This was demonstrated in the present study because the sample population with the least abundant variety of plants associated with wild *Capsicum* plants was in Los Gatos with the lowest precipitation, followed by San Bartolo with higher precipitation and Santiago with the highest.

In southern Arizona, U.S.A., where the vegetation is predominantly semi-desert grassland and mesquite woodland [35], Tewksbury *et al.* [36] found a greater association of wild plants of *C. annum* var. *aviculare* [Dierbach] D'arcy and Eshbaugh with seven species. These included *Celtis pallida* Torr., *Condalia globosa* Johnst., *Lycium andersonii* Gray, *Zizyphus obtusifolia* Hook, *Dodonea viscosa* Jacq., *Mimosa biuncifera* Benth., and *Prosopis velutina* Woot. They found that 78% of the plants were established under the canopies of fleshy-fruited shrub and tree species, while notably 58% of the *Capsicum* plants were found under just two species, desert hackberry (*Celtis pallida* Torr.) and netleaf hackberry (*Celtis reticulata* Torr.). A similar relationship has been documented for subtropical thorn scrubs in central Sonora, México, where wild *Capsicum* was 10 times more abundant under fleshy-fruited shrub [37]. In addition, Tewksbury *et al.* [36] also reported that wild *Capsicum* was not found in direct sunlight. Our study is in agreement with these authors, the distribution of *Capsicum* was determined by the micro environmental differences by different nurse-plants species or by nonrandom dispersal by *Capsicum* consumers. Specifically, our study showed that plants of wild *C. annuum* ecotypes in the populations were found to be associated to shrub or tree species, such as was reported by Laborde and Pozo [38] where they indicated that chili piquín was found under 1300 m.a.s.l., regularly in sites in association with shrubs plants where the environmental conditions such as humidity and luminosity are appropriate.

Leaf length of *Capsicum* plants from Santiago increased as relative humidity increased suggests that high morphometric variables are not necessarily related to environmental conditions, since leaf length values were higher in those plants from Los Gatos, where relative humidity was the lowest compared to the other sites. San Bartolo had high while Santiago intermediate values of relative humidity. In addition, root dry weight of plants collected in Santiago decreased as solar radiation increased. However, Santiago showed intermediate values of solar radiation compared to Los Gatos (the highest values) and San Bartolo (the lowest values). Our study showed that wild *Capsicum* plants were found under 700 m.a.s.l. which coincide with the reported by Laborde and Pozo [38] and Villalón-Mendoza *et al.* [34] where they stated that wild *Capsicum* species is commonly found with thorn scrubs at altitude limits at 600–800 m.a.s.l.

Medina-Martínez *et al.* [11] in a study of wild *C. annuum* in the northeast Mexico found that wild *Capsicum* can growth under high temperatures during summer season (up 40°C) with partial shade and were associated mainly with leguminous species. In a later study by also Medina-Martínez *et al.* [32] wild chili pepper populations were commonly found at intermountain and piedmont sites. They found that they grow mainly in vertisol and rendzins soil types, although less frequently in the later. The plants were found to be perennial with growth increasing with spring rains that produce fruits in summer and autumn to be commercialized by families in rural communities.

In the present study all wild *Capsicum* plants were found under shrubs and trees. The temperatures (20–30° C) of the autumn season (September, October and November) in the zone were conducive to wild *Capsicum* plants because flowering and seedling development improved and fruits production increased. The results are in agreement with the evidence showed by Heiser and Pickersgill [39] where they described that wild chilies identified as *Capsicum annuum* var. *glabriusculum*, commonly known as "chiltepines" are widely distributed in Mexico, especially under tree species of tropical deciduous forest, also it is possible found around field crops and to roadsides. Medina-Martínez *et al.* [32] stated that *C. annuum* var. *aviculare* grew favorably under clay-loam texture soils with pH of 7.5 and electrical conductivities between 0.5-1.0, with high organic matter content (3.5% on average) containing elements such as nitrogen, phosphorus and potassium. Our study showed that wild *Capsicum* plants were found in a range of temperature among 22 to 23° C, with maximum of 33°C, minimum of 13°C and average of 22.5° C which coincide with those reported by Medina-Martínez *et al.* [33].

Capsicum species occur in a wide range of different habitats with an average day temperature between 7 and 29°C, an annual precipitation between 300 and 4600 mmand a soil pH between 4.3 and 8.7 [40]. In general, *Capsicum* species are cold sensitive and grow best in well-drained, sandy or silt-loam soil [40].

In the present study, 70% of plants had significant morphometric differences between populations, while in fruits, 50% showed significant differences. It is important to note, that other studies of wild *Capsicum* have reported a high variability of morphometric traits such as main stem and foliage characteristics where the foliar covering or diameter was found to have a range of 0.60-1.05 m, in the plant height of 30–98 cm, in the leaves length of 1.9-4.2 cm and in leaf width of 1.1-2.3 cm and about the fruits production, high variability was appreciated in the precocity degree, fruit length and width and yield of fruits per plant [41]. The fruit length range of

1.1-2.5 cm and the fruit width was 0.5 at 1.0 cm [41]. In the same sense, Medina-Martínez *et al.* [32] reported a high variability between morphometric traits in chili piquín (*C. annunm* var. *aviculare*) with an average of 2.8 cm in leaf width, plant height of 2.0 m, length of petioles of 5–20 mm, fruit peduncle length of 1–2 cm and diameter of 0.5 mm, the fruit is a berry from 8–10 mm of length and 5–8 mm of width, with yellowish brown seeds of 2.5 mm of length. Because the fruit or pod, technically a berry, is the commodity of the pepper plant, fruit morphology flavor and pungency are the characteristics of most economic importance within the genus. A tremendous wealth of genetic variation is known with respect to fruit traits such as size, shape, color, and flavor, resulting in more than 50 commercially recognized pod types. The major pod types are described by Bosland [42], Andrews [43] and by Paran et al. [44].

Other studies in wild populations of *C. annuum* from northwest Mexico have found a high variation in morphometric traits such as fruit length (range 0.30-0.98 cm) and seed number (range 1–34) in same populations [16]. In other latitudes of the world, similar results have been reported. Shrilekha Misra *et al.* [45] reported that in 38 accessions of *C. annuum* collected from diverse locations in India, divergence of pooled characters ranged from 41–111 cm plant height, 6.62-45.39 cm^2 leaf surface area, 1.45-9.96 cm fruit length, 0.65-1.84 cm fruit diameter, 2.64-27.40 cm^2 fruit surface area, 0.36-4.447 mg fruit fresh weight and 0.14-0.96 mg fruit dry weight. Hernández-Verdugo *et al.* [46] reported high variability in 11 morphometric traits, except for main stem diameter which showed values between 1.1-1.8 cm in seven wild *Capsicum* populations in different habitats in Sinaloa, México. The measured morphometric traits were plant height (95–181 cm), plant width (68–175 cm), main stem length (21–61 cm), leaf width (1.4-3.3 cm), leaf length (3.5-5.6 cm), pedicel length (2.3-2.8 cm), fruit width (5.5-7.7 mm), fruit length (5.6-7.6 mm), number of seeds per fruit (11–17) and seed weight (1.9-2.7 mg) [46]. Some traits measured in the present study are between the range values with those found by Hernández-Verdugo *et al.* [46].

The results of our study show high morphometric variability between the populations of wild *C. annuum* in three sites near two reserve biospheres in Baja California Sur, Mexico. The phenotypic diversity and undoubtedly the genetic diversity of wild *Capsicum* in each of these populations are affected by geography, climate, ecology and human intervention. The trend of stem dry weight to decrease as evapotranspiration increased in those plants of Santiago suggests that evapotranspiration is an important climatic variable in the growth, production and yield of wild *Capsicum*. Higher evapotranspiration was found for plants measured in Los Gatos, followed by Santiago and San Bartolo. The main stem dry weight was higher in San

Bartolo plants followed by Santiago which showed the lowest values in those plants collected in Los Gatos but this sample population showed the higher values of evapotranspiration. Also, the maximum leaf width showed a trend to decrease as evapotranspiration increase in those plants collected in Los Gatos. According to Brown [47] an improved understanding of climate effects on the current structure of genetic diversity and morphometric variation within the species is important for efficient germplasm conservation and use.

In the present study, the significant differences found in population site morphometrics could be related to environmental condition(s) where the wild *Capsicum* populations are found. For example, the plants collected in two populations (San Bartolo and Santiago) near La Laguna reserve biosphere showed higher values in the majority of morphometric traits in both plants and fruits compared to Los Gatos probably because these populations are close to the Tropic of Cancer where the precipitation is higher. The Los Gatos population is close to the El Vizcaino reserve biosphere. Nevertheless, in spite of the lower amount of precipitation the wild plants collected in Los Gatos showed more vigor because length, area, average width and maximum leaf width were higher respect with respect to San Bartolo and Santiago plants. Leaf average width in those plants collected in Los Gatos increased as minimum temperature increased. Similarly, the leaf area showed a trend to increase as minimum temperature increase in Los Gatos. The results of both variables show that the range of temperature for better growth of this species is when temperature is higher than 13° C. Also, these differences could be an evidence that ecotype from Los Gatos differ genetically from the ecotypes collected in San Bartolo and Santiago; however, more studies related to genetic, physiology, botanical, and others topics are required. Evidently the differences in environmental conditions such as temperature, nutrient availability and altitude have an influence on plant growth [48]. In the present study, the micro-environmental conditions in the three different sample populations, such as temperature, photoperiod, light quality and nutrient availability suggest that they may be sufficiently distinct to have caused the observed differences in morphometric traits in both plants and fruits, also the mineral content of roots, stems and leaves of wild *Capsicum* plants may also pay a role. The mineral content in roots, stems and leaves is an important variable that influences the plant response under different environmental conditions. Our study showed that plants from Santiago had the higher values of Ca, K, Cu, Zn and P in roots, stems and leaves, higher values of Na in roots and stems, Fe in stems and leaves and Mg in leaves. Although plants from Santiago showed good nutrition condition, they did not necessarily have higher values of morphometric traits in both plants and

fruits; however, these plants showed higher values of main stem diameter and root dry weight, also in some morphometric traits in fruits such as peduncle length, fruit length and pulp/seeds ratio. Recently, research regarding the identification of hot pepper cultivars containing low Cadmium levels after growing on contaminated soil [49] and protective role of Selenium on pepper exposed to Cadmium stress during reproductive stage [50] have been reported. Cadmium and other non-essential and highly toxic elements to plants, can pose a human health risk throughout the food chain. Future work will be carried out to determine whether these cultivars are low or high Cd accumulation plants. This is essential if this crop is developed in the future as a commercial product for human consumption, since low Cd cultivars are preferred for human health reasons.

Conclusions

This is the first study evaluating the ecology and morphometric traits of both plants and fruits of wild *C. annuum* in Baja California Sur, Mexico. The results provide useful information regarding morphometric variation between wild *Capsicum* populations. This could prove valuable to future decision processes involved in the management and preservation of germplasm and genetic resources. The wild relatives of cultivated *C. annuum* are a valuable genetic resource that needs to be conserved. Probably, the populations of wild relatives of chili here in the Peninsula of Baja Calironia due to its geographic isolation maintain high levels of genetic, ecological variability, and are potentially useful genes for agriculture. Future studies are nneded that will evaluate *C. annuum* in the study area to investigate genetic differentiation for upcoming plant breeding efforts with *Capsicum*. There remain some areas of interest in the Peninsula that should be visited in the future, for example, Sierra of La Giganta in front of Loreto City, Sierra of Mulegé in front of Mulegé town, and other sites of the Region of the Cape in the southern part of the Baja California Peninsula. These areas should be a target for future data collection and investigation, including ethnobotanical studies, providing a seed sample bank that will be publicly available for research in plant improvement and for subsequent use in an inquiry into the domestication of *C. annuum*.

Methods

Ethics statement

The research conducted herein did not involve measurements with humans or animals. The study site is not considered a protected area. No protected or endangered or species were used in the course of carrying out this study, however, some special permissions need to be get at the Procuraduría Federal de Protección al Ambiente

(PROFEPA) at La Paz, Baja California Sur, México. *Capsicum annuum* used in the present study is not considered an endangered species and their use therefore had negligible effects on broader ecosystem functioning.

Sampling populations

Three populations (Figure 1) were located in three sites along Baja California Sur (B.C.S.), Mexico to identify wild *C. annuum* ecotypes. The three sample wild populations were selected based on information provided by local inhabitants in each municipality of Baja California Sur. This data of wild *Capsicum* plants was assessed in extensive field trips and respective interviews with communities and farmers located in wild areas, i.e. in Mulegé municipality, the population of Santa Lucia mountain with more abundance in wild *Capsicum* plants is the area called Los Gatos (The Cats) and surroundings. The sample populations were positioned geographically using a global positioning system (Garmin GPS Map 60Cx). One population was situated in the first site, which was in the municipality of Mulegé (Los Gatos Ranch) near the limit area of biosphere reserve El

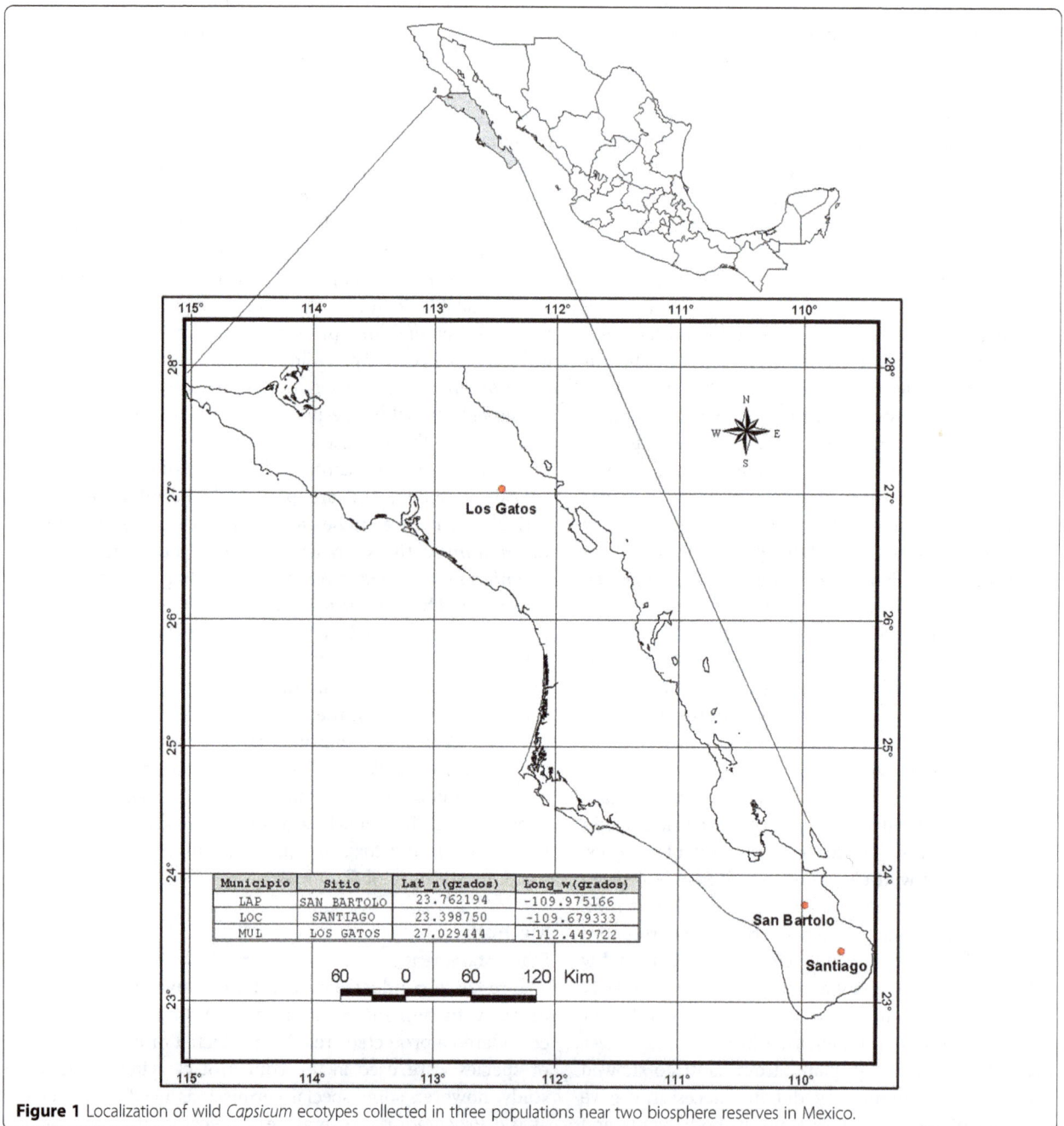

Municipio	Sitio	Lat_n(grados)	Long_w(grados)
LAP	SAN BARTOLO	23.762194	-109.975166
LOC	SANTIAGO	23.398750	-109.679333
MUL	LOS GATOS	27.029444	-112.449722

Figure 1 Localization of wild *Capsicum* ecotypes collected in three populations near two biosphere reserves in Mexico.

Vizcaino, B.C.S., México. The second population was located in a second site in the municipality of La Paz (San Bartolo town) and the third population of the third site was located in the municipality of Los Cabos (Santiago town) both near the area limit of the biosphere reserve La Laguna, B.C.S., México. Los Gatos is located in a semiarid zone of Baja California Sur, northwest of Mexico (27°01′46″ N, 112°26′59″ W), 680 meter above sea level (masl). Los Gatos is a wild *Capsicum* population surrounded by some cattle ranches, located in a small range just behind Santa Rosalía, B.C.S., at Santa Lucia Mountain, which joins the Sierra of Guadalupe to the south. This wild *Capsicum* population is located around the limits of El Vizcaino biosphere reserve, close to the highest hill called La Bandera. Below the Pacific slopes of the mid-peninsular range, the Central Desert stretches from 30°N to 26°N and encompasses the Vizcaino Desert and, to the south of the Madgalena Plain. The soils of this population are shallow, of recent formation and high rate of erosion, characterized as lithosol soils, with low organic matter, have no structure to be composed of unconsolidated material with high sand content. Are set on hills and mountain areas, where the type of vegetation is found of sarcocaule scrub. Are coarse textured and are associated with eutric regosols. San Bartolo is located in a subtropical zone of Baja California Sur, northwest of Mexico (23°45′43.9″ N, 109°58′30.6″ W), 526 masl. The wild *Capsicum* population of San Bartolo is located around the limits of La Laguna biosphere reserve, near Sierra of La Laguna lies below La Paz in the Cape Region. The range is called La Laguna after a mountain meadow that, according to natives, was once a lake. This wild *Capsicum* population is close to Arroyo (Dry River) of San Bartolo that it is large. This population is located in the east face of La Laguna Mountain, with high precipitation, with deep canyons and luxuriant growth found on many of these gradual eastern slopes. The soils of this population are predominantly eutric cambisol, a weakly developed mineral soils in unconsolidated materials, soil management affects moisture-holding capacity, the highest moisture contents is found in undistributed soils, which are related to low organic matter contents, medium to low porosity and low values of structural stability. Santiago is located in a subtropical zone of Baja California Sur, northwest of Mexico (23°23′55.5″ N, 109°40′45.6″ W), 226 masl. The wild *Capsicum* population of San Santiago is located around the limits of La Laguna biosphere reserve, near Sierra de La Laguna lies below La Paz in the Cape Region. The range is called La Laguna after a mountain meadow that, according to natives, was once a lake. This wild *Capsicum* population of Santiago is close to Arroyo (Dry River) of San Bernardo and Arroyo of San Dionisio, both are large. This population is located in the

southeast face of La Laguna Mountain, is steep, with deep canyons and luxuriant growth found on many of the more gradual eastern slopes. The soils of this population are dominated by eutric cambisol that with natural vegetation had the highest moisture-holding capacity, the highest rates of infiltration are found for natural vegetation soils, structural profile and porous system are more stable in unchanged soils. Figures 2, 3 and 4 shows the environmental conditions such maximum, minimum and average temperature (°C), precipitation (mm), evapotranspiration (mm), solar radiation (w m^{-2}) and relative humidity (%) of the three sample populations in a range of 73 years from 1939 to 2013 along January to December (monthly average). The meteorological observations were obtained during the study from an automated weather stations located at the study areas which are property of the National Institute of Forestry, Agricultural and Livestock Research (INIFAP) and from the National Weather Service (SMN) both institutions of the Secretary of Agriculture, Livestock, Rural Development, Fisheries and Food (SAGARPA) with coverture in all regions of Mexico.

Vegetation associated to wild *Capsicum* (*in-situ*)

In each sample population, two rectangles of 50 × 20 m (1000 m^2) were traced and each *Capsicum* plant were counted and identified in each rectangle. One square of 4 × 4 m (16 m^2) was traced around each *Capsicum* plant found and the vegetation associated was identified the family, common and scientific names.

Morphometric traits measured in plants (*in-situ*)

In each sample population, five wild *Capsicum* plants were selected completely randomized and the height (cm), plant coverage (m^2), main stem diameter (mm), as well as the height of the beginning of canopy (cm) were measured. We collected only five plants of each sample population since the Procuraduría Federal of Protección to the Ambiente (PROFEPA) authorized only the collection of a limited number of wild *Capsicum* plants and fruits. This species is perennial, however annual growth change yearly, thus this first study will be important for providing the baseline for future growth studies of this plant species. Plant coverage, plant height and height of beginning of canopy were measured using a metric tape of 5 m and main stem diameter was measured using a digital caliper (General No 143, General Tools®, Manufacturing Co., Inc., New York, USA) at a plant height of 0.20, 0.40 and 0.60 m and the result was averaged. The growth types of all *Capsicum* plants found in each sample population were recorded. The growth type was identified as two types, as erect (shrub type) or climbing (vine type).

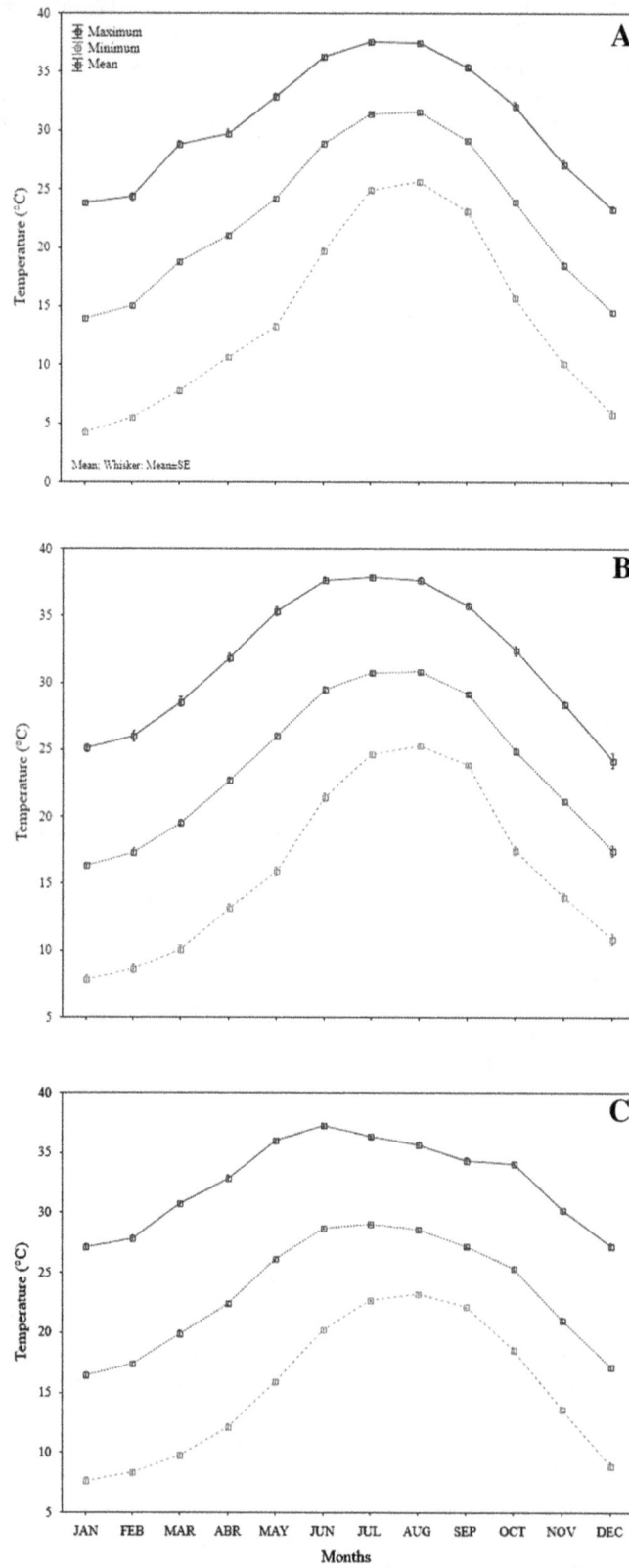

Figure 2 Maximum, minimum and mean temperature of three populations, Los Gatos **(A)**, San Bartolo **(B)** and Santiago **(C)** of wild *Capsicum* ecotypes collected near two biosphere reserves in Mexico.

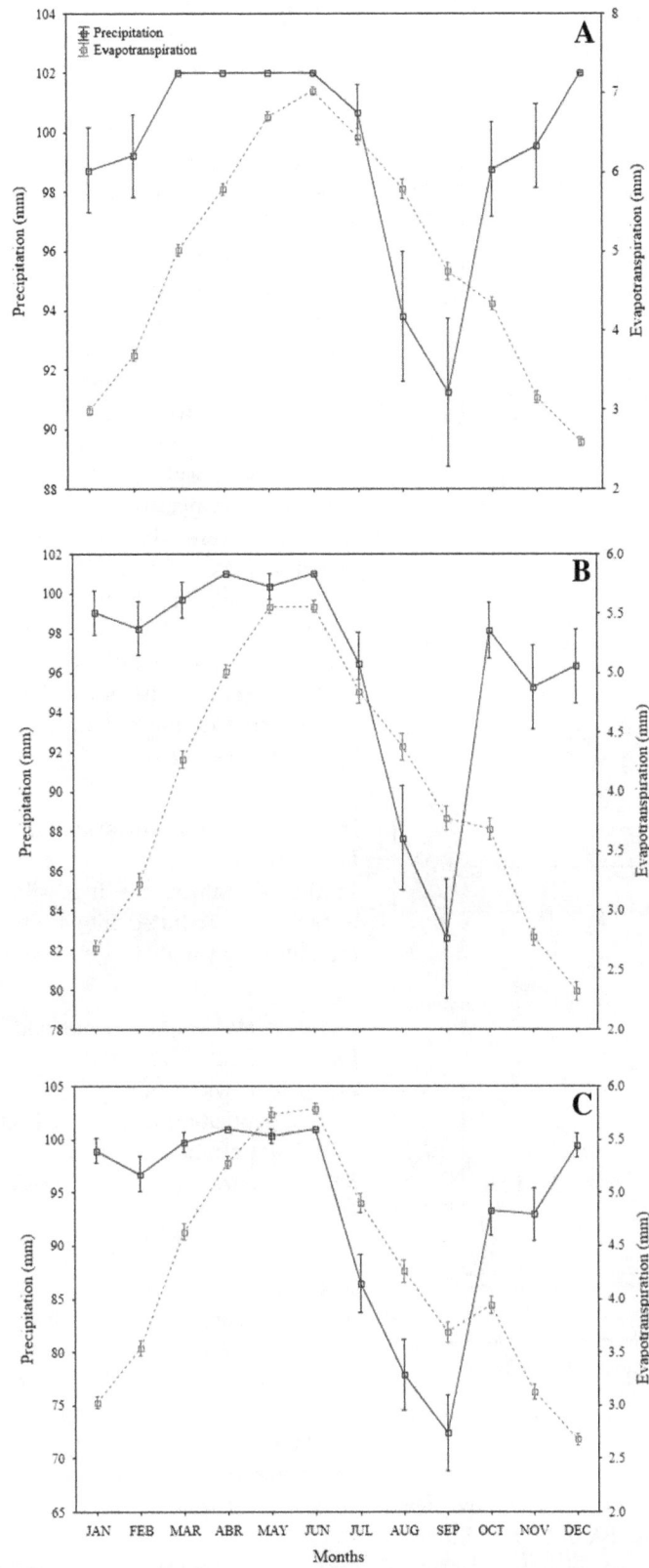

Figure 3 Precipitation and evapotranspiration of three populations, Los Gatos **(A)**, San Bartolo **(B)** and Santiago **(C)** of wild *Capsicum* ecotypes collected near two biosphere reserves in Mexico.

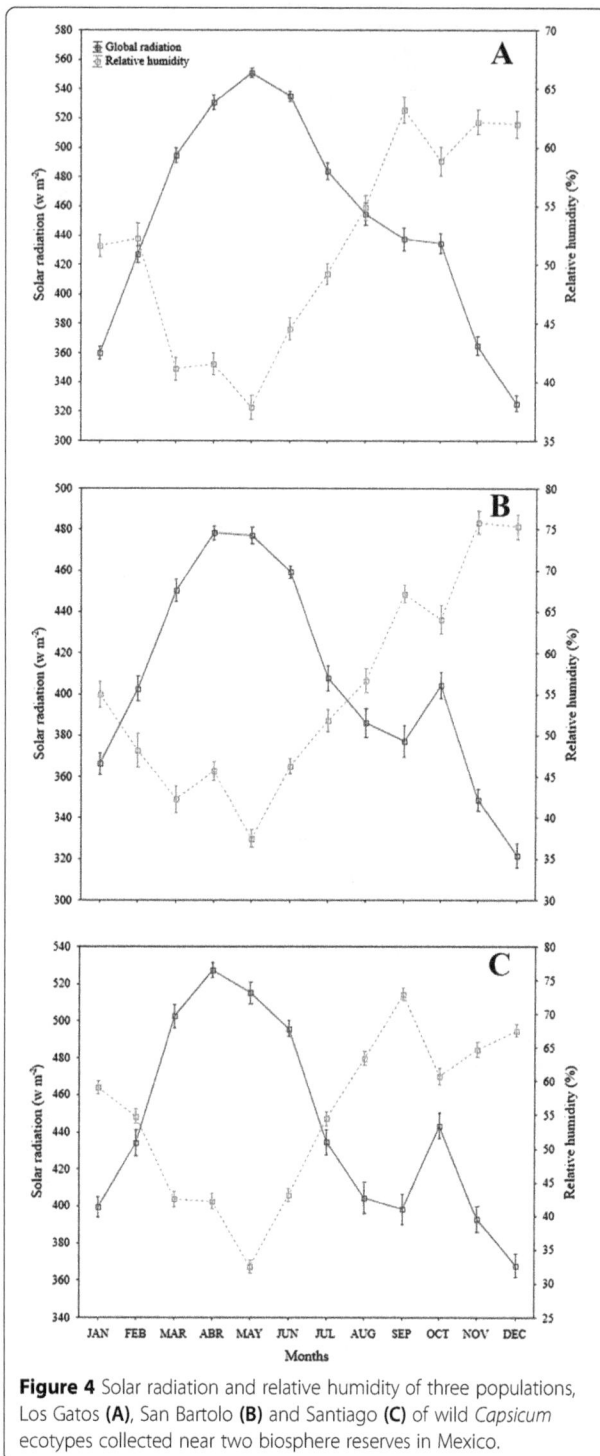

Figure 4 Solar radiation and relative humidity of three populations, Los Gatos **(A)**, San Bartolo **(B)** and Santiago **(C)** of wild *Capsicum* ecotypes collected near two biosphere reserves in Mexico.

Plants and fruits collection

Previously to realizing the collection, a specific permission needs to be granted by PROFEPA in La Paz, Mexico in order to collect wild *Capsicum* plants and fruits. These plants at present not considered endangered or protected species. However, for future plants and fruits collection of *Capsicum* and other species in the sample populations near both biosphere reserves will be sampled after attaining appropriate permissions contacting to Mr. Leonel Valerio Castro Santana, Federal Officer of PROFEPA in Baja California Sur. In each sample population, the five plants of wild *Capsicum* selected were collected and completely randomized (including roots). These plants were used for morphometric measurements. Each plant was considered as a replication. The collected plants were introduced in paper bags, labelled, stored in cardboard containers and moved to the laboratory of plant physiology at Centro de Investigaciones Biológicas del Noroeste, S.C. (CIBNOR®) at La Paz, México. Before the collection of each plant, the total fruits per plant were harvested and placed in paper bags, labelled and stored in a cardboard container and moved to the laboratory. At the same time, 400 mature fruits from different plants (without collecting) at each sample population were collected, introduced in paper bags, labelled and moved to the laboratory. Each group of 100 fruits was considered as one replication. We collected 400 mature fruits because PROFEPA authorized only the collection of this quantity of wild *Capsicum* fruits based on the criteria of the normativity for wild vegetation in Mexico considering criteria for conservation and management of resources.

Morphometric traits measured in collected plants and fruits (laboratory)

In the laboratory, the five wild *Capsicum* plants collected were separated into roots, leaves and stems and the following variables were measured:

Leaf area, leaf length, average and maximum width of leaf

Leaf area (cm^2), leaf length (cm), average (cm) and maximum (cm) width of leaf of each collected plant of each sample population that was collected was measured with a Li-Cor portable leaf area meter (Li-Cor®, modelo-Li-3000A, series Pam 1701, Li-Cor® Lincoln, Nebraska, USA).

Leaves, roots and stems dry weights

All leaves, roots and stems dry weights from each plant collected in each sample population were recorded. The leaves, roots and stems were placed in a pre-heated oven (Shel-Lab®, model Fx-5, serie-1000203) at 80°C, until constant weight, in order to obtain leaves (g), roots (g) and stems (g) dry weights which were obtained using a conventional scale (Ohaus®, model CT600-S, USA, series 18939).

In the laboratory, the 400 fruits harvested from each sample population and those fruits collected *in-situ* were separated into peduncle, seeds and fruit pulp and the following variables were measured:

Number of fruits per plant, peduncle length and fruit average fresh and dry weights

Each fruit collected from each collected plant were counted and recorded. The peduncle of each fruit was separated from the fruit and the length (cm) was recorded using a digital caliper (General No 143, General Tools®, Manufacturing Co., Inc., New York, USA.). Average fresh weight of fruit (g) was determined using a conventional scale (Ohaus®, model CT600-S, USA, series 18939) and average dry weight of fruit (g) were obtained when each group of fruits from each plant were placed in a pre-heated oven (Shel-Lab®, model Fx-5, serie-1000203) at 80°C, until constant weight.

Number of seeds per fruit, fruit length and width

The 400 mature fruits collected from each sample population were used to determine the number of seeds per fruit, length (mm) and width (mm) of fruit which were measured using a digital caliper (General No 143, General Tools®, Manufacturing Co., Inc., New York, USA.).

100 fruits dry weight, seeds and pulp dry weight of 100 fruits, 1000 seeds dry weight and pulp/seeds ratio

The 400 mature fruits collected from each sample population were separated in four groups of 100 fruits and fruits dry weight (g), seeds (g) and pulp (g) dry weight, pulp/seeds ratio and 1000 seeds dry weight (g) were measured. The dry weight were obtained when the fruits or seeds were introduced in a pre-heated oven (Shel-Lab®, model Fx-5, serie-1000203) at 80°C, until constant weight.

Mineral content of roots, stems and leaves

The mineral content in roots, stems and leaves is an important variable that influences the plant response under different environmental conditions. All roots, leaves and stems after being separated from the main plant were rinsed by dipping three times for a few seconds in distilled-deionised water before measuring dry weights. Separately roots, leaves and stems dried tissue were finely ground in a blender (Braun® 4–041 Model KSM-2) for mineral analysis. The Na, Ca, Mg, Mn, Fe, Cu, Zn, and K (all in g kg^{-1} dry-weight) content was determined by atomic absorption spectrophotometer (Shimadzu AA–660, Shimadzu®, Kyoto, Japan) after digestion with H_2SO_4, HNO_3, and $HClO_4$. Phosphorous (g kg^{-1} dry-weight) was estimated colorimetrically as phosphomolybdate blue complex method at 660 nm from the same extract.

Statistical analysis

Bartlett's test was performed on the data to test the homogeneity of variance. Data were analyzed using a fit model using a standard least squares means personality function and univariate and multivariate analysis of variance (ANOVA and MANOVA). All plants and fruits variables were analyzed for one way of classification, being sample population the study factor. The least significant differences were calculated using Tukey's HSD test ($p \leq$ 0.05) when the analysis of variance was significative. As a wild population, the coefficient of variation for each variable was considered. In all cases, differences among means were considered significant at $p \leq 0.05$. Single and multiple Pearson's correlation coefficients (r) at 95% confidence limits for independent variables (environmental conditions) and dependent variables measured in plants, fruits and seeds was determined. All analyses were done with Statistica software program v. 10.0 for Windows.

Availability of supporting data

The authors confirm that all data underlying the findings are fully available without restriction. All the relevant data that is needed to replicate this study and to draw the conclusions for this study is within the paper.

Competing interest

The authors have declared that they have no competing interests exist. The research was conducted in the absence of any commercial or financial relationships that could be construed as a potential conflict of interest.

Authors' contributions

Conceived and designed the experiments: BMA and ANG. Performed the experiments: BMA, EORP and ETD. Analyzed the data: BMA, MVCM and LGHM. Contributed reagents/materials/analysis tools/publication costs: EORP, ANG and ETD. Wrote, edited and revised the paper: BMA, MVCM and EORP. All authors read and approved the final manuscript. All authors agree that BMA to be accountable for all aspects of the work in ensuring that questions related to the accuracy or integrity of any part of the work are appropriately investigated and resolved.

Acknowledgments

This research was supported by grant AGROT1 from Centro de Investigaciones Biológicas del Noroeste, S.C. (CIBNOR®), and grant CB-2009-01 (0134460) from SEP-CONACYT. The first author thanks the support by project CONACYT number 245853 in the modality of short visits within the framework of the national program of sabbatical stay; CONACYT number 224216 in the modality of infrastructure consolidation The authors greatly appreciate the technical assistance of Álvaro González-Michel, Mario Benson-Rosas, Lidia Hirales-Lucero, Carmen Mercado-Guido, Margarito Rodríguez-Álvarez and Reymundo Domínguez-Cadena.

Author details

[1]Centro de Investigaciones Biológicas del Noroeste, S.C. La Paz, La Paz, Baja California Sur, México. [2]Universidad de Sonora, Hermosillo, Sonora, México.

References

1. O'Sullivan L, Jiwan MA, Daly T, O'Brien MN, Aherne SA. Bioaccessibility, uptake, and transport of carotenoids from peppers (Capsicum spp.) using the coupled in vitro digestion and human intestinal Caco-2 cell model. Journal of Agricultural. Food Chem. 2010;58:5374–9.
2. Pickersgill B. Relationships between weedy and cultivated forms in some species of chilli peppers (genus Capsicum). Evolution. 1971;25:683–91.
3. D'Arcy WG, Eshbaugh WH. New world peppers (Capsicum-Solanaceae) of north of Colombia. Baileya. 1974;19:93–103.

4. Bosland PW, Votava EJ. Peppers: vegetable and spice Capsicums. Wallingford, United Kingdom: CAB International; 2000. p. 204.

5. Hernández-Verdugo S, Dávila P, Oyama K. Síntesis del conocimiento taxonómico, origen y domesticación del género Capsicum. Bol Soc Bot Méx. 1999;64:65–84.

6. Pickersgill B, Heiser Jr CB, McNeill J. Numeral taxonomic studies on variation and domestication in some species of Capsicum. In: Hawkes JG, Lester RN, Skelding AD, editors. The biology and taxonomy of the Solanaceae. London: Academic; 1979. p. 679–700.

7. Votava EJ, Nabhan GP, Bosland PW. Genetic diversity and similarity revealed via molecular analysis among and within an in situ population and ex situ accessions of chiltepín (Capsicum annuum var. glabriusculum). Conserv Genet. 2002;3:123–9.

8. Pozo-Campodónico O, Montes HS, Redondo JE. Chile (Capsicum spp.). In: Ortega PR, Palomino HG, Castillo GF, González HVA, Livera MM, editors. Avances en el estudio de los recursos fitogenéticos de México, Sociedad Mexicana de Fitogenética. D.F.: A.C. Ciudad de México; 1991. p. 217–38.

9. Vázquez-Dávila MA. El amash y el pistoqué: un ejemplo de la etnoecología de los chontales de Tabasco. Méx Etnoecológica. 1996;3:59–63.

10. Almanza-Enríquez JG. El chile piquín (C. annuum L. var. avicularе Dierb. Estudio etnobotánico, biología y productividad. Universidad Autónoma de Nuevo León, Monterrey, N.L., México: Tesis de Licenciatura de la Facultad de Ciencias Biológicas; 1993. p. 72.

11. Medina-Martínez T, Villalón MH, Lara VM, Gaona GG, Trejo HL, Cardona EA. Informe técnico de proyecto 95/111. Universidad Autónoma de Tamaulipas SIREYES: Instituto de Ecología y Alimentos; 2000.

12. Medina-Martínez T. Reportaje. Ciudad Victoria, Tamaulipas, México: Periódico El Gráfico; 1997. p. 12.

13. González V. Reportaje sobre chile piquín. Cd. Victoria, Tamaulipas, México: Periódico "El Diario de Cd. Victoria"; 1999. p. 6-A.

14. de la León Luz JL, Domínguez-Cadena R, Domínguez-León M, Coria R, editors. Flora iconográfica de Baja California Sur II. S.C. La Paz, B.C.S., México: Centro de Investigaciones Biológicas del Noroeste; 2014. p. 278.

15. Encarnación-Dimayuga R. Medicina tradicional y popular de Baja California Sur. B.C.S., México: Secretaría de Educación Pública-Universidad Autónoma de Baja California Sur. La Paz; 1996. 122 p.

16. Hernández-Verdugo S, Luna-Reyes R, Oyama K. Genetic structure and differentiation of wild and domesticated populations of Capsicum annuum (Solanaceae) from Mexico. Plant Syst Evol. 2001;226:129–42.

17. Kraft KH, Luna-Ruiz JJ, Gepts P. A new collection of wild populations of Capsicum in Mexico and the southern United States. Genet Resour Crop Evol. 2013;60:225–32.

18. Gonzalez M, Bosland PW. Strategies for stemming genetic erosion of Capsicum germplasm in the Americas. Diversity. 1991;7:52–3.

19. Torres-Pacheco I, Garzón-Tiznado JA, Brown JK, Becerra-Flora A, Rivera-Bustamante RF. Detection and distribution of geminiviruses in Mexico and the southern United States. Phytopathology. 1996;86:1186–92.

20. Almanza-Enríquez JG, Maiti RK, Foroughbakhch PR, Cárdenas-Ávila ML, Núñez-González MA, Moreno-Limón S, et al. Bromatología del chile piquín (Capsicum annuum L. var. avicularе (Dierb.) D. & E.). Querétaro, Querétaro, México: Resúmenes XV Congreso Mexicano de Botánica; 2001.

21. Boukema RW. Allelism of genes controlling resistance to TMV in Capsicum. Euphytica. 1980;29:433–9.

22. Hernández-Verdugo S, Guevara-González RG, Rivera-Bustamente RF, Oyama K. Screening wild plants of Capsicum annuum for resistance to pepper huasteco virus (PHV): presence of viral DNA and differentiation among populations. Euphytica. 2001;122:31–6.

23. Hawkes JG. The diversity of crop plants. Cambridge, Massachusetts, USA: Harvard University Press; 1983. p. 184.

24. Burdon JJ, Jarosz AM. Wild relatives as sources of disease resistance. In: Brown AHD, Frankel OH, Marshall DR, Williams JT, editors. The use of plant genetic resources. Cambridge.: Cambridge University Press; 1989. p. 280–96.

25. Maiti RK, Almanza JG, Gutiérrez JL. Aspectos biológicos y productividad del chile piquín (C. annuum var. avicularе Dierb.) Resumen. Memorias del XV Congreso de Fitogenética. A.C. Monterrey, Nuevo, León, México: Sociedad Mexicana de Fitogenética; 1994. p. 262.

26. Almanza-Enríquez JG. Estudios ecofisiológicos, métodos de propagación y productividad del "chile piquín" (Capsicum annuum L. var. avicularе Dierb.) D. & E, Tesis de Maestría. Universidad Autónoma de Nuevo León, Monterrey, N.L., México: Facultad de Ciencias Biológicas; 1998.

27. Harlan JR. Crops and man. 2nd ed. Crop Science Society of America, Inc., Madison, Wisconsin, USA: American Society of Agronomy, Inc.; 1992. 283 p.

28. Casas A, Barbera G. Mesoamerican domestication and diffusion. In: Nobel PS, editor. Cacti: biology and uses. Los Angeles.: The University of California Press; 2002. p. 143–62.

29. Doebley J. Isozymic evidence and evolution of crop plants. In: Soltis ED, Soltis PM, editors. Isozymes in plant biology. Portland, Oregon.: Dioscorides; 1989. p. 165–91.

30. Oyama K, Hernández-Verdugo S, Sánchez C, González-Rodríguez A, Sánchez-Peña P, Garzón-Tiznado JA, et al. Genetic structure of wild and domesticated populations of Capsicum annuum (Solanaceae) from northwestern Mexico analyzed by RAPDs. Genet Resour Crop Evol. 2006;53:553–62.

31. Johnson DE. Applied multivariate methods for data analysis. Brooks Cole Publishing Company and International Thompson Publishing Company: Kansas State University; 1998. p. 566.

32. Medina-Martínez T, Villalón Mendoza H, Lara-Villalón M, Gaona-García G, Trejo-Hernández L, Cardona-Estrada A. El chile piquín del Noreste de México. Ciudad Victoria, Tamaulipas, México: Folleto técnico. Universidad Autónoma de Tamaulipas-Consejo Nacional de Ciencia y Tecnología-Universidad Autónoma de Nuevo León; 2000. p. 20.

33. Medina-Martínez T, Villalón MH, Carreón PA, Lara VM, Cardona EA, Gaona GG, Trejo HL. Chili piquin (Capsicum annuum) population and handling agro-forestry study in northeastern Mexico. In: Proceedings of the 16th International Pepper Conference. Tampico, Tamaulipas, México: Universidad Autónoma de Tamaulipas; 2002. p. 13–14.

34. Villalón-Mendoza H, Medina-Martínez T, Rodríguez del Bosque JL, Pozo-Campodónico O, Garza-Ocañas F, López de León R, Soto-Ramos JM, Lara-Villalón M, López-Aguillón R. Wild chilli pepper: a potential forest resource for sustainable management in northeastern México. In: Proceedings of the 16 th International Pepper Conference. Tampico, Tamaulipas, México: Universidad Autónoma de Tamaulipas; 2002. p. 15–16.

35. Brown DE. Biotic communities of the American South-west-United States and Mexico. Dessert Plants. 1982;41:1–342.

36. Tewksbury JJ, Nabhan GP, Norman D, Suzán H, Tuxill J, Donovan J. In situ conservation of wild chiles and their biotic associates. Conserv Biol. 1999;13:98–107.

37. Nabhan GP. Nurse-plant ecology of threatened plants in the U.S./ México borderlands. Pp. 377–383. In: Elias TS, Nelson J, editors. Conservation and management of rare endangered plants: proceedings of a California conference on the conservation and management of rare endangered plants. Sacramento, California, USA.: California Native Plant Society; 1987. p. 377–84.

38. Laborde CJA, Pozo CO. Presente y pasado del chile en México. México: Publicación especial No. 85. INIA; 1982. p. 80.

39. Heiser CB, Pickersgill B. Names for the bird peppers (Capsicum-Solanaceae). Baileya. 1975;19:151–3.

40. Simon JE, Chadwick AF, Craker LE. Herbs: an indexed bibliography, 1971–1980: the scientific literature on selected herbs and aromatic and medicinal plants of the temperate zone. Hamden: Connecticut, Archon Books; 1984. p. 770.

41. Acosta-Rodríguez GF, Luján-Favela M. Selection and characterization of plants in two populations of piquín pepper (Capsicum annuum L.) in Delicias, Chihuahua. In: Proceedings of the 16 th International Pepper Conference. Tampico, Tamaulipas, México: Universidad Autónoma de Tamaulipas; 2002. p. 16–17.

42. Bosland PW. Chiles: a diverse crop. Hort Technol. 1992;2:7–10.

43. Andrews J. The domesticated Capsicums. Newth ed. Austin, Texas, USA: University of Texas Press; 1995.

44. Paran I, Ben-Chaim A, Byoung-Cheorl K, Capsicums JM. Chapter 7. Genome mapping and molecular breeding in plants. In: Kole C, editor. Vegetables, vol. 5. Berlin Heidelberg: Springer; 2007. p. 209–26.

45. Misra S, Lal RK, Darokar MP, Khanuja SPS. Genetic variability in germplasm accessions of Capsicum annuum L. Am J Plant Sci. 2011;2:629–35.

46. Hernández-Verdugo S, Guevara-González RG, Rivera-Bustamante RF, Vázquez-Yanes C, Oyama K. Los parientes silvestres del chile (Capsicum spp.) como recursos genéticos. Bol Soc Bot Méx. 1998;62:171–81.

47. Brown AHD. Core collections. A practical approach to genetic resources management. Genome. 1989;31:818–24.

Baseline study of morphometric traits of wild Capsicum annuum growing near two biosphere reserves...

163

48. Grime JP, Hunt R. Relative growth rate: Its rage and adaptive significance in a local flora. J Ecol. 1975;63:393–422.

49. Xin J, Huang B, Liu A, Zhou W, Liao K. Identification of hot pepper cultivars containing low Cd levels after growing on contaminated soil: uptake and redistribution to the edible plant parts. Plant Soil. 2013;373:415–25. doi:10.1007/s11104-013-1805-y.

50. Mozafariyan M, Shekari L, Hawrylak-Nowak B, Kamelmanesh MM. Protective role of Selenium on pepper exposed to Cadmium stress during reproductive stage. Biol Trace Elem Res. 2014;160:97–107. doi:10.1007/s12011-014-0028-2.

Direct Contact – Sorptive Tape Extraction coupled with Gas Chromatography – Mass Spectrometry to reveal volatile topographical dynamics of lima bean (*Phaseolus lunatus* L.) upon herbivory by *Spodoptera littoralis* Boisd.

Lorenzo Boggia[1], Barbara Sgorbini[1], Cinzia M Bertea[2], Cecilia Cagliero[1], Carlo Bicchi[1], Massimo E Maffei[2] and Patrizia Rubiolo[1,2*]

Abstract

Background: The dynamics of plant volatile (PV) emission, and the relationship between damaged area and biosynthesis of bioactive molecules in plant-insect interactions, remain open questions. Direct Contact-Sorptive Tape Extraction (DC-STE) is a sorption sampling technique employing non adhesive polydimethylsiloxane tapes, which are placed in direct contact with a biologically-active surface. DC-STE coupled to Gas Chromatography – Mass Spectrometry (GC-MS) is a non-destructive, high concentration-capacity sampling technique able to detect and allow identification of PVs involved in plant responses to biotic and abiotic stresses. Here we investigated the leaf topographical dynamics of herbivory-induced PV (HIPV) produced by *Phaseolus lunatus* L. (lima bean) in response to herbivory by larvae of the Mediterranean climbing cutworm (*Spodoptera littoralis* Boisd.) and mechanical wounding by DC-STE-GC-MS.

Results: Time-course experiments on herbivory wounding caused by larvae (HW), mechanical damage by a pattern wheel (MD), and MD combined with the larvae oral secretions (OS) showed that green leaf volatiles (GLVs) [(*E*)-2-hexenal, (*Z*)-3-hexen-1-ol, 1-octen-3-ol, (*Z*)-3-hexenyl acetate, (*Z*)-3-hexenyl butyrate] were associated with both MD and HW, whereas monoterpenoids [(*E*)-β-ocimene], sesquiterpenoids [(*E*)-nerolidol] and homoterpenes (DMNT and TMTT) were specifically associated with HW. Up-regulation of genes coding for HIPV-related enzymes (Farnesyl Pyrophosphate Synthase, Lipoxygenase, Ocimene Synthase and Terpene Synthase 2) was consistent with HIPV results. GLVs and sesquiterpenoids were produced locally and found to influence their own gene expression in distant tissues, whereas (*E*)-β-ocimene, TMTT, and DMNT gene expression was limited to wounded areas.

Conclusions: DC-STE-GC-MS was found to be a reliable method for the topographical evaluation of plant responses to biotic and abiotic stresses, by revealing the differential distribution of different classes of HIPVs. The main advantages of this technique include: a) *in vivo* sampling; b) reproducible sampling; c) ease of execution; d) simultaneous assays of different leaf portions, and e) preservation of plant material for further "omic" studies. DC-STE-GC-MS is also a low-impact innovative method for *in situ* PV detection that finds potential applications in sustainable crop management.

(Continued on next page)

* Correspondence: patrizia.rubiolo@unito.it
[1]Department of Drug Science and Technology, University of Turin, Via P. Giuria 9, 10125 Turin, Italy
[2]Plant Physiology Unit, Department Life Sciences and Systems Biology, University of Turin, Via Quarello 15/A, 10135 Turin, Italy

(Continued from previous page)
Keywords: Direct Contact-Sorptive Tape Extraction (DC-STE), Gas Chromatography coupled with Mass Spectrometry (GC-MS), Herbivory-induced plant volatile (HIPV), *Phaseolus lunatus* L., *Spodoptera littoralis* Boisd., Plant-insect interactions, Herbivory, Green leaf volatiles (GLVs), Monoterpenoids, Sesquiterpenoids

Background

In the past ten years, the study of the interaction between larvae of the Mediterranean climbing cutworm (*Spodoptera littoralis* Boisd.) and leaves of the lima bean (*Phaseolus lunatus* L.) has provided evidence of both early and late events, and has been used as a model system to decipher plant-insect interactions [1-5]. Upon herbivory by *S. littoralis*, the lima bean responds, as do many other plants, with a cascade of events that lead to the activation of defense mechanisms. These mechanisms include the perception of molecular patterns or effectors of defense [6,7], mitogen-activated protein kinase (MAPK) activation, and protein phosphorylation [8,9], production of ethylene and jasmonates [10], expression of late defense response genes [11], and emission of herbivory-induced plant volatiles (HIPVs) [12,13].

Even if robotic mechanical wounding can simulate plant response similar to HIPV [4], the simple mechanical damage (MD) is not fully satisfactory to induce the same responses if not supported by the application of insect's oral secretions (OS) [14]. Despite the presence of several elicitors in *S. littoralis* OS (e.g., fatty acid conjugates) [7,15], it is not clear whether these factors originate from the salivary glands or other feeding-related organs, such as the ventral eversible gland [14,16]. However, the plant volatile (PV) blends emitted in response to herbivores differ markedly with different feeding modes [17-20].

In plant defensive strategies, the release of PVs plays multiple roles: direct deterrents against herbivores [21,22], attraction of natural enemies of the attacking herbivores [23-26], damage and disease long-distance signaling [27-30], and pathogen resistance priming [29-32]. Since volatiles are produced from several biosynthetic pathways, their qualitative and quantitative composition is the result of the concerted action of different pathways, triggered by multiple factors. To date, studies of the emission of PVs in response to herbivory have been limited to single organs or to the whole plant, either by destructive methods or by head-space analysis [33,34], and only one study analyzed PV gradients within a single leaf [35].

Direct Contact-Sorptive Tape Extraction (DC-STE) is a fast and easy-to-use sampling technique, developed to study the effect of cosmetic treatment on sebum composition, through *in vivo* sampling at the human skin surface [36,37]. The technique employs a thin flexible non-adhesive polydimethylsiloxane (PDMS) tape, which is placed directly in contact with a (biological) surface

for a fixed time (Figure 1). Bicchi et al. [38] showed that this technique can also be applied to plants to monitor PVs, in both surface-static headspace and direct-contact (DC) modes. In DC-STE, volatiles produced at the biological surface are concentrated in the apolar PDMS layer by sorption (a sampling approach based on the partition of a compound between the sample and the bulk of a polymeric retaining phase) in amounts depending on the compound polarity and volatility. While in headspace sampling (e.g. static and dynamic headspace, high concentration-capacity solid phase microextraction) sorption is applied to the plant surrounding air space [33], DC-STE interacts directly with leaf surfaces. In DC-STE, plant-air interaction equilibrium is eliminated thus limiting the number of phases involved with sampling to two (plant and PDMS) instead of three (plant, air and PDMS). In this study, a glass coverslip was placed just above the DC-STE tape in order to exclude PDMS – air interaction.

Compound recovery from PDMS is achieved either by thermal desorption and on-line transferred to the injector of a Gas Chromatography–Mass Spectrometry (GC-MS) system, or by liquid extraction with polar solvents. DC-STE can be used successfully for both qualitative and quantitative analyses [37] making DC-STE coupled with GC-MS an efficient approach to characterize the profile and dynamics of PV production in response to both biotic and abiotic stresses.

In this study, the use of DC-STE combined with GC-MS was applied *in vivo* for the first time to evaluate the dynamics of HIPV release, upon abiotic (MD) and biotic (herbivory wounding, HW) stresses, by using the model system *S. littoralis/P. lunatus*. Furthermore, MD was used in combination with *S. littoralis* OS (MDOS). Here we show that HIPVs are differentially produced in different parts of the wounded leaf, depending on the biotic or abiotic stress applied. The analytical method was compared to the expression of genes involved in HIPV biosynthesis, which showed the same HIPV topographical pattern.

Results

In response to herbivory, plants produce PVs, which can serve as direct deterrents [21] or to attract the herbivore's predators and parasitoids [23-26,39]. The dynamics of HIPV emission, and the relationship between damaged area and biosynthesis of bioactive molecules, remain open questions. An innovative *in vivo* strategy was here used to identify compounds actively related to

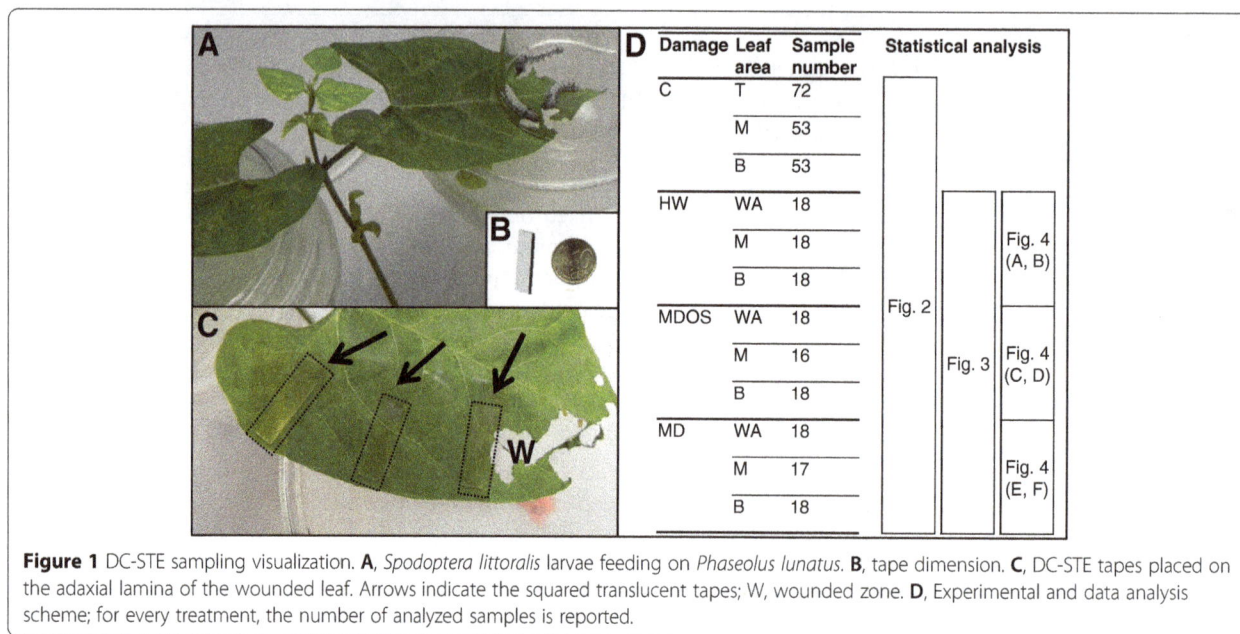

Figure 1 DC-STE sampling visualization. **A**, *Spodoptera littoralis* larvae feeding on *Phaseolus lunatus*. **B**, tape dimension. **C**, DC-STE tapes placed on the adaxial lamina of the wounded leaf. Arrows indicate the squared translucent tapes; W, wounded zone. **D**, Experimental and data analysis scheme; for every treatment, the number of analyzed samples is reported.

plant-insect interactions, employing a non-destructive high concentration-capacity sampling technique to capture volatiles from lima bean leaves after abiotic and biotic wounding.

DC-STE-GC-MS analysis discriminates herbivory from mechanical wounding

To analyze the topographical distribution of HIPVs, leaves from plants grown in a growth chamber treated with HW, MD and MDOS as well as control intact leaves were sampled with PDMS rectangular tapes ($4 \times 15 \times 0.2$ mm) placed in direct contact with leaves at specific distances from the damaged areas (0 cm, 1.5 cm, 3 cm) for different sampling times (2, 6, 24 h). Adaxial and abaxial leaf laminae were sampled in three different leaf portions: a) close to the damaged area (referred as the wounding zone, 0 cm); b) in the central portion (referred as the middle zone, 1.5 cm); and c) in the basal portion of the leaf (referred as the basal zone, 3 cm) (Figure 1). Preliminary trials showed no significant differences in PV results between adaxial and abaxial epidermises (data not shown). Analysis of camphor variation supports the repeatability of the method, accounting for 18.3% as relative standard deviation throughout the whole dataset.

Several PVs were identified by GC-MS analyses including green-leaf volatiles (GLVs, including aldehydes, alcohols and acetates), alkyl aldehydes, homoterpenes, mono- and sesquiterpenoids (Additional file 1). Because of the large number of samples (337), several Principal Component Analyses (PCA) were carried out; the best results were those obtained with logarithmic scaling as data pre-treatment [40].

Figure 2A reports the PCA (42% of explained variance) on the total dataset of samples, discriminating undamaged (controls) from damaged leaves. The damaged sample distribution in Figure 2A showed that HW and MD seemed divided into two different subsets, while application of OS to MD leaves produced intermediate results between them.

The resulting damage-related discriminant compounds included GLVs [(*E*)-2-hexenal, (*Z*)-3-hexen-1-ol, (*Z*)-3-hexenyl acetate, (*Z*)-3-hexenyl butyrate], a linoleic acid breakdown product (1-octen-3-ol), a monoterpene [(*E*)-β-ocimene], two homoterpenes [4,8-dimethyl-1,3,7-nonatriene (DMNT) and 4,8,12-trimethyl-1,3,7,11-tridecatetraene (TMTT)] and a sesquiterpenoid [(*E*)-nerolidol] (Figure 2B). These HIPVs were therefore used as variables for the subsequent PCA to explore the internal differences in the damaged leaf dataset. A better discrimination (about 71% of total variance explained) was obtained between HW and MD treatments, whereas MDOS samples showed a scattered pattern (Figure 3).

DC-STE-GC-MS determines and quantifies the topography of leaf HIPV production

The ability to discriminate between MD and HW highlights the potential of DC-STE-GC-MS as a reliable technique for *in vivo* HIPV monitoring. This ability was used to study the dynamics of volatile production as a function of topography in lima bean leaf responses to HW, MD and MDOS.

To visualize HIPV distribution, the damaged leaf dataset was divided into three different matrices, depending on the type of damage, each including a smaller but still

Figure 2 PCA representing the whole set of data. 337 samples are here plotted in PCA by using all compounds as variables. **A**, Control unwounded leaves (C) are well-separated from damaged leaves. HW and MD show a clear separation. MDOS produced intermediate patterns between HW and MD. **B**, Loading plot highlights the discriminant variables (blue circle). Compound legend: a, n-hexanal; b, (E)-2-hexenal; c, (Z)-3-hexen-1-ol; d, 1-octen-3-ol; e, 6-methyl-5-hepten-2-one; f, octanal; g, (Z)-3-hexenyl acetate; h, p-cymene; i, limonene; j, 2-ethyl hexanol; k, (E)-β-ocimene; l, 1-octanol; m, linalool; n, nonanal; o, DMNT; p, (Z)-3-hexenyl butyrate; q, decanal; r, tridecane; s, geranyl acetone; t, (E)-nerolidol; u, TMTT.

Figure 3 Damage sample dataset PCA. This analysis was done on the damaged samples with the variables selected in the first PCA. **A**, There is a clear distinction between HW (green squares) and MD (magenta circles) samples. Application of OS to MD (MDOS) produced scattered results. **B**, Loading plot. Compound legend: b, (E)-2-hexenal; c, (Z)-3-hexen-1-ol; d, 1-octen-3-ol; g, (Z)-3-hexenyl acetate; k, (E)-β-ocimene; o, DMNT; p, (Z)-3-hexenyl butyrate; t, (E)-nerolidol; u, TMTT.

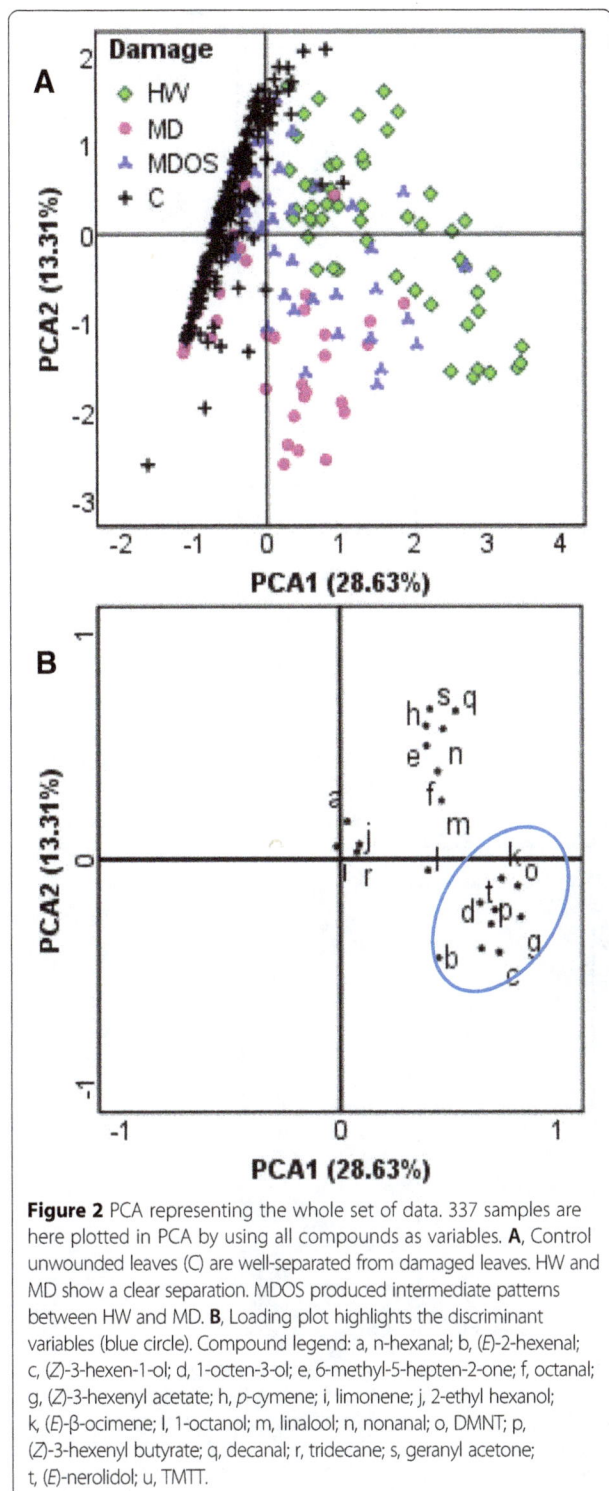

considerable number of samples (HW: 54 samples; MD: 53 samples; MDOS: 52 samples). PCA data processing was performed by using the discriminating variables identified above (GLVs, homoterpenes, mono- and sesquiterpenoids) with the aim of establishing a relationship between sampling time and leaf portion. A distinctive

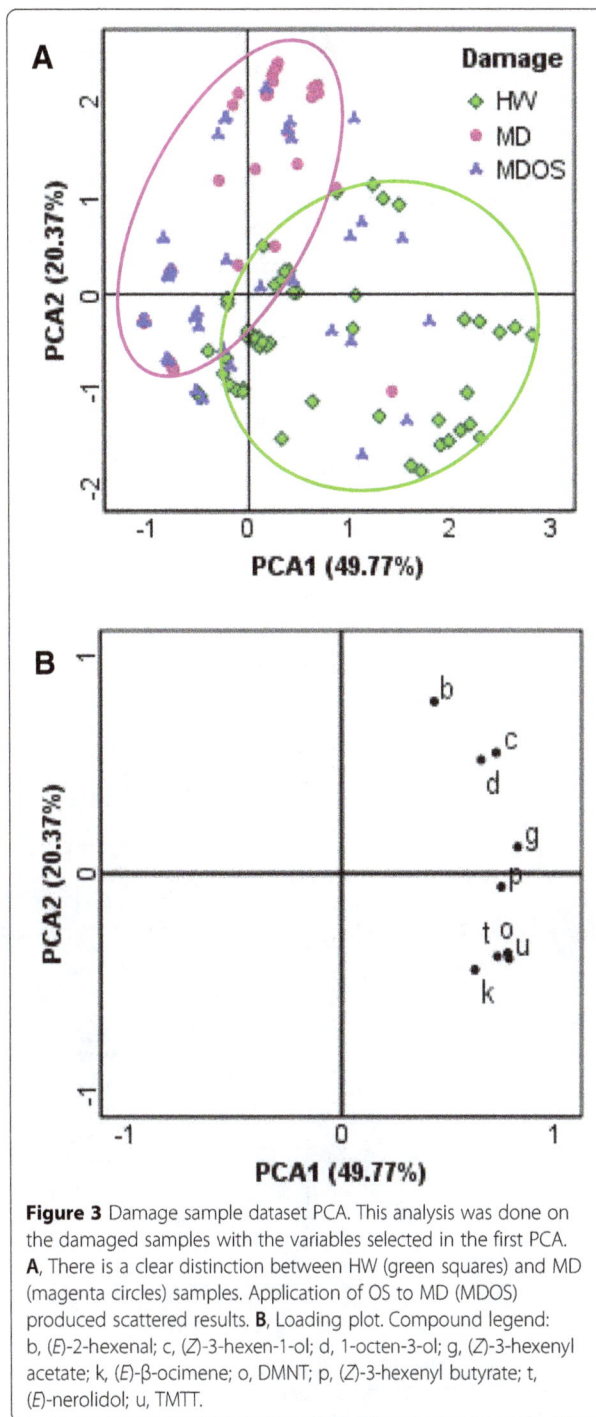

distribution of volatiles as a function of the damaged area was found for both HW and MDOS (Figure 4: A and C). Compared to controls, HW treated leaves showed a significantly higher production of the GLVs (E)-2-hexenal, (Z)-3-hexen-1-ol, (Z)-3-hexenyl acetate and of 1-octen-3-ol close to the HW damaged leaf portion (Figure 4B). (Z)-3-Hexenyl butyrate, DMNT, TMTT,

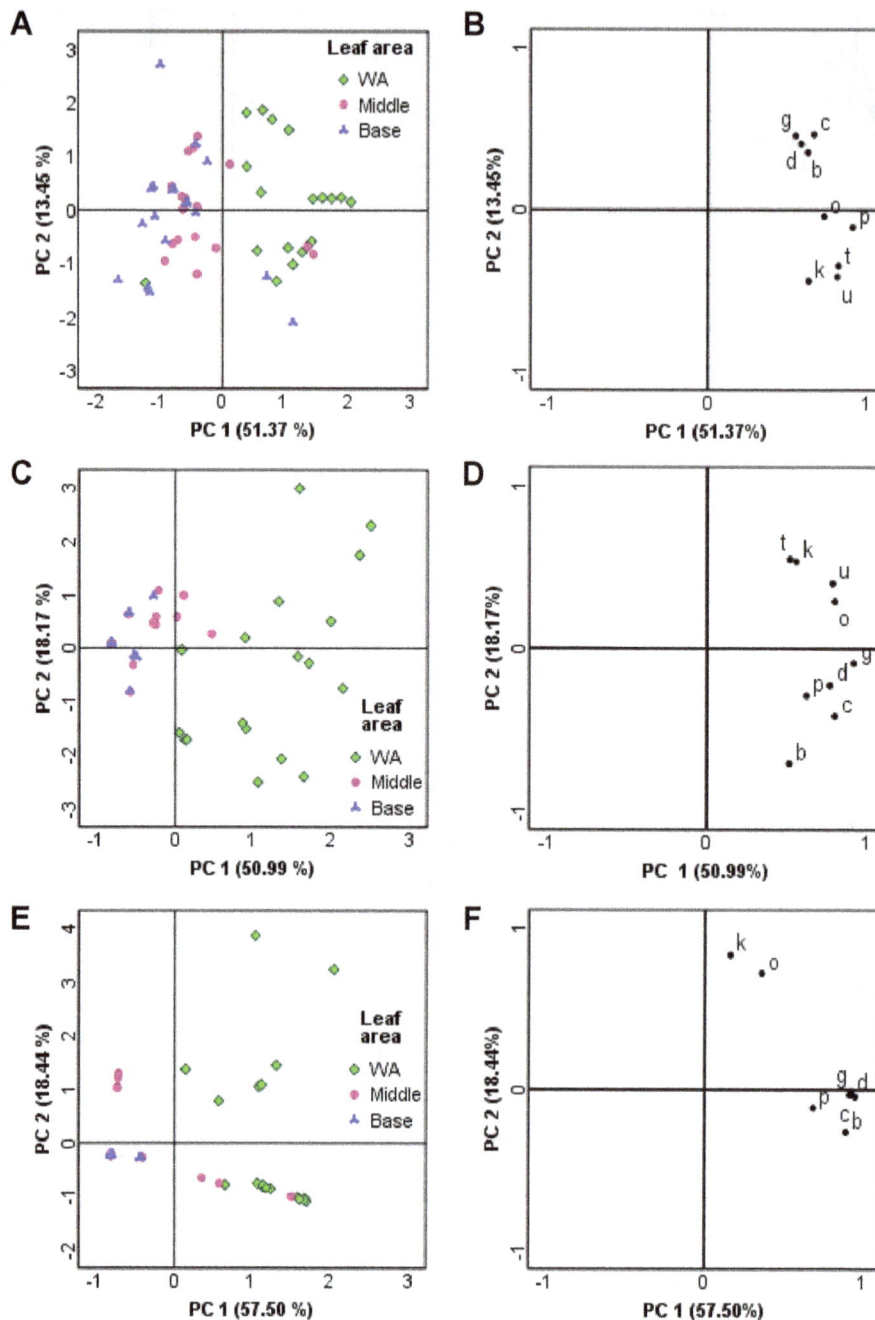

Figure 4 HIPV topography. HIPV topography is clearly shown in PCA score plots: wounded areas (WA) are in all cases clearly separated from other leaf portions. PCAs were carried out using: b, (*E*)-2-hexenal; c, (*Z*)-3-hexen-1-ol; d, 1-octen-3-ol; g, (*Z*)-3-hexenyl acetate; k, (*E*)-β-ocimene; o, DMNT; p, (*Z*)-3-hexenyl butyrate; t, (*E*)-nerolidol; u, TMTT. **A**, Score plot for HW leaves (54 samples) shows the distinction between WA samples (green squares) and other leaf portions. **B**, HW loading plot suggests that GLVs and terpenoids have the same localization in HIPV topographical distribution. **C**, MDOS leaves (52 samples) show a distribution similar to HW leaves (A). **D**, MDOS loading plot. **E**, MD score plot (53 samples) shows the same topographical distribution. **F**, MD loading plot shows a different distribution between GLVs and the terpenoid groups. The position of (k) and (o) and the absence of (t) and (u) suggest the non-significant role of terpenoids in MD reaction, unlike the HW and MDOS loading plots (B, D).

(*E*)-β-ocimene and (*E*)-nerolidol were produced in the same area, close to the HW zone, but also in distant leaf portions (Figure 4B). A similar pattern was found when MD plants were treated with OS (Figure 4: C and D).

In MD treated leaves, there was a clear distinction between the wounded area and the rest of the leaf (Figure 4E). However, only GLVs and 1-octen-3-ol were produced in wounded areas, while (*E*)-β-ocimene and

DMNT were not discriminant for the different leaf portions (Figure 4F).

The observation of the temporal differences showed that in all treatments GLVs were always produced early, whereas production of terpenoids and homoterpenes occurred later. In particular, PCAs highlighted some interesting differences in the temporal patterns between HW and other damages, with MDOS again showing intermediate values (Additional file 2).

A quantitative evaluation of the main damage-related compounds were carried out by combining in-tape camphor standardization with an external calibration by Gas Chromatography - Selected Ion Monitoring - Mass Spectrometry (GC-SIM-MS) for all types of damage, reaching a good linearity for every quantified HIPV (for quantitation parameters see Additional file 3).

In general, GLVs were the most abundant compounds in the damaged area (Table 1). (Z)-3-hexen-1-ol, (E)-2-hexenal, and 1-octen-3-ol reach rates of up to 100 ng/cm^2. (E)-β-ocimene and (E)-nerolidol were generally produced in smaller amounts far from the wounded zone; however, they were found to exceed 100 ng/cm^2 in the damage area. The homoterpenes, DMNT and TMTT, were mostly found in low quantities in HW-damaged leaves (Table 1).

Topographical gene expression analysis and DC-STE-GC-MS HIPV mapping

Because of the non-destructive DC-STE method of PV sampling, the different leaf sampled portions producing HIPVs could be used for gene expression analyses. Farnesyl Pyrophosphate Synthase (FPS) [41], P. lunatus Ocimene Synthase (PlOS) [10] and P. lunatus Terpene Synthase 2 (PlTPS2) [42] gene expressions were analyzed and compared to the results obtained by DC-STE for the related compounds . In addition, Lipoxygenase (LOX) [41] gene expression was analyzed, to assess any similarity with the observed high formation of GLVs.

Significantly higher expression of PlOS (Figure 5A) was in all cases coherent with the measured amount of the related compound (E)-β-ocimene (Figure 5B), with fold change values > 10 in the wounded zones of leaves treated by HW, MD or MDOS. Production of the homoterpene TMTT was associated with the gene expression pattern of PlTPS2 only for HW and MDOS treatments, whereas regulation of the gene was not comparable to the amount of the homoterpene upon MD treatment (Figure 5: C and D). Upregulation of FPS gene expression (Figure 5E) was consistent with (E)-nerolidol amount in HW and MDOS treatments (Figure 5F). Finally, the total GLV - production (Figure 5H) was in all cases higher in wounded zones, and consistent with LOX upregulation, in particular when referred to HW (Figure 5G). These results are fully supported by the

Kruskal-Wallis significance test (with Bonferroni adjustment, p < 0.017), as shown in Figure 5.

Discussion

One of the most challenging tasks in multitrophic interaction studies is the adoption of advanced analytical platforms that enable different analyses to be run simultaneously using different "omic" methodologies. DC-STE-GC-MS enabled to characterize the qualitative and quantitative topographical profile of leaf volatile emission upon herbivory, while evaluating at the same time the gene expression of the same sampled tissues.

In general, the HIPVs detected upon biotic and abiotic stresses in this study agree with those associated with biological damaging events [9,13,22,43] and with indirect plant defense [20,25,30,44].

The present results highlight the key role of the damaged area in HIPV production [35], with GLVs associated with both mechanical damage and herbivory, and monoterpenoids, sesquiterpenoids, and homoterpenes specifically associated with herbivory. In particular, MD treatment appears to be sufficient to induce higher amount of GLVs, including (E)-2-hexenal, (Z)-3-hexen-1-ol, 1-octen-3-ol, (Z)-3-hexenyl acetate and (Z)-3-hexenyl butyrate [20,25,32,45-48]. The DC-STE-GC-MS technique enabled GLVs to be determined qualitatively and to be quantified for further comparisons. Furthermore, the analysis revealed that some GLVs [(E)-2-hexenal, (Z)-3-hexen-1-ol and 1-octen-3-ol] are more intensively produced during MD than they are during other stresses. Conversely, (Z)-3-hexenyl acetate and (Z)-3-hexenyl butyrate are produced in higher amount in HW leaves, supporting their herbivory-induced production pattern [23,25,26,49]. GLVs are synthesized via the LOX pathway from C_{18} polyunsaturated fatty acids [50], which are cleaved to C_{12} and C_6 compounds by hydroperoxide lyases (HPL) [28]. Most plants have several isoforms of LOX [51], and a specific LOX that is essential to GLV formation has been identified in a few plant species [52]. In the present study, upregulation of LOX expression was evenly distributed throughout the leaf, although GLVs were mostly found in the wounded area. This discrepancy between gene expression and GLV production may be due, on the one hand, to the wide variety of roles played by LOX [53], and, on the other hand, to the effect of the GLVs on leaf tissues [30]. For instance, (Z)-3-hexenal in the vapor phase was taken up by Arabidopsis and converted into its alcohol and acetate in the cells. This scenario was further confirmed by the fact that the isotope ratios of alcohol and acetate were almost identical to that of (Z)-3-hexenal when ^{13}C-labeled (Z)-3-hexenal of a given isotope ratio was used for the exposure [54]. GLVs produced in the wounded zone may therefore influence expression of genes in unwounded tissues of the same leaf

Table 1 *Phaseolus lunatus* HIPV quantitation results

		(E)-2-hexenal	(Z)-3-hexen-1-ol	1-octen-3-ol	(Z)-3-hexenyl acetate	(E)-β-ocimene	DMNT	(Z)-3-hexenyl butyrate	(E)-nerolidol	TMTT
HW	WA	**185.4** (45.6)	**341.1** (67.5)	**152.8** (51.8)	**65.8** (11.3)	**274.3** (53.2)	**62.2** (8.0)	670.0 (539.7)	**237.7** (82.2)	**63.6** (23.6)
	M	nd	38.8 (9.9)	42.6 (12.6)	29.2 (10.7)	111.7 (38.5)	24.7 (5.8)	40.6 (31.0)	40.6 (30.1)	12.6 (6.3)
	B	nd	58.3 (31.5)	28.3 (10.0)	12.2 (2.3)	90.6 (49.1)	24.7 (12.2)	25.5 (22.1)	31.9 (24.0)	6.3 (4.6)
MDOS	WA	**354.6** (98.7)	**366.1** (95.6)	**195.5** (39.1)	**9.0** (1.2)	**68** (23.6)	**16.3** (3.7)	7.6 (5.8)	63.8 (40.5)	**16.2** (4.5)
	M	9.8 (9.8)	18.5 (12.8)	20.6 (8.4)	nd	**45.6** (19.3)	3.8 (1.7)	nd	nd	nd
	B	7.7 (7.7)	nd	8.2 (4.7)	nd	3.3 (1.8)	0.7 (0.7)	nd	nd	nd
MD	WA	**652.3** (157.7)	**1436.4** (220.0)	**435.1** (101.4)	**12.5** (1.7)	**277.6** (120.2)	2.5 (1.3)	**11.6** (5.5)	7.9 (7.9)	1.8 (1.8)
	M	49.8 (29.1)	71.4 (47.2)	51.3 (35.7)	1.4 (1.0)	26.9 (13.1)	nd	nd	nd	nd
	B	nd	4.1 (4.1)	2.4 (2.4)	nd	nd	nd	nd	nd	nd
Contr	T	3.0 (3.0)	nd	1.6 (1.1)	0.2 (0.2)	nd	0.6 (0.4)	nd	nd	nd
	M	4.0 (4.0)	3.6 (2.5)	6.9 (2.7)	nd	nd	nd	nd	15.2 (12.2)	nd
	B	13.9 (9.9)	4.0 (2.8)	2.5 (1.4)	0.4 (0.3)	0.3 (0.3)	nd	nd	nd	nd

Quantitative analysis of HIPVs produced by *Phaseolus lunatus* in different stress conditions, and topographical distribution of HIPVs. Results are expressed as ng/cm^2 (SEM). Reported results were submitted to ANOVA. Numbers in bold indicate statistical significance at the Tukey HSD test of the indicated leaf area (p < 0.05) in those cases in which ANOVA (treatments-control) was significant (p < 0.05). WA, wounded area; M, middle portion of the leaf; B, basal portion of the leaf; T, unwounded tip of the leaf; HW, herbivore wounding; MD, mechanical damage; MDOS, mechanical damage plus application of *Spodoptera littoralis* oral secretions; Contr, control undamaged leaves; nd, not detectable.

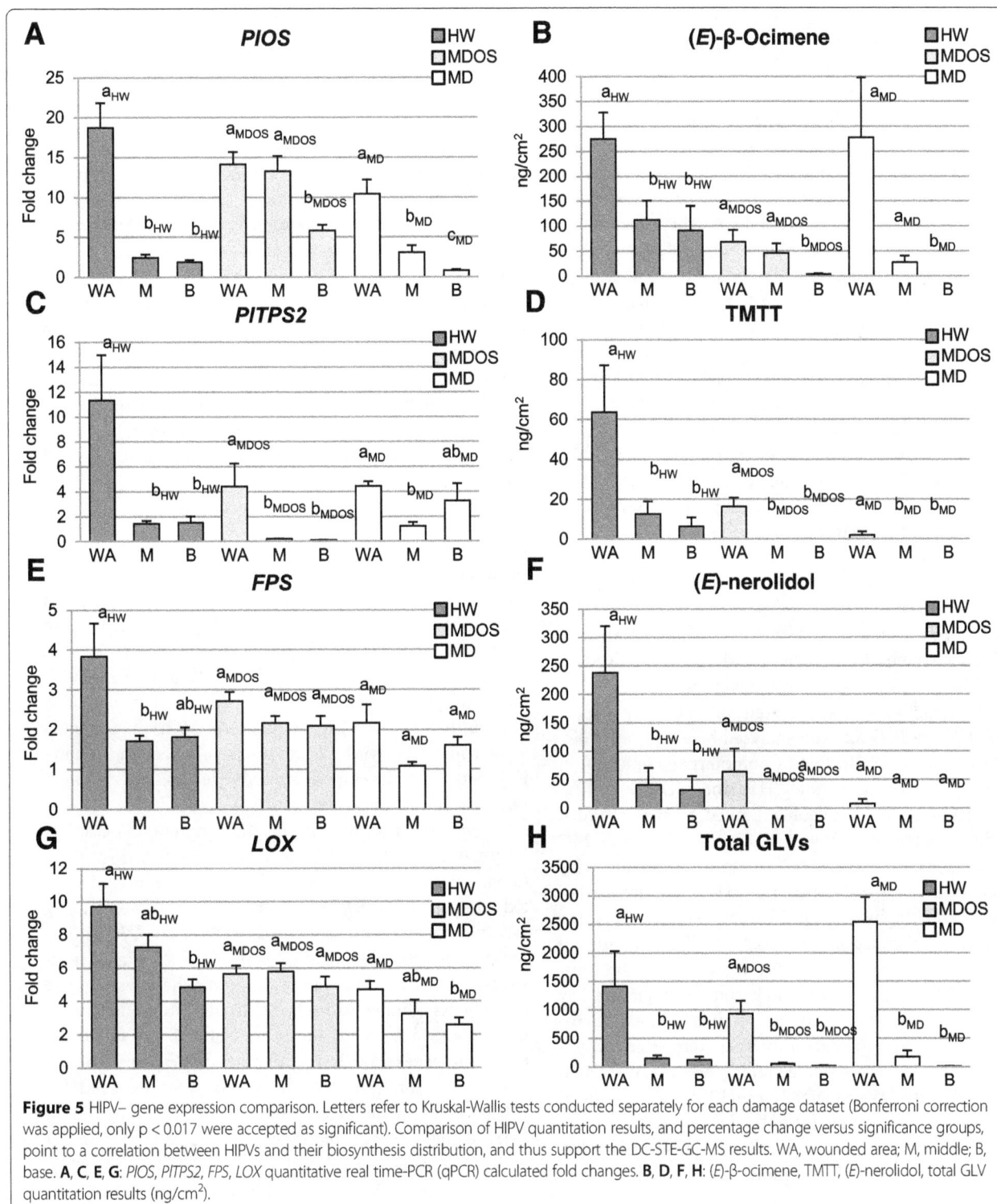

Figure 5 HIPV– gene expression comparison. Letters refer to Kruskal-Wallis tests conducted separately for each damage dataset (Bonferroni correction was applied, only p < 0.017 were accepted as significant). Comparison of HIPV quantitation results, and percentage change versus significance groups, point to a correlation between HIPVs and their biosynthesis distribution, and thus support the DC-STE-GC-MS results. WA, wounded area; M, middle; B, base. **A**, **C**, **E**, **G**: *PIOS*, *PITPS2*, *FPS*, *LOX* quantitative real time-PCR (qPCR) calculated fold changes. **B**, **D**, **F**, **H**: (*E*)-β-ocimene, TMTT, (*E*)-nerolidol, total GLV quantitation results (ng/cm^2).

because of GLV diffusion. In line with what Heil and Land have been recently reported [30], DC-STE sampling also highlights that GLVs play a central role in the so-called plant damage associated molecular pattern (DAMP). Indeed they seem to be essential to trigger gene expression required to prepare an adequate damage reaction in the surrounding tissues and organs. The MD related high GLV production could be explained with their well-known anti-microbial activity [32,55]. This is a resistance trait that is required during pathogen infection, which could occur after wounding [30], even without the herbivore interaction. Among monoterpenes, (*E*)-β-ocimene, a well-known damage-related HIPV [5,10,44] is a significant example of HIPV distribution. Its amount is limited to the damaged area in HW, while in MDOS (E)-β-ocimene also occurs distant from the wounded tissues. This different distribution agrees with the pattern of *PlOS* expression, demonstrating to produce almost exclusively (E)-β-ocimene when activated [56]; production is mainly located in the wounding area [45]. Transgenic *Arabidopsis*, transformed with the *PlOS* promoter *GUS* fusion constructs, shows that the activity is restricted to the wounded sites [10]. Lepidopteran caterpillars continuously remove leaf tissue after every bite, even if in a time longer than that one needed for the induction [57]. Conversely, application of OS to MD enables the elicitor to remain on the leaf longer, at least throughout the sampling time. This might explain why, in MDOS treated leaves, *PlOS* upregulation was observed in leaf areas distant from the damage.

Homoterpenes and sesquiterpenoids, such as DMNT, TMTT and (E)-nerolidol, are often associated with damage-related emission [5,22,26,58]; they have been studied as indirect defense mediators [25,39]. DMNT distribution is comparable to that of TMTT, and shows a general distribution from the damage zone throughout the leaf. However, their amount is higher in the wounded zone after both HW and MDOS treatment. The TPS enzymes have been found to be involved in DMNT and TMTT precursor production [42,58,59] and their products have been related to herbivory events [10,35,58]. The *PlTPS2* gene analyzed here showed a distribution comparable to that of homoterpene amount, in particular in leaves undergoing HW and MDOS.

Production of (E)-nerolidol is limited to the wounded zone, in particular in HW and MDOS damage, while MD does not seem to induce it. The lower amount of this compound in MDOS compared to HW is of interest because it shows the inability of OS alone to trigger the same HW-related leaf emission. Expression of *FPS* was found to be upregulated not only in damaged areas but also in leaf tissues distant from the wounding zone. *FPS* plays a key role in HIPV emission since its product, farnesyl pyrophosphate, is a basic precursor for sesquiterpenoid biosynthesis [13,60]. FPS is considered an important HW-related enzyme [43] and its inducibility by HIPVs has also been discussed and confirmed [61,62]. *FPS* upregulation was marked in HW leaves, underlining the relationship between herbivory and *FPS* activation [43].

Conclusions

The use of DC-STE-GC-MS provides a clearer picture of DAMP distribution in lima bean, by showing differential release of HIPV classes after different kinds of wounding. DAMPs, which are essential for airborne damage-signals, were found to be mainly related to disrupted tissues. The results confirm the role of HIPVs as DAMP signals and show their role as signals able to quickly spread in the surrounding environment of wounded areas. Upon herbivory a fast V_m depolarization is known to affect the whole damaged leaf, whereas calcium, potassium, ROS and NO responses are limited to the wounded zones. DC-STE-GC-MS results show that GLVs are released almost immediately and their emission is topographically in concomitance with early events such as V_m depolarization and calcium signaling, as previous data suggested [32,63-69].

The DC-STE-GC-MS results are in agreement with the present body of knowledge of plant damage recognition and reaction, and provide a better understanding of the dynamics of plant responses to damage. The main advantages of this technique compared to classical PV sampling methods are: a) *in vivo* sampling; b) ease of execution; c) simultaneous assays of different leaf portions, and d) preservation of plant material for further omic studies.

Methods

Plant and animal material

Feeding experiments were carried out using the lima bean (*Phaseolus lunatus* L. cv Ferry Morse var. Jackson Wonder Bush). Individual plants were grown from seed in plastic pots with quartz sand at 23°C and 60% humidity, using daylight fluorescent tubes at approximately $270 \; \mu E \; m^{-2} \; s^{-1}$ with a photophase of 16 h. Experiments were conducted with 12- to 16-day-old seedlings showing two fully-developed primary leaves, which were found to be the most responsive [1].

Spodoptera littoralis Boisd. (Lepidoptera, Noctuidae) larvae were kindly provided by R. Reist from Syngenta Crop. Protection Münchwilen AG, Switzerland, and were fed on an artificial diet comprising 125 g bean flour, 2.25 g ascorbic acid, 2.25 g ethyl 4-hydroxybenzoate, 750 μL formaldehyde, 300 mL distilled water and 20 g agar, previously solubilized in 300 mL of distilled water. The ingredients (Sigma-Aldrich, St. Louis, MO, USA) were mixed with a blender and stored at 4°C for not more than one week. With the exception of plant volatile (PV) collection (see below), plants were exposed for 2 h to third instar larvae reared from egg clutches in Petri dishes (9 cm diameter) in a growth chamber with 16 h photoperiod at 25°C and 60-70% humidity. The amount of herbivore damage was limited to 30% of leaf surface, as detected by ImageJ image analysis [4]. Feeding experiments were always performed between 1 and 3 p.m.

Collection of oral secretions

In order to evaluate the effect of *S. littoralis* oral secretions (OS), 5-day-old larvae were allowed to feed on lima bean leaves for 24 h. Regurgitation was caused by gently squeezing the larva with a forceps behind the head. OS was collected in glass capillaries connected to an evacuated sterile vial (peristaltic pump).

PV sampling setup

Biotic stress was caused by *S. littoralis* (HW); whereas abiotic stress was performed by mechanically damaging leaf tissues with a pattern wheel (MD). Furthermore abiotic and biotic stresses were connected by combining MD with *S. littoralis* oral secretions (MDOS). A large number of samples were analyzed (337) and multivariate methods were used to define discriminant variables (i.e., HIPVs) and to plot chemical and molecular topographical maps of leaf areas producing HIPVs in response to biotic and abiotic stress. In particular, the experiments were carried out in nine sampling steps, each representing a specific combination of type of damage (HW, MD, and MDOS) and sampling duration (2, 6, 24 h). For each sampling step, three biological replicates were analyzed, with 12 tapes for each. A control using two tapes was also sampled. HW was caused by *S. littoralis* caterpillars; the damaged area for each plant was as near as possible equal. MD was done by piercing the leaves manually with a pattern wheel. The damaged leaf area and the duration of time of the damaging mechanism were kept constant. The damage process in MDOS was similar to that in MD, with the addition on the wounded area of 10 µL of a solution 1:1 of *S. littoralis* OS and 5 mM MES (2-(N-morpholino)-ethane-sulphonic acid) buffer (pH 6.0). The OS quantity was assessed after several trials (from 0.5 to 10 µL) and was found the most appropriate to obtain reproducible experiments [43].

At the end of the sampling time, the tapes were removed and stored at –20°C. Leaves were cut into 3 parts (wounded area, middle, base) and stored at –80°C for further analyses.

Direct Contact–Sorptive Tape Extraction of PVs

Polydimethylsiloxane (PDMS) tapes (4 × 15 × 0.2 mm, *ca.* 33 mg) were placed on different areas of the adaxial and abaxial leaf lamina of *S. littoralis*-attacked and of control leaves. A glass coverslip was placed just above the DC-STE tape in order to exclude PDMS – air interaction. The quantitation of the collected PVs was obtained by an external standard at known concentration levels, being difficult to calculate an analyte recovery rate with DC-STE applied to *in vivo* plant matrices (unlike it was done in [37] with standards). Sampling was carried out in triplicate in the positions on the leaf shown in Figure 1, for the times reported above (2, 6, 24 h). Camphor (Sigma-Aldrich, Milan, Italy) was used as internal standard (I.S.) and was sorbed onto the tapes as proposed by Wang et al. [70] for Solid Phase Micro Extraction. Preliminary analysis with tapes with and without camphor I.S. were carried out to

verify any possible interference of camphor with lima bean PV production (Additional file 4). After sampling, the PDMS tapes were placed in thermal desorption tubes, stored in sealed vials, and submitted to automatic thermal desorption (see below). Sorption tapes were provided by the Research Institute for Chromatography (Kortrijk-Belgium).

GC-MS analysis

PDMS tape thermal desorption was carried out with a Thermal Desorption Unit (TDU) from Gerstel (Mülheima/d Ruhr, Germany). Analyses were driven automatically by an MPS-2 multipurpose sampler installed on an Agilent 7890 GC unit coupled to an Agilent 5975C MSD (Agilent, Little Falls, DE, USA). The TDU thermal desorption program was: from 30°C to 250°C (5 min) at 60°C/min in splitless flow mode, and transfer line at 300°C. A Gerstel CIS-4 PTV injector was used to cryofocus compounds thermally desorbed from the PDMS tapes, and inject them into the injector GC port. The PTV was cooled to –40°C using liquid CO_2; injection temperature: from –40°C to 250°C (5 min) at 12°C/s. The inlet was operating in the splitless mode. Helium was used as carrier gas at a flow rate of 1 mL/min. Column: HP5MS (30 m × 0.25 mm i.d. × 0.25 µm; Agilent Technologies). Temperature program: from –30°C (1 min) to 50°C at 50°C/min, then to 165°C at 3°C/min, then to 250°C (5 min) at 25°C/min. MS operated in EI mode at 70 eV with a mass range from 35 to 350 amu in full scan mode.

Quantitative Gas Chromatography – Selected Ion Monitoring – Mass Spectrometry analysis (GC-SIM-MS): appropriate amounts of 2-hexenal, 3-hexenol, 1-octen-3-ol, 3-hexenyl acetate, (Z)-3-hexenyl butyrate, 1-octen-3-ol, (E)-β-ocimene, 4,8-dimethyl-1,3,7-nonatriene and 4,8,12-trimethyl-1,3,7,11-tridecatetraene (Sigma-Aldrich, Milan, Italy) were diluted with cyclohexane (Sigma-Aldrich, Milan, Italy) to obtain nine different concentrations in the range 1 to 1000 µg/mL for each component. Calibration curves were constructed by analyzing the resulting standard solutions three times, by GC-MS in SIM mode, under the conditions reported above.

GC-MS data processing

Data were processed with Agilent MSD ChemStation *ver.* D.03.00.611 (Agilent Technologies). Components were identified by comparing their linear retention indices (I^Ts) (calculated *versus* a C_9-C_{25} hydrocarbon mixture) and their mass spectra to those of authentic samples, or by comparison with those present in commercially-available mass spectrum libraries (Wiley, Adams).

RNA extraction from lima bean leaves after HW, MD and MDOS

After each experiment, leaves were collected and immediately frozen in liquid nitrogen. Samples from time-course experiments were pooled so as to have a single pool of replicates for each stress condition (HW, MDOS, MD, undamaged leaves). Fifty mg of frozen leaf material were ground in liquid nitrogen with mortar and pestle. Total RNA was isolated using the Agilent Plant RNA Isolation Mini Kit (Agilent Technologies, Santa Clara, CA, US) and RNase-Free DNase set (Qiagen, Hilden, Germany). Sample quality and quantity were checked using the RNA 6000 Nano kit and the Agilent 2100 Bioanalyzer (Agilent Technologies), following the manufacturer's instructions. Quantification of RNA was also confirmed spectrophotometrically, using the NanoDrop ND-1000 (Thermo Fisher Scientific, Waltham, MA, US).

Quantitative real time–PCR (qPCR) reaction conditions and primers

First strand cDNA synthesis was run with 1 μg of total RNA and random primers, using the High-Capacity cDNA Reverse Transcription Kit (Applied Biosystems, Foster City, CA, US), and following the manufacturer's recommendations. Reactions were prepared by adding 1 μg of total RNA, 2 μL of 10X RT Buffer, 0.8 μL of 25X dNTPs mix (100 mM), 2 μL 10X RT random primer, 1 μL of Multiscribe™ Reverse Transcriptase, and nuclease-free sterile water to 20 μL. Reaction mixtures were incubated at 25°C for 10 min, 37°C for 2 h, and 85°C for 5 s.

The qPCR experiments were run on a Stratagene Mx3000P Real-Time System (La Jolla, CA, USA) using SYBR green I with ROX as an internal loading standard. The reaction mixture was 10 μL, comprising 5 μL of 2X Maxima™ SYBR Green qPCR Master Mix (Fermentas International, Inc, Burlington, ON, Canada), 0.5 μL of cDNA and 100 nM primers (Integrated DNA Technologies, Coralville, IA, US). Controls included non-RT controls (using total RNA without reverse transcription to monitor for genomic DNA contamination) and non-template controls (water template). Specifically, PCR conditions were the following: *P. lunatus* Actin1 (*PlACT1*), Farnesyl Pyrophosphate Synthase (*FPS*), Lipoxygenase (*LOX*) [41], *P. lunatus* Ocimene Synthase (*PlOS*) [10], *P. lunatus* Terpene Synthase 2 (*PlTPS2*) [42]: 10 min at 95°C, 45 cycles of 15 s at 95°C, 30 s at 55°C, and 30 s at 72°C, 1 min at 95°C, 30 s at 55°C, 30 s at 95°C. Fluorescence was read following each annealing and extension phase. All runs were followed by a melting curve analysis from 55°C to 95°C. The linear range of template concentration to threshold cycle value (Ct value) was determined by preparing a dilution series, using cDNA from three independent RNA extractions analyzed in three technical replicates. Primer efficiencies for all primer pairs were calculated using the standard curve method [71]. Two different reference genes (Actin1 (*PlACT1*) and the *18S* ribosomal RNA) were used to normalize the results of the qPCR. The best of the two

genes was selected using the Normfinder software [72]; the most stable gene was *PlACT1*. Primers used for qPCR were as described elsewhere [3,41,42] and are reported in Additional file 5.

All amplification plots were analyzed with the Mx3000P™ software to obtain Ct values. Relative RNA levels were calibrated and normalized with the level of *PlACT1* mRNA.

Statistical analyses

Analysis of variance (ANOVA) and the Tukey test were used to assess difference between treatments and control. For all other experiments, at least five samples per treatment group entered the statistical data analysis. PV chemical data are expressed as mean values ± standard error of the mean (SEM).

Principal Component Analysis (PCA) was used in three different steps, each targeting different discrimination (control-damage, different damage, different leaf areas). A log-transformation was used as GC-MS data pre-treatment [40]. Each PCA step was followed by a significance test for the discriminant compounds. To compare the different leaf areas, the Kruskal-Wallis test was applied to both chemical and gene expression data. Bonferroni adjustment (p/k; k = number of comparisons) was applied to protect against Type I Error [73,74].

All statistical data analyses were done using SPSS software for Windows.

Additional files

Additional file 1: PV DC-STE-GC-MS profile. The typical GC-MS profile of PVs captured by DC-STE on *Phaseolus lunatus* leaves wounded by *Spodoptera littoralis* obtained after thermal desorption of DC-STE. A compound table is also provided.

Additional file 2: PCA analysis of time-course experiments on different damage dataset. Score and loading plots presented in Figure 4 are here displayed taking into account sampling time of every sample (instead considering the distance from the wounded area).

Additional file 3: Parameters for the volatiles' quantitation. A table reports HIPV quantitation parameters obtained by Gas Chromatography – Selected Ion Monitoring – Mass Spectrometry (GC-SIM-MS).

Additional file 4: Camphor effects on *Phaseolus lunatus*. A table and a barchart report a comparison between PVs in plants analyzed with and without camphor preloading on tapes.

Additional file 5: Gene-specific primers used for quantitative real-time PCR. A table collecting GenBank accession number and sequences of every primer used.

Abbreviations

DAMP: Damage associated molecular pattern; DC-STE: Direct contact-sorptive tape extraction; DMNT: 4,8-dimethyl-1,3,7-nonatriene; GC: Gas chromatography; GLV: Green leaf volatile; HIPV: Herbivory-induced plant volatile; HW: Herbivore wounding; MD: Mechanical damage; MDOS: Mechanical damage with oral secretions; MS: Mass spectrometry; OS: Oral secretions; PDMS: Polydimethylsiloxane; PV: Plant volatile; TMTT: 4,8,12-trimethyl-1,3,7,11-tridecatetraene.

Competing interests

The authors declare that they have no competing interests.

Authors' contributions
LB and BS carried out the PV sampling and analyses. LB performed the statistical analysis. CMB and LB carried out the molecular genetic studies. CB, MEM and PR conceived the study, and participated in its design and coordination and helped to draft the manuscript. CC helped to draft the manuscript. All authors read and approved the final manuscript.

Authors' information
LB is a graduate student in the Doctorate School of Pharmaceutical and Biomolecular Sciences of the University of Turin. BS and CC are Assistant Professors of Pharmaceutical Biology in the Dept. of Drug Science and Technology, University of Turin. CB and PR are Professors of Pharmaceutical Biology in the Dept. of Drug Science and Technology, University of Turin. CMB is Associate Professor of Plant Physiology in the Dept. Life Sciences and Systems Biology, University of Turin. MEM is Professor of Plant Physiology in the Dept. Life Sciences and Systems Biology, University of Turin.

Acknowledgements
This work was partly supported by the Doctorate School of Pharmaceutical and Biomolecular Sciences of the University of Turin. The study was also carried out in the framework of the project "Studio di metaboliti secondari biologicamente attivi da matrici di origine vegetale" financially supported by the Ricerca Locale (Ex 60% 2013) of the University of Turin, Turin (Italy).

References

1. Maffei ME, Bossi S, Spiteller D, Mithöfer A, Boland W. Effects of feeding *Spodoptera littoralis* on lima bean leaves. I. Membrane potentials, intracellular calcium variations, oral secretions, and regurgitate components. Plant Physiol. 2004;134(April):1752–62.
2. Maffei ME, Mithöfer A, Boland W. Before gene expression: early events in plant-insect interaction. Trends Plant Sci. 2007;12:310–6.
3. Maffei ME, Mithöfer A, Arimura G, Uchtenhagen H, Bossi S, Bertea CM, et al. Effects of feeding *Spodoptera littoralis* on lima bean leaves. III. Membrane depolarization and involvement of hydrogen peroxide. Plant Physiol. 2006;140:1022–35.
4. Bricchi I, Leitner M, Foti M, Mithöfer A, Boland W, Maffei ME. Robotic mechanical wounding (MecWorm) versus herbivore-induced responses: early signaling and volatile emission in Lima bean (*Phaseolus lunatus* L.). Planta. 2010;232:719–29.
5. Mithöfer A, Wanner G, Boland W. Effects of Feeding *Spodoptera littoralis* on Lima Bean Leaves II. Continuous Mechanical Wounding Resembling Insect Feeding Is Sufficient to Elicit Herbivory-Related Volatile Emission. Plant Physiol. 2005;137(March):1160–8.
6. Bonaventure G, VanDoorn A, Baldwin IT. Herbivore-associated elicitors: FAC signaling and metabolism. Trends Plant Sci. 2011;16:294–9.
7. Maffei ME, Arimura G-I, Mithöfer A. Natural elicitors, effectors and modulators of plant responses. Nat Prod Rep. 2012;29:1288–303.
8. Arimura G-I, Maffei ME. Calcium and secondary CPK signaling in plants in response to herbivore attack. Biochem Biophys Res Commun. 2010;400:455–60.
9. Arimura G-I, Ozawa R, Maffei ME. Recent advances in plant early signaling in response to herbivory. Int J Mol Sci. 2011;12:3723–39.
10. Arimura G, Köpke S, Kunert M, Volpe V, David A, Brand P, et al. Effects of feeding *Spodoptera littoralis* on lima bean leaves: IV. Diurnal and nocturnal damage differentially initiate plant volatile emission. Plant Physiol. 2008;146:965–73.
11. Wu J, Baldwin IT. New insights into plant responses to the attack from insect herbivores. Annu Rev Genet. 2010;44:1–24.
12. Baldwin IT. Plant volatiles. Curr Biol. 2010;20:R392–7.
13. Maffei ME, Gertsch J, Appendino G. Plant volatiles: production, function and pharmacology. Nat Prod Rep. 2011;28:1359–80.
14. Zebelo S, Piorkowski J, Disi J, Fadamiro H. Secretions from the ventral eversible gland of *Spodoptera exigua* caterpillars activate defense-related genes and induce emission of volatile organic compounds in tomato *Solanum lycopersicum*. BMC Plant Biol. 2014;14:140.
15. Mori N, Yoshinaga N. Function and evolutionary diversity of fatty acid amino acid conjugates in insects. J Plant Interact. 2011;6:103–7.
16. Zebelo SA, Maffei ME. The ventral eversible gland (VEG) of *Spodoptera littoralis* triggers early responses to herbivory in *Arabidopsis thaliana*. Arthropod Plant Interact. 2012;6:543–51.
17. Van Poecke RMP, Dicke M. Indirect defence of plants against herbivores: using *Arabidopsis thaliana* as a model plant. Plant Biol (Stuttg). 2004;6:387–401.
18. Leitner M, Boland W, Mithöfer A. Direct and indirect defences induced by piercing-sucking and chewing herbivores in *Medicago truncatula*. New Phytol. 2005;167:597–606.
19. Turlings TCJ, Ton J. Exploiting scents of distress: the prospect of manipulating herbivore-induced plant odours to enhance the control of agricultural pests. Curr Opin Plant Biol. 2006;9:421–7.
20. Maffei ME. Sites of synthesis, biochemistry and functional role of plant volatiles. South African J Bot. 2010;76:612–31.
21. Zebelo SA, Bertea CM, Bossi S, Occhipinti A, Gnavi G, Maffei ME. *Chrysolina herbacea* modulates terpenoid biosynthesis of *Mentha aquatica* L. PLoS One. 2011;6, e17195.
22. Mithöfer A, Boland W. Plant defense against herbivores: chemical aspects. Annu Rev Plant Biol. 2012;63:431–50.
23. De Moraes CM, Lewis WJ, Paré PW, Alborn HT, Tumlinson JH. Herbivore-infested plants selectively attract parasitoids. Nature. 1998;393(June):570–3.
24. Ponzio C, Gols R, Pieterse CMJ, Dicke M. Ecological and phytohormonal aspects of plant volatile emission in response to single and dual infestations with herbivores and phytopathogens. Funct Ecol. 2013;27:587–98.
25. Mäntylä E, Alessio GA, Blande JD, Heijari J, Holopainen JK. From plants to birds: higher avian predation rates in trees responding to insect herbivory. PLoS One. 2008;3, e2832.
26. Clavijo McCormick A, Irmisch S, Reinecke A, Boeckler GA, Veit D, Reichelt M, et al. Herbivore-induced volatile emission in black poplar: regulation and role in attracting herbivore enemies. Plant Cell Environ. 2014;37(8):1909–23.
27. Dicke M, Baldwin IT. The evolutionary context for herbivore-induced plant volatiles: beyond the "cry for help". Trends Plant Sci. 2010;15:167–75.
28. Engelberth J, Alborn HT, Schmelz EA, Tumlinson JH. Airborne signals prime plants against insect herbivore attack. Proc Natl Acad Sci U S A. 2004;101:1781–5.
29. Yi H-S, Heil M, Adame-Alvarez RM, Ballhorn DJ, Ryu C-M. Airborne induction and priming of plant defenses against a bacterial pathogen. Plant Physiol. 2009;151:2152–61.
30. Heil M, Land WG. Danger signals - damaged-self recognition across the tree of life. Front Plant Sci. 2014;5(October):578.
31. Ballhorn DJ, Kautz S, Schädler M. Induced plant defense via volatile production is dependent on rhizobial symbiosis. Oecologia. 2013;172:833–46.
32. Scala A, Allmann S, Mirabella R, Haring MA, Schuurink RC. Green leaf volatiles: a plant's multifunctional weapon against herbivores and pathogens. Int J Mol Sci. 2013;14:17781–811.
33. Tholl D, Boland W, Hansel A, Loreto F, Röse USR, Schnitzler J-P. Practical approaches to plant volatile analysis. Plant J. 2006;45:540–60.
34. Bicchi C, Maffei M. The Plant Volatilome: Methods of Analysis. Methods Mol Biol. 2012;918:289–310.
35. Köllner TG, Lenk C, Schnee C, Köpke S, Lindemann P, Gershenzon J, et al. Localization of sesquiterpene formation and emission in maize leaves after herbivore damage. BMC Plant Biol. 2013;13:15.
36. Sandra P, Sisalli S, Adao A, Lebel M, Le Fur I. Sorptive Tape Extraction — A Novel Sampling Method for the *in vivo* Study of Skin. LCGC Eur. 2006;19:33–9.
37. Sgorbini B, Ruosi MR, Cordero C, Liberto E, Rubiolo P, Bicchi C. Quantitative determination of some volatile suspected allergens in cosmetic creams spread on skin by direct contact sorptive tape extraction-gas chromatography–mass spectrometry. J Chromatogr A. 2010;1217:2599–605.
38. Bicchi C, Cordero C, Liberto E, Rubiolo P, Sgorbini B, Sandra P. Sorptive tape extraction in the analysis of the volatile fraction emitted from biological solid matrices. J Chromatogr A. 2007;1148:137–44.
39. Heil M. Indirect defence via tritrophic interactions. New Phytol. 2008;178:41–61.
40. Brereton RG. Chemometrics for Pattern Recognition. Chichester, UK: John Wiley & Sons, Ltd; 2009.
41. Ozawa R, Bertea CM, Foti M, Narayana R, Arimura G-I, Muroi A, et al. Exogenous polyamines elicit herbivore-induced volatiles in lima bean leaves: involvement of calcium, H2O2 and Jasmonic acid. Plant Cell Physiol. 2009;50:2183–99.
42. Brillada C, Nishihara M, Shimoda T, Garms S, Boland W, Maffei ME, et al. Metabolic engineering of the C16 homoterpene TMTT in *Lotus japonicus* through overexpression of (E, E)-geranyllinalool synthase attracts generalist and specialist predators in different manners. New Phytol. 2013;200:1200–11.

43. Bricchi I, Occhipinti A, Bertea CM, Zebelo SA, Brillada C, Verrillo F, et al. Separation of early and late responses to herbivory in Arabidopsis by changing plasmodesmal function. Plant J. 2013;73:14–25.

44. Zhang P-J, Zheng S-J, Van Loon JJ, Boland W, David A, Mumm R, et al. Whiteflies interfere with indirect plant defense against spider mites in Lima bean. Proc Natl Acad Sci U S A. 2009;106:21202–7.

45. Arimura G-I, Matsui K, Takabayashi J. Chemical and molecular ecology of herbivore-induced plant volatiles: proximate factors and their ultimate functions. Plant Cell Physiol. 2009;50:911–23.

46. Heil M, Lion U, Boland W. Defense-inducing volatiles: in search of the active motif. J Chem Ecol. 2008;34:601–4.

47. Holopainen JK, Gershenzon J. Multiple stress factors and the emission of plant VOCs. Trends Plant Sci. 2010;15:176–84.

48. De Boer J, Snoeren T, Dicke M. Predatory mites learn to discriminate between plant volatiles induced by prey and nonprey herbivores. Anim Behav. 2005;69:869–79.

49. D'Auria JC, Pichersky E, Schaub A, Hansel A, Gershenzon J. Characterization of a BAHD acyltransferase responsible for producing the green leaf volatile (Z)-3-hexen-1-yl acetate in *Arabidopsis thaliana*. Plant J. 2007;49:194–207.

50. Grechkin A. Recent developments in biochemistry of the plant lipoxygenase pathway. Prog Lipid Res. 1998;37:317–52.

51. Joo Y-C, Oh D-K. Lipoxygenases: potential starting biocatalysts for the synthesis of signaling compounds. Biotechnol Adv. 2012;30:1524–32.

52. Christensen SA, Nemchenko A, Borrego E, Murray I, Sobhy IS, Bosak L, et al. The maize lipoxygenase, ZmLOX10, mediates green leaf volatile, jasmonate and herbivore-induced plant volatile production for defense against insect attack. Plant J. 2013;74:59–73.

53. Liavonchanka A, Feussner I. Lipoxygenases: occurrence, functions and catalysis. J Plant Physiol. 2006;163:348–57.

54. Matsui K, Sugimoto K, Mano J, Ozawa R, Takabayashi J. Differential metabolisms of green leaf volatiles in injured and intact parts of a wounded leaf meet distinct ecophysiological requirements. PLoS One. 2012;7, e36433.

55. Heil M. Herbivore-induced plant volatiles: targets, perception and unanswered questions. New Phytol. 2014;204:297–306.

56. Degenhardt J, Köllner TG, Gershenzon J. Monoterpene and sesquiterpene synthases and the origin of terpene skeletal diversity in plants. Phytochemistry. 2009;70:1621–37.

57. Schittko U, Preston CA, Baldwin IT. Eating the evidence? *Manduca sexta* larvae can not disrupt specific jasmonate induction in *Nicotiana attenuata* by rapid consumption. Planta. 2000;210:343–6.

58. Tholl D, Sohrabi R, Huh J-H, Lee S. The biochemistry of homoterpenes. Common constituents of floral and herbivore-induced plant volatile bouquets. Phytochemistry. 2011;72:1635–46.

59. Arimura G, Garms S, Maffei M, Bossi S, Schulze B, Leitner M, et al. Herbivore-induced terpenoid emission in *Medicago truncatula*: concerted action of jasmonate, ethylene and calcium signaling. Planta. 2008;227:453–64.

60. Dewick PM. The biosynthesis of C5-C25 terpenoid compounds. Nat Prod Rep. 1997;14:111–44.

61. Ueda H, Kikuta Y, Matsuda K. Plant communication. Mediated by individual or blended VOCs? Plant Signal Behav. 2012;7:222–6.

62. Arimura G, Ozawa R, Shimoda T, Nishioka T, Boland W, Takabayashi J. Herbivory-induced volatiles elicit defence genes in lima bean leaves. Nature. 2000;406:512–5.

63. Arimura G, Ozawa R, Horiuchi J, Nishioka T, Takabayashi J. Plant–plant interactions mediated by volatiles emitted from plants infested by spider mites. Biochem Syst Ecol. 2001;29:1049–61.

64. Kost C, Heil M. Herbivore-induced plant volatiles induce an indirect defence in neighbouring plants. J Ecol. 2006;94:619–28.

65. Allmann S, Halitschke R, Schuurink RC, Baldwin IT. Oxylipin channelling in *Nicotiana attenuata*: lipoxygenase 2 supplies substrates for green leaf volatile production. Plant Cell Environ. 2010;33:2028–40.

66. Hatanaka A. The biogeneration of green odour by green leaves. Phytochemistry. 1993;34:1201–18.

67. Farag MA, Paré PW. C6-green leaf volatiles trigger local and systemic VOC emissions in tomato. Phytochemistry. 2002;61:545–54.

68. Matsui K. Green leaf volatiles: hydroperoxide lyase pathway of oxylipin metabolism. Curr Opin Plant Biol. 2006;9:274–80.

69. Zebelo SA, Matsui K, Ozawa R, Maffei ME. Plasma membrane potential depolarization and cytosolic calcium flux are early events involved in tomato (*Solanum lycopersicon*) plant-to-plant communication. Plant Sci. 2012;196:93–100.

70. Wang Y, O'Reilly J, Chen Y, Pawliszyn J. Equilibrium in-fibre standardisation technique for solid-phase microextraction. J Chromatogr A. 2005;1072:13–7.

71. Pfaffl MW. A new mathematical model for relative quantification in real-time RT-PCR. Nucleic Acids Res. 2001;29, e45.

72. Andersen CL, Jensen JL, Ørntoft TF. Normalization of Real-Time Quantitative Reverse Transcription-PCR Data: A Model-Based Variance Estimation Approach to Identify Genes Suited for Normalization, Applied to Bladder and Colon Cancer Data Sets. Cancer Res. 2004;64:5245–50.

73. Bland JM, Altman DG. Multiple significance tests: the Bonferroni method. Br Med J. 1995;310:170.

74. Strassburger K, Bretz F. Compatible simultaneous lower confidence bounds for the Holm procedure and other Bonferroni-based closed tests. Stat Med. 2008;27:4914–27.

Dissection of the style's response to pollination using transcriptome profiling in self-compatible (*Solanum pimpinellifolium*) and self-incompatible (*Solanum chilense*) tomato species

Panfeng Zhao[1,2], Lida Zhang[2] and Lingxia Zhao[1,2*]

Abstract

Background: Tomato (*Solanum lycopersicum*) self-compatibility (SC) is defined as self-pollen tubes that can penetrate their own stigma, elongate in the style and fertilize their own ovules. Self-incompatibility (SI) is defined as self-pollen tubes that are prevented from developing in the style. To determine the influence of gene expression on style self-pollination, a transcriptome-wide comparative analysis of SC and SI tomato unpollinated/pollinated styles was performed using RNA-sequencing (RNA-seq) data.

Results: Transcriptome profiles of 24-h unpollination (UP) and self-pollination (P) styles from SC and SI tomato species were generated using high-throughput next generation sequencing. From the comparison of SC self-pollinated and unpollinated styles, 1341 differentially expressed genes (DEGs) were identified, of which 753 were downregulated and 588 were upregulated. From the comparison of SI self-pollinated and unpollinated styles, 804 DEGs were identified, of which 215 were downregulated and 589 were upregulated. Nine gene ontology (GO) terms were enriched significantly in SC and 78 GO terms were enriched significantly in SI. A total of 105 enriched Kyoto Encyclopedia of Genes and Genomes (KEGG) pathways were identified in SC and 80 enriched KEGG pathways were identified in SI, among which "Cysteine and methionine metabolism pathway" and "Plant hormone signal transduction pathway" were significantly enriched in SI.

Conclusions: This study is the first global transcriptome-wide comparative analysis of SC and SI tomato unpollinated/pollinated styles. Advanced bioinformatic analysis of DEGs uncovered the pathways of "Cysteine and methionine metabolism" and "Plant hormone signal transduction", which are likely to play important roles in the control of pollen tubes growth in SI species.

Keywords: Tomato, Self-incompatibility, Self-compatibility, Style, Transcriptome

Background

In flowering plants, the male organ of the flower is the stamen and the female organ of the flower is pistil. The stamen comprises an anther generating pollen grains and a filament supporting the anther. The pistil comprises the stigma, the style and the ovary. Pollination is a process of pollen-pistil interaction during which pollen adheres, hydrates, and germinates on the stigma, the pollen tube elongates on an active extracellular matrix in the style and finally transports male gametes (sperm cells) to the ovary, releasing it into ovules to complete fertilization [1]. Mate selection is crucial to successful reproduction and species survival [2]. Self-compatibility (SC) and self-incompatibility (SI) are the two predominant forms of mate selection. SC is defined as self-pollen that can penetrate its own pistil and fertilize its own ovules [1]; SI is where self-pollen is prevented from developing on the pistil [3].

Tomatoes (*Solanum lycopersicum*) are one of the most important vegetable crops in the world, and possess

* Correspondence: lxzhao@sjtu.edu.cn
[1]Joint Tomato Research Institute, School of Agriculture and Biology, Shanghai Jiao Tong University, Shanghai 200240, China
[2]Plant Biotechnology Research Center, School of Agriculture and Biology, Shanghai Jiao Tong University, Shanghai 200240, China

genetic diversities in fruit color, size, and mating system. In particular, the mating systems play key roles to control the reproductive habits between intra-/interspecies in tomatoes. Generally, color-fruited species such as *Solanum lycopersicum*, *S. pimpinellifolium* and *S. neorickii* are SC species, while some green-fruit species, such as *S. habrochaites* and *S. chilense*, are SI species [4]. However, the growth of pollen tubes within styles differs between SI and SC species. Pollen growth is arrested at the middle style in SI species, but not in SC. Some models were proposed for growth behavior of pollen tubes within styles that are related to pollen factors such as F-box protein and pistil factor of RNase [5,6]; however, the mechanism controlling the growth of pollen tubes remains unclear in tomatoes.

The transcriptome is the sum of all the RNA transcription for specific cells in a certain functional condition, including mRNAs, non-coding RNAs (ncRNA) and small RNAs [7,8]. RNA-Seq is a deep-sequencing technology [7,9] that has many advantages compared with Serial Analysis of Gene Expression (SAGE) [10], Expressed Sequence Tag (EST) [11], cDNA-amplified fragment length polymorphism (AFLP) [12], DNA microarrays [13] and massively parallel signature sequencing (MPSS) [14]. RNA-seq has already been widely used for transcriptome research in *Miscanthus sinensis* [15], tomato [16], *Wolfiporia cocos* [17], *Hevea brasiliensis* [18], *Populus tomentosa* [19], *Lolium rigidum* [20] and wheat [21]. It has also been applied to study pollination in maize [22,23], and to study SC/SI in *Citrus clementina* [24], lemon [25] and *Leymus chinensis* [26]. To understand what occurs after pollination in the styles of tomatoes of different mating types at the transcriptome level, we compared the transcription profiles differences between tomato SI and SC species. The results provide valuable information for understanding the growth behavior of pollen tubes within styles.

At present, research into tomato SC and SI has mainly concentrated on the S-RNase aspect, with no comprehensive transcriptome-level studies. Thus, to the best of our knowledge, this is the first study to perform comparative transcriptome analyses of SC and SI tomato unpollinated/pollinated styles using RNA-seq. The results of RNA-seq were analyzed by mapping, differential gene expression analysis, GO and pathway analysis. The results revealed comprehensive information concerning SI and SC, and provided clues to the molecular mechanisms of SI and SC.

Results

Summary of RNA-seq datasets

SC unpollination/self-pollination (SCUP/SCP) and SI unpollination/self-pollination (SIUP/SIP) styles (total of 12 samples) were performed RNA-seq. The raw sequence

data yielded approximately 3.0 gigabases (GB) per sample and more than 96% of the raw read pairs obtained had a quality score of $\geq$ Q20. Total raw read pairs among the 12 samples ranged from 15 to 18 million. By later removing reads containing adapters, reads containing poly-N and low-quality reads from the raw data, high-quality read pairs were obtained. The number of high-quality read pairs among the 12 samples ranged from 14 to 17 million (about 98% of the raw read pairs). Approximately 90% of the high-quality read pairs from the SC samples and 70% of the SI samples could be mapped to the tomato reference genome sequence. In addition, unmapped read pairs ranged from 1 to 5 million and multiple mapped read pairs ranged from about 0.30% to 0.50% of mapped read pairs among the 12 samples (Table 1).

Differential gene expression profiles of unpollinated (UP) and self-pollinated (P) styles in SC and SI, and hierarchical cluster analysis

To quantify the expression levels of the transcripts, HT-seq was used to count the read numbers mapped to each gene, based on the 34,726 genes of the tomato reference genome. These data were then normalized to reads per kilobase of exon region in a given gene per million mapped reads (RPKM) values, which were calculated based on the length of the gene and read count mapped to this gene. The RPKM values for each gene are listed in Additional file 1. To determine differential expression genes (DEGs) of UP and P styles in SC and SI, we screened for DEGs between UP and P styles in SC, and between UP and P styles in SI using the following criteria: Log_2 fold-change (FC) > 1 or $Log_2 FC < -1$ and P-value < 0.05. We identified 1341 DEGs between UP and P styles in SC, and 804 DEGs between UP and P styles in SI (Additional file 2). Of these DEGs, 753 genes were downregulated and 588 genes were upregulated after self-pollination in SC; 215 genes were downregulated and 589 genes were upregulated after self-pollination in SI (Figure 1). We used hierarchical cluster analysis to compare the DEGs between UP and P styles in SC, between UP and P styles in SI, and the similarity of the expression patterns of the three biological replicates (Figure 1).

GO annotation of all DEGs in SCP *vs*. SCUP and SIP *vs*. SIUP

To identify the functions of thee DEGs, we performed gene ontology (GO) analysis. A total of 798 DEGs of SC comparing UP and P styles were assigned GO annotations and 525 DEGs of SI comparing UP and P styles were assigned GO annotations. GO has three ontologies: molecular function, cellular component and biological process. In many cases, one gene was annotated with multiple GO terms. The GO terms of 798 DEGs of SCP

Table 1 Statistics of raw and mapped read pairs from RNA-seq analysis of SC unpollination/self-pollination (SCUP/SCP) and SI unpollination/self-pollination (SIUP/SIP) styles

Sample ID	Raw read pairs	High-quality read pairs	High-quality Percent	Mapped read pairs	Mapped Percent	Unmapped read pairs	Multi-mapped read pairs	Multi-mapped Percent
SCP1	17000933	15215933	89.50%	13817410	90.80%	1398523	65920	0.50%
SCP2	16374027	14680679	89.66%	13485391	91.90%	1195288	59339	0.50%
SCP3	17667649	15893802	89.96%	14489321	91.20%	1404481	67431	0.50%
SCUP1	18248702	16320337	89.43%	14747316	90.40%	1573021	48233	0.30%
SCUP2	17346914	15557760	89.69%	14145517	90.90%	1412243	59543	0.40%
SCUP3	18986356	17021427	89.65%	15362024	90.30%	1659403	56730	0.40%
SIP1	15510971	13879490	89.48%	9431478	68.00%	4448012	32428	0.30%
SIP2	16845976	15163409	90.01%	10608995	70.00%	4554414	37544	0.40%
SIP3	16920459	15154474	89.56%	10396040	68.60%	4758434	43009	0.40%
SIUP1	17664280	15847493	89.71%	10885898	68.70%	4961595	29071	0.30%
SIUP2	17752773	15880716	89.45%	11004025	69.30%	4876691	31678	0.30%
SIUP3	18253204	16435260	90.04%	11232677	68.30%	5202583	31212	0.30%

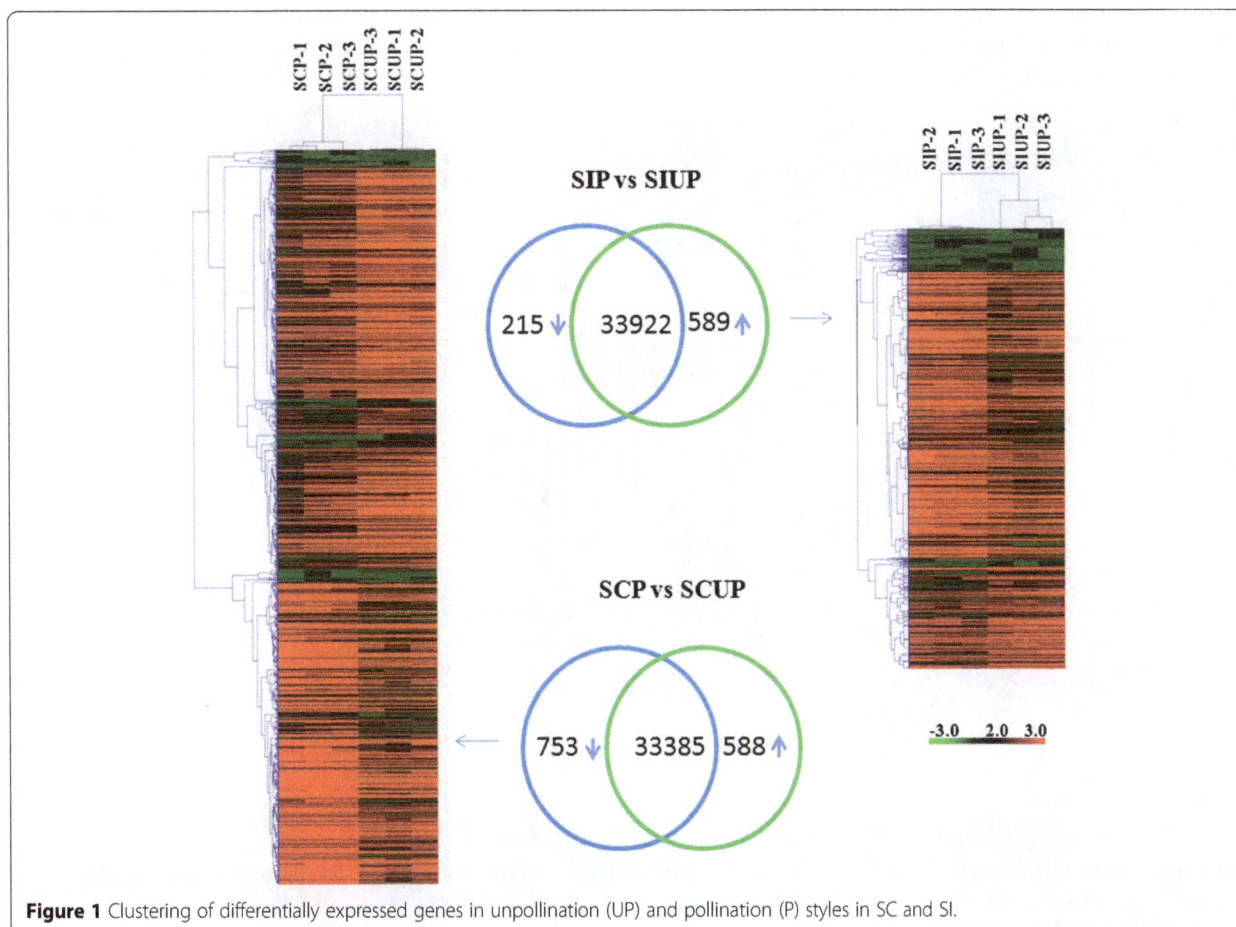

Figure 1 Clustering of differentially expressed genes in unpollination (UP) and pollination (P) styles in SC and SI.

vs. SCUP styles were categorized into 42 main functional groups belonging to the three categories and the GO terms of 525 DEGs of SIP *vs.* SIUP styles were categorized into 41 main functional groups belonging to the three categories (Figure 2).

Comparative analysis of GO terms assigned to SCP *vs.* SCUP DEGs and those assigned to SIP *vs.* SIUP DEGs

To better understand the distribution of gene functions at the macro level, the GO function classification of the DEGs in SCP *vs.* SCUP styles and SIP *vs.* SIUP styles were analyzed using the WEGO online tool. The comparative analysis showed that DEGs in SCP *vs.* SCUP styles and SIP *vs.* SIUP styles shared broad similarities in the proportion of genes in the three main categories, but differences were detected in many subcategories (Figure 2). Most GO terms of DEGs in SCP *vs.* SCUP styles and SIP *vs.* SIUP styles were categorized into the same biological processes, cellular components and molecular functions. Most GO subcategories terms were detected in both of SCP *vs.* SCUP styles and SIP *vs.* SIUP styles; however, GO subcategory terms, including membrane-enclosed lumen, organelle part, molecular transducer, transcription regulator, biological regulation, developmental process, multicellular organismal process, pigmentation, reproduction, reproductive process and response to stimulus showed significant (P-value < 0.05) differences in counts between

SCP *vs.* SCUP styles and SIP *vs.* SIUP styles. These results suggested that despite certain mechanisms of SC and SI appear to be conserved, the regulation mechanisms appear to be different between these two reproductive systems.

GO enrichment analysis of all DEGs in SCP *vs.* SCUP and SIP *vs.* SIUP

Significantly enriched GO terms were identified using singular enrichment analysis (SEA). The results showed that nine GO terms were significant in DEGs of SCP *vs.* SCUP based on a P-value < 0.05 and the false discovery rate (FDR) < 0.05 cutoffs (Figure 3A), which comprised two, three and four terms for the cellular components, molecular functions, biological processes categories, respectively. Seventy-eight GO terms were significant in DEGs of SIP *vs.* SIUP based on a P-value < 0.05 and the FDR < 0.05 cutoffs (Figure 3B, only 9), which comprised eight and 70 terms for the molecular functions and biological processes categories, respectively. The detailed results of the SCP *vs.* SCUP and SIP *vs.* SIUP Go enrichment analysis are presented in Additional file 3.

KEGG pathway mapping of all DEGs in SCP *vs.* SCUP and SIP *vs.* SIUP

To further investigate the influence of the DEGs on pathways, statistical pathway enrichment analysis of DEGs in SCP *vs.* SCUP and SIP *vs.* SIUP were performed

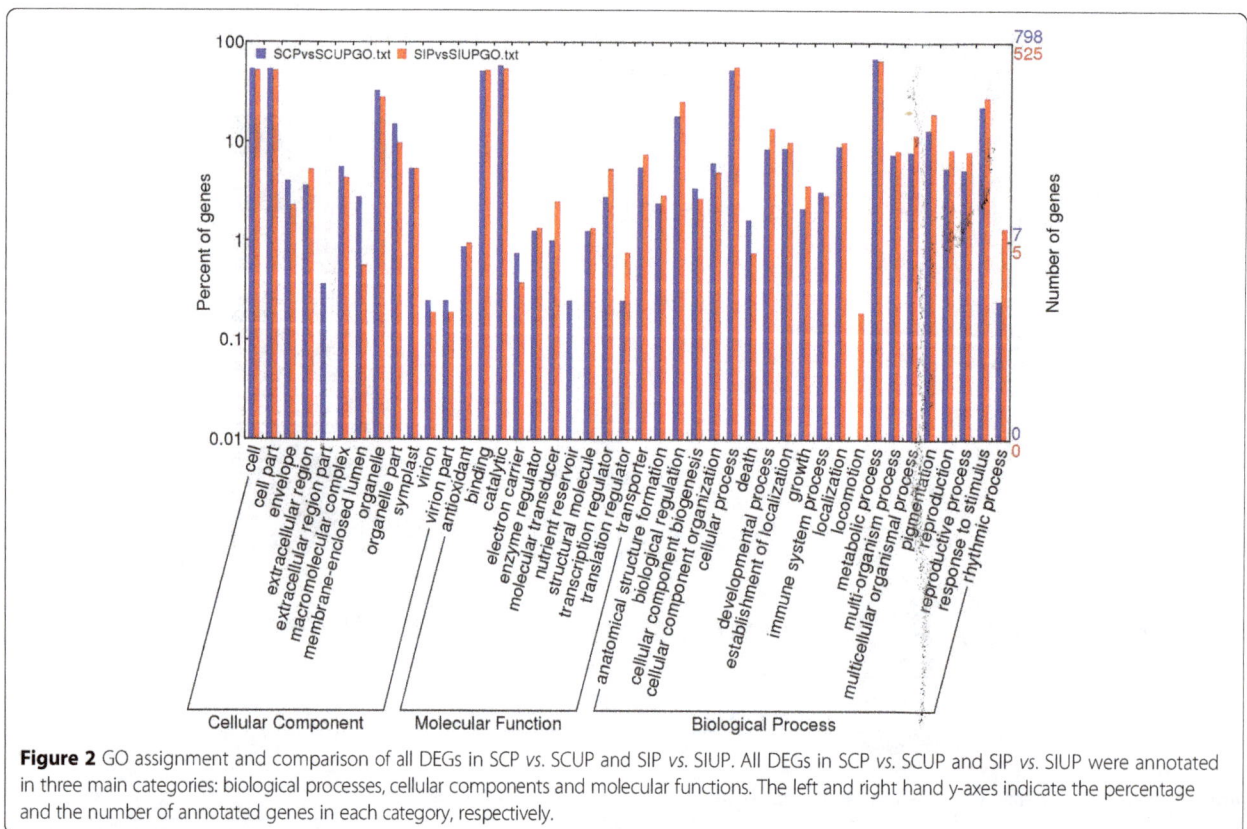

Figure 2 GO assignment and comparison of all DEGs in SCP *vs.* SCUP and SIP *vs.* SIUP. All DEGs in SCP *vs.* SCUP and SIP *vs.* SIUP were annotated in three main categories: biological processes, cellular components and molecular functions. The left and right hand y-axes indicate the percentage and the number of annotated genes in each category, respectively.

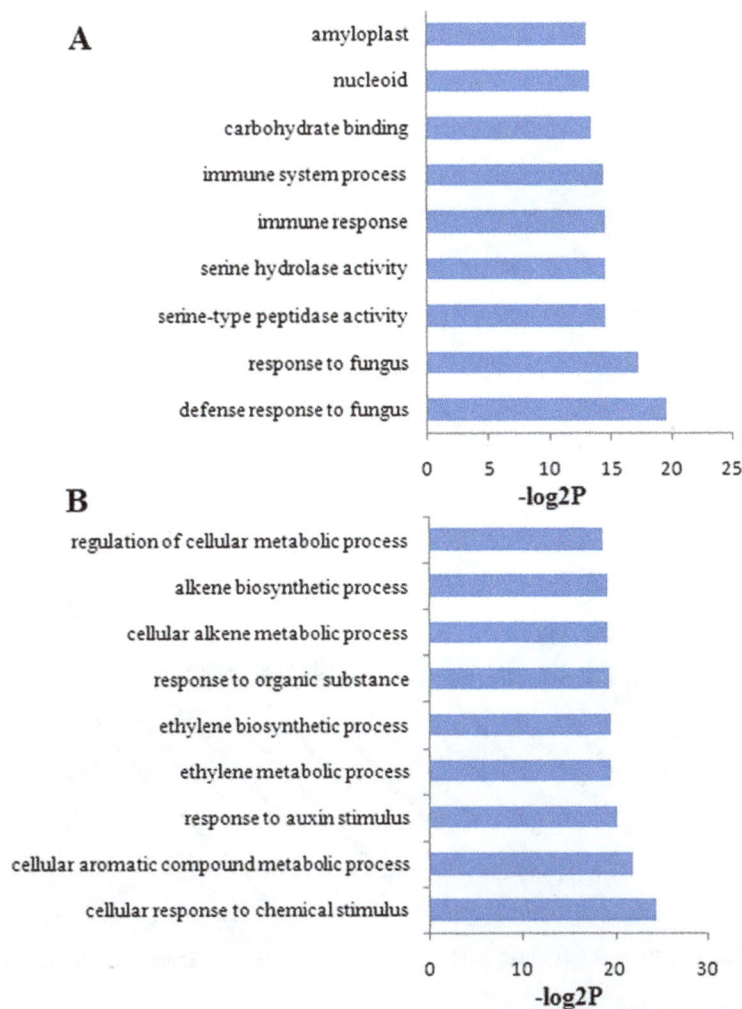

Figure 3 Significant gene ontology analysis of DEGs in SCP *vs.* SCUP and SIP *vs.* SIUP. **A.** Significant GO terms of SCP *vs.* SCUP; **B.** Significant GO terms of SIP *vs.* SIUP (The first nine significant GO terms). P-value < 0.05 and FDR < 0.05 for all significant GO terms.

based on KEGG database, using Fisher's exact test. The DEGs of SCP *vs.* SCUP were enriched in 105 KEGG metabolic pathways and the DEGs of SIP *vs.* SIUP were enriched in 80 KEGG metabolic pathways (Additional file 4). The top ten KEGG metabolic pathways and three P-value < 0.05 metabolic pathways of the DEGs in SCP *vs.* SCUP are shown in Figure 4A. Among these 105 pathways of SCP *vs.* SCUP, those containing the greatest numbers of DEGs transcripts were "Metabolic pathways" (containing 111 DEGs) and "Biosynthesis of secondary metabolites" (containing 75 DEGs). Other GO terms associated with higher numbers of DEGs were "Starch and sucrose metabolism" (16 DEGs), "Plant hormone signal transduction" (16 DEGs), "Biosynthesis of amino acids" (15 DEGs), "Carbon metabolism" (15 DEGs), "Plant-pathogen interaction" (12 DEGs), "Phenylpropanoid biosynthesis" (11 DEGs), "Glycolysis/Gluconeogenesis" (nine DEGs), and "Amino sugar and nucleotide sugar metabolism" (eight

DEGs); The pathways of "Biosynthesis of secondary metabolites", "Biotin metabolism", "Brassinosteroid biosynthesis" and "Degradation of aromatic compounds" had P-values < 0.05 (Figure 4A). For SIP *vs.* SIUP, of 13 KEGG metabolic pathways were identified. The top 11 KEGG metabolic pathways and two P-value < 0.05 metabolic pathways of DEGs in SIP *vs.* SIUP are shown in Figure 4B. Among the 80 pathways of SIP *vs.* SIUP, those containing the greatest numbers of DEGs were "Metabolic pathways" (69 DEGs), "Biosynthesis of secondary metabolites" (40 DEGs), "Plant hormone signal transduction" (22 DEGs), "Plant-pathogen interaction" (10 DEGs), "Starch and sucrose metabolism" (9 DEGs), "Biosynthesis of amino acids" (nine DEGs), "Phenylpropanoid biosynthesis" (nine DEGs), "Carbon metabolism" (eight DEGs), "Pentose and glucuronate interconversions" (eight DEGs), "Phenylalanine metabolism" (seven DEGs). The pathways of "Cysteine and methionine metabolism", "Plant hormone signal transduction", "Pentose

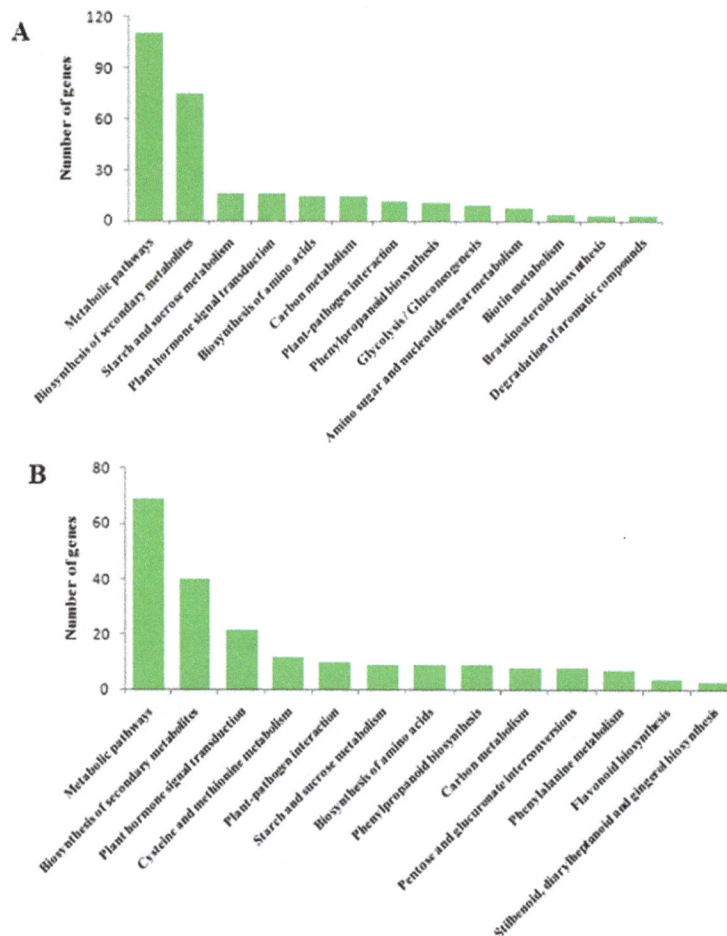

Figure 4 Pathway enrichment analysis of DEGs in SCP *vs.* SCUP and SIP *vs.* SIUP based on KEGG. **A**. Enriched pathways in SCP *vs.* SCUP; **B**. Enriched pathways in SIP *vs.* SIUP.

and glucuronate interconversions", "Flavonoid biosynthesis" and "Stilbenoid, diarylheptanoid and gingerol biosynthesis" all had P-values < 0.05 (Figure 4B). In addition, the pathways of "Cysteine and methionine metabolism" and "Plant hormone signal transduction" were significant pathways in DEGs of SIP *vs.* SIUP, based on a P-value < 0.05 and the FDR < 0.05 cutoffs (Figure 4B). The detailed results of the SIP *vs.* SIUP significant pathways enrichment analysis are presented in Figures 5 and 6.

"Cysteine and methionine metabolism" is the ethylene biosynthesis pathway, which was significantly enriched in the SIP *vs.* SIUP analysis. DEGs were enriched in the step of O-Acetyl-L-serine conversion to L-Cysteine, L-Homocysteine conversion to L-Methionine, L-Methionine conversion to S-adenosyl-L-methionine (AdoMet), AdoMet conversion to 1-aminocyclopropane-1-carboxylate (ACC) and ACC production ethylene (Figure 5). L-Methionine conversion to AdoMet was the first step of ethylene biosynthesis, AdoMet conversion to ACC was the rate-limiting step in ethylene biosynthesis and ACC production ethylene was the last steps for ethylene biosynthesis. Plant hormone

signal transduction is very important to hormone-instigated biochemical changes during plant growth, development, and environmental information processing pathways, which were also significantly enriched in the SIP *vs.* SIUP comparison. DEGs were also enriched in Auxin signal transduction, Abscisic acid (ABA) signal transduction, Ethylene signal transduction, Jasmonic acid (JA) signal transduction and Salicylic acid (SA) signal transduction (Figure 6).

Significant pathways enrichment analysis showed that cysteine and methionine metabolism and plant hormone signal transduction were the most important pathways in SIP *vs.* SIUP comparison, and plant hormone signal transduction was the key biological event. All the plant hormone signaling pathways pointed to it and the significant pathway of "Cysteine and methionine metabolism" also (Figure 7). This evidence indicated that plant hormone signal transduction plays important roles in tomato SI.

Discussion

RNA-seq is a powerful tool that can provide a global overview of gene expression at the transcriptome level. With

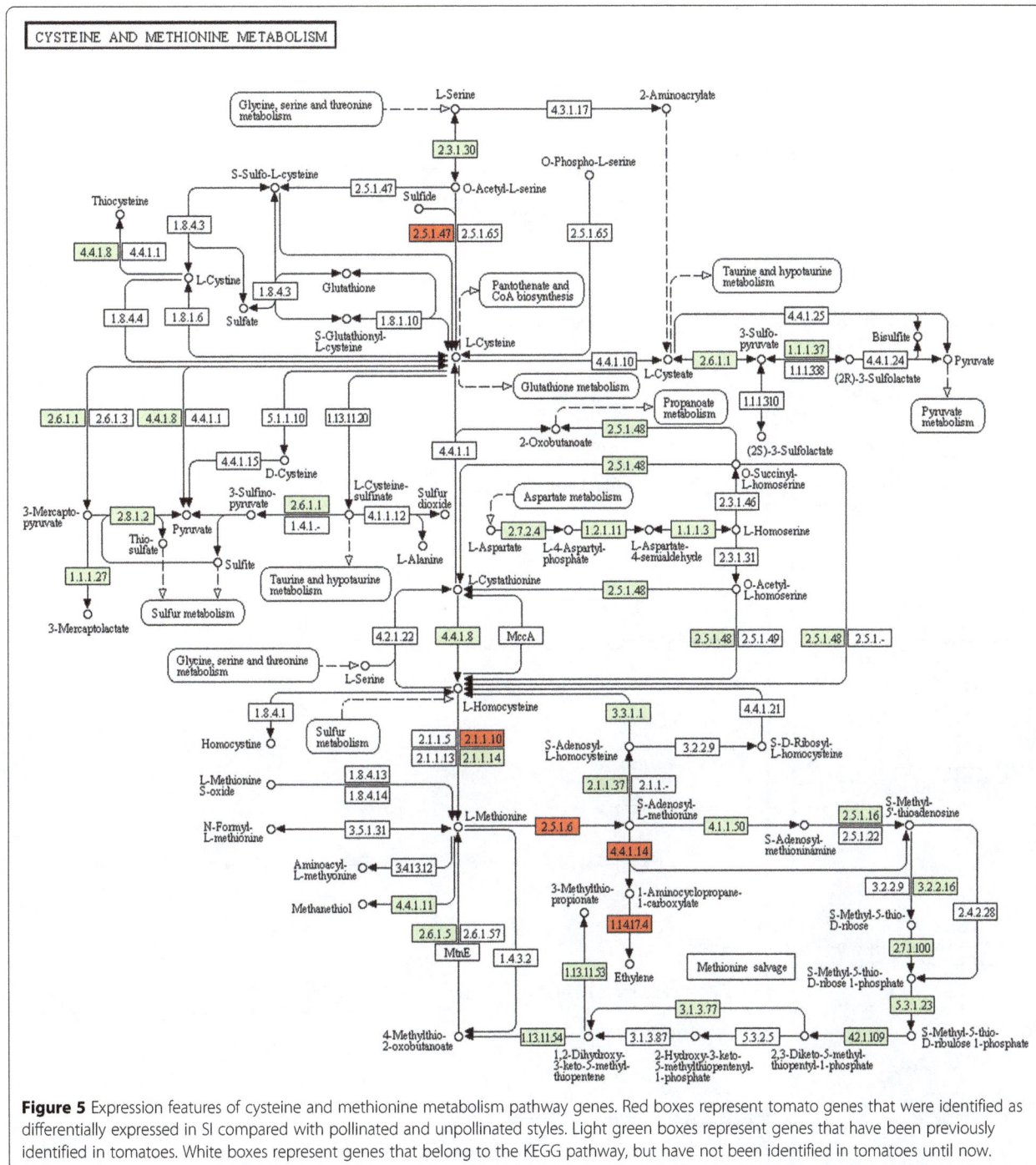

Figure 5 Expression features of cysteine and methionine metabolism pathway genes. Red boxes represent tomato genes that were identified as differentially expressed in SI compared with pollinated and unpollinated styles. Light green boxes represent genes that have been previously identified in tomatoes. White boxes represent genes that belong to the KEGG pathway, but have not been identified in tomatoes until now.

reductions in sequencing costs and the advance of technologies, RNA-seq will become more accessible to researchers to identify and track the expression changes of all genes [7]. The present study identified 1341 significant (P-value < 0.05) DEGs after comparing UP and P styles in SC and 804 significant (P-value < 0.05) DEGs in the comparison of UP and P styles in SI, using RNA-seq analysis. The total number of gene changes demonstrated that SC self-pollination and SI self-pollination

are complex processes. This finding is consistent with other plant pollination studies. For example, 1025 differentially expressed genes were potentially involved in the pollination response and SI mechanisms in sheepgrass [26]. In a comparison of pollinated and unpollinated stigmas with styles, 4785 DEGs were identified in SI lemon [25]. These data demonstrate the complex nature of the transcriptome changes in SC self-pollination and SI self-pollination.

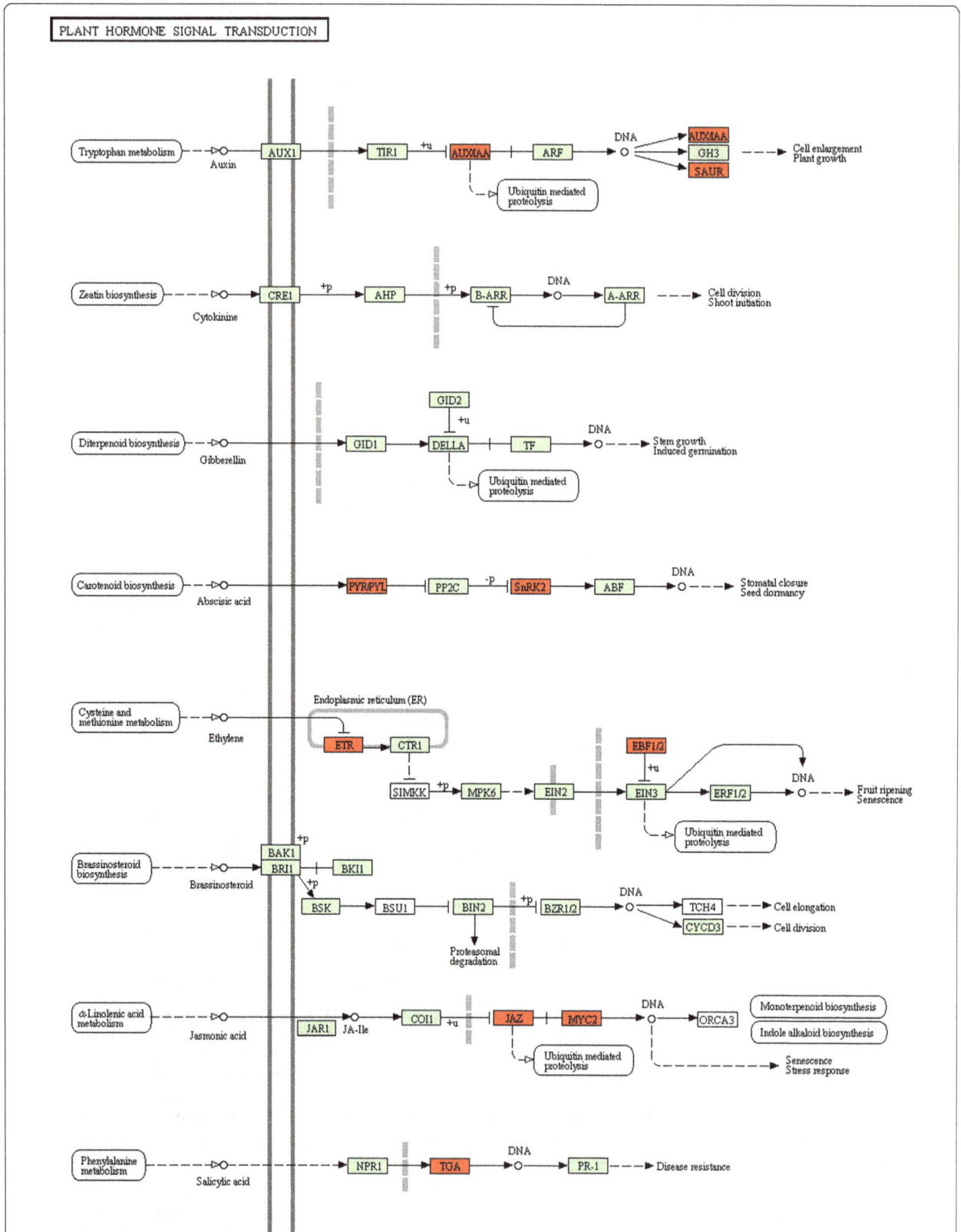

Figure 6 Expression features of plant hormone signal transduction pathway genes. Red boxes represent tomato genes that were identified as differentially expressed in SI compared with pollinated and unpollinated styles. Light green boxes represent genes that have been previously identified in tomatoes. White boxes represent genes that belong to the KEGG pathway, but have not been identified in tomatoes until now.

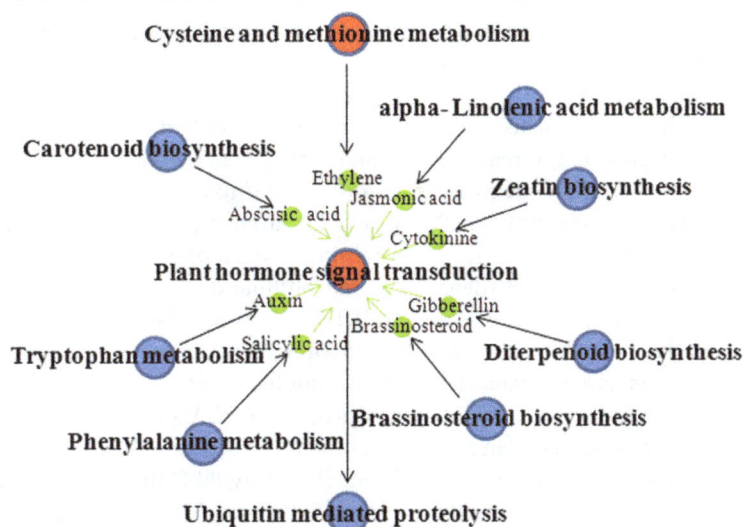

Figure 7 Significant pathways enrichment analysis and interaction network of SIP *vs.* SIUP based on KEGG. Red circles represent significantly enriched pathways.

Pollination shares striking similarities with fungal infection in terms of biological responses and processes that result in cell death [27,28]. Our transcriptome GO enrichment analysis identified several significant GO terms involved in pathogen invasion responses, such as defense response to fungus, response to fungus, immune response, and immune system process in the SCP *vs.* SCUP comparison. This result is consistent with other plant pollination studies, such as in Arabidopsis [29,30] and rice [31]. However, GO terms involved in stimuli and hormones were the most important of the 78 significant GO terms in the SIP *vs.* SIUP comparison.

Pollination leads to senescence of petunia corollas by inducing many hormonal, physiological, and molecular changes [32]. Ethylene is a gaseous plant hormone with a wide range of effects on plant growth and development [33]. Ethylene is synthesized from L-Methionine via the intermediates AdoMet and ACC (Figure 5) [34-36]. AdoMet is made from L-Methionine by the enzyme S-adenosylmethionine synthase (SAM), representing the first step of ethylene biosynthesis (Figure 5). 1-aminocyclopropane-1-carboxylate synthase (ACS) gene family members and 1-aminocyclopropane-1-carboxylate oxidase (ACO) gene family members are two important enzymes for ethylene biosynthesis. ACS catalyzes the conversion of AdoMet to ACC, which is the rate-limiting step in ethylene biosynthesis. ACO then catalyzes the conversion of ACC to ethylene (Figure 5) [37]. After SI self-pollination, one SAM gene (S-adenosyl-methionine synthase 2-like) (Solyc10g083970), five ACS gene family members (Solyc00g095760, Solyc08g081550, Solyc08g008100, Solyc08g081540, Solyc00g095860) and four ACO gene family members (Solyc02g036350,

Solyc07g026650, Solyc07g049530, Solyc07g049550) were significantly upregulated, which indicated that SI self-pollination is associated with results in significant upregulation of ethylene biosynthesis related genes and ethylene production. It has been reported that ethylene biosynthesis is induced by pollination in petunias [38]. After SC self-pollination, although the pathway of "Cysteine and methionine metabolism" was not a significant enrichment pathway in the SCP *vs.* SCP comparison, two ACS gene family members (Solyc08g081540, Solyc00g095860) and one ACO gene family member (Solyc07g049530) were significantly upregulated, which indicated that SC self-pollination results in some upregulation of ethylene biosynthesis of partly related genes. The above results suggest that SI self-pollination induces more ethylene production than SC self-pollination.

Plant hormone signal transduction is very important to hormone triggered biochemical changes [39]. Plant hormone signal transduction plays an important role in pollination of petunias pollination; for example, RNA-seq revealed that plant hormone signal transduction-related KEGG pathways were enriched in petunia corollas when comparing pollinated and unpollinated samples [32]. After SI self-pollination, plant hormone signal transduction-related KEGG pathways were significantly enriched in the SIP *vs.* SIUP comparison, but not after SC self-pollination (Figure 6). This result indicated that plant hormone signal transduction might play an important role in tomato SI. Plants recognize and transduce the ethylene signal via ethylene receptors (ETR) [40] in the ethylene signal transduction pathway (Figure 6) [41]. We identified two ethylene receptors, LeETR6 (Solyc06g053710) and tETR (Solyc09g089610), which were significantly upregulated in the SIP *vs.* SIUP comparison, both of which mapped to the plant hormone signal transduction KEGG pathway. LeETR2 (Solyc07g056580) was the only ethylene receptor identified from the SCP *vs.* SCUP comparison, and significantly downregulated in P styles compared with UP styles. This protein also mapped to the plant hormone signal transduction KEGG pathway, which was not a significantly enriched pathway in the SCP *vs.* SCUP comparison. The above results indicated that SI self-pollination not only involves the induction of ethylene production, but also enhanced the perception ethylene. Although SC self-pollination may involve some enhancement of ethylene production, the ability to perceive ethylene was weakened by the significant downregulation of LeETR2. Plant responses to ethylene initiates with ethylene binding to ETRs and terminates in a transcription cascade of plant-specific transcription factors families, especially the ethylene-insensitive protein 3 (EIN3/EIL) and ethylene-responsive transcription factor (ERF). EIN3 protein is a key transcription factor for mediating the expression of ethylene-regulated genes and morphological responses. EIN3 interacts physically with the Ein3-binding f-box protein1/2 (EBF1/EBF2) and is ultimately and quickly degraded through a ubiquitin/proteasome pathway mediated by the SCF complex, which comprises a RING-box protein 1 (RBX1), Cullin 1 (Cul1), S-phase kinase-associated protein 1 (Skp1), F-box protein (F-box) [42,43]. We identified one EBF1/2 (Solyc07g008250) from the SC and two EBF1/2 (Solyc07g008250, Solyc12g009560) from the SI, both of which were significantly upregulated in P compared with UP styles. In addition, we also identified one Skp1 (Solyc01g111640) and one Cul1 (Solyc01g067120) from SI, which were significantly upregulated in P compared with UP styles. This result indicated that key transcription factor EIN3 was negatively regulated by targeting EIN3 it for degradation through the ubiquitin/proteasome pathway after SI self-pollination, but not in SC pollination.

A previous study demonstrated that auxin was significantly increased after compatible pollination and ethylene was strongly increased after incompatible pollination [44,46]. The last step of indole-3-acetic acid (IAA) biosynthesis is performed by aldehyde dehydrogenase. We identified one aldehyde dehydrogenase (aldehyde dehydrogenase family 2 member B4, Solyc08g068190) from SC that was significantly upregulated in P compared with UP styles and one aldehyde dehydrogenase (aldehyde dehydrogenase family 3 member H1-like, Solyc06g060250) from SI that was significantly downregulated in P compared with UP styles. This result is consistent with the results of the previous study. Auxin is likely to be directly or indirectly involved in pollen-pistil recognition and pollen tube elongation in Nicotiana [45] and might have an important role in the SI response in plants such as *Theobroma cacao* [46], *Petunia hybrida* [47] and *Olea europaea* [48]. Auxins regulate plant growth and development by a complex signal transduction network [49], which was included in the significantly enriched KEGG pathways of plant hormone signal transduction KEGG in the SIP *vs.* SIUP comparison. Auxin influx carrier (AUX1 LAX family) is a polar auxin transporter in cells that is involved in attaining a hormone maximum (Figure 6) [50]. After SC self-pollination, LAX2 protein (auxin influx carrier, AUX1 LAX family) (Solyc01g111310) was significantly downregulated. Auxins alter three major gene families: auxin/indole-3-acetic acid (Aux/IAA), GH3 and small auxin-up RNA (SAUR) to direct plant growth and development (Figure 6) [49,51]. Aux/IAA gene families: IAA1 (Solyc09g083280), IAA2 (Solyc06g084070), IAA3 (Solyc09g065850), IAA19 (Solyc03g120380), IAA22 (Solyc06g008580), IAA26 (Solyc03g121060), IAA35 (Solyc07Vg008020) and IAA36 (Solyc06g066020) were significantly upregulated in the SIP *vs.* SIUP comparison, and only IAA2 (Solyc06g084070), IAA29 (Solyc08g021820) and IAA 35 (Solyc07g008020) were significantly upregulated in the SCP *vs.* SCUP comparison. For the GH3 gene families, only one probable indole-3-acetic acid-amido synthetase GH3.1-like gene (Solyc02g092820) was significantly upregulated in the SCP *vs.* SCUP comparison. For the SAUR gene families, small auxin-up protein 58 (Solyc06g053260), auxin-induced protein 10A5-like (Solyc03g033590), uncharacterized LOC101249064 (Solyc03g124020) and uncharacterized LOC101254455 (Solyc12g009280) were significantly upregulated, and auxin-induced protein 15A-like (Solyc01g110570) and auxin-induced protein 10A5-like (Solyc01g110560) were significantly downregulated in the SIP *vs.* SIUP comparison. Only auxin-induced protein 15A-like (Solyc09g009980) and indole-3-acetic acid-induced protein ARG7-like (Solyc04g081250) were significantly upregulated in the SCP *vs.* SCUP comparison. These results indicated that although auxin was strongly increased after compatible pollination, because the auxin influx carrier (AUX1 LAX family) (Solyc01g111310) was significantly downregulated, fewer auxin-responsive genes showed altered expressions. During SC pollination, the auxin influx carrier (AUX1 LAX family) was not affected, resulting in many auxin-responsive genes showing altered expression after incompatible pollination. A previous study indicated that auxin influx carriers (AUX1 LAX family) were involved in auxin-ethylene interactions in *Arabidopsis thaliana* [52]; however, whether auxin influx carriers (AUX1 LAX family) are also involved in auxin-ethylene interactions in tomato SI is unknown.

Ethylene and JA, as well as ABA and auxin, have direct or indirect interactions [32], but the roles of JA and ABA in tomato pollination, especially in SI self-pollination, were unknown. ABA is a phytohormone that acts in seed dormancy, plant development and environmental stress. The carotenoid biosynthesis pathway is an ABA biosynthesis pathway (Figure 6) that was enriched in SC and SI. Endogenous ABA levels are regulated by both ABA biosynthesis and ABA catabolism: xanthoxin dehydrogenase is a key enzyme for ABA biosynthesis and ABA 8′-hydroxylase is a key enzyme for ABA catabolism [53,54]. Xanthoxin dehydrogenase (Solyc12g056600) was significantly upregulated in SC not in SI and ABA 8′-hydroxylase 1-like (CYP707A2, Solyc08g005610) was significantly upregulated in SI but not in SC, which indicated that endogenous ABA levels increased in SC and decreased in SI styles during pollination. Pyrabactin resistance/pyrabactin resistance-like (PYR/PYL) family is an ABA receptor that is very important to ABA recognition and signaling [55,56]. We identified two genes of the PYR/PYL family: ABA receptor PYL8-like (Solyc03g007310) from SI and ABA receptor PYL6-like (Solyc06g050500) from SC. PYL8-like was significantly downregulated in SI and PYL6-like was significantly upregulated in SC styles during pollination, which indicated that the ability to perceive ABA was weakened in SI and enhance in SC. A previous study showed that PYR/PYLs are negative regulatory receptors, whereby ABA binds to PYR/PYLs, which in turn binds to type 2C protein phosphatases (PP2Cs) to inhibit PP2Cs. SNF1-related protein kinase subfamily 2 (SnRK2) is located downstream of PP2Cs and is negatively regulated by PP2Cs (Figure 6). SnRK2 (serine/threonine-protein kinase SAPK3-like, Solyc08g077780) was upregulated in SI (in which PP2Cs are not inhibited) and an SnRK2 (serine/threonine-protein kinase SAPK7-like, Solyc05g056550) was downregulated in SC, wherePP2Cs are inhibited. In addition, SnRK2s can phosphorylate b-ZIP transcription factors, which bind to the ABA-responsive element to activate ABA-responsive genes. Phosphorylated b-ZIP transcription factors are important to active ABA-responsive genes [57]. One b-ZIP transcription factor (Solyc10g076920) was significantly downregulated in the SCP *vs.* SCUP comparison, but not in the SIP *vs.* SIUP comparison. This indicated that ABA might have important regulatory roles in SI. Jasmonates are phytohormones that are essential for plant development and survival, and can induce jasmonate ZIM-domain proteins (JAZs) to be degraded through the ubiquitin/proteasome pathway, mediated by the SCF[COI1] complex. In addition, JAZs negatively regulate MYC2, which is a key jasmonate responses transcriptional activator [58]. We identified a JAZ (jasmonate ZIM-domain protein 1, Solyc12g009220) and a transcription factor MYC2 (Solyc08g076930), both of which were both significantly upregulated in the SIP *vs.* SIUP comparison. The TGA family comprises key transcription factors of the salicylic acid (SA)-mediated signal transduction pathway [59]. After SI self-pollination, TGA family transcription factor (Solyc10g080410) was significantly upregulated.

Conclusions

This is the first global transcriptome-wide comparative analysis of styles from SC and SI tomatoes using a high-throughput RNA-seq. The enriched GO term analysis of the identified DEGs showed that nine GO terms were significantly enriched in the SCP *vs.* SCUP comparison and 78 GO terms were significantly enriched in the SIP *vs.* SIUP comparison. The ethylene biosynthesis pathway of the cysteine and methionine metabolism pathway and the plant hormone signal transduction pathway play an important role in tomato SI. Further GO and KEGG analyses showed that SI self-pollination induced more ethylene production and catabolism of ABA, and SC self-pollination induced more auxin production and ABA biosynthesis. Moreover, the phytohormones ethylene, auxin and ABA play important roles by plant hormone signal transduction in tomato SI.

Methods
Plant materials

Tomato seeds of *S. chilense* (LA0130, SI) and *S. pimpinellifolium* (LA1585, SC) were obtained from the Charles Rick Tomato Genetics Resource Center (UC, Davis http:// tgrc.ucdavis.edu/index.aspx). The seeds were germinated in peat pellets and seedlings with three to four leaves were grown on medium containing the perlite: peat (1:1) under a thermoperiod of 26/20°C (day/night) in a greenhouse. Plants were supplied with a commercial fertilizer every week. During flowering, 24 h UP and P styles (containing stigmas) (Additional file 5) were collected from *S. chilense* (LA0130) (SIUP/SIP) and *S. pimpinellifolium* (LA1585) (SCUP/SCP), respectively, and immediately frozen in liquid nitrogen and stored at −80°C for RNA extraction. Three biological replicates of each sample were collected and used for RNA extraction.

RNA extraction and deep sequencing

Total RNA was extracted from each sample using an RNAprep pure Plant Kit (Tiangen, Beijing, China), according to the manufacturer's protocol. The RNA concentration of each sample was measured using a NanoDrop 2000 (Thermo Scientific, Waltham, MA, USA). The RNA quality was assessed using an Agilent2200 (Agilent Technologies, Santa Clara, CA, USA).

The sequencing library for each RNA sample was prepared using a TruseqTM RNA sample prep Kit (Illumina, San Diego, CA, USA), following the manufacturer's protocol. Briefly, mRNA was purified using poly-T oligo-attached magnetic beads (Invitrogen,Carlsbad, CA, USA) from 5 μg total RNA. The mRNA was fragmented, and the RNA fragments were reverse transcribed and amplified to double-stranded cDNA. Index adapters were then ligated to the cDNA according to the protocol of the TruseqTM RNA sample prep Kit (Illumina). The library was quantified using a TBS-380 mini-fluorometer (Picogreen, Cohasset, MA, USA). The clustering of the index-coded samples was performed on a cBot Cluster Generation System, using a TruSeq PE Cluster Kit v3-cBot-HS (Illumina), according to the manufacturer's instructions. After cluster generation, the library preparations were sequenced on an Illumina Hiseq 2500 platform and a sequence length of 2*101 bp paired-end reads were generated.

Filtering raw reads and mapping

The raw reads were pass-filtered using the Trimmomatic tool [60] and then used for mapping. The reference tomato genome and gene model annotation files were downloaded from the genome website (http://solgenomics.net/) directly. The paired-end clean reads were aligned to the reference tomato genome using Tophat [61] and the mapped reads were counted with using HT-seq [62].

Identification of DEGs

Gene expression levels were estimated as RPKM [63]. Differential expression analysis of SCUP/SCP groups and SIUP/SIP groups was performed using the DESeq R package (1.10.1), which provides statistical routines for determining differential expression in digital gene expression data using a model based on the negative binomial distribution. After statistical analysis, the DEGs were identified using significance analysis by t-tests, with a P-value < 0.05 and at least two-fold changes (either up- or downregulation) being considered significant.

GO analysis

The blast2go [64] program was used to obtain GO annotations for all identified genes. GO functional classification was performed using the WEGO online tool [65] to gain an understanding of the distribution of gene functions at the macro level. GO is the key functional classification of NCBI, which was applied to analyze the functions of the DEGs [66,67]. GO enrichment analysis of DEGs was implemented using SEA [68], in which Fisher's exact test and a χ^2 test were used to classify the GO categories; the FDR was calculated to correct the P-value [69,70]. P-values for the GOs of all the DEGs were computed. The significant GO terms were defined as having a P-value < 0.05 and an FDR < 0.05.

Pathway analysis

KEGG is a database resource for understanding high-level functions and utilities of biological systems, such as cells,

organisms and ecosystems, from molecular-level information, especially large-scale molecular datasets generated by genome sequencing and other high-through put experimental technologies (http://www.genome.jp/kegg/). KEGG pathway analysis was used to identify the significant pathways involving the DCEGs [71-73]. Fisher's exact test and a χ^2 test were used to identify significant pathways (P-value < 0.05 and FDR < 0.05) [74-76]. We used the KEGG Orthology Based Annotation System (KOBAS) software to test the statistical enrichment of DEGs in KEGG pathways.

Availability of supporting data

The data sets supporting the results of this article are available in the Gene Expression Omnibus repository under accession no GSE67654 (http://www.ncbi.nlm.nih.gov/geo/query/acc.cgi?acc=GSE67654) [77].

Additional files

Additional file 1: Table S1. RNA-seq of data of all counts for SI and SC compared with self-pollinated and unpollinated styles.

Additional file 2: Table S2. List of differentially expressed genes for SI and SC compared with self-pollinated and unpollinated styles.

Additional file 3: Table S3. GO analysis of differentially expressed genes for SI and SC compared with self-pollinated and unpollinated styles.

Additional file 4: Table S4. Pathway analysis differentially expressed genes for SI and SC compared with self-pollinated and unpollinated styles.

Additional file 5: Figure S1. The structure of the tomato pistil. Red lines show the cutting position of a style containing a stigma.

Abbreviations

SC: Self-compatibility; SI: Self-incompatibility; ncRNA: Non-coding RNAs; SAGE: Serial Analysis of Gene Expression; EST: Expressed sequence tag; MPSS: Massively parallel signature sequencing; RNA-seq: mRNA sequencing; GO: Gene ontology; GB: Gigabases; SCUP: SC self-pollination; SCP: SC self-pollination; SIUP: SI unpollination; SIP: SI self-pollination; UP: Unpollination; P: Pollination; RPKM: Reads per kilobase of exon region in a given gene per million mapped reads; DEGs: Differentially expressed genes; FC: Fold-change; SEA: Singular enrichment analysis; FDR: False discovery rate; KEGG: Kyoto encyclopedia of genes and genomes; AdoMet: S-adenosyl-L-methionine; ACC: 1-aminocyclopropane-1-carboxylate; ABA: Abscisic acid; ACO: 1-aminocyclopropane-1-carboxylate oxidase; ACS: 1-aminocyclopropane-1-carboxylate synthase; SAM: S-adenosyl methionine synthase; ETR: Ethylene receptor; EIN3/EIL: Ethylene-insensitive protein 3; ERF: Ethylene-responsive transcription factor; EBF1/EBF2: Ein3-binding f-box protein1/2; IAA: Indole-3-acetic acid; AUX1 LAX: Auxin influx carrier; Aux/IAA: Auxin/indole-3-acetic acid; SAUR: Small auxin-up RNA; JA: Jasmonic acid; RBX1: RING-box protein 1; Cul1: Cullin 1; Skp1: S-phase kinase-associated protein 1; F-box: F-box protein; PYR/PYL: Pyrabactin resistance/pyrabactin resistance-like; PP2Cs: Type 2C protein phosphatases; SnRK2: SNF1-related protein kinase subfamily 2; JAZs: Jasmonate ZIM-domain protein; KOBAS: KEGG Orthology Based Annotation System.

Competing interests

The authors declare that they have no competing interests.

Authors' contributions

ZP participated in the experimental design, collected the material, performed the RNA extraction, participated in the bioinformatics analyses and wrote the manuscript. ZL participated in the bioinformatics analyses. ZL designed and wrote this manuscript. All authors read and approved the final manuscript.

Acknowledgements
We thank the Charles Rick Tomato Genetics Resource Center at the University of California Davis for supplying the tomato seeds for this study. The research was supported by the Key Technology Research and Development Program of Shanghai Science and Technology Committee (No. 13391901202 and No. 14JC1403400), the National Natural Science Foundation of China (No. 31071810) and the China National '863' High-Tech Program (No. 2011AA100607).

References

1. Lord EM, Russell SD. The mechanisms of pollination and fertilization in plants. Annu Rev Cell Dev Biol. 2002;18:81–105.

2. Samuel MA, Tang W, Jamshed M, Northey J, Patel D, Smith D, et al. Proteomic analysis of Brassica stigmatic proteins following the self-incompatibility reaction reveals a role for microtubule dynamics during pollen responses. Mol Cell Proteomics. 2011. Doi: 10.1074/mcp.M111.011338.

3. Kear PJ, McClure B. How did flowering plants learn to avoid blind date mistakes? Self-incompatibility in plants and comparisons with non self rejection in the immune response. Adv Exp Med Biol. 2012;738:108–23.

4. Li W, Chetelat RT. A pollen factor linking inter- and intraspecific pollen rejection in tomato. Science. 2010;330(6012):1827–30.

5. Qiao H, Wang F, Zhao L, Zhou J, Lai Z, Zhang Y, et al. The F-box protein AhSLF-S2 controls the pollen function of S-RNase-based self-incompatibility. Plant Cell. 2004a;16:2307–22.

6. Franklin-Tong VE, Franklin FC. Gametophytic self-incompatibility inhibits pollen tube growth using different mechanisms. Trends Plant Sci. 2003;8:598–605.

7. Wang Z, Gerstein M, Snyder M. RNA-Seq: a revolutionary tool for transcriptomics. Nat Rev Genet. 2009;10(1):57–63.

8. Costa V, Angelini C, De Feis I, Ciccodicola A. Uncovering the complexity of transcriptomes with RNA-Seq. J Biomed Biotechnol. 2010;2010:853916.

9. Haas BJ, Zody MC. Advancing RNA-Seq analysis. Nat Biotechnol. 2010;28:421–3.

10. Veleulescu VE, Zhang L, Zhou W, Vogelstein J, Basrai MA, Bassett Jr DE, et al. Characterization of the yeast transcriptome. Cell. 1997;88:243–51.

11. Adams MD, Kelley JM, Gocayne JD, Dubnick M, Polymeropoulos MH, Xiao H, et al. Complementary DNA sequencing: expressed sequence tags and human genome project. Science. 1991;252:1651–6.

12. Brugmans B, del Fernandez CA, Bachem CW, van Os H, van Eck HJ, Visser RG. A novel method for the construction of genome wide transcriptome maps. Plant J. 2002;31:211–22.

13. Schena M, Shalon D, Davis RW, Brown PO. Quantitative monitoring of gene expression patterns with a complementary DNA microarray. Science. 1995;270:467–70.

14. Brenner S, Johnson M, Bridgham J, Golda G, Lloyd DH, Johnson D, et al. Gene expression analysis by massively parallel signature sequencing (MPSS) on microbead arrays. Nat Biotechnol. 2000;18(6):630–4.

15. Swaminathan K, Chae WB, Mitros T, Varala K, Xie L, Barling A, et al. A framework genetic map for Miscanthus sinensis from RNAseq-based markers shows recent tetraploidy. BMC Genomics. 2012. doi: 10.1186/1471-2164-13-142.

16. Koeniga D, Jiménez-Gómez JM, Kimura S, Fulop D, Chitwood DH, Headland LR, et al. Comparative transcriptomics reveals patterns of selection in domesticated and wild tomato. Proc Natl Acad Sci U S A. 2013;110:E2655–62.

17. Shu S, Chen B, Zhou M, Zhao X, Xia H, Wang M. De novo sequencing and transcriptome analysis of Wolfiporia cocos to reveal genes related to biosynthesis of triterpenoids. PLoS One. 2013;8(8), e71350.

18. Salgado LR, Koop DM, Pinheiro DG, Rivallan R, Le Guen V, Nicolás MF, et al. De novo transcriptome analysis of Hevea brasiliensis tissues by RNA-seq and screening for molecular markers. BMC Genomics. 2014;15:236. doi:10.1186/1471-2164-15-236.

19. Liao W, Ji L, Wang J, Chen Z, Ye M, Ma H, et al. Identification of glutathione S-transferase genes responding to pathogen infestation in populus tomentosa. Funct Integr Genomics. 2014;14:517–29.

20. Gaines TA, Lorentz L, Figge A, Herrmann J, Maiwald F, Ott MC, et al. RNA-Seq transcriptome analysis to identify genes involved in metabolism-based diclofop resistance in Lolium rigidum. Plant J. 2014;78:865–76.

21. Li A, Liu D, Wu J, Zhao X, Hao M, Geng S, et al. mRNA and small RNA transcriptomes reveal insights into dynamic homoeolog regulation of allopolyploid heterosis in nascent hexaploid wheat. Plant Cell. 2014;26:1878–900.

22. Xu XH, Chen H, Sang YL, Wang F, Ma JP, Gao XQ, et al. Identification of genes specifically or preferentially expressed in maize silk reveals similarity and diversity in transcript abundance of different dry stigmas. BMC Genomics. 2012;13:294. doi: 10.1186/1471-2164-13-294.

23. Li G, Wang D, Yang R, Logan K, Chen H, Zhang S, et al. Temporal patterns of gene expression in developing maize endosperm identified through transcriptome sequencing. Proc Natl Acad Sci U S A. 2014;111:7582–7.

24. Caruso M, Merelo P, Distefano G, Malfa SL, Lo Piero AR, Tadeo FR, et al. Comparative transcriptome analysis of stylar canal cells identifies novel candidate genes implicated in the self-incompatibility response of citrus clementina. BMC Plant Biol. 2012;12:20. doi:10.1186/1471-2229-12-20.

25. Zhang S, Ding F, He X, Luo C, Huang G, Hu Y. Characterization of the 'Xiangshui' lemon transcriptome by de novo assembly to discover genes associated with self-incompatibility. Mol Genet Genomics. 2015;290(1):365–75.

26. Zhou Q, Jia J, Huang X, Yan X, Cheng L, Chen S, et al. The large-scale investigation of gene expression in Leymus chinensis stigmas provides a valuable resource for understanding the mechanisms of poaceae self-incompatibility. BMC Genomics. 2014. doi: 10.1186/1471-2164-15-399.

27. van Doorn WG, Woltering EJ. Physiology and molecular biology of petal senescence. J Exp Bot. 2008;59(3):453–80.

28. Tintor N, Ross A, Kanehara K, Yamada K, Fan L, Kemmerling B, et al. Layered pattern receptor signaling via ethylene and endogenous elicitor peptides during Arabidopsis immunity to bacterial infection. Proc Natl Acad Sci U S A. 2013;110(15):6211–6.

29. Tung CW, Dwyer KG, Nasrallah ME, Nasrallah JB. Genome-wide identification of genes expressed in arabidopsis pistils specifically along the path of pollen tube growth. Plant Physiol. 2005;138:977–89.

30. Swanson R, Clark T, Preuss D. Expression profiling of arabidopsis stigma tissue identifies stigma-specific genes. Sex Plant Reprod. 2005;18:163–71.

31. Li M, Xu W, Yang W, Kong Z, Xue Y. Genome-wide gene expression profiling reveals conserved and novel molecular functions of the stigma in rice (Oryza sativa L.). Plant Physiol. 2007;144:1797–812.

32. Broderick SR, Wijeratne S, Wijeratn AJ, Chapin LJ, Meulia T, Jones ML. RNA-sequencing reveals early, dynamic transcriptome changes in the corollas of pollinated petunias. BMC Plant Biol. 2014. doi:10.1186/s12870-014-0307-2.

33. Wilkinson JQ, Lanahan MB, Clark DG, Bleecker AB, Chang C, Meyerowitz EM, et al. A dominant mutant receptor from Arabidopsis confers ethylene insensitivity in heterologous plants. Nat Biotechnol. 1997;15(5):444–7.

34. Yang SF, Hoffman NE. Ethylene biosynthesis and its regulation in higher plants. Annu Rev Plant Physiol. 1984;35:155–89.

35. Kende H. Enzymes of ethylene biosynthesis. Plant Physiol. 1989;91(1):1–4.

36. Bleecker AB, Kende H. Ethylene: a gaseous signal molecule in plants. Annu Rev Cell Dev Biol. 2000;16:1–18.

37. Chae HS, Kieber JJ. Eto Brute? Role of ACS turnover in regulating ethylene biosynthesis. Trends Plant Sci. 2005;10(6):291–6.

38. Singh A, Evensen KB, Kao TH. Ethylene synthesis and floral senescence following compatible and incompatible pollinations in Petunia inflata. Plant Physiol. 1992;99(1):38–45.

39. Bowler C, Chua NH. Emerging themes in plant signal transduction. Plant Cell. 1994;6:1529–41.

40. Schaller GE, Bleecker AB. Ethylene-binding sites generated in yeast expressing the Arabidopsis EtRI K7gene. Science. 1995;270:1809–11.

41. Sakai H, Hua J, Chen QG, Chang C, Medrano LJ, Bleecker AB, et al. ETR2 is an ETR1-like gene involved in ethylene signaling in Arabidopsis. Proc Natl Acad Sci U S A. 1998;95:5812–7.

42. Potuschak T, Lechner E, Parmentier Y, Yanagisawa S, Grava S, Koncz C, et al. EIN3-dependent regulation of plant ethylene hormone signaling by two arabidopsis F box proteins: EBF1 and EBF2. Cell. 2003;115(6):679–89.

43. Guo H, Ecker JR. Plant responses to ethylene gas are mediated by SCF (EBF1/EBF2)-dependent proteolysis of EIN3 transcription factor. Cell. 2003;115(6):667–77.

44. Holden MJ, Marty JA, Singh-Cundy A. Pollination-induced ethylene promotes the early phase of pollen tube growth in Petunia inflata. J Plant Physiol. 2003;160(3):261–9.

45. Chen D, Zhao J. Free IAA in stigmas and styles during pollen germination and pollen tube growth of Nicotiana tabacum. Physiol Plant. 2008;134:202–15.

46. Hasenstein KH, Zavada MS. Auxin modification of the incompatibility response in Theobroma cacao. Physiol Plant. 2001;112:113–8.

47. Kovaleva L, Zakharova E. Hormonal status of the pollen-pistil system at the progamic phase of fertilization after compatible and incompatible pollination in Petunia hybrida L. Sex Plant Reprod. 2003;16:191–6.

48. Solfanelli C, Bartolini S, Vitagliano C, Lorenzi R. Immunolocalization and quantification of IAA after self- and free pollination in Olea europaea L. Sci Hortic. 2006;110:345–51.

49. Khan S, Stone JM. Arabidopsis thaliana GH3.9 influences primary root growth. Planta. 2007;226(1):21–34.

50. Grebe M, Friml J, Swarup R, Ljung K, Sandberg G, Terlou M, et al. Cell polarity signaling in Arabidopsis involves a BFA-sensitive auxin influx pathway. Curr Biol. 2002;12(4):329–34.

51. Jain M, Kaur N, Tyagi AK, Khurana JP. The auxin-responsive GH3 gene family in rice (Oryza sativa). Funct Integr Genomics. 2006;6(1):36–46.

52. Vandenbussche F, Petrásek J, Zádníková P, Hoyerová K, Pesek B, Raz V, et al. The auxin influx carriers AUX1 and LAX3 are involved in auxin-ethylene interactions during apical hook development in Arabidopsis thaliana seedlings. Development. 2010;137(4):597–606.

53. Gonzalez-Guzman M, Apostolova N, Belles JM, Barrero JM, Piqueras P, Ponce MR, et al. The short-chain alcohol dehydrogenase ABA2 catalyzes the conversion of xanthoxin to abscisic aldehyde. Plant Cell. 2002;14:1833–46.

54. Okamoto M, Kuwahara A, Seo M, Kushiro T, Asami T, Hirai N, et al. CYP707A1 and CYP707A2, which encode abscisic acid 8′-hydroxylases, are indispensable for proper control of seed dormancy and germination in Arabidopsis. Plant Physiol. 2006;141:97–107.

55. Nishimura N, Hitomi K, Arvai AS, Rambo RP, Hitomi C, Cutler SR, et al. Structural mechanism of abscisic acid binding and signaling by dimeric PYR1. Science. 2009;326:1373–9.

56. Park SY, Fung P, Nishimura N, Jensen DR, Fujii H, Zhao Y, et al. Abscisic acid inhibits type 2C protein phosphatases via the PYR/PYL family of START proteins. Science. 2009. doi: 10.1126/science.1173041.

57. Fujii H, Zhu JK. Arabidopsis mutant deficient in 3 abscisic acid-activated protein kinases reveals critical roles in growth, reproduction, and stress. Proc Natl Acad Sci U S A. 2009;106:8380–5.

58. Chini A, Fonseca S, Fernandez G, Adie B, Chico JM, Lorenzo O, et al. The JAZ family of repressors is the missing link in jasmonate signalling. Nature. 2007;448:666–71.

59. Zhou JM, Trifa Y, Silva H, Pontier D, Lam E, Shah J, et al. NPR1 differentially interacts with members of the TGA/OBF family of transcription factors that bind an element of the PR-1 gene required for induction by salicylic acid. Mol Plant Microbe Interact. 2000;13:191–202.

60. Bolger AM, Lohse M, Usadel B. Trimmomatic: a flexible trimmer for Illumina sequence data. Bioinformatics. 2014;30(15):2114–20.

61. Trapnell C, Roberts A, Goff L, Pertea G, Kim D, Kelley DR, et al. Differential gene and transcript expression analysis of RNA-seq experiments with TopHat and Cufflinks. Nat Protoc. 2012;7(3):562–78.

62. Anders S, Pyl PT, Huber W. HTSeq-a Python framework to work with high-throughput sequencing data. Bioinformatics. 2015;31(2):166–9.

63. Mortazavi A, Williams BA, McCue K, Schaeffer L, Wold B. Mapping and quantifying mammalian transcriptomes by RNASeq. Nat Methods. 2008;5:621–8.

64. Conesa A, Gotz S, Garcia-Gomez JM, Terol J, Talon M, Robles M. Blast2GO: a universal tool for annotation, visualization and analysis in functional genomics research. Bioinformatics. 2005;21:3674–6.

65. Ye J, Fang L, Zheng HK, Zhang Y, Chen J, Zhang ZJ, et al. WEGO: a web tool for plotting GO annotations. Nucleic Acids Res. 2006;34:W293–7.

66. Gene Ontology Consortium. The Gene Ontology (GO) project in 2006. Nucleic Acids Res. 2006;34(Database issue):D322–6.

67. Ashburner M, Ball CA, Blake JA, Botstein D, Butler H, Cherry JM, et al. Gene ontology: tool for the unification of biology. The Gene Ontology Consortium. Nat Genet. 2000;25:25–9.

68. Du Z, Zhou X, Ling Y, Zhang Z, Su Z. AgriGO: a GO analysis tool kit for the agricultural community. Nucleic Acids Res. 2010;38:W64–70.

69. Benjamini Y, Hochberg Y. Controlling the false discovery rate: a practical and powerful approach to multiple testing. J Royal Sta Soc Ser B. 1995;57(1):289–300.

70. Pawitan Y, Michiels S, Koscielny S, Gusnanto A, Ploner A. False discovery rate, sensitivity and sample size for microarray studies. Bioinformatics. 2005;21:3017–24.

71. Wang K, Singh D, Zeng Z, Coleman SJ, Huang Y, Savich GL, et al. MapSplice: accurate mapping of RNA-seq reads for splice junction discovery. Nucleic Acids Res. 2010;38(18), e178.

72. Kanehisa M, Goto S. KEGG: kyoto encyclopedia of genes and genomes. Nucleic Acids Res. 2000;28:27–30.

73. Joshi-Tope G, Gillespie M, Vastrik I, D'Eustachio P, Schmidt E, de Bono B, et al. Reactome: a knowledgebase of biological pathways. Nucleic Acids Res. 2005;33:D428–32.

74. Kanehisa M, Goto S, Kawashima S, Okuno Y, Hattori M. The KEGG resource for deciphering the genome. Nucleic Acids Res. 2004;32 (Database issue):D277–80.

75. Yi M, Horton JD, Cohen JC, Hobbs HH, Stephens RM. WholePathwayScope: a comprehensive pathway-based analysis tool for high-throughput data. BMC Bioinformatics. 2006;7:30. doi: 10.1186/1471-2105-7-30.

76. Draghici S, Khatri P, Tarca AL, Amin K, Done A, Voichita C, et al. A systems biology approach for pathway level analysis. Genome Res. 2007;17:1537–45.

77. Zhao P, Zhang L, Zhao L. Dissection of the style's response to pollination using transcriptome profiling in self-compatible (Solanum pimpinellifolium) and self-incompatible (Solanum chilense) tomato species. Gene Expression Omnibus. http://www.ncbi.nlm.nih.gov/geo/query/acc.cgi?acc=GSE67654.

Characterization of *Brachypodium distachyon* as a nonhost model against switchgrass rust pathogen *Puccinia emaculata*

Upinder S Gill[1†], Srinivasa R Uppalapati[1,2†], Jin Nakashima[1] and Kirankumar S Mysore[1*]

Abstract

Background: Switchgrass rust, caused by *Puccinia emaculata*, is an important disease of switchgrass, a potential biofuel crop in the United States. In severe cases, switchgrass rust has the potential to significantly affect biomass yield. In an effort to identify novel sources of resistance against switchgrass rust, we explored nonhost resistance against *P. emaculata* by characterizing its interactions with six monocot nonhost plant species. We also studied the genetic variations for resistance among *Brachypodium* inbred accessions and the involvement of various defense pathways in nonhost resistance of *Brachypodium*.

Results: We characterized *P. emaculata* interactions with six monocot nonhost species and identified *Brachypodium distachyon* (Bd21) as a suitable nonhost model to study switchgrass rust. Interestingly, screening of *Brachypodium* accessions identified natural variations in resistance to switchgrass rust. *Brachypodium* inbred accessions Bd3-1 and Bd30-1 were identified as most and least resistant to switchgrass rust, respectively, when compared to tested accessions. Transcript profiling of defense-related genes indicated that the genes which were induced in Bd21after *P. emaculata* inoculation also had higher basal transcript abundance in Bd3-1 when compared to Bd30-1 and Bd21 indicating their potential involvement in nonhost resistance against switchgrass rust.

Conclusion: In the present study, we identified *Brachypodium* as a suitable nonhost model to study switchgrass rust which exhibit type I nonhost resistance. Variations in resistance response were also observed among tested *Brachypodium* accessions. *Brachypodium* nonhost resistance against *P. emaculata* may involve various defense pathways as indicated by transcript profiling of defense related genes. Overall, this study provides a new avenue to utilize novel sources of nonhost resistance in *Brachypodium* against switchgrass rust.

Keywords: Brachypodium, Switchgrass, *Puccinia emaculata*, Nonhost resistance

Background

Switchgrass (*Panicum virgatum* L.) is considered a potential biofuel crop by the United States Department of Energy (DOE) [1]. Switchgrass can grow on marginal lands with low-input agriculture and without many crop management practices. Due to extensive root systems and clumping growth patterns, switchgrass can provide protection against soil erosion and also acts as an excellent habitat for wildlife [2]. Since switchgrass is perennial and a monoculture crop, it can become more susceptible to pathogens and insects. However, to date, very limited information is available on diseases of switchgrass [3]. Among diseases of switchgrass, switchgrass rust, caused by *Puccinia emaculata*, is economically very important since it has the potential to significantly affect biomass yield. *P. emaculata* is a biotrophic fungal pathogen and is widely distributed in switchgrass growing regions of North America with a moderate to high incidence of infection [4-9]. Genetic variations for rust resistance exist in natural populations of switchgrass and have been studied in detail [6,8]. In general, lowland switchgrass cultivars such as Alamo and Kanlow are moderately resistant to *P. emaculata* compared to upland cultivars such as Summer and Cave-in-Rock [8]. Variations also exist in the virulence of urediniospores collected from

* Correspondence: ksmysore@noble.org
†Equal contributors
[1]Plant Biology Division, The Samuel Roberts Noble Foundation, Ardmore, Oklahoma 73401, USA
Full list of author information is available at the end of the article

different sources [10]. Urediniospores of *P. emaculata* collected from ornamental switchgrass were found to have greater virulence than urediniospores collected from agronomic switchgrass plots [10]. These variations in virulence of wind-borne rust urediniospores pose a great threat to monoculture of switchgrass varieties in new geographical areas. Genetic variations in switchgrass germplasm can be exploited to find sources of host resistance, but host resistance is generally less durable due to the fact that variations also exist in rust pathogen isolates.

Nonhost resistance (NHR), on the other hand, is a form of durable resistance shown by all members of a plant species against all isolates of a specific pathogen [11]. NHR response in the plant may not lead to any visual symptoms (type I), or it can be associated with visible symptoms (type II), depending on the host-pathogen interaction [12]. During nonhost interactions, the first line of defense involves preformed physical and chemical barriers such as surface topology, cytoskeleton, antimicrobial compounds and secondary metabolites [12-14]. The importance of wax layers on leaf surfaces has been described specifically for NHR against fungal pathogens where epicuticular wax affects fungal pre-infection structures [15,16]. An inducible defense response is often triggered if the primary line of defense is breached by the pathogen [14]. Generally, conserved elicitor molecules, often called microbe- or pathogen-associated molecular patterns (MAMPS or PAMPs), are sensed by plant plasma membrane receptors to trigger basal or NHR response [17]. In certain situations, if a nonhost species is closely related to a host species for a particular pathogen, the NHR response is often associated with hypersensitive (HR) response [18,19].

The study of NHR against biotrophic rust pathogens, which usually infect via urediniospores and pass through a set of defined developmental stages, is potentially more informative because resistance at each stage of development can be precisely defined [20]. NHR against rusts typically happens before haustoria formation during pre-penetration events or due to restrictive fungal growth in the substomatal cavity [21]. In some cases, post-haustorial resistance of rust fungi is also observed [22]. Rust diseases of cereals and other grasses are mainly caused by rust fungi belonging to the genus *Puccinia*, and it is considered the most economically destructive genus of biotrophic fungi [23]. NHR mechanisms against rust fungi of wheat have been studied using divergent species such as Arabidopsis and broad-bean (*Vicia faba* L.), and also closely related species such as *Brachypodium distachyon*, barley and rice [20,22,24-26]. Among these species, rice is the only monocot known so far which is immune to all rust pathogens and shows an active NHR response against

cereal rust pathogens by involving hydrogen peroxide production and callose depositions [24]. Quantitative trait loci (QTL) analysis of NHR in barley against *P. triticina* (wheat leaf rust) identified map locations similar to genes conferring partial resistance to *P. hordei*, a pathogen of barley [26].

Brachypodium, which is considered a model plant species for the study of some members of the family *Poaceae*, is a host to the rust pathogen *P. brachypodii*. Variations exist among *Brachypodium* accessions for resistance against *P. brachypodii*, and QTLs for resistance have been identified [27,28]. Since its acceptance of *Brachypodium* as a model species by the scientific community at the start of the 21st century, a variety of genetic and genomic resources have been developed, such as T-DNA insertion lines, efficient genetic transformation and sequencing of the whole genome [29-31]. *Brachypodium* has been used as a nonhost to study plant diseases caused by a variety of plant pathogens including cereal rusts, *Fusarium* head blight of wheat, rice blast, powdery mildew and *Barley stripe mosaic virus* (BSMV) [24,32-36]. *Brachypodium* accessions show large variations in NHR response against cereal rust pathogens [33]. In some instances, sporulating pustules of *P. striiformis* (a wheat pathogen) appeared on a few of the tested *Brachypodium* accessions [33]. Variations in resistance were also reported in *Brachypodium* against *P. graminis* f. sp. *tritici*, *lolii* and *phlei-pratensis*, where many of the tested accessions showed sporulating pustules against *P. graminis* f. sp. *lolii* and *phlei-pratensis* [34]. Using genetic mapping populations of *Brachypodium* ecotypes, the inheritance of variations against *P. striiformis* f. sp. *tritici* were studied [24]. Genetic analysis indicated a relatively simple inheritance of NHR in *Brachypodium*, including single gene segregation in one of the families [24].

Here we present our results on identification of *Brachypodium* as a suitable plant to study nonhost resistance against *P. emaculata* and detailed characterization of *Brachypodium-P. emaculata* nonhost interactions involving *Brachypodium* inbred accessions. The transcript level of plant defense-related genes was also studied to understand genetic variation for resistance among these accessions.

Results

Identification of an appropriate nonhost monocot plant species to study NHR against *P. emaculata*

To identify a suitable monocot nonhost model system for *P. emaculata*, we screened several monocot plants from BEP (Bambusoideae, Ehrhartoideae and Pooideae) and PACCMAD (Panicoideae, Arundinoideae, Chloridoideae, Centothe-coideae, Micrairoideae, Aristidoideae and Danthonioideae) clades (Additional file 5). As reported previously, switchgrass cv. Summer is a susceptible host to *P. emaculata* [8]. As expected, switchgrass cv. Summer was infected with rust urediniospores and showed disease symptoms in the form of sporulating pustules under controlled (Figure 1a, b) and natural conditions (Figure 1c). *P. emaculata* germ tubes failed to recognize the host surface, followed by a lack of oriented growth of germ tube and appressoria formation on the abaxial or adaxial leaf surfaces of corn, sorghum or foxtail millet belonging to the PACCMAD clade (Figure 2a, b, c). However, on sorghum leaf surfaces, few urediniospore germ tubes were able to show oriented growth similar to that of the host plant, switchgrass (Figure 2c). Oriented growth of rust spore germ tubes was also noticed on the leaf surfaces of both barley and rice, but the appressoria were developed only in the case of rice (Figure 2d, e).

On *Brachypodium* accession Bd21, a commonly used accession for which the genome sequence is available [31] and belonging to the BEP clade, *P. emaculata* germ tubes showed oriented growth perpendicular to the long axis of the epidermal cells (Figures 2f and 3a). The oriented growth of germ tubes indicates recognition of topographic and chemical signals on the host surface. Furthermore, the germ tubes encountered stomata and formed appressoria on *Brachypodium* (Figure 3b, e). Further infection occurred by formation of a penetration peg which is presumably followed by fungal hyphae growth in mesophyll cells (Figure 3c, f). In some cases, microscopic evaluation also revealed hypersensitive cell death at the site of fungal penetration (Figure 3d). However, disease symptoms

in the form of urediniospore-containing rust pustules or hypersensitive resistance response were not visually observed on *Brachypodium* leaves (Figure 3g) thus representing type I NHR as proposed previously [12]. These results suggest that *Brachypodium* would be a suitable model plant among the monocot plants tested to identify signaling components to NHR against *P. emaculata*.

Variability of *Brachypodium* inbred accessions for resistance against *P. emaculata*

P. emaculata was able to penetrate *Brachypodium* line Bd21, but failed to produce disease symptoms. Therefore, we decided to study genetic variations for NHR by investigating various *Brachypodium* germplasm lines. A total of 38 *Brachypodium* germplasm lines were tested for their response to *P. emaculata* urediniospore inoculation. Out of 38 lines, 32 lines representing different geographical locations in the world with unknown heterozygosity were procured from USDA-GRIN. Six inbred accessions, Bd21, Bd21-3, Bd3-1, Bd18-1, Bd29-1 and Bd30-1, were developed by single-seed descent to increase homozygosity and minimum variations within each line [37]. None of the tested 38 lines showed any disease susceptibility to *P. emaculata* (data not shown). Six *Brachypodium* inbred accessions which were considered to have minimum heterogeneity were further evaluated microscopically for the

Figure 1 Switchgrass cv. Summer infected with switchgrass rust. Switchgrass cv. Summer plants were inoculated by spraying *P. emaculata* urediniospores at a concentration of 10^5 spores per milliliter under controlled conditions. Two weeks after inoculation, dark brown rust pustules containing dikaryotic urediniospores appeared on leaf **(a)** and stem **(b)** of switchgrass plants. **(c)** A switchgrass plant in a field severely infested with rust under natural conditions.

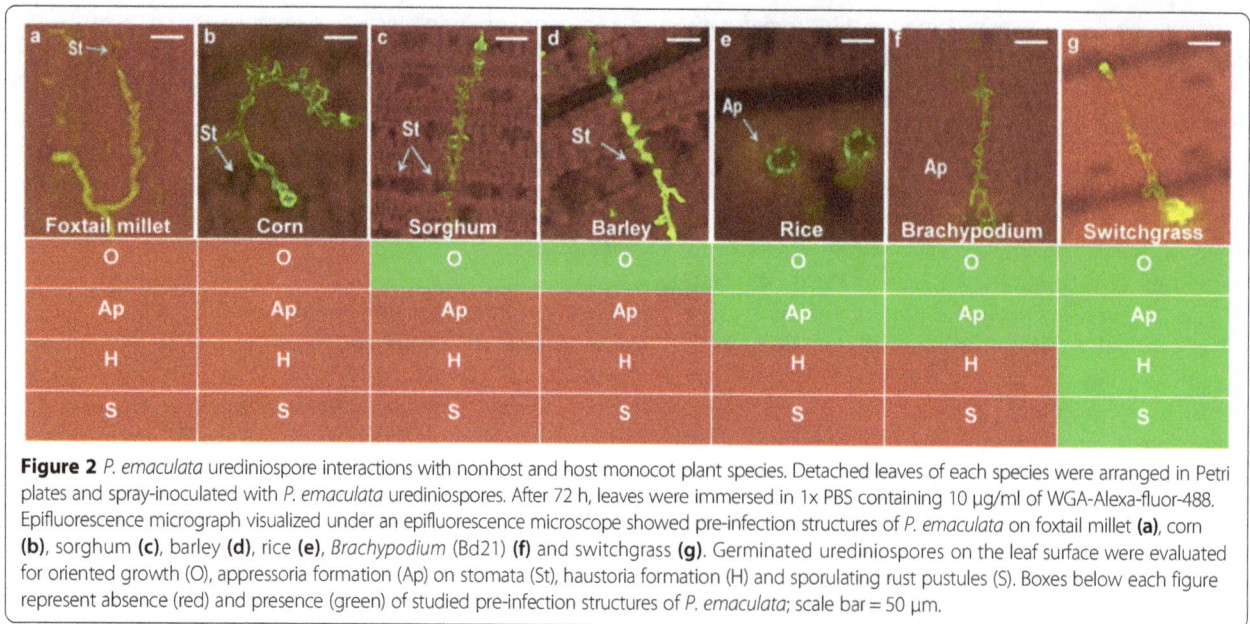

Figure 2 *P. emaculata* urediniospore interactions with nonhost and host monocot plant species. Detached leaves of each species were arranged in Petri plates and spray-inoculated with *P. emaculata* urediniospores. After 72 h, leaves were immersed in 1x PBS containing 10 μg/ml of WGA-Alexa-fluor-488. Epifluorescence micrograph visualized under an epifluorescence microscope showed pre-infection structures of *P. emaculata* on foxtail millet (**a**), corn (**b**), sorghum (**c**), barley (**d**), rice (**e**), *Brachypodium* (Bd21) (**f**) and switchgrass (**g**). Germinated urediniospores on the leaf surface were evaluated for oriented growth (O), appressoria formation (Ap) on stomata (St), haustoria formation (H) and sporulating rust pustules (S). Boxes below each figure represent absence (red) and presence (green) of studied pre-infection structures of *P. emaculata*; scale bar = 50 μm.

development of pre-infection structures of *P. emaculata* upon inoculation (Figure 4a). Microscopic evaluations were conducted on 3–4 weeks old plants by studying at least 10 leaves per genotype pooled from three individual plants to represent one biological replication. For each replication, minimum of 100 interaction sites were scored. One way Analysis Of Variance (ANOVA) indicated significant differences among accessions for all tested parameters. Our results indicated a germination rate of more than 80% of *P. emaculata* urediniospores on the leaf surface of each

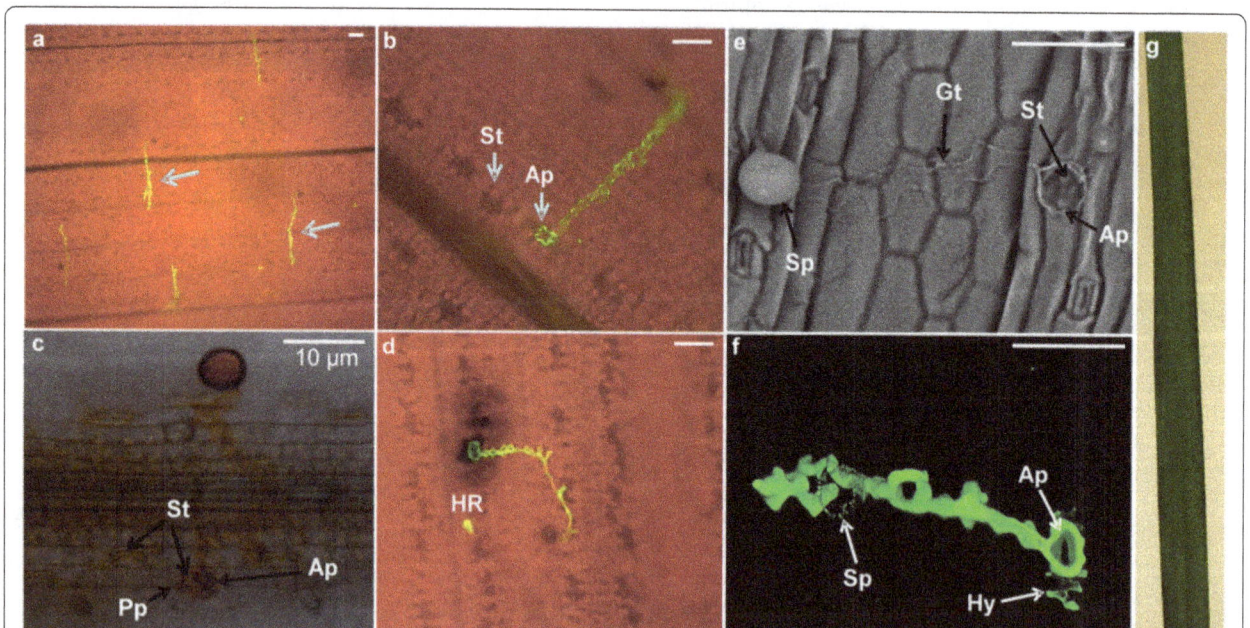

Figure 3 *P. emaculata* urediniospore interactions with *Brachypodium*. Epifluorescence micrograph showing pre-infection structures after germination of *P. emaculata* urediniospores (Sp) on *Brachypodium* (Bd21) leaf surface in the form of the oriented growth pattern of germ tubes (Gt) (**a**) and formation of appressoria (Ap) on stomatal (St) openings (**b**). (**c**) 3, 3'-diaminobenzidine (DAB)-stained fungal structures showing a penetration peg (Pp) originating from the appressorium that helps to push through the closed stomata for entry into the intercellular space within the host leaf. (**d**) Localized cell death in the form of hypersensitive (HR) response at the site of infection. (**e**) Scanning electron microscope (SEM) image showing the formation of pre-infection structures after urediniospore germination on the leaf surface of *Brachypodium*. (**f**) Confocal microscope image showing intercellular hyphal (Hy) growth inside mesophyll cells. (**g**) *Brachypodium* leaf at two weeks after infection with rust urediniospores. No visible disease symptoms appeared on *Brachypodium*. Scale bar = 50 μm or unless specified.

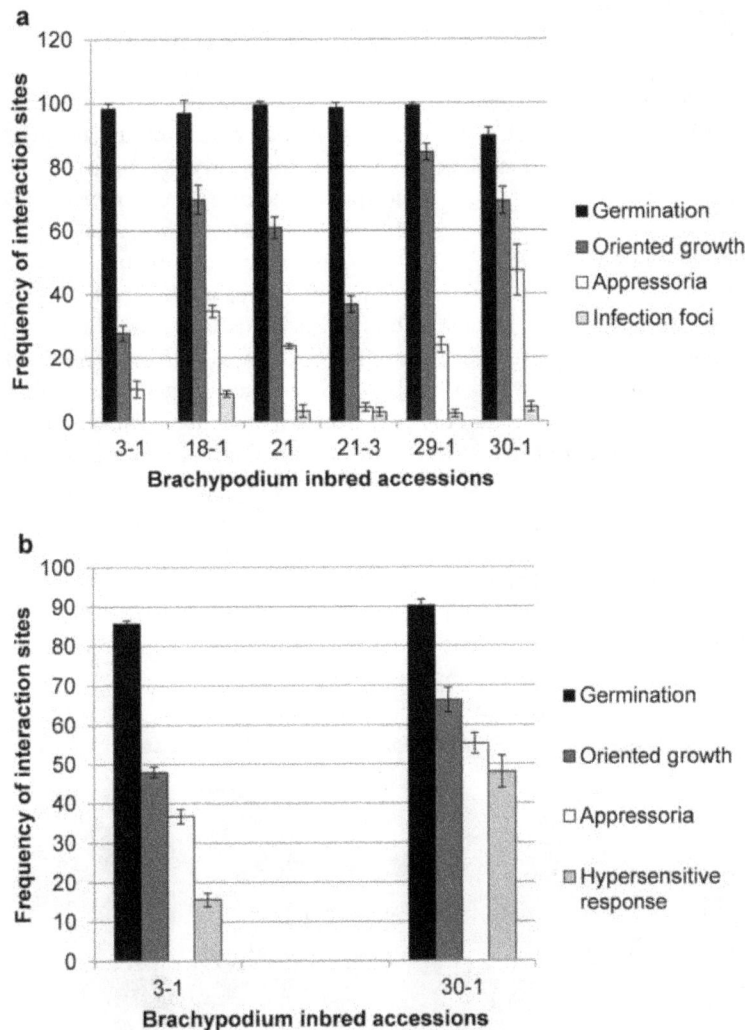

Figure 4 Genetic variations for rust pre-infection structures on *Brachypodium* inbred accessions. Detached leaves of accessions Bd3-1, Bd18-1, Bd21, Bd21-3, Bd29-1 and Bd30-1 were spray-inoculated with *P. emaculata* urediniospores followed by staining with WGA-Alexa-fluor-488, 72 h after inoculation. Stained fungal structures were visualized and quantified by epifluorescence microscope. Percent of *P. emaculata* urediniospore germination, germ tubes showing oriented growth, appressoria formation on stomatal apertures and infection foci were measured on six *Brachypodium* accessions **(a)**. One way Analysis of variance (ANOVA) indicated differences among accessions for all parameters (*p*-values 0.0015, 0.0000000008, 0.0000001 and 0.0001 for germination, oriented growth, appressoria and HR, respectively). **(b)** Bd3-1 and Bd30-1 pre-infection structures, including microscopic evaluation of hypersensitive response (HR). Two tailed Student's *t*-test (*p*-values 0.010, 0.0007, 0.0005 and 0.0002 for germination, oriented growth, appressoria and HR, respectively) indicate significant differences among both accessions for all tested parameters. It is important to note that *P. emaculata* failed to sporulate on any of the accessions tested. For statistical analysis, data from three biological replications was used. Error bars indicate standard deviation from mean.

tested inbred accession (Figure 4a). Oriented growth of urediniospore germ tubes was lowest in Bd3-1 (27.8%) and highest in Bd29-1 (84.6%), followed by Bd8-1 (69.7) and Bd30-1 (69.3%) (Figure 4a). Appressoria formation was also highest in Bd30-1 (47.4%) (Figure 4a). Surprisingly, no infection foci were noticed in Bd3-1, whereas infection foci were formed in all other tested inbred accessions (Figure 4a). Infection foci are the sites of infection in mesophyll cells after the fungus penetrates via stomata. Appressoria formation and formation of infection foci are the important steps before colonization of the

fungus and/or to trigger elicitor-induced defense response. These results suggest that Bd3-1 is more resistant to *P. emaculata* compared to other inbred accessions tested, while Bd30-1 was somewhat less resistant to *P. emaculata*. Based on the variation in appressoria and infection foci formation, Bd3-1 and Bd30-1 were further evaluated for more comprehensive analyses of these variations (Figure 4b). Data was collected from four separate experiments with eight replications each. Because rust infection foci were not detected in Bd3-1 in previous analyses, both Bd3-1 and Bd30-1 accessions were tested for the presence of hypersensitive cell death at the site of penetration using a light microscope. The percentage of germinating spores showing oriented growth, appressoria formation and infection sites with hypersensitive cell death was higher in Bd30-1 than in Bd3-1 (Figure 4b). Overall, our results indicated an enhanced resistance response in Bd3-1 against *P. emaculata* compared to Bd30-1.

Transcript profiling of defense-related genes in *P. emaculata*-inoculated Bd21

To identify the role of known defense genes against switchgrass rust, transcript profiles of 21 representative genes involved in various plant defense pathways were studied. A quantitative real-time PCR (qRT-PCR) method was used to measure transcript level changes of defense-related genes in Bd21 at 0 h (hours), 12 h, 24 h, 48 h and 72 h after *P. emaculata* inoculations (Figure 5). Transcripts of mock inoculated samples collected at 12 h, 24 h, 48 h and 72 h was used as control and the fold change ratio between *P. emaculata* inoculated vs mock was used for analysis. Transcripts of most genes involved in Ethylene

(ET), Salicylic Acid (SA) and Jasmonic Acid (JA) biosynthesis or signaling pathways were induced at different time-points after pathogen inoculation depending upon their role in plant defense. For example, transcription factors that play a role in the ethylene signaling pathway, *ERF1* (Ethylene Response Factor 1) and *ERF3* (Ethylene Response Factor 3), were induced more than two folds in first 24 hours after inoculation (hai) and later maintained till 72 hai with slight reduction in expression (Figure 5). Transcript level of *ACO1* (1-aminocyclopropane-1-carboxylic acid oxidase) was also slightly induced with similar pattern to *ERF1* and *ERF3*. *ACO1* is an ethylene biosynthetic pathway gene and is up-regulated in response to pathogen infection [38]. Among genes encoding enzymes involved in JA biosynthesis such as *LOX2* (Lipoxygenase 2), *OPR3* (12-oxophytodienoate reductase 3) and *AOS* (Allene Oxide Synthase), only *OPR3* was induced by up to two folds at 48 hai (Figure 5). Another gene, *MKK3* which is involved in JA signaling also showed similar trend as *OPR3*. Interestingly, *VSP1* (vegetative storage protein 1), a gene known to be induced by JA, was not induced

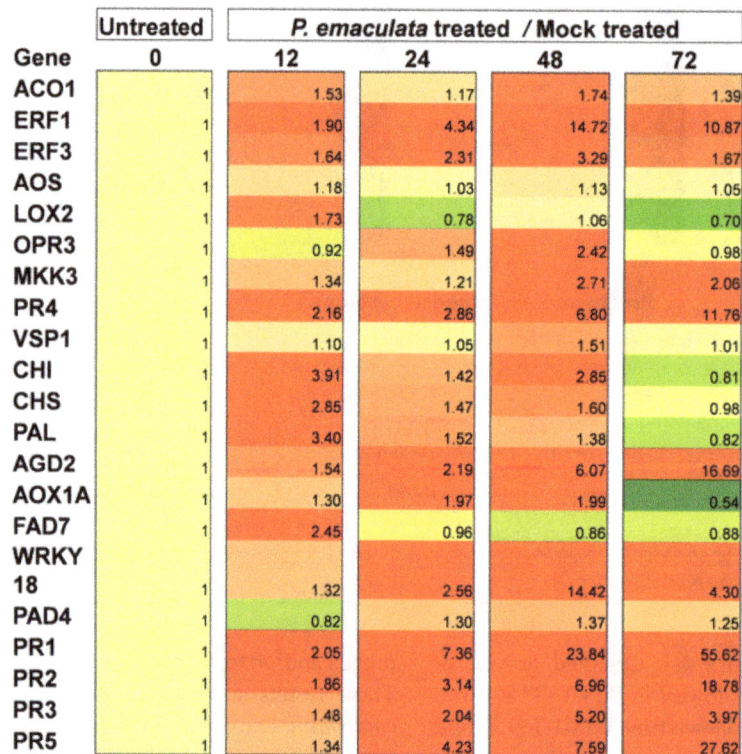

Figure 5 Heat map of transcript profile of defense-related genes in *P. emaculata* inoculated Bd21 leaves. Transcript profiling of defense-related genes in Bd21 leaf tissue collected at 0, 12, 24, 48 and 72 hours after inoculation (hai) with *P. emaculata* or after treatment with water containing 0.05% Tween20 (mock). For each treatment, leaves of at least five plants were pooled for each replication with a total of three replications per treatment. Relative quantification (in fold change) of *P. emaculata* and mock treated samples was calculated in relation to 0 hai and given a value of one. For each time-point, change in gene expression was calculated by measuring ratios of fold change between *P. emaculata* inoculated and mock inoculated samples. Intensity of red and green color indicates the extent of upregulation and downregulation, respectively, with respect to the gene expression at 0 hai which has been normalized to 1 (yellow). Numerical values of fold change are given in parenthesis in each box. 0, 12, 24, 48 and 72 represent hai. Three technical replicates were used for each sample.

appreciably at any tested time point compared to mock inoculations. Among SA signaling pathway genes, transcripts of AGD2 (Aberrant Growth Defects 2) and AOX1a (Alternative Oxidase) were highly induced by 16.7 fold at 72 hai and ~2 fold at 48 hai, respectively. Strikingly, all tested pathogenesis-related proteins encoding genes such as PR1, PR2, PR3, PR4 and PR5 were significantly induced in response to P. emaculata with upward trend in transcript abundance until 72 hai (Figure 5). Among these, PR1 was the highly induced gene with 55 fold increase at 72 hai (Figure 5). No significant differences in transcript level changes of FAD7 (omega-3 Fatty Acid Desaturase) and PAD4 (PhytoAlexin Deficient 4) were observed at tested time-points except FAD7 was induced by two fold at 12 hai. WRKY18 was induced up to two fold at 24 hai with highest induction of 14 fold at 72 hai (Figure 5). Genes involved in secondary metabolism, especially during the early steps of phenylpropanoid pathway, such as CHI (Chalcone Isomerase), CHS (Chalcone Synthase) and PAL (Phenylalanine Ammonia-Lyase) were induced by two fold mainly at 12 hai but the induction was below two fold in later time points (Figure 5). Overall, our data suggest that transcripts of many tested plant defense-related genes were induced in Bd21 leaves at various time-points after P. emaculata inoculation.

Basal transcript profiles of plant defense-related genes in Bd3-1, Bd21 and Bd30-1

Transcript profiling of Bd21 upon P. emaculata inoculation showed induction of most defense-related genes indicating their potential involvement in NHR against P. emaculata. In order to explain the variations in resistance in Bd3-1 and Bd30-1, we extended our analysis by conducting a basal transcript profiling of tested defense-related genes in Bd3-1 and Bd30-1 in relation to Bd21. It was surprising to see that some of the genes which were induced after P. emaculata inoculations in Bd21 had inherently high basal transcript levels in Bd3-1 which also have more penetration resistance (Figure 6). Transcript levels of ERF1, OPR3, VSP1, AGD2, AOX1A and PR5 were more than two folds in Bd3-1 compared to Bd21, whereas, transcript levels of rest of the tested genes in Bd3-1 were either less than two fold or comparable to Bd21 (Figure 6). On the other hand, Bd30-1 did not show higher transcript abundance compared to Bd21 for most of the tested genes except CHS and AOX1A which showed more than two fold increase (Figure 6). Interestingly, some of the genes such as VSP1 and AGD2, which had higher transcript abundance in Bd3-1, showed two fold decrease in transcript abundance in Bd30-1 (Figure 6). These results correspond with our phenotypic evaluation in which Bd3-1 was inherently more resistant to P. emaculata infection when compared to Bd21 and Bd30-1.

Discussion

Switchgrass rust, caused by P. emaculata, is an important disease of switchgrass, but not much attention has been given to this disease so far. Although variations in host resistance have been reported among switchgrass germplasm, those variations have not been exploited so far in switchgrass breeding [8]. In an effort to identify sources of NHR against switchgrass rust, we tested six different monocot plant species belonging to the BEP and PACCMAD clades for their response to P. emaculata inoculation (Additional file 5). From this analysis, we determined that Brachypodium, where the P. emaculata can successfully penetrate, is a suitable model to study nonhost interactions and to identify novel sources of resistance against P. emaculata. The inability of P. emaculata to successfully penetrate other monocot species tested, such as foxtail millet, corn, sorghum and barley, could be due to the absence of biochemical, topographical and thigmotropic signals from these nonhost species (Figure 2). These signals provide important cues to germinating spores for successful penetration by recognizing the host surface [39]. Based on our results we can speculate that these signals were recognized by germinating urediniospores on leaf surfaces of rice and Brachypodium to form appressoria (Figures 2e, f and 3). In some instances, intercellular fungal hyphal growth was also observed between mesophyll cells, but the presence of haustoria was not confirmed in Brachypodium (Figure 3f). Occasionally, Brachypodium also exhibited HR-related cell death at the site of rust infection which is only visible under the microscope (Figure 3d). Similar observations were also reported earlier for Puccinia species which are pathogenic to wheat, and the growth of pre-infection structures was far greater on Brachypodium than on rice [24]. Additionally, so far none of the rust pathogens are able to breach NHR imparted by rice [22]. Wheat stripe rust fungus P. striiformis and wheat stem rust fungus P. graminis f. sp. tritici showed successful colonization and sporulating pustules on some of the Brachypodium accessions [24,33]. Interestingly, in our study, P. emaculata failed to produce successful disease establishment in the form of rust pustules on the Brachypodium accessions tested. The failure of P. emaculata to colonize on other cereal species can be better explained by studying the evolutionary relationship among different species of genus Puccinia. An internal transcribed spacer (ITS) primer-based phylogenetic tree placed P. emaculata closer to P. asparagi, an asparagus rust pathogen, than other cereal rusts [8]. Detailed phylogenetic information is needed to study the evolution of P. emaculata and to identify the precise relationship of P. emaculata with other rust fungi to explain variability in host range.

Variations in pre-infection structures among different Brachypodium accessions can be classified under type I NHR (Figure 4). However, in later stages, intercellular

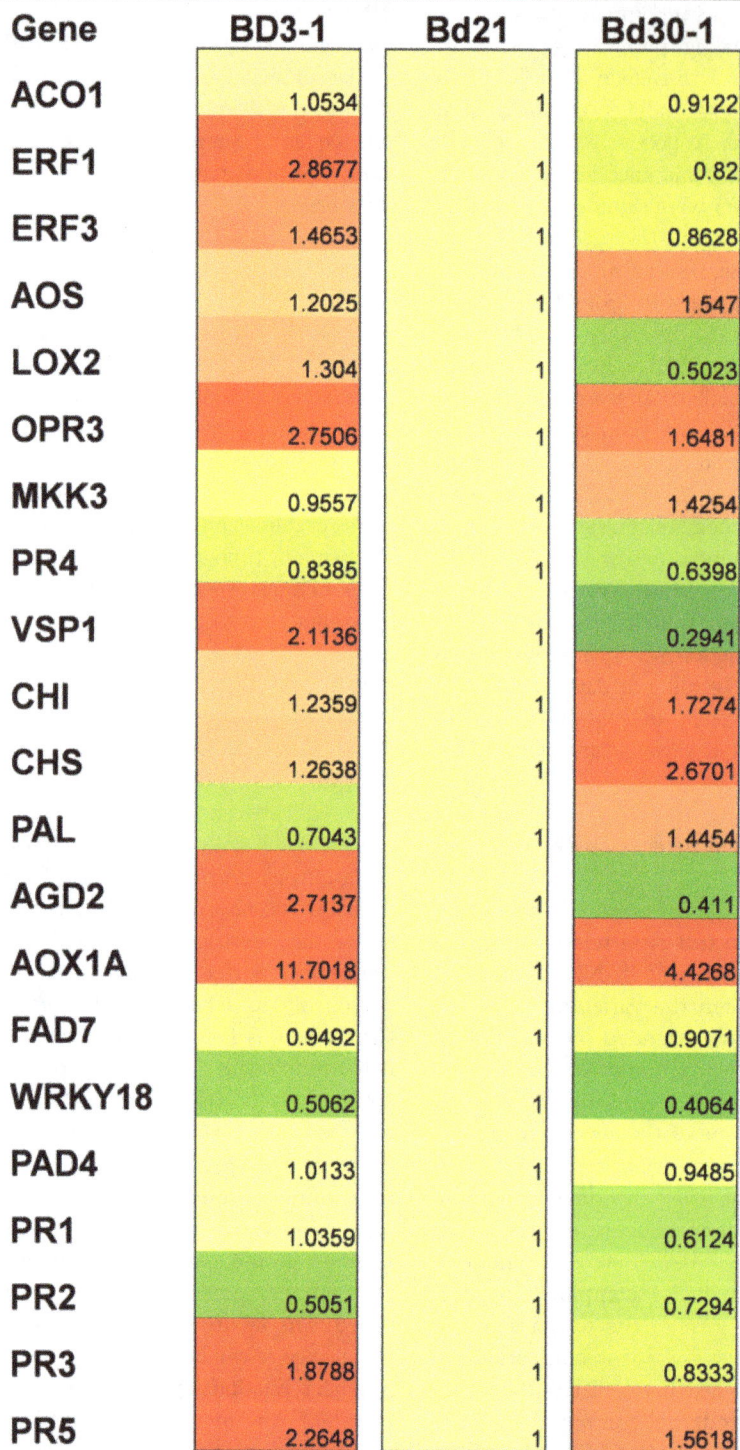

Figure 6 Heat map of basal transcript profile of defense-related genes among Bd3-1, Bd21 and Bd30-1. Transcript profile of defense-related genes was estimated from leaves of Bd3-1, Bd21 and Bd30-1. For each sample, leaves of at least five plants were pooled with three biological replicates per sample. Relative transcript levels were calculated in Bd3-1 and Bd30-1 by keeping transcript levels of Bd21 constant. Intensity of red and green color indicate extent of upregulation and downregulation, respectively, with respect to the gene expression of Bd21 normalized to 1 (yellow). Numerical values of fold change are given in parenthesis in each box. Three technical replicates were used for each sample.

growth of fungal hyphae in mesophyll cells followed by hypersensitive response is observed in all tested *Brachypodium* accessions also indicating a strong apoplastic defense response. Among tested accessions, Bd3-1 and Bd30-1 were most resistant and least resistant to switchgrass rust, respectively (Figure 4). Similar to our results, it has been previously reported that Bd3-1 was more resistant than Bd30-1 against *P. graminis* f. sp. *lolii, phleipratensis* and *tritici* [34]. A genetic mapping population followed by QTL analyses can be used in future to capture genetic variation among these accessions.

To understand the role of known defense genes in NHR of *Brachypodium* against switchgrass rust, we conducted a transcript profiling of defense-related genes in Bd21 leaves at various time-points after *P. emaculata* inoculations (Figure 5). Several genes involved in JA, ET and SA pathways were induced in *P. emaculata*-inoculated leaves of Bd21 (Figure 5). JA and ET pathways are usually involved in defense against necrotrophic pathogens, whereas the SA pathway is involved in defense against biotrophic pathogens [40]. In Arabidopsis, *AOX1a* expression is controlled by H_2O_2 signaling, SA application and a pathway involving *EDS1* and *PAD4* [41]. AOX acts as an antioxidant to combat excessive reactive oxygen species production during HR response during defense against pathogens [42]. *AOX1a* was also induced at 48 to 72 h after rust inoculation in Bd21 (Figure 5) indicating its potential involvement in post-penetration resistance which is often associated with HR response. Another protein, AGD2, which is involved in lysine biosynthesis, acts as a negative regulator of plant defense against biotrophic pathogens [43,44]. *AGD2* was induced up to 16 fold in Bd21 after pathogen inoculation (Figure 5). Ethylene response factors encoding genes, *ERF1* and *ERF2* were also induced after rust inoculation which was surprising because these genes are often involved in resistance against necrotrophic pathogens. Role of ethylene response factors has been reported previously in *Medicago truncatula* for resistance against *R. solani* [45]. Similarly, a JA biosynthetic pathway gene, *OPR3* was also induced in Bd21 in response to rust inoculation (Figure 5). JA biosynthesis and elicitation of JA response upon wounding and stress are associated with increased transcriptional activity of this gene [46]. Contrary to *OPR3*, *VSP1* was not induced after pathogen inoculation (Figure 5). *VSP1* is generally induced by JA, and a mutant with constitutive expression of *VSP1* shows enhanced resistance to pathogens [47,48]. However, in the present context, lack of induction of *VSP1* against biotrophic pathogen makes sense because of its involvement in JA signaling. However, induction of *OPR3* and its role in defense against a biotrophic rust pathogen is difficult to explain. Induction of transcript levels of phenylpropanoid pathway genes (*PAL, CHI* and *CHS*) during early phases of infection was also noticed (Figure 5).

Phenylpropanoid pathway compounds are generally involved in plant defense by acting as barriers against infection or as signaling molecules [49]. Induced transcripts of all tested *PR* genes with *P. emaculata* inoculation was interesting but not surprising due to perhaps their involvement in plant defense against invading pathogens [50].

To test if the defense-related genes which were induced in Bd21 in response to *P. emaculata* inoculation are also responsible for variations in resistance among *Brachypodium* inbred accessions, we extended our analyses by studying the basal transcript abundance of tested genes in Bd3-1 and Bd30-1 relative to Bd21. Surprisingly, the genes such as *ERF1, OPR3, AGD2* and *AOX1A* which were induced in response to *P. emaculata* were also showed more than two fold basal transcript abundance in Bd3-1 relative to Bd21 (Figure 6). Coincidently, Bd3-1 was also selected as the most resistance inbred accession among tested accessions. It could be possible that some or all of these genes were involved in high penetration resistance in Bd3-1 against *P. emaculata*. Additionally, out of five *PR* genes which were induced in Bd21 after rust inoculation, only *PR3* and *PR5* showed two fold transcript abundance in Bd3-1 relative to Bd21 or Bd30-1 (Figure 6). *PR3* encodes endochitinase which could act against fungal pathogens [50]. In Arabidopsis, *PR5* (a thaumatin) gene is induced in response to the SA pathway against biotrophic pathogens [51]. Previously, the transcript level of *PR5* was shown to be associated with incompatible interactions of wheat with wheat stripe rust [52]. Considering these studies, higher transcript levels of *PR3* and *PR5* in Bd3-1 (Figure 6) may be directly correlated with enhanced resistance against switchgrass rust.

Conclusion

We characterized switchgrass rust interactions with six monocot nonhost species and identified *Brachypodium* as a suitable nonhost model to study switchgrass rust. Analyses of *Brachypodium*-switchgrass rust interactions suggest type I NHR responses exhibited by *Brachypodium*. Genetic variation in resistance was reported among *Brachypodium* accessions against switchgrass rust. Among tested accessions, Bd3-1 exhibited more resistance against *P. emaculata*. These variations were further characterized at the molecular level by studying the transcript profiling of representative defense-related genes after *P. emaculata* inoculations in Bd21 and by measuring the basal transcript levels of these genes in a few accessions. Transcript profiling indicated involvement of various defense pathways in NHR imparted by *Brachypodium* against *P. emaculata*. Overall, the current study provides an avenue to identify novel sources of resistance against switchgrass rust by utilizing the extensive genomic and genetic resources available for *Brachypodium*.

Methods

Plant material and growth conditions

Brachypodium inbred accessions Bd21, Bd21-3, Bd3-1, Bd18-1, Bd29-1 and Bd30-1 were kindly provided by Dr. David Garvin, ARS-USDA. Thirty-two *Brachypodium* germplasm lines were procured from USDA-GRIN (United States Department of Agriculture-Germplasm Resources Information Network) (Additional file 1). Seeds of selected monocots, corn, sorghum, barley, rice and foxtail millet were procured from Drs. Xin Ding and Malay Saha at The Samuel Roberts Noble Foundation, Ardmore, Oklahoma, USA. Switchgrass cv. Summer, which is susceptible to *P. emaculata*, was used for multiplication of urediniospores [8]. *Brachypodium*, switchgrass and selected monocots belonging to the PACCMAD and BEP clades were planted and grown in a greenhouse with daytime and nighttime temperatures of 22°C and 18°C, respectively. Plant inoculations with rust were conducted in a biosafety level 2 room and kept in growth chambers at 29/22°C day/night temperature, 16 h photoperiod, 90% relative humidity and photon flux density 150–200 $\mu molm^{-2}\,s^{-1}$.

Switchgrass rust inoculation and screening

Urediniospores of *P. emaculata* were originally collected from switchgrass fields in Ardmore, Oklahoma [8]. Plant material was inoculated with freshly collected urediniospores from controlled inoculations in a growth chamber. Urediniospore suspension at a concentration of 10^5 spores/ml was prepared in water containing 0.05% Tween-20. Plants were inoculated with spray inoculation using an artist's airbrush (Paasche Airbrush Company, Chicago, Illinois, USA) followed by misting with distilled water. For mock inoculation, plants were sprayed with water containing 0.05% Tween-20. Both *P. emaculata* and mock inoculated plants were kept at 22°C under dark conditions for 16 hours before putting them back in a growth chamber under the above described environmental conditions. Rust-inoculated plants were screened after two weeks for disease susceptibility. For Bd21 time course experiment, similar inoculation procedure and environmental conditions were followed.

Microscopic evaluation/screening

Microscopic evaluation of rust urediniospore pre-infection structures was conducted on detached leaves. Minimum of 10 newly emerged leaves for each genotype were clipped from 3–4 week old plants and arranged on wet paper towels in Petri plates to represent one biological replicate. Statistical analyses of microscopic observations was conducted by performing one way ANOVA and Student's *t*-test. Spray-inoculation of urediniospore suspension was followed as described in the previous section. Sprayed leaves were kept under the same environmental conditions described in the previous section. Microscopic

screening experiment was conducted at least three times for consistency. Microscopic evaluations were conducted 72 hours after rust inoculation. For microscopic evaluations, leaves were first immersed in 1X PBS (phosphate-buffered saline) buffer containing 0.05% Tween-20 and 10 μg/ml wheat germ agglutinin conjugated with Alexa-fluor-488 (Invitrogen, Carlsbad, California, USA) for 10 minutes followed by three washings with 1X PBS buffer [16]. Leaf samples were arranged on a glass slide with a cover slip on top and visualized under an Olympus BX41 epifluorescence microscope (Olympus Corporation, Tokyo, Japan) and/or a Leica TCS SP2 AOBS confocal laser scanning microscope (Leica Microsystems CMS GmbH, Mannheim, Germany) with UV excitation. To evaluate percent germination, urediniospores which formed germ tubes >10 μm were considered germinated at 24 hours after inoculation. Germinated spores were evaluated for oriented growth and appressoria formation on top of stomatal opening. More than 100 spores were evaluated from five independent leaves of each genotype for each replication. Scanning electron microscopy was conducted using Hitachi TM3000, a tabletop scanning electron microscope (Hitachi High-Technologies Corporation, Tokyo, Japan) to analyze and score the pre-infection structures of *P. emaculata*.

DAB (3,3'-diaminobenzidine) staining

For DAB staining, infected leaves were placed in a freshly prepared solution of DAB-HCl (pH 3.8) at a concentration of 1 mg ml^{-1} for eight hours at room temperature. Leaf chlorophylls were removed with 95% ethanol before visualizing under a light microscope as described by Ishiga et al. [53].

RNA isolation and quantitative real-time PCR

Leaf tissue was collected from 3–4 week old *Brachypodium* plants. Leaves of five plants were pooled for each treatment per replicate and minimum of three biological replicates were used for RNA isolation. Collected leaf samples were used for total RNA isolation by the hot phenol/guanidinium thiocyanate method (TRIzol® Reagent, Invitrogen, Carlsbad, California, USA). First-strand cDNA was synthesized from 2 μg of total RNA using the SuperScript® III First-Strand Synthesis System (Invitrogen, Carlsbad, California, USA). Quantitative real-time PCR (RT-qPCR) analysis was conducted on CFX Connect™ Real-Time PCR Detection System (Bio-Rad, Hercules, California, USA) and Applied Biosystems 7900HT Fast Real-Time PCR system (Life Technologies, Grand Island, New York, USA) by following the manufacturer's instructions. For RT-qPCR analyses, three biological replicates per treatment and three experimental replicates per sample were used. Data was analyzed by software DataAssist™ v3.01 (Life Technologies, Grand Island, New York, USA) by calibrating to housekeeping control gene, *Ubiquitin*. Relative quantification was measured using $2^{-\Delta\Delta CT}$ method [54]. PCR efficiencies for each PCR well were calculated by using software, LinRegPCR [55]. Average PCR efficiency for each primer pair is given in Additional file 6. Relative quantification (in fold change) of transcripts was estimated in reference to Bd21 with selected value of one (Additional file 3 and Additional file 4). Detail of selected defense-related genes and their primer sequences is given in Additional file 2.

Additional files

Additional file 1: Table S1. List of *Brachypodium* accessions procured from United States Department of Agriculture-Germplasm Resources Information Network (USDA-GRIN) and tested for disease reaction against *P. emaculata*.

Additional file 2:Table S2. *Brachypodium* defense-related genes and their primer sequences used for quantitative RT-PCR analysis.

Additional file 3: Table S3. Relative quantification (RQ) and P-value of defense related genes in Bd21 leaf tissue at 0, 12, 24, 48 and 72 hours after *P.emaculata* inoculation.

Additional file 4: Table S4. Relative quantification (RQ) and P-value of basal transcript level of defense related genes in Bd3-1 and Bd30-1 relative to Bd21.

Additional file 5: Taxonomy tree of tested monocot species. Seven monocot species, *Zea mays, Sorghum bicolor, Panicum virgatum, Setaria italica, Oryza sativa, Brachypodium distachyon* and *Hordeum vulgare*, were tested for nonhost/host resistance against switchgrass rust pathogen *P. emaculata*. Taxonomic information is based on the NCBI (National Center for Biotechnology Information) database.

Additional file 6: Table S5. Average PCR efficiencies and correlation coefficient of each primer pair used for qRT-PCR. Data points with PCR efficiencies of more than 1.8 were used for analysis.

Abbreviations

DOE: Department of energy; NHR: Nonhost resistance; HR: Hypersensitive response; QTL: Quantitative trait loci; BSMV: Barley stripe mosaic virus; ET: Ethylene; JA: Jasmonic acid; SA: Salicylic acid.

Competing interests

The authors declare that they have no competing interests.

Authors' contributions

KM, SU and UG designed research; SR, UG and JN performed research; UG and SR analyzed data; UG and KM wrote the paper. All authors read and approved the final manuscript.

Acknowledgments

We thank Dr. David Garvin (USDA) for providing us the seeds of *Brachypodium* inbred lines Bd3-1, Bd18-1, Bd21, Bd21-3, Bd29-1 and Bd30-1; Drs. X. S. Ding and Malay Saha for providing seeds of barley, rice sorghum and maize; Drs. Prasun Ray and Chengke Liu for their help with confocal microscopy; Dr. Hee-Kyung Lee for help with qRT-PCR; Drs. Prasun Ray and Prasanna Kankanala for critical reading of the manuscript; and Jackie Kelley for editing the manuscript. This work was supported through a grant to K.S.M from NSF-EPSCoR (EPS-0814361) and in part by The Samuel Roberts Noble Foundation.

Author details

^{1}Plant Biology Division, The Samuel Roberts Noble Foundation, Ardmore, Oklahoma 73401, USA. ^{2}Current address: Biologicals and Fungicide Discovery, DuPont Crop Protection, Newark DE 19711, USA.

References

1. Bouton J. Improvement of switchgrass as a bioenergy crop. In: Vermerris W, editor. Genetic Improvement of Bioenergy Crops. New York: Springer Verlag; 2008. p. 295–308.
2. Black JA. The epidemiology of *Puccinia emaculata* (rust) in switchgrass and evaluation of the mycoparasite *Sphaerellopsis filum* as a potential biological control organism for switchgrass rust, MS thesis. Knoxville: The University of Tennessee; 2012.
3. Stewart A, Cromey M. Identifying disease threats and management practices for bioenergy crops. Curr Opin Environ Sustain. 2011;3:75–80.
4. Frazier T, Shen Z, Zhao B. First report of *Puccinia emaculata* infection on switchgrass in Virginia. Dis Notes. 2013;97:424.
5. Gravert CE, Munkvold GP. Fungi and diseases associated with cultivated switchgrass in Iowa. J Iowa Acad Sci. 2002;109:30–3.
6. Gustafson DM, Boe A, Jin Y. Genetic variation for *Puccinia emaculata* infection in switchgrass. Crop Sci. 2003;43:755–9.
7. Hirsch RL, TeBeest DO, Bluhm BH, West CP. First report of rust caused by *Puccinia emaculata* on switchgrass in Arkansas. Plant Dis. 2010;94:381.
8. Uppalapati SR, Serba DD, Ishiga Y, Szabo LJ, Mittal S, Bhandari HS, et al. Characterization of the rust fungus, *Puccinia emaculata*, and evaluation of genetic variability for rust resistance in switchgrass populations. Bioenergy Res. 2013;6:458–68.
9. Zale J, Freshour L, Agarwal S, Sorochan J, Ownley BH, Gwinn KD, et al. First report of rust on switchgrass (*Panicum virgatum*) caused by *Puccinia emaculata* in Tennessee. Plant Dis. 2008;92:1710.
10. Li Y, Windham M, Trigiano R, Windham A, Ownley B, Gwinn K, et al. Cultivar-specific interactions between switchgrass and *Puccinia emaculata*. Phytopathol. 2009;99:S72.
11. Heath MC. Nonhost resistance and nonspecific plant defenses. Curr Opin Plant Biol. 2000;3:315–9.
12. Mysore KS, Ryu CM. Nonhost resistance: how much do we know? Trends Plant Sci. 2004;9:97–104.
13. Senthil-Kumar M, Mysore KS. Nonhost Resistance against bacterial pathogens: Retrospectives and prospects. Annu Rev Phytopathol. 2013;51:407–27.
14. Thordal-Christensen H. Fresh insights into processes of nonhost resistance. Curr Opin Plant Biol. 2003;6:351–7.
15. Tsuba M, Katagiri C, Takeuchi Y, Takada Y, Yamaoka N. Chemical factors of the leaf surface involved in the morphogenesis of *Blumeria graminis*. Physiol Mol Plant Pathol. 2002;60:51–7.
16. Uppalapati SR, Ishiga Y, Doraiswamy V, Bedair M, Mittal S, Chen J, et al. Loss of abaxial leaf epicuticular wax in *Medicago truncatula* irg1/palm1 mutants results in reduced spore differentiation of anthracnose and nonhost rust pathogens. Plant Cell. 2012;24:353–70.
17. Zipfel C, Robatzek S. Pathogen-associated molecular pattern-triggered immunity: veni, vidi...? Plant Physiol. 2010;154:551–4.
18. Schulze-Lefert P, Panstruga R. A molecular evolutionary concept connecting nonhost resistance, pathogen host range, and pathogen speciation. Trends Plant Sci. 2011;16:117–25.
19. Gill US, Lee S, Mysore KS. Host versus nonhost resistance: Distinct wars with similar arsenals. Phytopathol. 2015. http://dx.doi.org/10.1094/PHYTO-11-14-0298-RVW
20. Mellersh DG, Heath MC. An investigation into the involvement of defense signaling pathways in components of the nonhost resistance of *Arabidopsis thaliana* to rust fungi also reveals a model system for studying rust fungal compatibility. Mol Plant-Microbe Interact. 2003;16:398–404.
21. Heath MC. Resistance of plants to rust infection. Phytopathol. 1981;71:971–4.
22. Ayliffe M, Devilla R, Mago R, White R, Talbot M, Pryor A, et al. Non-host resistance of rice to rust pathogens. Mol Plant-Microbe Interact. 2011;24:1143–55.
23. Hooker AL. The genetics and expression of resistance in plants to rusts of the genus *Puccinia*. Annu Rev Phytopathol. 1967;5:163–82.
24. Ayliffe M, Singh D, Park R, Moscou M, Pryor T. Infection of *Brachypodium distachyon* with selected grass rust pathogens. Mol Plant-Microbe Interact. 2013;26:946–57.
25. Cheng Y, Zhang H, Yao J, Wang X, Xu J, Han Q, et al. Characterization of non-host resistance in broad bean to the wheat stripe rust pathogen. BMC Plant Biol. 2012;12:96.

26. Jafary H, Albertazzi G, Marcel TC, Niks RE. High diversity of genes for nonhost resistance of barley to heterologous rust fungi. Genetics. 2008;178:2327–39.

27. Barbieri M, Marcel TC, Niks RE. Host status of false brome grass to the leaf rust fungus *Puccinia brachypodii* and the stripe rust fungus *P. striiformis*. Plant Dis. 2011;95:1339–45.

28. Barbieri M, Marcel TC, Niks RE, Francia E, Pasquariello M, Mazzamurro V, et al. QTLs for resistance to the false brome rust *Puccinia brachypodii* in the model grass *Brachypodium distachyon* L. Genome. 2012;55:152–63.

29. Bragg JN, Wu J, Gordon SP, Guttman MA, Thilmony RL, Lazo GR, et al. Generation and characterization of the Western Regional Research Center *Brachypodium* T-DNA Insertional Mutant Collection. PLoS One. 2012;7:e41916.

30. Vogel J, Hill T. High-efficiency Agrobacterium-mediated transformation of *Brachypodium distachyon* inbred line Bd21-3. Plant Cell Rep. 2008;27:471–8.

31. The International Brachypodium Initiative. Genome sequencing and analysis of the model grass *Brachypodium distachyon*. Nature. 2010;463:763–8.

32. Cui Y, Lee MY, Huo N, Bragg J, Yan L, Yuan C, et al. Fine mapping of the *Bsr1* barley stripe mosaic virus resistance gene in the model grass *Brachypodium distachyon*. PLoS One. 2012;7:e38333.

33. Draper J, Mur LAJ, Jenkins G, Ghosh-Biswas GC, Bablak P, Hasterok R, et al. *Brachypodium distachyon*: A new model system for functional genomics in grasses. Plant Physiol. 2001;127:1539–55.

34. Figueroa M, Alderman S, Garvin DF, Pfender WF. Infection of *Brachypodium distachyon* by formae speciales of *Puccinia graminis*: Early infection events and host-pathogen incompatibility. PLoS One. 2013;8:e56857.

35. Peraldi A, Beccari G, Steed A, Nicholson P. *Brachypodium distachyon*: a new pathosystem to study *Fusarium* head blight and other *Fusarium* diseases of wheat. BMC Plant Biol. 2011;11:100–14.

36. Routledge APM, Shelley G, Smith JV, Talbot NJ, Draper J, Mur L. *Magnaporthe grisea* interactions with the model grass *Brachypodium distachyon* closely resemble those with rice (*Oryza sativa*). Mol Plant Pathol. 2004;5:253–65.

37. Vogel J, Garvin D, Leong O, Hayden D. Agrobacterium-mediated transformation and inbred line development in the model grass *Brachypodium distachyon*. Plant Cell Tiss Organ Cult. 2006;84:199–211.

38. Ciardi JA, Tieman DM, Lund ST, Jones JB, Stall RE, Klee HJ. Response to *Xanthomonas campestris* pv. vesicatoria in tomato involves regulation of ethylene receptor gene expression. Plant Physiol. 2000;123:81–92.

39. Hoch HC, Staples RC, Whitehead B, Comeau J, Wolf ED. Signaling for growth orientation and cell differentiation by surface topography in uromyces. Science. 1987;235:1659–62.

40. Glazebrook J. Contrasting mechanisms of defense against biotrophic and necrotrophic pathogens. Annu Rev Phytopathol. 2005;43:205–27.

41. Ho LH, Giraud E, Uggalla V, Lister R, Clifton R, Glen A, et al. Identification of regulatory pathways controlling gene expression of stress-responsive mitochondrial proteins in Arabidopsis. Plant Physiol. 2008;147:1858–73.

42. Mur LAJ, Kenton P, Lloyd AJ, Ougham H, Prats E. The hypersensitive response; the centenary is upon us but how much do we know? J Exp Bot. 2008;59:501–20.

43. Hudson AO, Singh BK, Leustek T, Gilvarg C. An Il-Diaminopimelate Aminotransferase Defines a Novel Variant of the Lysine Biosynthesis Pathway in Plants. Plant Physiol. 2006;140:292–301.

44. Song JT, Lu H, Greenberg JT. Divergent roles in *Arabidopsis thaliana* development and defense of two homologous genes, *ABERRANT GROWTH AND DEATH2* and *AGD2-LIKE DEFENSE RESPONSE PROTEIN1*, encoding novel aminotransferases. Plant Cell. 2004;16:353–66.

45. Anderson JP, Lichtenzveig J, Gleason C, Oliver RP, Singh KB. The B-3 ethylene response factor MtERF1-1 mediates resistance to a subset of root pathogens in *Medicago truncatula* without adversely affecting symbiosis with rhizobia. Plant Physiol. 2010;154:861–73.

46. Turner JG, Ellis C, Devoto A. The jasmonate signal pathway. Plant Cell (Suppl). 2002;14:S153–64.

47. Benedetti CE, Xie D, Turner JG. Coi1-dependent expression of an Arabidopsis vegetative storage protein in flowers and siliques and in response to coronatine or methyl jasmonate. Plant Physiol. 1995;109:567–72.

48. Ellis C, Turner JG. The Arabidopsis mutant cev1 has constitutively active jasmonate and ethylene signal pathways and enhanced resistance to pathogens. Plant Cell. 2001;13:1025–33.

49. Dixon RA, Achnine L, Kota P, Liu C, Reddy M, Wang L. The phenylpropanoid pathway and plant defense-a genomics perspective. Mol Plant Pathol. 2002;3:371–90.

50. Van Loon LC, Rep M, Pieterse CMJ. Significance of inducible defense-related proteins in infected plants. Annu Rev Phytopathol. 2006;44:135–62.

51. Thomma BP, Eggermont K, Penninckx IA, Mauch-Mani B, Vogelsang R, Cammue BP, et al. Separate jasmonate-dependent and salicylate-dependent defense-response pathways in Arabidopsis are essential for resistance to distinct microbial pathogens. Proc Natl Acad Sci U S A. 1998;95:15107–11.

52. Wang X, Tang C, Deng L, Cai G, Liu X, Liu B, et al. Characterization of a pathogenesis-related thaumatin-like protein gene *TaPR5* from wheat induced by stripe rust fungus. Physiologia Plantarum. 2010;139:27–38.

53. Ishiga Y, Uppalapati SR, Ishiga T, Elavarthi S, Martin B, Bender CL. The phytotoxin coronatine induces light-dependent reactive oxygen species in tomato seedlings. New Phytol. 2009;181:147–60.

54. Livak KJ, Schmittgen TD. Analysis of relative gene expression data using real-time quantitative PCR and the $2^{-\Delta\Delta C_T}$ method. Methods. 2001;25:402–8.

55. Ruijter JM, Ramakers C, Hoogaars WMH, Karlen Y, Bakker O, van den Hoff MJB, et al. Amplification efficiency: linking baseline and bias in the analysis of quantitative PCR data. Nucleic Acids Res. 2009;37(6):e45.

X-ray micro-computed tomography in willow reveals tissue patterning of reaction wood and delay in programmed cell death

Nicholas James Beresford Brereton[1*], Farah Ahmed[2], Daniel Sykes[2], Michael Jason Ray[3], Ian Shield[4], Angela Karp[4] and Richard James Murphy[5]

Abstract

Background: Variation in the reaction wood (RW) response has been shown to be a principle component driving differences in lignocellulosic sugar yield from the bioenergy crop willow. The phenotypic cause(s) behind these differences in sugar yield, beyond their common elicitor, however, remain unclear. Here we use X-ray micro-computed tomography (μCT) to investigate RW-associated alterations in secondary xylem tissue patterning in three dimensions (3D).

Results: Major architectural alterations were successfully quantified in 3D and attributed to RW induction. Whilst the frequency of vessels was reduced in tension wood tissue (TW), the total vessel volume was significantly increased. Interestingly, a delay in programmed-cell-death (PCD) associated with TW was also clearly observed and readily quantified by μCT.

Conclusions: The surprising degree to which the volume of vessels was increased illustrates the substantial xylem tissue remodelling involved in reaction wood formation. The remodelling suggests an important physiological compromise between structural and hydraulic architecture necessary for extensive alteration of biomass and helps to demonstrate the power of improving our perspective of cell and tissue architecture. The precise observation of xylem tissue development and quantification of the extent of delay in PCD provides a valuable and exciting insight into this bioenergy crop trait.

Keywords: Willow, Biofuel, X-Ray micro-computational tomography, Programmed-cell-death, Reaction wood

Background

Dedicated bioenergy crops have the potential to provide a sustainable and carbon neutral replacement to petroleum based liquid transport fuels. However, the glucose rich cell walls of dedicated bioenergy crops (such as willow or *Miscanthus* in the UK) are generally recalcitrant to deconstruction, requiring high amounts of energy and severe chemical pretreatment before the glucose can be released in a form suitable for fermentation. To overcome this barrier, research efforts worldwide have been directed towards understanding the natural variation of cell wall recalcitrance in dedicated bioenergy crops.

The basis of genotype-specific variation in recalcitrance was recently identified in the fast-growing biomass crop willow (*Salix sp.*) as genetic variation in a natural response to *gravity*, known as the "reaction wood" (RW) response [1]. RW formation in trees is characterised by major alterations in xylem cell development and tissue patterning in the stem in response to displacement from vertical, either through the perception of gravity or mechanical load. These changes are polarized across the stem with the "upper" side of the stem termed Tension Wood (TW) and the "bottom" side termed Opposite Wood (OW). Despite being recognised as a key determinant of glucose yield, many aspects of this trait, and specifically how the trait

* Correspondence: Nicholas.Brereton@UMontreal.ca
[1]Institut de recherche en biologie végétale, Université de Montréal, Montreal, QC H1X 2B2, Canada
Full list of author information is available at the end of the article

differs between genotypes to result in such large alterations to glucose release yields, remains a mystery.

General reaction wood tissue patterning and development

The majority of tree biomass develops from the vascular cambium, the ring of differentiating cells between the bark and the inner/secondary xylem. The proportion of the secondary xylem to the biomass of the stem varies with age and genotype, but is roughly 85-90% [2]. Most angiosperms, such as willow (*Salix* sp.), have a degree of specialisation within the secondary xylem, with fibre cells predominantly delivering the structural demands of the organism, vessel elements comprising purely hydraulic architecture and ray parenchyma cells thought to mostly serve as storage elements. This increased tissue complexity and diversity of function is distinct from the more ancient gymnosperms, where tracheids serve both functions.

Further specialisation has evolved in a smaller number (<50%) of woody angiosperms [3] where gelatinous fibres (g-fibres) can form on the TW side of secondary xylem, in a stem displaced from vertical, in order to return the apical meristem to vertical and increase the mechanical strength of the stem. The structural re-enforcement of fibre cells with an extra cell wall layer (the gelatinous layer or g-layer) is developed at the expense of the fibre cell lumen, and thus chould be accompanied by a deleterious reduction in water conductance in TW. A positive correlation between fibre cell lumen and xylem water capacity has been observed by Pratt *et al.* [4]. Even though there is this large change in cell structure upon RW formation, most g-fibre forming angiosperms, unlike gymnosperms [5,6], are thought to maintain their efficient water translocation, although the mechanism of how this is achieved is unclear.

During secondary xylem development from the vascular cambium, normal fibre cells undergo a very strictly controlled apoptosis, the end result being long tube-like cells with thick secondary cell walls and no protoplast. How the process of programmed-cell-death (PCD) is altered in TW development is poorly established in terms of evidence, but it has been suggested in several reviews [7,8] that PCD is delayed in certain species of poplar, with this delay hypothesised as being necessary to accommodate g-layer biosynthesis.

X-Ray micro-computed tomography (μCT)

X-Ray μCT has been used increasingly as a powerful method for plant anatomical assessment mainly driven by its value in the timber industry for evaluation of wood quality. Recent published studies, while low resolution in terms of the current state-of-the-art, show how this non-destructive technology can be used systematically to identify the presence or absence of rameal traces, i.e. irregularities relating to branching such as knots, in oak [9]. High resolution X-Ray μCT has, over the past decade, been presented as a potentially valuable method for quantitative investigation of plant anatomy in numerous studies, and more recently wood anatomy. Stuppy *et al.* [10] demonstrated how 3D architecture could be rendered (at a relatively poor linear resolution of 50 μm) in a diverse range of plants including sections of palm, oak, pineapple, a tulip flower and inflorescence of *Leucospermum tottum*. Exclusively in wood, broad tissue 3D models have been rendered of sections of: beech, oak, spruce heartwood, Douglas fir, loblolly pine, teak and eucalyptus (as well as non-woody Arabidopsis) [11-13]. Most recently Broderson *et al.* have established a range of tools useful for assessment of 3D xylem structre using X-Ray μCT [14,15].

Variation in reaction wood

Juvenile willow genotypes (3 month old) grown under greenhouse conditions only exhibit fully mature field (3 year old trees, 7 year root stock) lignocellulosic sugar yield phenotype if tipped to induce RW [1], demonstrating the significance of variation in RW response to wood development as well as the constant RW inducing conditions of field environments. It seems likely that the high sugar release yields achieved from willow and poplar biomass is due to abundance of the cellulose rich g-layers in TW tissue of RW (which are always present to some extent in short rotation coppice (SRC) willow stem sections) and that sugar release yield variation between genotypes is therefore due to variation in g-fibre abundance. Evidence for this is absent to date and, surprisingly, some genotypes of willow that do not significantly increase in sugar release upon RW induction did have increased g-fibre abundance [1]. This suggests that variation in RW might extend beyond g-fibre abundance alone. Traditional sectioning and microscopy fall short of providing a means of robust quantification of RW tissue patterning on a whole tree level as a transverse section or several transverse sections may not be representative due to the irregular nature of wood growth. To overcome these limitations, an approach to larger scale 3D tissue assessment was devised in the hope further resolving the nature of the RW response.

To test the hypothesis that tissue patterning alters significantly upon RW induction; we used 3D X-Ray μCT to directly assess wood architecture in willow trees after being grown vertically and *tipped* at 45°.

Methods
Plant cultivation and RW induction

Six short rotation coppice willow cuttings (cultivar Resolution – pedigree: (*S. viminalis.* x (*S. viminalis.* x *S. schwerinii* SW930812)) x (*S. viminalis.* x (*S. viminalis.* x *S.*

schwerinii 'Quest'))) were planted in 12 l pots with 10 l of growing medium consisting of $^1/_3$ vermiculite, $^1/_3$ sharp sand and $^1/_3$ John Innes No.2 compost, by volume. Trees were then grown under a 16 h (23°C) day cycle and an 8 h (18°C) night cycle for 12 weeks. After 6 weeks of growth all stems from all trees were tied to a supporting bamboo cane at regular intervals and three of the trees tipped at a 45° angle to the horizontal (three left growing vertically as controls). All trees were checked every two days and tied to maintain controlled growth orientation, either 45° or vertical. After 12 weeks of growth (and 6 weeks of differential treatment) tree stem biomass was harvested.

Fixation, sectioning, staining and microscopy
Upon stem harvest an eight cm section of the measured middle of all the stems from each tree was debarked (for ease of 2° xylem specific analysis) and "fixed" in FAA (formaldehyde 3.7%, acetic acid 5% – ethanol 47.5%). The fixation step is crucial to maintain cell contents for downstream 3D X-ray μCT. Sections were then cut into four cm sections, one was air dried for 3–5 days before X-ray μCT, whilst the other was used for sectioning (using a sledge microtome to 25 μm) and histochemisty before then being used for destructive basic density assessment. Sections were stained with 1% safranin O (aq) as an unspecific cell wall counterstain, 1% chlorazol black E (in methoxyethanol) to stain g-layers [1,16] or with 1% Coomassie to highlight the remnants of cell content. As fibre cell length can often be greater than 1 mm (and sections for microscopy were limited to 25 μm depth) efforts were made to compare these partial cell images to 3D data. Further comparative analysis was conducted using cell wall and cell contents auto-fluorescence confocal microscopy by Z-stacking with high resolution for closer comparison of cell content fragments to 3D images. Excitation and emission wavelengths were 488 nm and 500-700 nm respectively [17].

Basic density assessment
The basic density of wood was assessed using traditional methods [18]. Here, 2-4 cm wood sections were vacuum infiltrated with water before green volume was measured via water displacement and wood oven dry weight was measured after drying over night at 105°C.

X-ray μCT scanning
The scans were performed using a Nikon Metrology HMX ST 225. The samples were scanned using a tungsten reflection target, at an accelerating voltage of 160 kV and current of 180 μA using a 500 ms exposure time (giving a scan time of 25 minutes). No filters were used and 3,142 projections were taken over a 360° rotation. The voxel size of the resulting dataset had linear resolution of 9 μm. A common piece of willow wood, cut from NW of control trees and treat in the same manner as the samples, was used as a reference standard and scanned alongside each sample. For 2D images, high voxel intensity, and therefore greater X-Ray attenuation, is visible as lighter regions whereas regions of low voxel intensity are visible as darker regions.

3D image processing
The 3D volumes were reconstructed using CT Pro (Nikon Metrology, Tring, UK) and TIFF stacks exported using VG Studio Max (Volume Graphics GmbH, Heidelberg, Germany) (see Additional file 1). Drishti [19] was used to generate a 3D rendering and analysis of ROI. A standardised transfer function was designed and applied to each 3D ROI in isolation to allow comparison. ROI were also saved as individual 2D Tiff files and MatLab (MATLAB 6.1, The MathWorks Inc., Natick, MA, R2012b) was used to collate data from all files and produce histograms of voxel intensity distribution. The reference standard was used to normalise voxel intensity and allow direct comparison of collated data between samples.

Results
Density and G-layer verification
The basic density of 2–4 cm long stem middle segments from 3 month old willow (cultivar Resolution) was assessed after trees had been tipped for 6 weeks of their growth or grown vertically as controls. Debarked stem segments from control trees had an average basic density of 195 kg/m^3 which was significantly (t-test p < 0.001) increased in similar segments from RW induced trees, to 275 kg/m^3 (Figure 1A). Stems were then sectioned and stained which confirmed that RW induction had successfully produced g-fibres (Figures 1B). RW induced trees had abundant g-fibre production with clear transverse polarisation aligned with the vector of gravitational stimulus ("upper" stem during tipping).

μCT scanning and voxel intensity/distribution of regions of interest
Reconstruction of μCT scans from each of the six stems were made allowing generation of ~1500 (2 MB) images each. These images were then stacked and rendered into a 3D volume where the stem segment is reproduced *in silico* down to a voxel (a 3D pixel) representing an *in planta* linear resolution of 9 μm. Clear increases in X-ray attenuation (represented by voxel intensity) were visible at the lateral part of the stem, corresponding anatomically to the vascular cambium, elongating secondary xylem and maturing secondary xylem tissue (Figure 2). In TW, this region of increased voxel intensity was greatly extended, also from the periphery of the stem, and with transverse polarisation aligned with the vector of gravitational stimulus (Figure 2 *top three panels*).

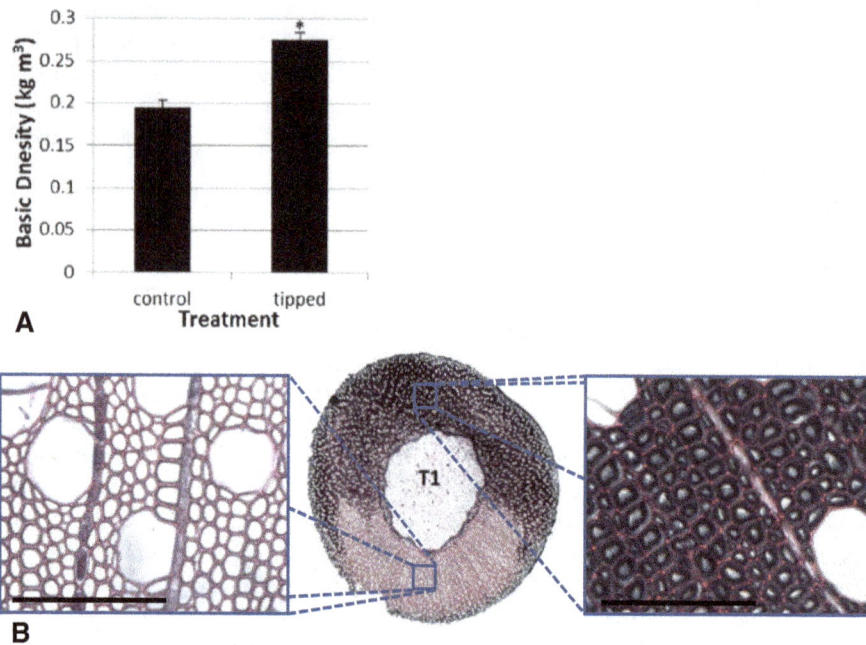

Figure 1 Reaction wood impact on basic density and 2D xylem architecture. A Basic density of debarked willow cultivar Resolution after 3 months of growth either unperturbed or including 6 weeks of RW induction (tipping at 45° from vertical). n = 3 trees. **B** Transverse middle stem section (25 μm) of a RW induced tree stained with safranin O (red – nonspecific staining the cell wall) and chlorazol black (black – specifically staining the g-layer of g-fibres). Panels: OW (left) and TW (right) are included with scale bar = 100 μm. *$p < 0.05$ (Students t-test).

The 3D Regions Of Interest (ROI) were then isolated *in silico* representing: TW, OW and NW (Figure 2). These ROI were assessed for voxel intensity and spatial distribution. Using MatLab, histograms of the voxel intensity for each ROI were plotted (n ranging from 3–14 million voxels for each ROI) (Figure 3). The distribution of voxels was consistent between OW and NW but distinct for TW. The voxels for a given ROI were then each counted into one of 26 bins of relative greyscale intensity from 0–50000 (0–1999, 2000-3999...) with numbers of voxels in each bin expressed as a proportion of total ROI voxels. The relative bin greyscale intensity was then normalised against a small segment of willow used as a common internal standard for each scan. When the normalised average voxel intensity of each scan was compared, TW was the only tissue to be significantly increased (Figure 3B, t-test $p < 0.05$).

Treatment specific tissue patterning/architectural patterning

As well as quantifying average voxel intensity, voxels can be binned according to intensity *in silico,* this can be applied to each rendered volume using a common transfer function as part of the image processing [19] to quantify (and view) voxel groups of similar intensity. In this way it was possible to isolate the vessel elements within each tissue type to compare architectural changes generated by RW induction (Figure 4).

The vessel frequency was consistent between the OW and NW ROI (averaging 37 vessel elements) but reduced in the TW ROI (averaging 30 vessel elements). However, vessel *volume* was significantly increased by over 50% in the TW ROI (Figures 3 and 4). The total vessel surface area (per cm^3) can also be quantified after isolation using the common transfer function; in tension wood the vessel surface area to vessel volume ratio drops well below an average of 0.9-0.95 to that of 0.64 (the largest cell type present in the stem).

Quantification of delayed programmed cell death

The variation in X-ray attenuation, and so voxel intensity, observed in RW induced trees was clearly aligned with TW but did *not* correspond to g-fibre presence in the tissue (Figure 5). This is not surprising on a cell by cell basis as resolution was not great enough to distinguish individual fibre cells (but was sufficient to distinguish vessel elements) despite the fact that each voxel was resolved to 9 μm.

Interestingly, the pattern of increased voxel intensity followed that of developing xylem before the termination of fibre cell maturation and completion of PCD. The post-cambial cells, from the secondary xylem elongation stage to the onset of autophagy where the protoplast and cytoplasmic contents are still retained, was visible as a circle surrounding the stem present in both control and RW induced trees.

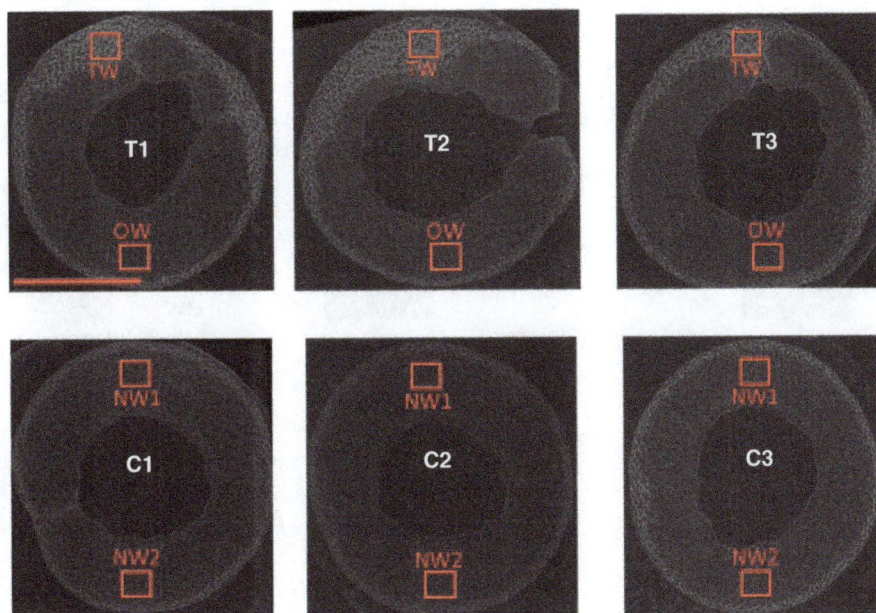

Figure 2 2D transverse X-Ray CT scans. A single representative image from the stack reconstructed from the X-ray CT scanning of each stem segment. Each tree is either RW induced (T1, T2 and T3) or a control grown without induction (C1, C2 and C3). Regions of interest assessed for voxel intensity and distribution, TW, OW and NW are highlighted in red. High voxel intensity, and therefore 554 great X-Ray attenuation, is visible as lighter regions whereas regions of low voxel intensity are visible as darker regions. Scale bar = 4mm.

The delay of PCD often referred to as associated with TW can be seen by light microscopy, but is inadequately represented due to the transverse sectioning process. By using direct coomassie staining and Z-stacked confocal microscopy of 25 μm sections the extension of cell life can be roughly observed in TW (Figure 5B). Fibre cells are sheared during the sectioning process leaving only a proportion of the cell contents/remnants visible so that, whilst the irregular nature of this extended tissue patterning is evident, quantification is difficult. By assessing this tissue patterning without sectioning, by X-ray μCT, the extent of this irregularity was revealed directly (Figures 2 and 5).

Discussion

RW response has been identified as a principle cause of variation in enzymatic saccharification yields in willow, yet understanding of the tissue architectural and cellular remodelling associated with RW has typically been limited to classical sectioning and microscopy. Here, RW stem remodelling was explored using μCT following the theory that such widespread alterations, with accompanying extensive influence on saccharification yields, would likely effect enough change in X-ray attenuation as to be amenable to more direct quantification.

Density, G-layer verification and distribution of voxel intensity

The tipping of trees at 45° and maintaining this angle by restraint is designed as an *analytical* technique for studying RW. RW induction is not optimised to produce large amounts of TW but to deliver a stimulus in a consistent and controlled manner allowing transferable analysis of the response. This constant, known magnitude of stimulus is crucial to such studies as, in field-grown willow trees, g-fibres can always be seen in transverse sections willow material. The explanation for this is that trees in the field are constantly exposed to some degree of RW inducing stimulus from the environment but of an ever-changing intensity and from varying vectors in the form a wind speed, land incline and/or internal growth stresses [1]. A number of common morphological alterations have been reported to occur in both Poplar and willow upon development under increased RW inducing conditions, either gravitational or thigmomorphogenic in nature. A common result is more compact growth, with reduced stem height, increased diameter and increased density [20,21]. These make sense from an architectural standpoint as, under conditions such as high wind speeds, the structure of a smaller, wider stem will reduce the stress a stem is exposed to. The degree to which such changes vary between different varieties has been less well studied.

The utilisation of the RW response is an attractive way to increase sugar accessibility in willow due to being part of natural plant physiology, and so unlikely to negatively impact plant integrity. In fact, large increases in sugar yield have been reported without any detriment to biomass yield (although only in pot trials) even though plant size was reduced [1,22]. An increase in density

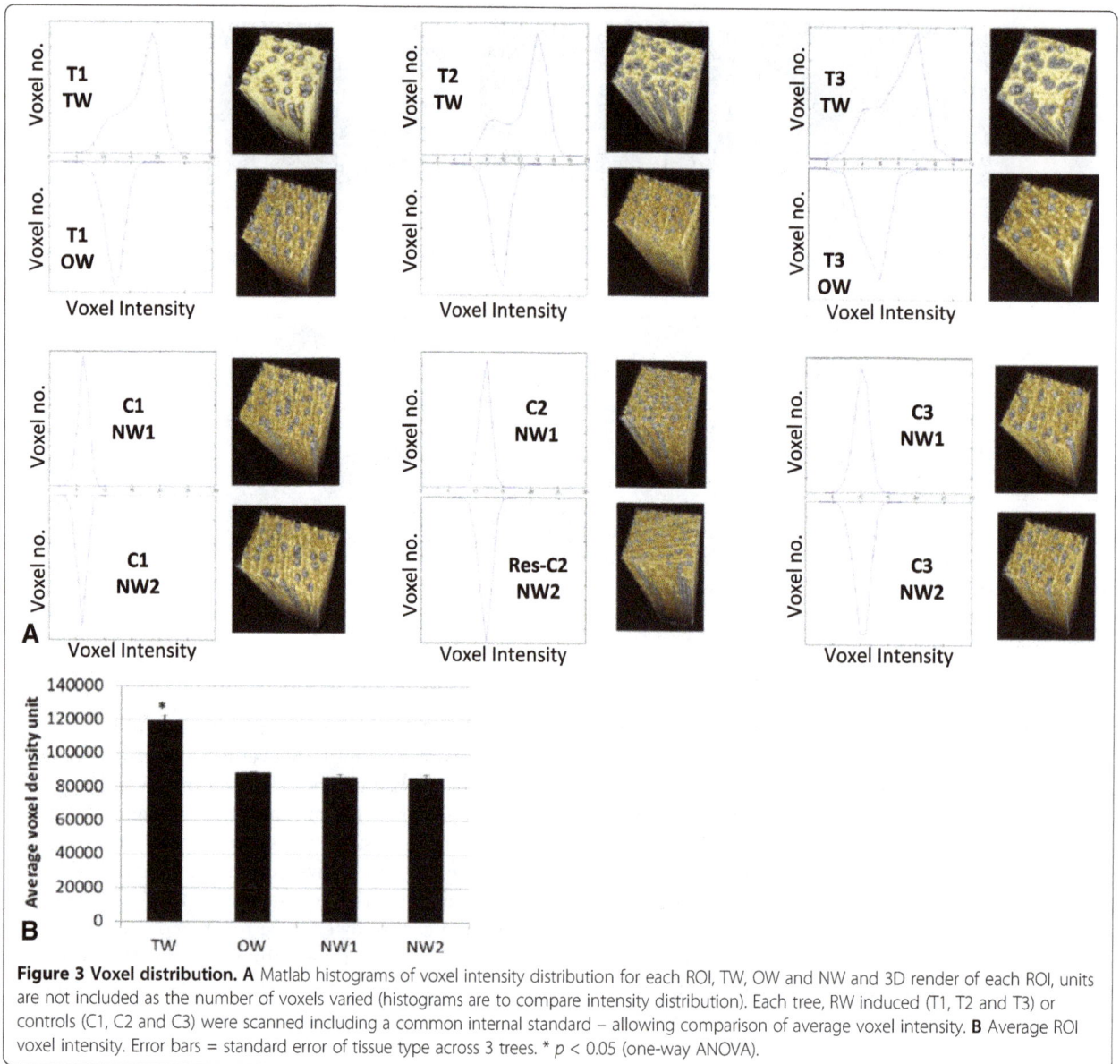

Figure 3 Voxel distribution. A Matlab histograms of voxel intensity distribution for each ROI, TW, OW and NW and 3D render of each ROI, units are not included as the number of voxels varied (histograms are to compare intensity distribution). Each tree, RW induced (T1, T2 and T3) or controls (C1, C2 and C3) were scanned including a common internal standard – allowing comparison of average voxel intensity. **B** Average ROI voxel intensity. Error bars = standard error of tissue type across 3 trees. * $p < 0.05$ (one-way ANOVA).

straightforwardly describes this phenomenon and speaks to the extent of the changes in biomass structure elicited during RW formation. The density observed in the pot grown trees here (195 and 275 kg/m^3) is very low when compared to that of mature willow ($\sim$300 – 500 kg/m^3) [23], but not surprising for juvenile, debarked wood. The increase in density associated with RW induction may be due to increased g-fibres, which substantially increase in abundance upon induction (Figure 1), as the extra cell wall layer replaced fiber lumen void space. Whilst g-fibre abundance was clearly increased, g-fibre enriched TW was not visible in the µCT scans, this is likely due to the linear resolution of the scans which was just short of a fibre cell width (once voxel bleeding, a localised overlap of signal, is accounted for) at $\sim$10 µm as well as due to the

nature of the extra cell wall layer (g-layer), which did not greatly attenuate X-rays being almost entirely composed of cellulose. However, clear differences in X-ray attenuation associated with RW induction were observable in 3D.

Treatment specific tissue patterning/architectural patterning

Broad secondary xylem tissue remodelling occurs during RW formation. An increase in vessel length but severe decrease in vessel frequency in tension wood of young inclined stems was recorded in poplar [24] and consequently total vessel volume should be reduced as a product. Our data agrees with this reduction in vessel frequency but also measures the volume of vessels in relation to other

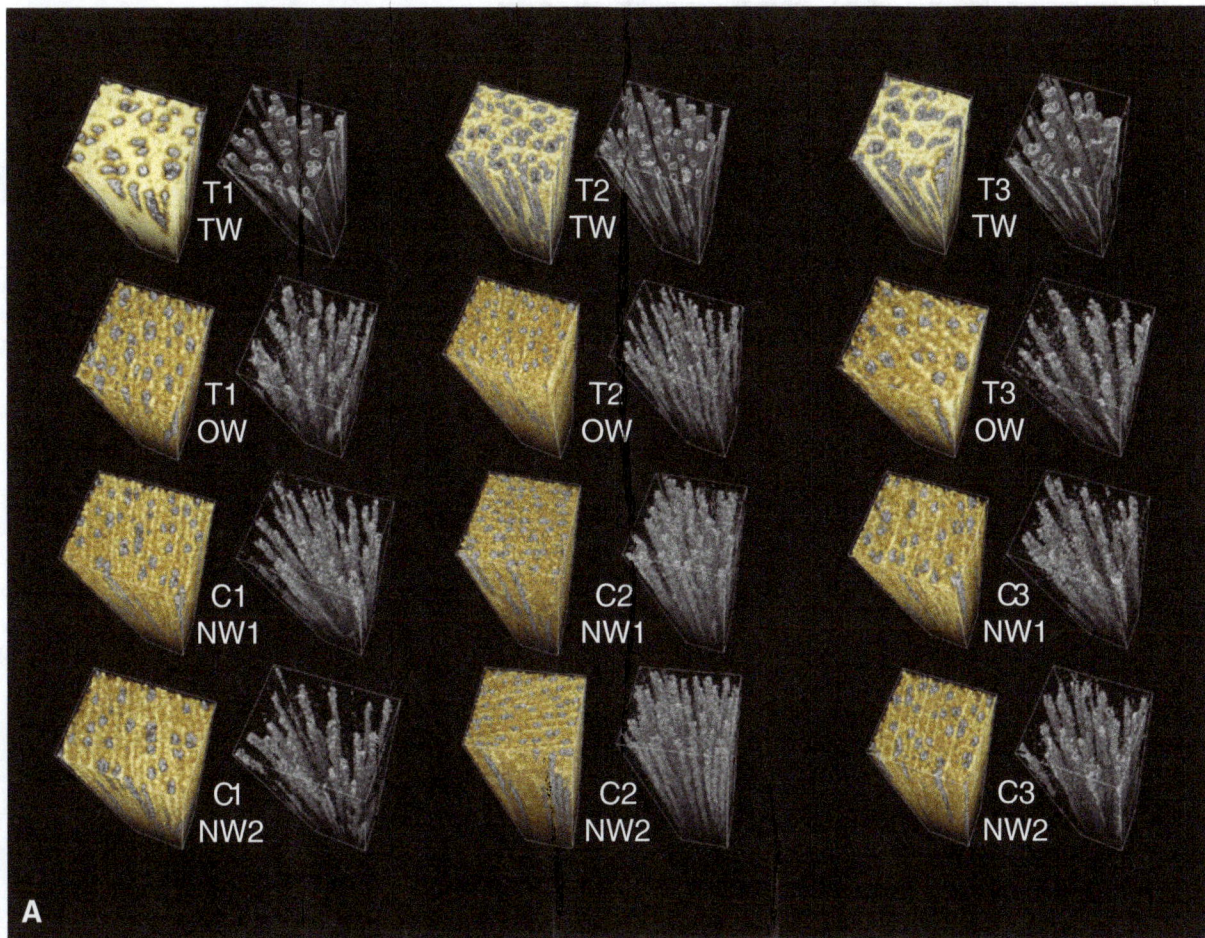

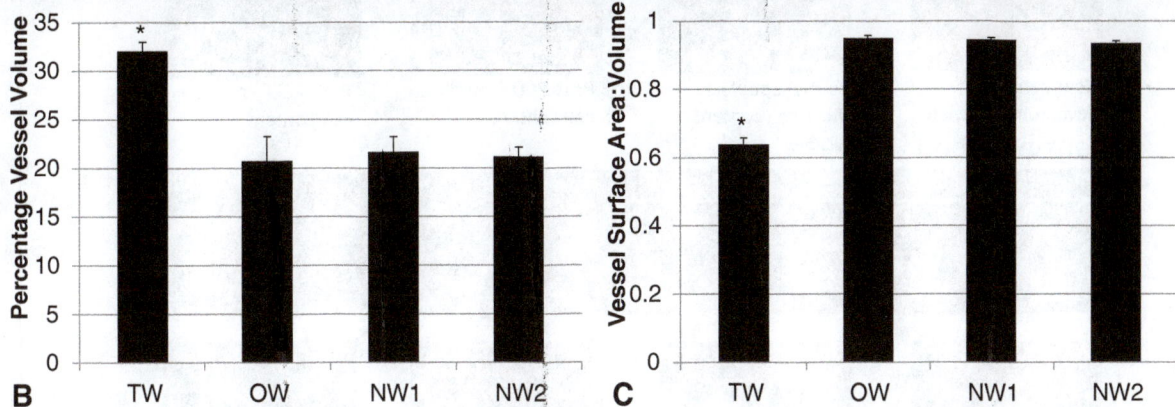

Figure 4 3D xylem architecture. A 3D render of each ROI (TW, OW or NW) from X-ray CT scans of RW induced trees (tipped T1, T2 and T3) or controls (C1, C2 and C3). The 3D ROI render on the right after the common vessel specific transfer function was applied *in silico*. **B** Total volume of vessels as a percentage of each ROI was averaged for each tissue. **C** Vessel surface area:volume ratio of each ROI was averaged for each tissue. Error bars = standard error of tissue type across 3 trees. * $p < 0.05$ (one-way ANOVA).

cell types in the xylem. As can be seen in Figure 4, the total volume of vessels is greatly increased in tension wood of the willow variety Resolution, even though the frequency is reduced. From this we can speculate that there is no penalty to trees grown in high RW inducing conditions due to limitations in water transport capacity. This increase in vessel volume:surface area ratio, in certain parts of tension wood, may represent a mechanism by which this maintenance of conductivity is achieved and may also reflect the penalty associated with such

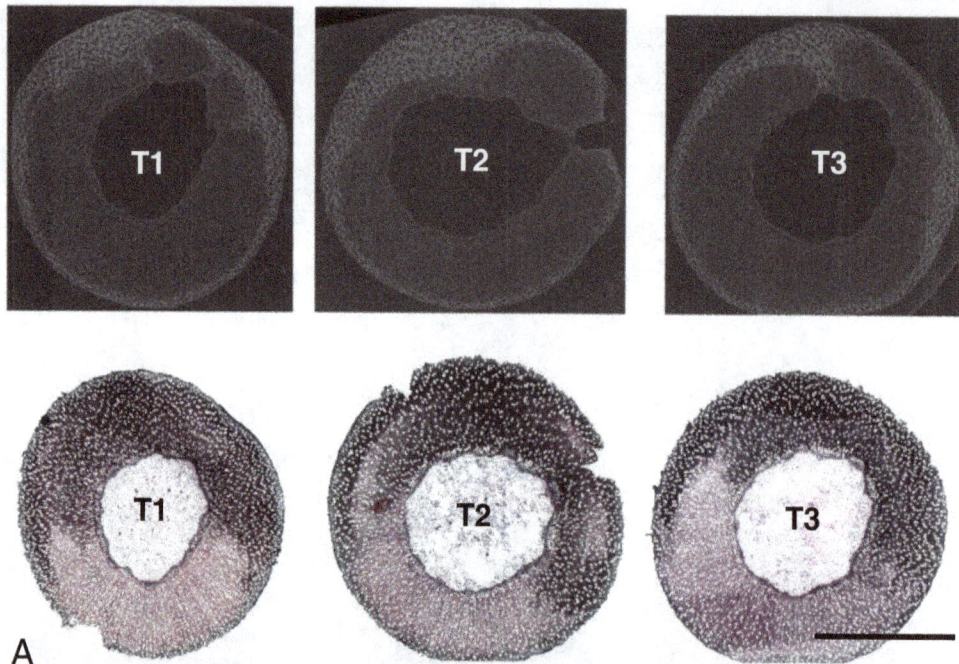

A
Opposite Wood

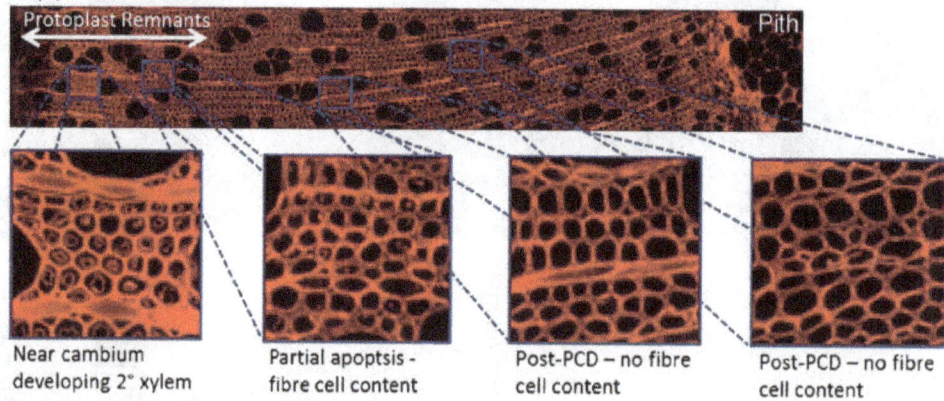

B
Tension Wood

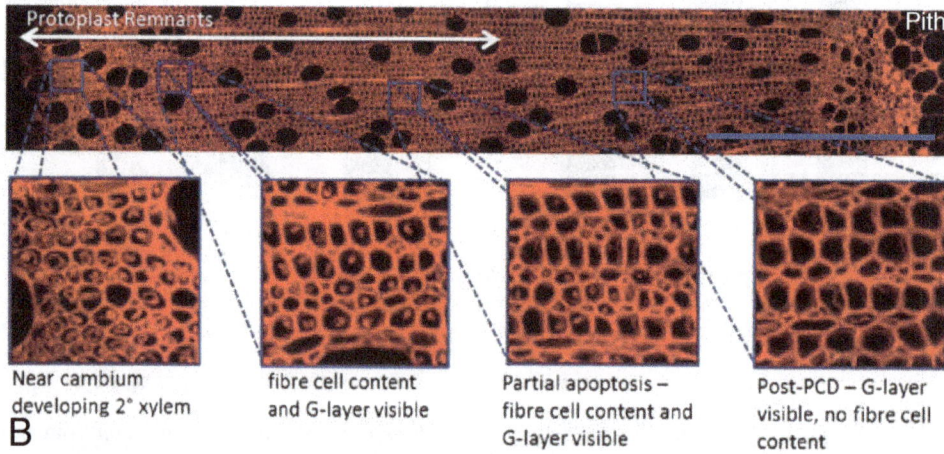

Figure 5 (See legend on next page.)

structural change if an increase in volume is required due to a reduction in efficiency of the new vessel structure.

The increase in relative fibre cell frequency is structurally necessary to either bring the stem back to vertical or help tolerate the increased load bestowed by the displaced stem. This remodelling is made at the expense of vessel number, yet a reduction in the water transporting capacity would be detrimental to plant fitness. It is not then surprising to see alterations to vessel dimensions to mitigate this penalty. Jourez *et al.* [24] also found that solitary vessels in TW, whilst less circular, had a greater external diameter (2 μm more) and greater length (10 μm more). When these increases are envisaged in 3D, the volume of vessels is likely to be larger. This would agree with another of their findings that mean lumen of TW vessels is larger (5%) than OW. Remodelling resulting in such large increases in vessel volume suggests that the TW form may not be as efficient at water transport but is still effective as well as permitting a greatly increased structural function. They also discuss the variability of such measurements in different species and we would re-iterate that these changes are likely to be species and variety specific.

This trade-off between mechanical support and water conductivity is recognised in conifers as compression wood has reduced k_s (specific water conductivity) [5,6]. Unlike angiosperms, gymnosperms, a more ancient phylum in evolutionary terms, do not have specialised vessel cells so the homogenous tracheids play the role of bestowing large structural modification without loss of plant integrity alone. Gartner *et al.* [25] found that *Quercus ilex* (holm Oak) TW elicits large scale modification for mechanical support *without* impairment to water conductivity (specific conductivity, k_s). Interestingly for Gartner *et al.* [25], whilst vessel area was similar, vessel frequency was actually increased in tension wood – the reverse physiological solution to that implied by the tissue modifications here in willow but with the same outcome. The lesson from nature here is that complex interdependent relationships exist between biomass mechanics support and water conductivity, or importantly from a bioenergy perspective, between lignocellulosic sugar yield (as driven by RW) and water-use-efficiency (of great importance for crop sustainability).

A reduction in vessel lumen area in poplar tension wood has been well documented [20,21,26] and is in contradiction to the data revealed here for willow. Whilst this may be a point of distinction between willow and poplar, the ROI specifically investigated, or novel method of their assessment here, may also be the source of this disparity. The ROI were selected specifically as the regions of variation in terms of X-ray attenuation which were located at the periphery/lateral side of the stem (Figure 2). We can see that at this periphery tissue architecture is distinct from more medial/older wood of lower intensity, in a manner which is aligned with g-fibre orientation (the "upper" part of the stem) but not overlapping with g-layers. This variation within TW and lack of g-fibre associated increase in intensity was surprising. A major aim of the technique was to be able to separate fibres from g-fibres, therefore providing a valuable approach in quantifying g-fibre abundance in 3D (hopefully affording more accuracy than multiple transverse sections). Although resolution of the X-ray μCT fell short of such separation there was sufficient variation, associated with our treatment, to suggest TW tissue variation beyond g-layer presence alone. This led us to question what other TW tissue modifications might underlie such stark treatment specific differences as well as which might impact vessel size during development.

Quantification of delayed programmed cell death

Although published evidence appears absent to date, it is well recognised that PCD is delayed in willow and poplar tension wood as fibre cell protoplast/cellular remnants can be observed (by microscopy) as present long after the completion of fibre cell maturation, apoptosis and degradation of cytoplasmic remnants in normal or opposite wood [7,8]. Overlapping RW formation and PCD EST libraries also strongly indicate alteration of "normal PCD" in xylem development of tension wood [27,28]. It is widely speculated that perhaps this delay occurs to accommodate biosynthesis of the g-layer from the plasma membrane, the extra internal cell wall layer of g-fibres, as the cellulose microfibrils would have to be produced after the establishment of the secondary cell wall.

Traditional methods of assessing cell *viability* are difficult to perform quantitatively in woody tissue as the process of transverse sectioning is such a destructive process, impairing methods such as NBT staining (for superoxide) and Tunnel Staining (for DNA degradation as resulting from autolysis). Whilst these methods are powerful in smaller model systems or when targeting limited numbers of cells, the assessment of larger scale tissue patterning in crops, such as the polarised patterning in wood as a result of RW induction, requires improved techniques for assessment and confidence.

Assessment is further made difficult by the natural asymmetry in wood growth (lack of perfectly uniform growth/ cell development) as can be seen in Figures 1, 2 and 5. Hence, when investigating developmental variation due to RW formation a quantitative method, encompassing the variation across a given point in the stem but also the variation along the stem, is of substantial value. X-ray μCT provided a surprisingly powerful method for such assessment of PCD in 3D. Unlike histochemical or immunohistochemical 2D assessment of transverse sections where some amount of cell content is lost through processing, 3D analysis (if cell content is fixed/preserved as here) reveals the stark difference between fibre cells having completed PCD and those still developing or undergoing PCD. This is, in retrospect, predictable as the comparison in 3D is between "empty" fibre cells of secondary xylem and "solid" fibre cells retaining cell content/protoplast should reveal substantial difference in density. It is interesting to note that lines of increased voxel intensity, and so increased X-ray attenuation, are also visible leading from the pith of the wood to the periphery. Although these are not well resolved, they could potentially represent the ray parenchyma, which are also cells where cell content would be present and preserved.

Validation of the state of cell development in 2D (compared to single images of x-ray scans used for 3D rendering) was made using coomassie staining (data not shown) and z-stacked confocal microscopy (Figure 5B) of section made by traditional sectioning. It should be made clear that the nature of this assessment is by no means an assessment of *viability*, as methodology such as NBT staining but more an assessment of architectural variation which effects x-ray attenuation and, as such, is not limited to quantification of PCD.

One of the lessons evident from recent advances in biomass composition and enzymatic saccharification assessment is that above ground stem biomass varies between stems and throughout a stem and consequently should be considering in 3D. RW formation is a compelling illustration of this as it forms in localised positions across and along the stem at different times throughout biomass development, so that, if the genetics of a crop variety or the *net* composition of a feedstock for a biofuel process is to be elucidated, a holistic model of assessment is not only of great benefit but a necessity.

Conclusions

Alterations to tissue patterning due to RW induction were visible in 3D using X-ray μCT. These changes

describe the compromise or trade-off between hydraulic architecture and mechanical support. As the effectiveness of the functional interplay is likely to vary in a genotype specific manner, this RW remodelling would interestingly link dedicated bioenergy crop sustainability and yield via WUE and cell wall sugar accessibility. Greater resolution is needed to distinguish g-fibres and, as a more general property of the technique, the higher the resolution the more biological complexity can be investigated as there seems to be no "lower limit" to the patterning in nature. If resolution can be very slightly increased while the macro scale of biological samples is maintained then X-ray μCT could join contemporary *in vivo* imaging techniques such as GFP-fusions.

Variation in RW response has been established and identified as being of major importance to glucose yields from willow biomass, however, the root cause of this variation remains unclear. Evidence of variation in g-layer structure has been shown to exist between species [29]. Further work should focus on quantitatively assessing the degree to which, if any, delay in PCD varies between genotypes known to differ in RW response in terms of glucose yield. If the basis for the delay in PCD is in fact due to a necessity of maintaining the protoplast for g-layer construction, variation in the extent of this delay may affect the composition, structure and/or abundance of the layer, and consequently be of key importance to lignocellulosic biofuel yields.

Availability of supporting data
The video supporting the results of this article is available in the BMC-series YouTube channel repository, http://youtu.be/CxWR10gdwQc.

Additional file

> **Additional file 1: 3D video rendered from X-Ray μCT scans of willow, cultivar Resolution.** Stem segment (2 cm length) was air dried for several months without fixation, bark was retained. Void volume of the wood was coloured in blue to highlighted vessel element lumens.

Competing interests
The authors declare that they have no competing interests.

Authors' contributions
NJBB designed the study. NJBB and MJR designed and performed the plant growth trials and sample preparation. FA, DS and NJBB performed the X-ray CT experiments and interpreted the data. NJBB and MJR drafted the manuscript. AK is overall leader of the BSBEC BioMaSS project and RJM leads the sub programme of which this work is a part. NJBB, MJR, IS, FA, DS, AK and RJM conceived the study, commented on the results and contributed to the manuscript. All authors read and approved the final manuscript.

Acknowledgements
We are grateful for the financial support for this research from the BBSRC Sustainable Bioenergy Centre (BSBEC), working within the BSBEC BioMASS (http://www.bsbec-biomass.org.uk/) Programme (Grant BB/G016216/1) as well as the Natural History Museum. The authors would like to thank Rodriguez Geraldes and Volker Behrends for their assistance with data processing software.

Author details

[1]Institut de recherche en biologie végétale, Université de Montréal, Montreal, QC H1X 2B2, Canada. [2]Micro-CT Lab, Imaging and Analysis Centre, Natural History Museum, London SW7 5BD, UK. [3]Department of Chemistry, Imperial College, London SW7 2AZ, UK. [4]Department of AgroEcology, Rothamsted Research, Harpenden, Herts AL5 2JQ, UK. [5]Centre for Environmental Strategy, University of Surrey, Guildford, Surrey GU2 7XH, UK.

References

1. Brereton NJB, Ray MJ, Shield I, Martin P, Karp A, Murphy RJ. Reaction wood - a key cause of variation in cell wall recalcitrance in willow. Biotechnol Biofuels 2012, 5:38.
2. Serapiglia MJ, Cameron KD, Stipanovic AJ, Smart LB. Analysis of Biomass Composition Using High-Resolution Thermogravimetric Analysis and Percent Bark Content for the Selection of Shrub Willow Bioenergy Crop Varieties. BioEnergy Res. 2009;2(1–2):1–9.
3. Fisher JB, Stevenson JW. Occurrence of Reaction Wood in Branches of Dicotyledons and Its Role in Tree Architecture. Bot Gaz. 1981;142(1):82–95.
4. Pratt RB, Jacobsen AL, Ewers FW, Davis SD. Relationships among xylem transport, biomechanics and storage in stems and roots of nine Rhamnaceae species of the California chaparral. New Phytologist. 2007;174(4):787–98.
5. Spicer R, Gartner BL. Hydraulic properties of Douglas-fir (Pseudotsuga menziesii) branches and branch halves with reference to compression wood. Tree Physiol. 1998;18(11):777–84.
6. Spicer R, Gartner BL. How does a gymnosperm branch (Pseudotsuga menziesii) assume the hydraulic status of a main stem when it takes over as leader? Plant Cell Environ. 1998;21(10):1063–70.
7. Bollhoner B, Prestele J, Tuominen H. Xylem cell death: emerging understanding of regulation and function. J Exp Bot. 2012;63(3):1081–94.
8. Courtois-Moreau CL, Pesquet E, Sjodin A, Muniz L, Bollhoner B, Kaneda M, et al. A unique program for cell death in xylem fibers of Populus stem. Plant J. 2009;58(2):260–74.
9. Colin F, Mothe F, Freyburger C, Morisset JB, Leban JM, Fontaine F. Tracking rameal traces in sessile oak trunks with X-ray computer tomography: biological bases, preliminary results and perspectives. Trees-Struct Funct. 2010;24(5):953–67.
10. Stuppy WH, Maisano JA, Colbert MW, Rudall PJ, Rowe TB. Three-dimensional analysis of plant structure using high-resolution X-ray computed tomography. Trends Plant Sci. 2003;8(1):2–6.
11. Steppe K, Cnudde V, Girard C, Lemeur R, Cnudde JP, Jacobs P. Use of X-ray computed microtomography for non-invasive determination of wood anatomical characteristics. J Struct Biol. 2004;148(1):11–21.
12. Trtik P, Dual J, Keunecke D, Mannes D, Niemz P, Stahli P, et al. 3D imaging of microstructure of spruce wood. J Struct Biol. 2007;159(1):46–55.
13. Mayo SC, Chen F, Evans R. Micron-scale 3D imaging of wood and plant microstructure using high-resolution X-ray phase-contrast microtomography. J Struct Biol. 2010;171(2):182–8.
14. Brodersen CR, Lee EF, Choat B, Jansen S, Phillips RJ, Shackel KA, et al. Automated analysis of three-dimensional xylem networks using high-resolution computed tomography. New Phytologist. 2011;191(4):1168–79.
15. Brodersen CR, McElrone AJ, Choat B, Matthews MA, Shackel KA. The Dynamics of Embolism Repair in Xylem: In Vivo Visualizations Using High-Resolution Computed Tomography. Plant Physiol. 2010;154(3):1088–95.
16. Robards AW, Purvis MJ. Chlorazol Black E as a Stain for Tension Wood. Biotech Histochem. 1964;39(5):309–15.
17. Donaldson L. Softwood and Hardwood Lignin Fluorescence Spectra of Wood Cell Walls in Different Mounting Media. IAWA J. 2013;34(1):3–19.
18. Biermann CJ. Handbook of pulping and papermaking. 2nd ed. San Diego: Academic Press; 1996.
19. Limaye A. Drishti, A Volume Exploration and Presentation Tool. Proc Spie 2012, 8506. Developments in X-Ray Tomography VIII, 85060X (October 17, 2012); doi:10.1117/12.935640.
20. Kern KA, Ewers FW, Telewski FW, Koehler L. Mechanical perturbation affects conductivity, mechanical properties and aboveground biomass of hybrid poplars. Tree Physiol. 2005;25(10):1243–51.
21. Pruyn ML, Ewers BJ, Telewski FW. Thigmomorphogenesis: changes in the morphology and mechanical properties of two Populus hybrids in response to mechanical perturbation. Tree Physiol. 2000;20(8):535–40.
22. Brereton NJB, Pitre FE, Ray MJ, Karp A, Murphy RJ. Investigation of tension wood formation and 2,6-dichlorbenzonitrile application in short rotation coppice willow composition and enzymatic saccharification. Biotechnol Biofuels. 2011;4:13.
23. Chave J, Coomes D, Jansen S, Lewis SL, Swenson NG, Zanne AE. Towards a worldwide wood economics spectrum. Ecol Lett. 2009;12(4):351–66.
24. Jourez B, Riboux A, Leclercq A. Anatomical characteristics of tension wood and opposite wood in young inclined stems of poplar (Populus euramericana cv 'Ghoy'). Iawa J. 2001;22(2):133–57.
25. Gartner BL, Roy J, Huc R. Effects of tension wood on specific conductivity and vulnerability to embolism of Quercus ilex seedlings grown at two atmospheric CO2 concentrations. Tree Physiol. 2003;23(6):387–95.
26. A.B. W. The Reaction Anatomy of Arborescent Angiosperms. In: Maria Moors Cabot Foundation for Botanical Research. New York: Academic Press; 1964.
27. Moreau C, Aksenov N, Lorenzo MG, Segerman B, Funk C, Nilsson P, Jansson S, Tuominen H. A genomic approach to investigate developmental cell death in woody tissues of Populus trees. Genome Biol 2005,6(4):R34.
28. Andersson-Gunneras S, Mellerowicz EJ, Love J, Segerman B, Ohmiya Y, Coutinho PM, et al. Biosynthesis of cellulose-enriched tension wood in Populus tremula: global analysis of transcripts and metabolites identifies biochemical and developmental regulators in secondary wall biosynthesis. (vol 45, pg 144, 2005). Plant J. 2006;46(2):349.
29. Prodhan AKMA, Funada R, Ohtani J, Abe H, Fukazawa K. Orientation of microfibrils and microtubules in developing tension-wood fibres of Japanese ash (Fraxinus mandshurica var. japonica). Planta. 1995;196(3):577–85.

Permissions

The contributors of this book come from diverse backgrounds, making this book a truly international effort. This book will bring forth new frontiers with its revolutionizing research information and detailed analysis of the nascent developments around the world.

We would like to thank all the contributing authors for lending their expertise to make the book truly unique. They have played a crucial role in the development of this book. Without their invaluable contributions this book wouldn't have been possible. They have made vital efforts to compile up to date information on the varied aspects of this subject to make this book a valuable addition to the collection of many professionals and students.

This book was conceptualized with the vision of imparting up-to-date information and advanced data in this field. To ensure the same, a matchless editorial board was set up. Every individual on the board went through rigorous rounds of assessment to prove their worth. After which they invested a large part of their time researching and compiling the most relevant data for our readers.

The editorial board has been involved in producing this book since its inception. They have spent rigorous hours researching and exploring the diverse topics which have resulted in the successful publishing of this book. They have passed on their knowledge of decades through this book. To expedite this challenging task, the publisher supported the team at every step. A small team of assistant editors was also appointed to further simplify the editing procedure and attain best results for the readers.

Apart from the editorial board, the designing team has also invested a significant amount of their time in understanding the subject and creating the most relevant covers. They scrutinized every image to scout for the most suitable representation of the subject and create an appropriate cover for the book.

The publishing team has been an ardent support to the editorial, designing and production team. Their endless efforts to recruit the best for this project, has resulted in the accomplishment of this book. They are a veteran in the field of academics and their pool of knowledge is as vast as their experience in printing. Their expertise and guidance has proved useful at every step. Their uncompromising quality standards have made this book an exceptional effort. Their encouragement from time to time has been an inspiration for everyone.

The publisher and the editorial board hope that this book will prove to be a valuable piece of knowledge for researchers, students, practitioners and scholars across the globe.

List of Contributors

Hongbo Cao
Key Laboratory of Horticultural Plant Biology (Ministry of Education), Huazhong Agricultural University, 430070 Wuhan, Hubei, China
College of Horticulture, Agricultural University of Hebei, 071001 Baoding, Hebei, China

Jiangbo Wang
Key Laboratory of Horticultural Plant Biology (Ministry of Education), Huazhong Agricultural University, 430070 Wuhan, Hubei, China
Present address: College of Plant Science, Tarim University, 843300 Alar, China

Xintian Dong
College of Horticulture, Agricultural University of Hebei, 071001 Baoding, Hebei, China

Yan Han
College of Horticulture, Agricultural University of Hebei, 071001 Baoding, Hebei, China

Qiaoli Ma
Key Laboratory of Horticultural Plant Biology (Ministry of Education), Huazhong Agricultural University, 430070 Wuhan, Hubei, China

Yuduan Ding
Key Laboratory of Horticultural Plant Biology (Ministry of Education), Huazhong Agricultural University, 430070 Wuhan, Hubei, China

Fei Zhao
Key Laboratory of Horticultural Plant Biology (Ministry of Education), Huazhong Agricultural University, 430070 Wuhan, Hubei, China

Jiancheng Zhang
Key Laboratory of Horticultural Plant Biology (Ministry of Education), Huazhong Agricultural University, 430070 Wuhan, Hubei, China
Present address: Shanxi Agricultural University, 030801 Taigu, Shanxi, China

Haijiang Chen
College of Horticulture, Agricultural University of Hebei, 071001 Baoding, Hebei, China

Qiang Xu
Key Laboratory of Horticultural Plant Biology (Ministry of Education), Huazhong Agricultural University, 430070 Wuhan, Hubei, China

Juan Xu
Key Laboratory of Horticultural Plant Biology (Ministry of Education), Huazhong Agricultural University, 430070 Wuhan, Hubei, China

Xiuxin Deng
Key Laboratory of Horticultural Plant Biology (Ministry of Education), Huazhong Agricultural University, 430070 Wuhan, Hubei, China

Silvia Santopolo
Dipartimento di Biologia e Biotecnologie "C. Darwin", Sapienza Università di Roma, Piazzale Aldo Moro 5, 00185 Rome, Italy

Alessandra Boccaccini
Dipartimento di Biologia e Biotecnologie "C. Darwin", Sapienza Università di Roma, Piazzale Aldo Moro 5, 00185 Rome, Italy
Istituto Pasteur Fondazione Cenci Bolognetti, Dipartimento di Biologia e Biotecnologie "C. Darwin", Sapienza Università di Roma, Piazzale Aldo Moro 5, 00185 Rome, Italy

Riccardo Lorrai
Dipartimento di Biologia e Biotecnologie "C. Darwin", Sapienza Università di Roma, Piazzale Aldo Moro 5, 00185 Rome, Italy
Istituto Pasteur Fondazione Cenci Bolognetti, Dipartimento di Biologia e Biotecnologie "C. Darwin", Sapienza Università di Roma, Piazzale Aldo Moro 5, 00185 Rome, Italy

Veronica Ruta
Dipartimento di Biologia e Biotecnologie "C. Darwin", Sapienza Università di Roma, Piazzale Aldo Moro 5, 00185 Rome, Italy

Davide Capauto
Dipartimento di Biologia e Biotecnologie "C. Darwin", Sapienza Università di Roma, Piazzale Aldo Moro 5, 00185 Rome, Italy

Emanuele Minutello
Dipartimento di Biologia e Biotecnologie "C. Darwin", Sapienza Università di Roma, Piazzale Aldo Moro 5, 00185 Rome, Italy

Giovanna Serino
Dipartimento di Biologia e Biotecnologie "C. Darwin", Sapienza Università di Roma, Piazzale Aldo Moro 5, 00185 Rome, Italy

Paolo Costantino
Dipartimento di Biologia e Biotecnologie "C. Darwin", Sapienza Università di Roma, Piazzale Aldo Moro 5, 00185 Rome, Italy

Paola Vittorioso
Dipartimento di Biologia e Biotecnologie "C. Darwin", Sapienza Università di Roma, Piazzale Aldo Moro 5, 00185 Rome, Italy
Istituto Pasteur Fondazione Cenci Bolognetti, Dipartimento di Biologia e Biotecnologie "C. Darwin", Sapienza Università di Roma, Piazzale Aldo Moro 5, 00185 Rome, Italy

Lingye Su
Beijing Key Laboratory of Grape Sciences and Enology and CAS Key Laboratory of Plant Resources, Institute of Botany, Chinese Academy of Sciences, Beijing 100093, China
University of Chinese Academy of Sciences, Beijing 100049, China

Zhanwu Dai
INRA, Institut des Sciences de la Vigne et du Vin, UMR 1287 Ecophysiologie et Génomique Fonctionnelle de la Vigne (EGFV), 210 Chemin de Leysotte, 33882 Villenave d' Ornon, France

Shaohua Li
Beijing Key Laboratory of Grape Sciences and Enology and CAS Key Laboratory of Plant Resources, Institute of Botany, Chinese Academy of Sciences, Beijing 100093, China
Key Laboratory of Plant Germplasm Enhancement and Specialty Agriculture, Wuhan Botanical Garden, Chinese Academy of Sciences, Wuhan 430074, China

Haiping Xin
Key Laboratory of Plant Germplasm Enhancement and Specialty Agriculture, Wuhan Botanical Garden, Chinese Academy of Sciences, Wuhan 430074, China

Desheng Liu
Department of Chemistry and Biochemistry, Miami University, Oxford, OH 45056, USA

Christopher A Makaroff
Department of Chemistry and Biochemistry, Miami University, Oxford, OH 45056, USA

Bhakti Prinsi
Dipartimento di Scienze Agrarie e Ambientali - Produzione, Territorio, Agroenergia (DISAA), Università degli Studi di Milano, Via Celoria, 2, 20133 Milano, Italy

Luca Espen
Dipartimento di Scienze Agrarie e Ambientali - Produzione, Territorio, Agroenergia (DISAA), Università degli Studi di Milano, Via Celoria, 2, 20133 Milano, Italy

Yong Jia
School of Agriculture, Food and Wine, University of Adelaide, Adelaide 5005, Australia

Darren CJ Wong
School of Agriculture, Food and Wine, University of Adelaide, Adelaide 5005, Australia
Present address: Wine Research Center, Faculty of Land and Food Systems, University of British Columbia, Vancouver V6T 1Z4BC, Canada

Crystal Sweetman
School of Agriculture, Food and Wine, University of Adelaide, Adelaide 5005, Australia
Present address: School of Biological Sciences, Flinders University, GPO Box 2100, Adelaide 5001, Australia

John B Bruning
School of Biological Sciences, University of Adelaide, Adelaide 5005, Australia

Christopher M Ford
School of Agriculture, Food and Wine, University of Adelaide, Adelaide 5005, Australia

John D Liu
Department of BioSciences, Rice University, Houston, TX 77005, USA

Danielle Goodspeed
Department of BioSciences, Rice University, Houston, TX 77005, USA
Current Address: Department of Obstetrics and Gynecology, Baylor College of Medicine, Houston, TX 77030, USA

Zhengji Sheng
Department of BioSciences, Rice University, Houston, TX 77005, USA

Baohua Li
Department of Plant Sciences, University of California, Davis, CA 95616, USA

Yiran Yang
Department of BioSciences, Rice University, Houston, TX 77005, USA

Daniel J Kliebenstein
Department of Plant Sciences, University of California, Davis, CA 95616, USA
DynaMo Centre of Excellence, Department of Plant and Environmental Sciences, Faculty of Science, University of Copenhagen, Thorvaldsensvej 40, 1871 Frederiksberg C, Denmark

Janet Braam
Department of BioSciences, Rice University, Houston, TX 77005, USA

Shancen Zhao
Centre for Soybean Research, Partner State Key Laboratory of Agrobiotechnology, The Chinese University of Hong Kong, Shatin, New Territories, Hong Kong
BGI-Shenzhen, Main Building, Beishan Industrial Zone, Yantian District, Shenzhen 518083, China

Fengya Zheng
Centre for Soybean Research, Partner State Key Laboratory of Agrobiotechnology, The Chinese University of Hong Kong, Shatin, New Territories, Hong Kong

Weiming He
BGI-Shenzhen, Main Building, Beishan Industrial Zone, Yantian District, Shenzhen 518083, China

Haiyang Wu
BGI-Shenzhen, Main Building, Beishan Industrial Zone, Yantian District, Shenzhen 518083, China

Shengkai Pan
Centre for Soybean Research, Partner State Key Laboratory of Agrobiotechnology, The Chinese University of Hong Kong, Shatin, New Territories, Hong Kong

Hon-Ming Lam
Centre for Soybean Research, Partner State Key Laboratory of Agrobiotechnology, The Chinese University of Hong Kong, Shatin, New Territories, Hong Kong

Suong T T Nguyen
Centre for Plant Science, School of Environmental and Life Sciences, The University of Newcastle, Newcastle, NSW 2308, Australia

David W McCurdy
Centre for Plant Science, School of Environmental and Life Sciences, The University of Newcastle, Newcastle, NSW 2308, Australia

Péter Szövényi
Institute of Evolutionary Biology and Environmental Studies, University of Zurich, Zurich, Switzerland
Institute of Systematic Botany, University of Zurich, Zurich, Switzerland
Swiss Institute of Bioinformatics, Quartier Sorge-Batiment Genopode, Lausanne, Switzerland
MTA-ELTE-MTM Ecology Research Group, ELTE, Biological Institute, Budapest, Hungary

Eftychios Frangedakis
Department of Plant Sciences, University of Oxford, South Parks Rd, Oxford, UK
Graduate School of Science, University of Tokyo, 7-3-1 Hongo, Bunkyo-ku, Tokyo 113 0033, Japan

Mariana Ricca
Institute of Evolutionary Biology and Environmental Studies, University of Zurich, Zurich, Switzerland
Swiss Institute of Bioinformatics, Quartier Sorge-Batiment Genopode, Lausanne, Switzerland

Dietmar Quandt
Nees-Institut für Biodiversität der Pflanzen, University of Bonn, Meckenheimer Allee 170, D – 53115 Bonn, Germany

Susann Wicke
Nees-Institut für Biodiversität der Pflanzen, University of Bonn, Meckenheimer Allee 170, D – 53115 Bonn, Germany
Institute for Evolution and Biodiversity, University of Muenster, Huefferstr. 1, 48149 Muenster, Germany

Jane A Langdale
Department of Plant Sciences, University of Oxford, South Parks Rd, Oxford, UK

Savithri U Nambeesan
Department of Plant Biology, Miller Plant Sciences, University of Georgia, Athens, GA 30602, USA
Department of Horticulture, University of Georgia, Athens, GA 30602, USA

Jennifer R Mandel
Department of Plant Biology, Miller Plant Sciences, University of Georgia, Athens, GA 30602, USA
Department of Biological Sciences, University of Memphis, Memphis, TN 38152, USA

John E Bowers
Department of Plant Biology, Miller Plant Sciences, University of Georgia, Athens, GA 30602, USA

Laura F Marek
North Central Regional Plant Introduction Station, Iowa State University/ USDA-ARS, Ames, IA 50014, USA

Daniel Ebert
Department of Botany, University of British Columbia, Vancouver, BC V6T 1Z4, Canada

Jonathan Corbi
Department of Plant Biology, Miller Plant Sciences, University of Georgia, Athens, GA 30602, USA
Department of Crop and Soil Sciences, University of Georgia, Athens, GA 30602, USA

Loren H Rieseberg
Department of Botany, University of British Columbia, Vancouver, BC V6T 1Z4, Canada

Steven J Knapp
Department of Plant Sciences, University of California, Davis, CA 95616, USA

John M Burke
Department of Plant Biology, Miller Plant Sciences, University of Georgia, Athens, GA 30602, USA

Bernardo Murillo-Amador
Centro de Investigaciones Biológicas del Noroeste, S.C. La Paz, La Paz, Baja California Sur, México

Edgar Omar Rueda-Puente
Universidad de Sonora, Hermosillo, Sonora, México

Enrique Troyo-Diéguez
Centro de Investigaciones Biológicas del Noroeste, S.C. La Paz, La Paz, Baja California Sur, México

Miguel Víctor Córdoba-Matson
Centro de Investigaciones Biológicas del Noroeste, S.C. La Paz, La Paz, Baja California Sur, México

Luis Guillermo Hernández-Montiel
Centro de Investigaciones Biológicas del Noroeste, S.C. La Paz, La Paz, Baja California Sur, México

Alejandra Nieto-Garibay
Centro de Investigaciones Biológicas del Noroeste, S.C. La Paz, La Paz, Baja California Sur, México

Lorenzo Boggia
Department of Drug Science and Technology, University of Turin, Via P. Giuria 9, 10125 Turin, Italy

Barbara Sgorbini
Department of Drug Science and Technology, University of Turin, Via P. Giuria 9, 10125 Turin, Italy

Cinzia M Bertea
Plant Physiology Unit, Department Life Sciences and Systems Biology, University of Turin, Via Quarello 15/A, 10135 Turin, Italy

Cecilia Cagliero
Department of Drug Science and Technology, University of Turin, Via P. Giuria 9, 10125 Turin, Italy

Carlo Bicchi
Department of Drug Science and Technology, University of Turin, Via P. Giuria 9, 10125 Turin, Italy

Massimo E Maffei
Plant Physiology Unit, Department Life Sciences and Systems Biology, University of Turin, Via Quarello 15/A, 10135 Turin, Italy

Patrizia Rubiolo
Department of Drug Science and Technology, University of Turin, Via P. Giuria 9, 10125 Turin, Italy
Plant Physiology Unit, Department Life Sciences and Systems Biology, University of Turin, Via Quarello 15/A, 10135 Turin, Italy

Panfeng Zhao
Joint Tomato Research Institute, School of Agriculture and Biology, Shanghai Jiao Tong University, Shanghai 200240, China
Plant Biotechnology Research Center, School of Agriculture and Biology, Shanghai Jiao Tong University, Shanghai 200240, China

Lida Zhang
Plant Biotechnology Research Center, School of Agriculture and Biology, Shanghai Jiao Tong University, Shanghai 200240, China

Lingxia Zhao
Joint Tomato Research Institute, School of Agriculture and Biology, Shanghai Jiao Tong University, Shanghai 200240, China
Plant Biotechnology Research Center, School of Agriculture and Biology, Shanghai Jiao Tong University, Shanghai 200240, China

Upinder S Gill
Plant Biology Division, The Samuel Roberts Noble Foundation, Ardmore, Oklahoma 73401, USA

Srinivasa R Uppalapati
Plant Biology Division, The Samuel Roberts Noble Foundation, Ardmore, Oklahoma 73401, USA
Biologicals and Fungicide Discovery, DuPont Crop Protection, Newark DE 19711, USA

Jin Nakashima
Plant Biology Division, The Samuel Roberts Noble Foundation, Ardmore, Oklahoma 73401, USA

Kirankumar S Mysore
Plant Biology Division, The Samuel Roberts Noble Foundation, Ardmore, Oklahoma 73401, USA

Nicholas James Beresford Brereton
Institut de recherche en biologie végétale, Université de Montréal, Montreal, QC H1X 2B2, Canada

Farah Ahmed
Micro-CT Lab, Imaging and Analysis Centre, Natural History Museum, London SW7 5BD, UK

Daniel Sykes
Micro-CT Lab, Imaging and Analysis Centre, Natural History Museum, London SW7 5BD, UK

Michael Jason Ray
Department of Chemistry, Imperial College, London SW7 2AZ, UK

Ian Shield
Department of AgroEcology, Rothamsted Research, Harpenden, Herts AL5 2JQ, UK

Angela Karp
Department of AgroEcology, Rothamsted Research, Harpenden, Herts AL5 2JQ, UK

Richard James Murphy
Centre for Environmental Strategy, University of Surrey, Guildford, Surrey GU2 7XH, UK

www.ingramcontent.com/pod-product-compliance
Lightning Source LLC
Chambersburg PA
CBHW080631200326
41458CB00013B/4587